TELEVISION SERVICING

TELEVISION SERVICING

Alex Levy
THOMAS A. EDISON VOCATIONAL HIGH SCHOOL
JAMAICA, NEW YORK

Murray Frankel
SERVIDEO ELECTRONIC MAINTENANCE CORPORATION

McGRAW-HILL BOOK COMPANY, INC.
New York Toronto London 1959

TELEVISION SERVICING

Library of Congress Catalog Card Number: 58–9997

II

Preface

Television, although not an entirely new field, is nevertheless still very much in the growing stage. The millions of television sets now in use are serviced mainly by technicians who grew up with the industry. By servicing innumerable receivers, these men have learned to recognize symptoms as caused by a particular malperforming component and are therefore able to remedy most receiver troubles. In the occasional case where the trouble is obscure, it must be located by trial-and-error methods. However, the great length of time needed to acquire such a familiarity with the many types of receivers makes this an impractical and almost impossible method for the student or beginning serviceman. There is therefore a need for a definite and universal trouble-shooting procedure which is easily learned. The authors, being closely associated with both the practical and teaching ends of television, have developed such a procedure and present it in a method based on sound principles of learning.

The book applies the same approach to the field of television that has proven so successful in *Elements of Radio Servicing* by William Marcus and Alex Levy, published by the McGraw-Hill Book Company, Inc. Basic circuits are described and servicing procedures formulated for them. The procedures are then adapted to commonly used variations of the basic circuits. The authors believe that once the reader has mastered the basic procedures, he will be able to apply them to any variation he may encounter. In the presentation, the emphasis is placed on practical repair work, theory being added when necessary for comprehension of the subject matter.

This book can be used as a school text, as it is designed as a reference manual in a course devoted to television servicing. Summary charts, together with suitable questions, are given at the end of each chapter.

The beginning repairman will also find it useful for self-study purposes. Even the experienced bench serviceman will find methods of improving his techniques and increasing his speed and therefore his earnings.

These are the aims and purposes of *Television Servicing*.

Murray Frankel
Alex Levy

Contents

TELEVISION SERVICING

1

Functional Sections of a Television Receiver

When an electrician is called to a home to service a condition where some circuits are dead but others are normal, as shown by the fact that some lights are working, there is no need for him to check the main fuse. He starts by checking the nonworking branch circuits. Similarly, in a television receiver, the breakdown of one component usually stops the operation of one section, leaving others functioning. The functioning and nonfunctioning sections are therefore the first clue to the location of the defective component. Since a television receiver, by its very nature, includes both sight and sound effects, a natural division of its functions is to describe the faulty conditions as they affect the picture and the sound. Another important consideration in service analysis is whether the screen of the television receiver lights up at all. Any description of faulty operation of a television receiver should therefore include answers to the following questions:

1. Does the face of the picture tube light up?
2. Is the sound absent or, if present, distorted in some way?
3. Is the picture absent or, if present, distorted in some way?

These three questions and their answers channel the serviceman's thinking into the main sections of the television receiver. These functional sections are now described, together with the symptoms which result from the failure of each.

The CRT and Deflection Section. In a television receiver, the screen on which the picture is viewed is the face of a large vacuum tube, known as a cathode-ray tube (abbreviated CRT), or picture tube. Some of the important components of the picture tube are illustrated in Fig. 1-1. In the neck of the tube near the base is a cathode. When heated, this emits electrons in exactly the same manner as the cathode in any vacuum tube. Surrounding the cathode and extending out from it are a group of electrodes which shape the emitted electrons into a beam aimed at the screen of the tube. The cathode, together with the electrodes near it, is called an electron gun.

The inside face of the picture tube is coated with a phosphorescent material. This becomes the screen. The phosphor glows, or fluoresces, when struck by electrons. The electron beam from the gun is focused to a narrow spot for maximum efficiency, and so a small dot of light is produced in the center of the screen, as shown in the diagram.

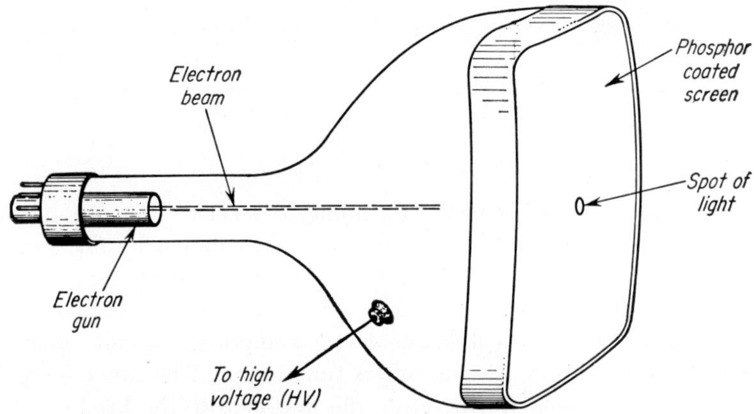

Fig. 1-1. The cathode-ray tube which reproduces the picture.

The distance from the gun to the screen is quite large, 24 in. or more, depending on the size of the picture tube. In order to propel the beam over such a long distance, a high voltage of approximately 12,000 to 16,000 volts is required. This is labeled HV in the diagram.

As we have just seen, the electron beam causes a dot of light in the center of the screen. To light up the entire screen, the beam is rapidly deflected back and forth (horizontally) and up and down (vertically) across the face of the screen. The moving electrons in the beam, just like moving electrons in a wire, exhibit magnetic effects and will interact with, or be deflected by, a magnet. The deflecting magnetic fields are furnished by two sets of windings which fit around the neck of the tube in a component known as the yoke. This is illustrated in Fig. 1-2, which also shows important subdivisions of the deflection and CRT sections.

One set of windings in the yoke is fed by the horizontal deflection section. This energizes the yoke to pull the beam magnetically back and forth in a horizontal plane. This occurs at a rapid rate, 15,750 times per second. At the same time, the other set of windings fed by the vertical section pulls the beam up and down in a vertical direction at a much slower rate, 60 times per second. The result is that the beam traverses the screen in a series of horizontal lines slanted slightly downward because of the pull of the vertical system, and each line is slightly below the preceding one because of the same downward pull. The moving beam creates lines of light across the screen, and since the entire field is covered in a comparatively short time ($\frac{1}{60}$ sec) and then repeated, persistence of vision makes it appear that the entire screen is lighted at once. Actually, at any one instant of time, the beam is lighting only one spot somewhere on the face of the screen.

In most modern receivers, the horizontal system also furnishes energy to operate the high-voltage supply. This is indicated by the high-voltage block in Fig. 1-2.

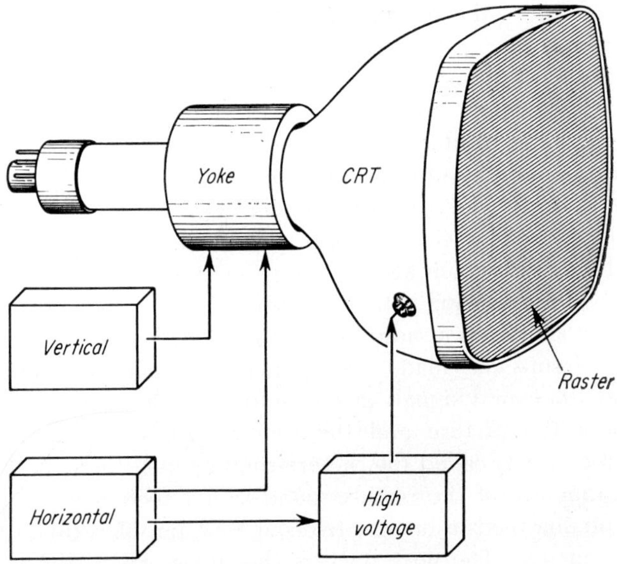

Fig. 1-2. The deflection and high-voltage section that produces the raster.

The lines of light covering the screen are known as the raster. Note that the raster is entirely independent of any signal sent out from the transmitting station. In a normally operating receiver, the raster, or fully lighted screen, should be seen on a channel setting where no station is operating. In a sharply focused large-screen receiver, the individual raster lines are easily seen with the unaided eye, especially when the receiver is on an unused channel.

The picture tube, deflection yoke, vertical and horizontal deflection circuits, and the high-voltage circuit form our first group of functional sections. Together, they produce the raster, and unless we have a normal raster, we cannot have normal reception results. For example, assume a weak vertical deflection section. The beam cannot be fully deflected for a full raster in the vertical direction, and the resulting picture appears compressed into a narrow band across the screen. Furthermore, if the vertical deflection is completely inoperative, the picture is compressed into a single line across the center of the screen.

Similarly, a weak or inoperative horizontal deflection system tends to reduce the width of the picture. However, reference to Fig. 1-2 shows that the horizontal circuits also provide energy to the high-voltage supply, and this failure therefore cuts down on the high voltage as well. Lack of high voltage stops electron flow in the picture tube, resulting in a blank screen.

Our first question regarding the condition of a television screen was, "Does the screen of the picture tube light?" The answer is the first step in a television-servicing procedure. The serviceman checks for the presence of a normal full raster. If this appears, our first functional section is in the clear. The picture tube, deflection, and high-voltage circuits are operating, and the trouble lies elsewhere in the receiver.

The Picture-Sound Signal Section. The channel allotted to a television transmitter is 6 Mc wide. In this frequency spread, spaced 4.5 Mc apart, the transmitter broadcasts two signals, one for the picture information and the other for sound which accompanies the picture. The two signals are picked up by the receiving antenna and fed to a superheterodyne receiver which is very similar to a radio broadcast receiver. One difference is that the tuning circuits are made broad enough to accommodate both the picture and the sound signals at one setting of the tuning dial. Since it handles both the picture and the sound signals, this section of the television receiver is called the picture-sound signal section.

A block diagram of the picture-sound signal section is shown in Fig. 1-3. The antenna feeds a one- or two-stage r-f amplifier tuned to a given channel frequency. The next stage is the mixer where the picture and sound carriers are mixed with the output of the local oscillator to produce two i-f signals. By the nature of superheterodyne action, the picture and sound i-f signals are still 4.5 Mc apart. The r-f, mixer, and oscillator stages are usually combined in a small subchassis called the tuner, or "front end."

From the tuner, the signal is fed to the i-f amplifier. This is usually a three-stage unit as indicated in the diagram. The i-f amplifier carries both the picture and sound i-f signals, and since these are 4.5 Mc apart, its tuning must be broad enough to handle this range. Broadly tuned circuits are characterized by low gain, thus necessitating three and sometimes four stages of i-f amplification.

The next stage is the second detector. In the intercarrier sound system used in most television receivers, the second detector serves a twofold function. One is to demodulate the picture i-f signal and feed it to the picture tube through a stage known as the video amplifier. The other is to provide a nonlinear circuit to produce the beat formed by the picture and sound i-f signals. This beat will be at 4.5 Mc and will carry the sound modulation. In effect, the original sound carrier has been converted to a new i-f signal at 4.5 Mc by dual superheterodyne action.

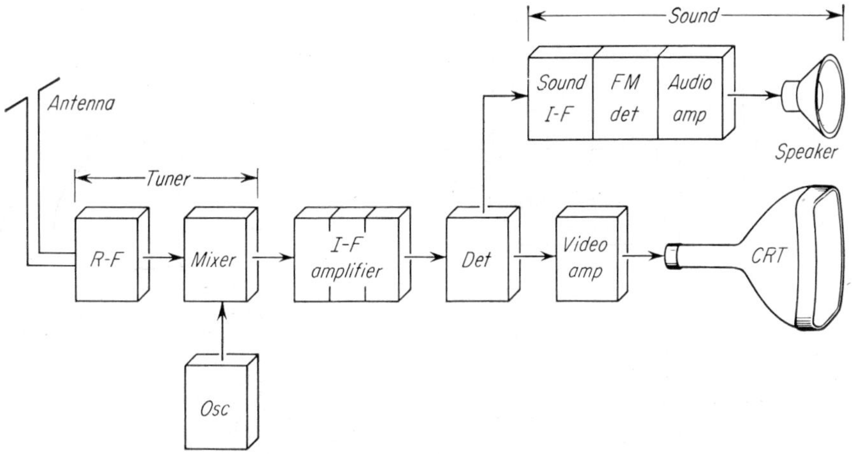

Fig. 1-3. Picture-sound signal chain.

The 4.5-Mc sound i-f signal is fed to a 4.5-Mc i-f amplifier, then to a demodulator, and finally to an audio amplifier and loudspeaker. For simplicity, all these stages have been combined into one block in Fig. 1-3, and called the sound section.

The stages preceding the sound section carry the sound signal, and these should rightly be called the picture-sound stages. Common usage, however, is to add the video-amplifier and picture-tube circuits and call the entire unit the picture-signal chain. A diagram showing this interpretation of the picture and sound section is given in Fig. 1-4. The deflection and high-voltage section have been included in this diagram.

Before studying symptoms caused by breakdown in the picture and sound sections, it would be of advantage to describe how the picture is formed on the screen. A simplified over-all view of the television system follows.

At the transmitter, the scene being televised is focused onto a screen made up of a light-sensitive surface. Light striking this surface produces an electric charge at the point of impact, the charge being dependent on the intensity of the light of the scene being viewed. Deflection generators, very similar to those used for the receiving picture tube, send an electron beam across this surface, picking up the light charges, and thus producing a current which varies with the intensity of the light. In this way, the current changes record the light changes encountered as the beam traverses a given line on the light-sensitive screen. This is the picture, or video signal, which is used to modulate the transmitting carrier, and so form the modulated r-f wave which is broadcast. At the same time, a

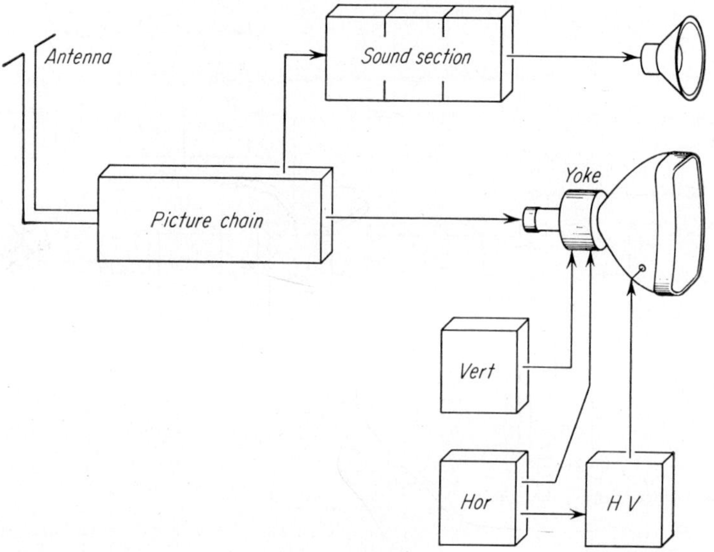

Fig. 1-4. Combined picture, sound, and raster sections.

microphone picks up the sound on the television stage. The sound signal is used to modulate a second transmitting carrier which is at a frequency 4.5 Mc higher than the picture carrier. Both carriers are transmitted simultaneously using the same antenna at the transmitting site.

The receiving antenna picks up both transmitted waves. These are tuned and amplified in the early stages of the picture-sound signal chain. At the detector, the modulated sound carrier is separated and sent to the sound section. The video signal is recovered from the modulated picture carrier and fed to the picture tube after it has been amplified in the video amplifier.

In the picture tube, as previously explained, the raster is produced by the deflection and high-voltage section. When the video signal is added, its effect is to strengthen and weaken the electron beam, thereby brightening or darkening portions of each line of light as it is drawn, proportionately to the strength of the video signal. This re-creates, on the picture tube, the line being scanned at the transmitter. With the deflection circuits at the receiver and transmitter operating in unison, or synchronization as it is called, the complete television scene appears on the face of the picture tube. At the same time, the loudspeaker reproduces the ac companying sound through the receiver's sound system.

An inoperative stage in any portion of the picture chain will stop the video signal and there will be no picture on the screen. However, the operative deflection circuit is still producing a raster so that the symptom "no picture—normal raster" indicates trouble in the picture chain. We can also have the condition of normal raster and picture but no sound. This proves that the complete picture chain is operative, and therefore the break is probably in the sound section.

The Sync Section. It was previously mentioned that the deflection generators at the transmitter and receiver must be in step. To stress the importance of this item, let us suppose that both sets of generators are working at the same rates but that they are out of step by half a line. This is a small timing error, since it represents about $\frac{1}{30{,}000}$ sec. Now, while the beam at the transmitter is scanning the left side of any one line, the beam at the receiver will be reproducing the same line on the right side of the screen. The result will be as though a photograph of the scene being transmitted were cut in half, and the two halves were transposed at the receiver. If the rates of speed of the deflection generators were also to vary, the reproduced picture would be a meaningless jumble of light and dark areas.

In order to keep both deflection generators in step, or synchronization, the transmitted signal includes synchronization information together with

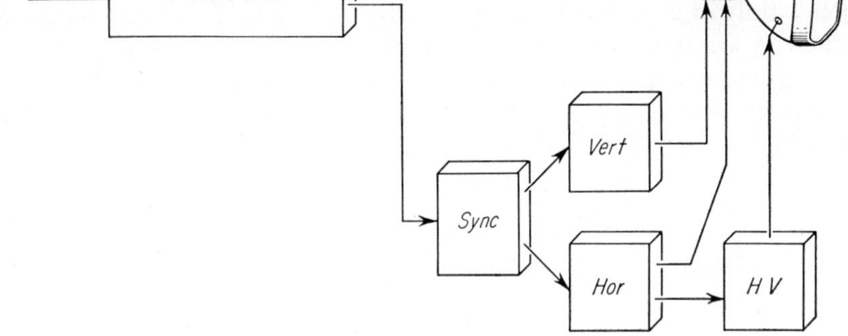

Fig. 1-5. Location of the sync section in a television receiver.

the signal. This is in the form of a short series of pulses transmitted at the end of each line. These keep the horizontal deflection circuits in step. A longer series of pulses at the end of each field is transmitted for the vertical system. These added signals are called synchronizing—or, more simply, sync pulses. In the receiver, the sync pulses are separated from the rest of the signal, amplified, and then used to lock the deflection generators at the receiver in step with those at the transmitter. The portion of the receiver in which this is done is called the sync section. It is shown in block-diagram form in Fig. 1-5.

The deflection generators in the receiver, like all generators, tend to drift somewhat from their correct frequency. The sync signals are meant to correct this tendency. If the sync pulses are absent because of a fault in the sync section, the picture will tend to drift up and down, or slip sideways, or do both at once as the deflection generators drift out of step with those at the transmitter. Therefore, an unstable picture, or a meaningless display of moving black and white areas on the screen is the symptom which points to trouble in the sync section. Raster and sound are normal.

Low-voltage Power Supply. In common with all electronic equipment, the television receiver has a power supply whose function is to light the tubes and furnish a d-c potential for the plate and screen circuits. In the television receiver, this is called the low-voltage power supply in order to distinguish it from the high-voltage supply, which furnishes 12,000 to 16,000 volts of d-c potential to attract the beam in the picture tube. The low-voltage power supply furnishes approximately 250 volts at the B plus point to operate the tubes in the receiver. Since the power supply is furnishing operating voltages to all sections of the receiver, its failure stops the entire set. There will be no raster, no picture, and no sound. A complaint that the receiver is completely dead, therefore, points to the power supply as the cause.

In our discussion so far, we have been considering the results of complete failure of a section. Of course, partial failure such as low gain or distortion in an amplifier will bring on some malfunction of the receiver. In many cases, the resulting symptom may be quite different from that caused by the complete failure of the stage. These conditions will be described more fully in succeeding chapters.

SUMMARY

Diagram of the Functional Sections of a Television Receiver. A diagram showing all the main functional sections of a television receiver is given in Fig. 1-6.

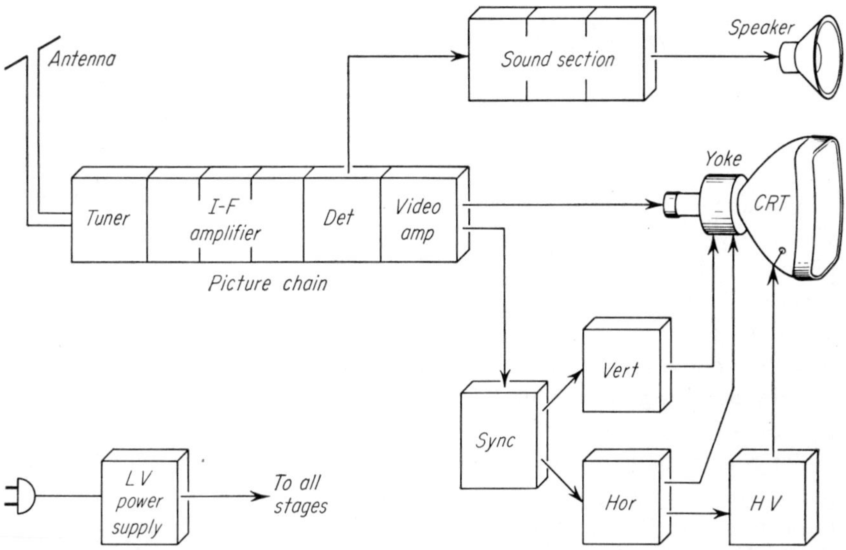

Fig. 1-6. Functional sections of a television receiver.

Section Symptom Chart. Table 1-1 shows the results to raster, picture, and sound when each of the functional sections becomes inoperative.

Table 1–1

Inoperative Section	*Resulting Symptom*		
	Raster (Screen Illumination)	*Picture Information*	*Sound*
Low-voltage power supply	No	No	No (set is dead)
Deflection:			
Vertical	White line	Compressed	Yes
Horizontal	No	No	Yes
Picture	Yes	No	Yes or no
Sound	Yes	Yes	No
Sync	Yes	Scrambled	Yes

2

Scope of the Book

As was pointed out in the preceding chapter, a basic approach to television servicing is to study the nature of the complaint as it affects raster, picture, and sound, and to determine from these symptoms which of the functional sections of the receiver is at fault. The section is then subjected to a servicing procedure designed to localize the trouble to the defective component. This book provides such a procedure.

Field Servicing and Bench Servicing. Generally, a television service job starts in the consumer's home as the result of a call that his receiver is not functioning properly. The man sent on the call verifies the complaint, decides on the section which is at fault, checks tubes, and makes adjustments in that section. Sometimes, by means of these fast checks, the fault is located, normal reception is restored, and the service job is completed.

However, the fault is sometimes more elusive, and the set must be removed from its cabinet so that test instruments may be applied to make a more thorough examination before the fault is found. All this work may be done in the customer's home, but generally, in the interests of greater efficiency, it is better at some point in a long servicing job to take the chassis to the repair shop, and finish the job on the service bench. In television repairs, work done in the customer's home is known as field servicing, and work taken to the shop is called bench servicing.

No exact line of demarcation can be drawn between field and bench servicing, because the point at which the field serviceman decides to bring the set to the bench is a variable dependent on his own working conditions. However, in general, most service organizations make it a practice to limit field servicing to checking tubes and making adjustments in the section that seems to be at fault. This reduces the field serviceman's equipment, allows him to handle more calls, and permits a concentration of expensive equipment permanently installed for thorough testing on the service bench.

This book will follow the same practice in its organization. A field-servicing procedure covering tubes and adjustments is given first, reserving the more detailed bench-servicing procedures for the chapters that follow.

Presentation of Field-servicing Material. Chapter 3 describes the field-servicing procedures to be applied for various troubles encountered in television-service work. These are listed in an illustrated trouble index. Each symptom is described by word and photograph and analyzed as to the functional section involved. This is followed by a discussion of the probable causes and the field-servicing procedures to be used. If this does not restore the receiver to normal operation, reference should be made to the subsequent chapter where the more thorough bench-servicing procedure is given.

Presentation of Bench-servicing Material. The material for the detailed servicing analysis of each functional section is organized as follows:

1. Quick check to confirm that the trouble is in this section.
2. Localizing checks to place the trouble in a subsection.
3. Function of the subsection or section.
4. Basic diagram of the subsection or section.
5. Description of how the circuit works, including functions and values of components.
6. Location and appearance of the section for identification purposes.
7. Normal test data.
8. Step-by-step procedure to locate the defective component which is causing complete failure of the section.
9. Other symptoms which result from weakness, or partial malfunction of the section.
10. Step-by-step servicing procedures for these symptoms.
11. Variations from the basic circuit, and necessary adaptations of the servicing procedures to fit these variations.
12. Information relating to the choice of suitable replacement components, and notes as to their installation.
13. Adjustment procedures.
14. Summary of tests, including procedural charts to be followed in tracing various symptoms to their cause.

In all cases, simple practical tests as performed by servicemen on the bench are presented. These tests are based on years of practical servicing experience. The test equipment consists of a multimeter of the 20,000 ohms per volt or vacuum-tube type, an a-m signal generator, and an oscilloscope. It is assumed that the student is familiar with the multimeter, but instructions on the calibration and use of the generator and oscilloscope are given when needed in the text.

3

Field Servicing and Trouble Index

Location Diagrams. When the trouble symptom points to a functional section as being at fault, the next step in servicing is to identify and locate that section on the chassis. Although varying layout schemes are used in different receivers, the need for short leads between associated components makes it possible to recognize a general pattern in this regard.

The section diagrams introduced earlier show a straight-line signal chain from antenna to picture tube and loudspeaker in order to point out signal flow and function. The same signal chain must be followed on the chassis, but the signal flow lines must be bent around to fit the sections on the square chassis commonly used for television receivers. A diagram illustrating the signal flow from section to section on a typical chassis is shown in Fig. 3-1.

The receiver signal chain begins with the tuner. This is easily identified on most receivers as a small subchassis mounted on the main chassis, or outboard from it. The channel-selector switch is part of the tuner. Locate the tuner in Fig. 3-1, and note the short signal flow line to the i-f and detector stages. In a functional-section diagram, signal from the detector divides to feed the video amplifier and sound section. Note the placement of these blocks in the diagram as to their proximity to the detector for short signal leads. At the video amplifier, the signal splits again, this time to feed the picture tube and sync section. Note that the video amplifier is conveniently located for short leads to the picture-tube socket and to the sync section. The signal flow splits again at the sync section to lock the vertical and horizontal circuits which drive the deflection yoke. Finally, the horizontal block feeds the high-voltage stage which provides the high potential needed for the picture tube. The low-voltage power supply fills up the unused corner of the chassis.

Although other receivers may not follow this exact layout scheme, a knowledge of the signal flow lines helps in an understanding of the variations. This makes it possible to identify sections on an unfamiliar chassis, when a location diagram is not available.

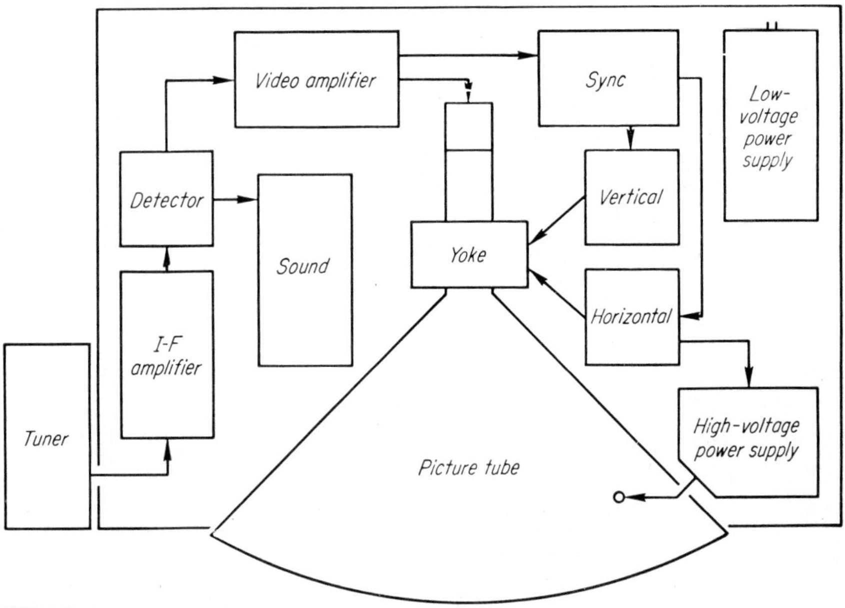

Fig. 3-1. Signal flow lines on a chassis.

Location diagrams of receivers are often found pasted in the back of the cabinet as an aid in field servicing. They are called tube-location charts. Tubes are usually labeled by type number and functional purpose, like 6AU6—second i-f amplifier. Service notes almost always include location charts and may be obtained from receiver manufacturers at little or no cost. In case they are needed when working on a receiver for which a location chart is not available, further tips on identifying functional sections are given later in this book in the detailed servicing chapters for the sections.

CHECKING TUBES IN FIELD SERVICING

A typical tube-location chart as prepared by a manufacturer is shown in Fig. 3-2. In this receiver, the chassis stands vertically with the picture tube extending through a hole in the center. Printed-circuit wiring boards are grouped around the tube. In receivers of this type, the functional sections are easily recognized since each printed-circuit board usually contains an entire section or subsection. In receivers using conventional wiring, the functional sections are identified from the given functional names of the tubes.

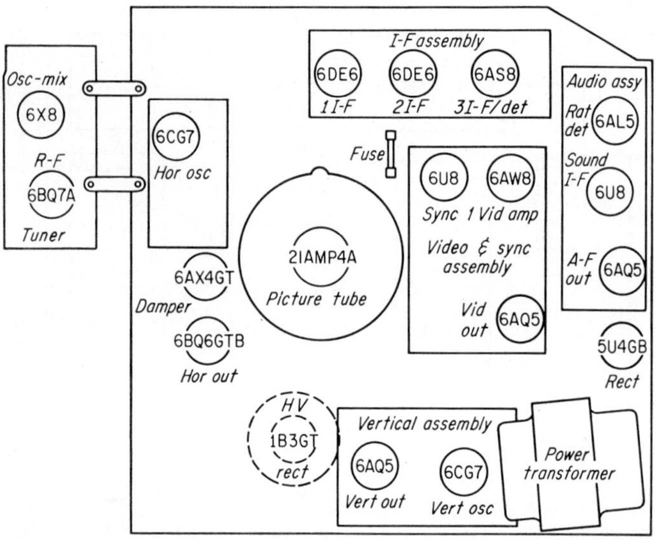

Fig. 3-2. Typical tube-location chart.

Examine the diagram and note that, except for the comparatively large space between the tuner and i-f strip, the sections and subsections are arranged for close leads similar to the arrangement given in Fig. 3-1. Locate the detector tube in the i-f assembly and see how the video amplifier, sync, and sound sections are in close proximity. See if you can draw in signal flow lines as was done in the earlier diagram. Note that the diagram also shows the location of a fuse. Changing fuses is an important item in field servicing.

Tubes Used in the Functional Sections. The functional sections usually contain several tubes, each one of which has a name descriptive of its subsection. For a servicing classification these names are more important than the tube type numbers, since there is less variation. The following is a listing of tube names usually found in the various sections:

Low-voltage power-supply section:
  Rectifier: one or two tubes, or dry-plate selenium rectifiers.
Picture chain: tuner.
  R-f amplifier.
  Mixer-oscillator.

Picture chain: i-f amplifier and detector.

There are usually three similar i-f amplifier tubes and one detector. Sometimes the detector is combined with the last i-f amplifier in a dual tube as is the case in Fig. 3-2. Sometimes it is a crystal.

Picture chain: video amplifier.

One or two video-amplifier tubes.

Sync section:

One or two tubes variously called clippers, separators, amplifiers, or inverters.

Deflection section: vertical.

Vertical oscillator.

Vertical output (both functions are sometimes combined in one dual-purpose tube).

Deflection section: horizontal.

Horizontal oscillator.

Phase detector or automatic frequency control (afc).

Horizontal output.

Damper.

High-voltage rectifier. This tube is often enclosed in an r-f cage. This is indicated by dotted lines, as shown for the 1B3-GT tube in Fig. 3-2.

Sound section:

Sound or audio i-f amplifier.

Discriminator, ratio detector or audio detector.

Audio or sound amplifier.

Audio or sound output.

General Notes on Checking Tubes. When the observed symptom points to a functional section, an early step in the field-servicing procedure calls for a check of the tubes in that section. Tubes in television receivers normally become quite hot in service; so allow them to cool down while studying the tube-location chart for the types used, and their location. Portable tube checkers are generally not accurate enough for television work; so the tube-checking method favored by most field servicemen is to carry a large stock of replacement tubes in a tube caddy and to check operation of the receiver with new tubes.

When removing the old tube from a socket in an inaccessible or blind corner, if it is of the seven- or nine-pin miniature type, pull it straight out, and note the position of break in the pin circle. Holding the new tube with the break in the same position aids in replacing the tube in its socket. It is advisable to use a pin straightener on all miniature tubes before placing them in their sockets. Certain tubes in a receiver are often enclosed in closely fitting metal shells, called goats. When these are found, make sure to replace the goat after changing the tube. When selecting a replacement tube from your caddy, make sure to choose the exact type number called for on the tube-location chart. For example, if the chart calls for a 6SN7-GT type, do not use the 6SN7-GTB. The latter will probably work, but it has different characteristics which may cause a failure at a later date.

Finding the Open Heater in a Series Filament String. Many receivers make use of a heater circuit where all the heaters are connected in series and fed directly from the line. This is the same circuit so widely used in small a-c/d-c radio receivers. In circuits of this type, if one of the tube heaters should open, the entire line is open, none of the tubes light, and the set is dead. In the radio receiver, this is an easily found trouble, since there are only five tubes to check. In a television receiver, however, there may be fourteen or more tubes, and finding the defective one by the tube-substitution method may be a long process, although it works.

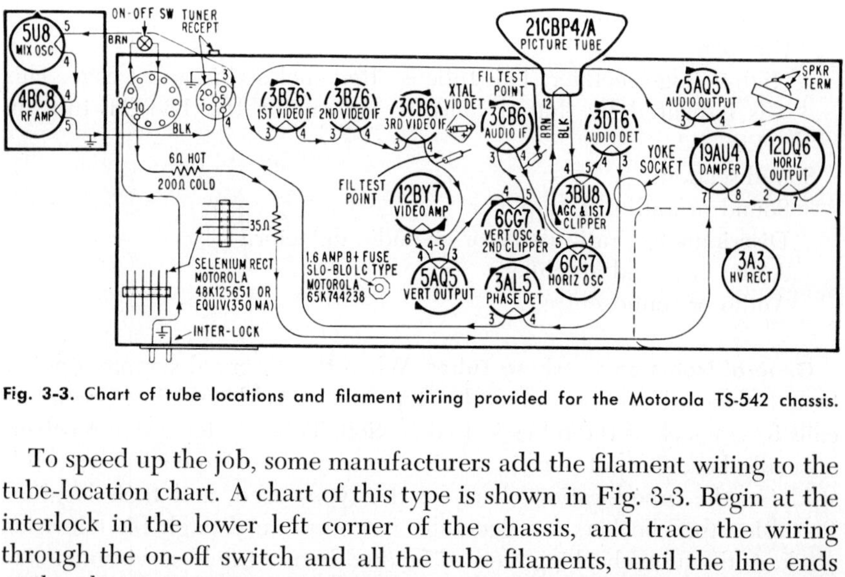

Fig. 3-3. Chart of tube locations and filament wiring provided for the Motorola TS-542 chassis.

To speed up the job, some manufacturers add the filament wiring to the tube-location chart. A chart of this type is shown in Fig. 3-3. Begin at the interlock in the lower left corner of the chassis, and trace the wiring through the on-off switch and all the tube filaments, until the line ends at the chassis connection in the tuner. Note the two filament test points provided near the video i-f and audio i-f tubes. The test points are in the form of a piece of hookup wire coming up through the chassis near the tube socket. These are used for checking sections of the filament line. The method employed uses an ohmmeter, and the plug is disconnected from the outlet.

The checking procedure can be explained as follows: If an ohmmeter or continuity checker were to be placed across the two contacts of the interlock with all good tubes in the circuit and the switch on, there would be a reading. With a dead tube somewhere in the line, the circuit is open, and the meter will not read. Since one of the line terminals goes to chassis, clip one of the meter terminals to chassis, and touch the other to the filament test point near the sound i-f tube. If this reads open, the break is in this portion of the line. Now starting at the ground connection in the tuner, and tracing the line through the tuner-tube filaments backward against the arrowheads, we go through the phase detector, sound detector, agc–first clipper, picture tube, and horizontal oscillator filaments to the first test point. One of these tubes is open, and they must be checked individually. Note that the diagram gives the pin numbers of the filaments, so that the tubes can be checked quickly with the ohmmeter.

If there was a reading at the first test point near the sound i-f tube, the tubes mentioned above are all good and do not have to be checked. The ohmmeter is next connected to the second test point near the third video i-f tube. If it reads open now, the break is between the two test points. Referring to the chart, the break must be in the video amplifier, vertical output, vertical oscillator–second clipper, or audio i-f tubes. These are now checked individually.

A continuity reading at the second filament test point places the trouble in one of the remaining tubes. Note that there are two resistors and the on-off switch in this section of the circuit. If one of these is open, the filament circuit will be open at this point, and no tubes will light. These components are also checked with the ohmmeter.

Not all receiver manufacturers provide the two filament test points. When they are not given, the filament contacts on the picture-tube socket will make convenient test points if the picture tube is roughly in the center of the filament string. Note that, in the circuit of Fig. 3-3, assuming that there are no other test points, the No. 1 pin of the picture tube is connected toward the grounded end, and the No. 12 pin is connected toward the hot end. Now, when checking for an open filament, an ohmmeter check from chassis to the No. 1 pin of the picture tube would indicate, by a continuity reading, that all tubes in this half of the circuit are good or, by an open reading, that one of them is bad. The number of tubes to be checked is thus cut roughly in half.

TROUBLE INDEX

The following pages contain a list of trouble symptoms. In each case, the symptom is described together with an explanation of the probable section at fault. A field-servicing procedure is given in detail, and if this does not clear up the trouble, reference is made to the chapter which gives the bench-servicing procedure.

A receiver is generally considered as having two sets of controls. One is a set of operating controls on the front panel for the consumer to change stations, control the volume of sound, adjust the picture to his individual preferences, etc. The other set is known as service, or rear-apron, controls. These are on the back of the chassis (rear apron) or on the picture-tube assembly and are adjustments of a semipermanent nature in that, once properly set, they do not require readjustment except to compensate for aging, changing of tubes, etc. These are the adjustments that are reset by the field serviceman. The description of the adjustments together with the adjustment procedure will be interspersed in the trouble index at those points where the symptom calls for adjustment.

In all cases in the text, throughout the field- and bench-servicing procedures, it is assumed that there is only one fault in the receiver. Of course, in actual practice there may be two or more, but in learning a servicing procedure, it is advisable to work on the most obvious symptom first. After that is cleared up, any other fault that remains is worked on as though it were a new job.

TROUBLE SYMPTOMS: PICTURE-TUBE ACCESSORIES

There is a group of symptoms which are caused by incorrect mechanical setup of the picture-tube accessories. These mechanical adjustments have to do with tilt, brightness, centering, and focusing. The problems are encountered when installing a new set, when installing a new picture tube, or when normal aging or vibration causes a need for resetting.

SYMPTOM: Tilted picture.

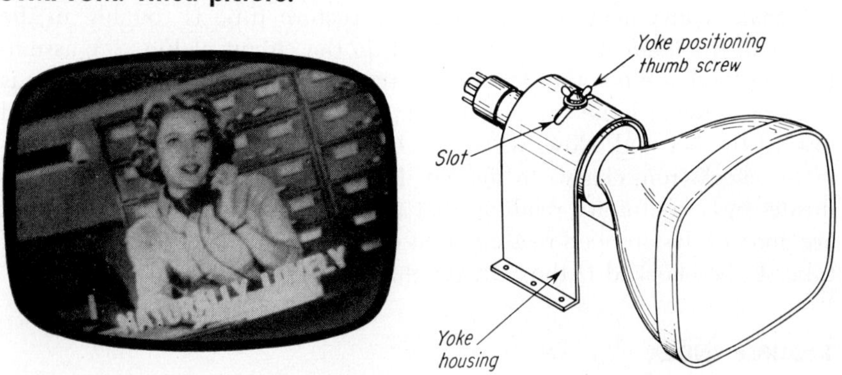

Fig. 3-4. A tilted raster and the yoke-positioning screw that corrects the condition.

This symptom is caused by incorrect positioning of the deflection yoke. The photograph in Fig. 3-4 shows an uncommon degree of tilt so as to make the condition more obvious. Generally, a slight degree of tilt goes unnoticed until written matter appears on the screen. Then the usual complaint is that writing or printed matter seems to slant uphill or down.

SERVICING PROCEDURE: To correct the condition, examine the deflection yoke to see how it is fastened. Generally, it is by means of a thumbscrew, as shown in Fig. 3-4. Loosen the screw and rotate the yoke until the raster lines are horizontal. This can be judged most easily at the top or bottom of the mask opening. Make sure that the yoke does not slide back but remains snugly forward against the bell of the picture tube. Finally tighten the wing nut in position.

SYMPTOM: Neck shadow.

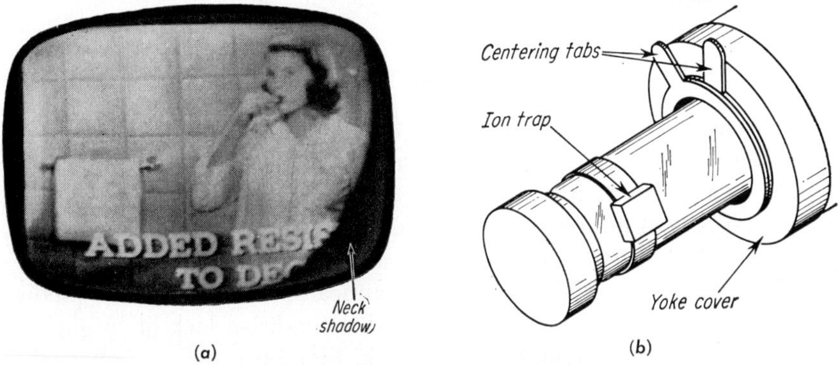

Fig. 3-5. Adjustment of the centering device is a cure for neck shadow.

The identifying feature for this symptom is the rounded dark area in one corner of the screen as shown in Fig. 3-5*a*. This happens when an off-center electron stream from the gun is partially blocked from the screen by the neck of the picture tube.

SERVICING PROCEDURE: First check that the yoke is snugly against the bell of the tube, and then look for the centering device. A common type is a pair of tabs like those shown in Fig. 3-5*b*. These are rotated singly or together, until the raster fills the screen.

After centering, it is usually necessary to readjust the ion trap and the vertical and horizontal size controls. Procedures for these adjustments are given later in this section. Other types of centering devices are described in Chap. 10.

SYMPTOM: Insufficient brightness.

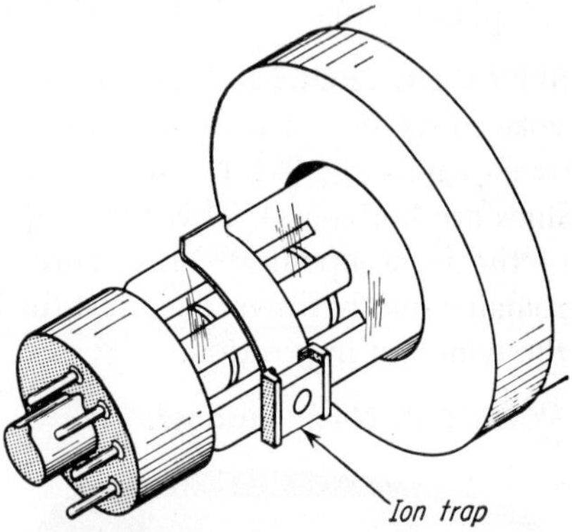

Fig. 3-6. Lack of brightness may be caused by a misadjusted ion trap.

The symptom must not be confused with a weak picture. On a normally operating receiver, turn the contrast down and the brightness up. This is a weak picture, and although the customer may say it lacks brightness, the screen is well illuminated. Now turn the contrast up and the brightness down. Note that the picture is too dark, portions that should be gray are black, and only very bright portions appear. Note also that the picture remains in focus. The serviceman calls this condition lack of brightness. An illustration of this symptom is given in Fig. 3-6.

In sets over one year old, the condition may be caused by a layer of soot over the face of the picture tube and the inside of the safety glass. Aside from this factor, it can be caused by a misadjusted ion trap, a weak picture tube, or trouble in the brightness circuit. In some receivers, it may also be caused by a weak video-amplifier tube.

SERVICING PROCEDURE:

Step 1. Turn the receiver off and shine your flashlight directly on the face of the receiver. Layers of soot will be immediately obvious, if present. If the picture tube and safety glass require cleaning, refer to the manufacturer's service notes to see if instructions are given for removing the safety glass, or if the CRT must be removed for the purpose.

Step 2. Turn the station selector to an unused channel, and rotate the brightness control through its range. As the brightness control is advanced, and the screen becomes more illuminated, take note of the raster lines. If these seem to increase in size and lose definition, the trouble is a condition known as "blooming." In severe cases, the raster lines disappear entirely at the full on position of the control. The servicing procedure for blooming is given later in the trouble index.

Step 3. If the lines remain in focus, but the screen is barely illuminated at the full on position of the brightness control, adjust the ion trap by moving it slowly forward and backward while rotating it at the same time. If the brightness comes up, reduce the setting of the brightness control and again adjust the ion trap for maximum brightness. Remember that some picture tubes do not use an ion trap, in which case this step is omitted.

Step 4. If the ion-trap adjustment shows no improvement, the other possibilities must be investigated. A weak picture tube is checked by connecting a CRT booster to it. If the brightness is improved by increasing the heater voltage in this way, the picture tube is weak. If the booster makes no improvement, try a new video-amplifier tube. If it is still inadequate, check the brightness circuit as described in Chap. 6.

SYMPTOM: Blurred picture.

(a)

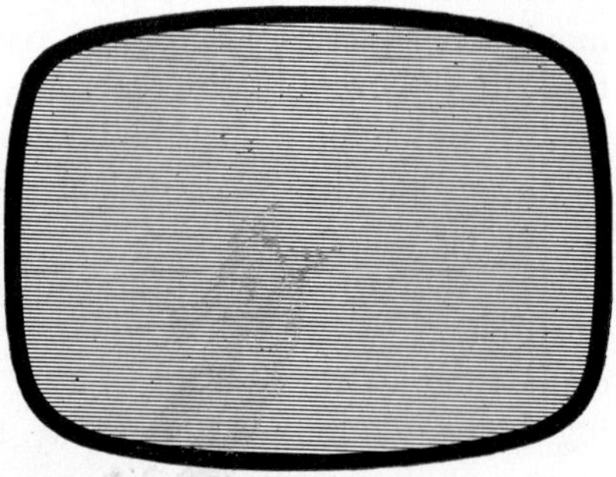

(b)

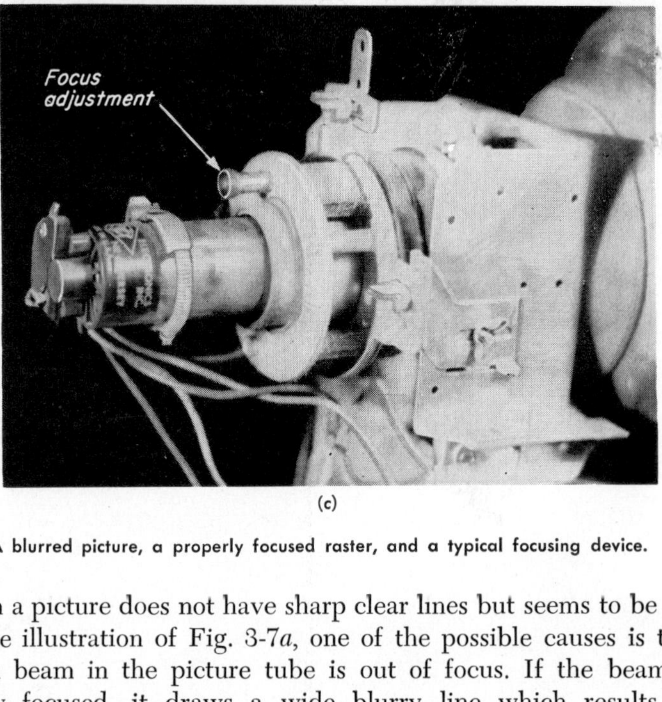

(c)

Fig. 3-7. A blurred picture, a properly focused raster, and a typical focusing device.

When a picture does not have sharp clear lines but seems to be blurred as in the illustration of Fig. 3-7*a*, one of the possible causes is that the electron beam in the picture tube is out of focus. If the beam is not properly focused, it draws a wide blurry line which results in the blurred picture. There are many devices for focusing the beam in the picture tube, a common one being the focalizer unit of Fig. 3-7*c*. The beam is focused by means of the screw shown in the drawing. Other focusing systems are described in Chap. 10.

The blurred picture may also be caused by insufficient bandwidth in the i-f or video amplifiers. In the servicing procedure which follows, the first step is to determine whether the cause is focusing or bandwidth, and then to continue accordingly.

SERVICING PROCEDURE: Tune the receiver to an unused channel, and adjust the brightness control for a well-lighted screen. Look for the raster lines. If they are sharp and well defined as in Fig. 3-7*b*, the beam is in focus and the bandwidth is at fault. Pull the receiver and refer to Chap 9 for the procedure for lack of definition or the blurred picture.

If the raster lines cannot be clearly seen, adjust the focusing device for sharp clear lines. It is sometimes difficult to do this over the entire screen. In this case adjust for clear lines over the center portion, which is normally the center of interest. A small amount of blurring is permissible at the outer edges of the screen. When proper focus is attained, reduce the setting of the brightness control, and retouch the ion trap for maximum brightness.

The above procedure may be applied to all types of focusing adjustments. In the case of picture tubes with electrostatic focusing, refer to the receiver service notes for recommendations as to how to change voltage at the No. 6 pin for focusing purposes. With these tubes, it is sometimes difficult to maintain good focus after the final touchup is made with the ion trap for maximum brightness. If this happens, try reversing the polarity of the ion trap by sliding it off the neck of the tube, turning it over, and replacing it. Then readjust the ion trap for maximum brightness.

ADDITIONAL ADJUSTMENTS: Width shims and raster corrector magnets. Some picture tubes are equipped with one or both of two additional accessories: the width shim and raster corrector magnets. Their locations are shown on the diagram of Fig. 3-8. The width shim is a brass or aluminum cylinder fitting snugly over the neck of the tube, and arranged so that it can be slipped in and out of the yoke assembly. Inserting it further causes the raster to become narrower; sliding it back adds to the width; complete insertion may cause the tube to go entirely dark. The shim should be adjusted so that the raster is just outside the mask opening.

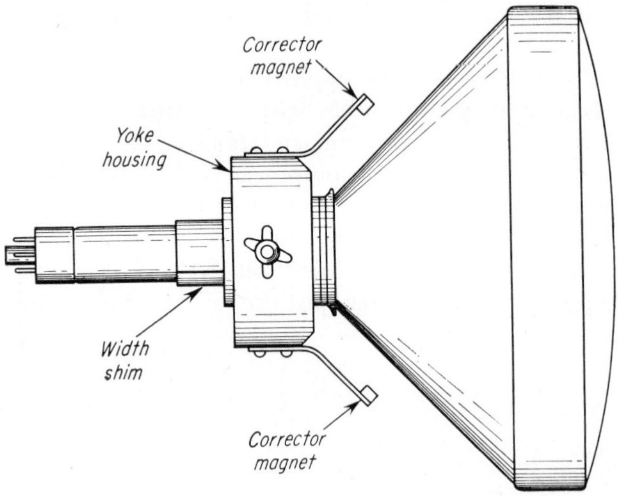

Fig. 3-8. Top view of picture-tube assembly, showing the location of the width shim and corrector magnets.

The corrector magnets are mounted on the yoke housing, one on each side of the picture tube. The supporting rods may be bent so as to bring the magnets closer to the sides of the picture tube, or farther away. In some tubes, the sides of the raster tend to bend, causing a distortion or pulling most noticeable on vertical lines in the picture near the edges. Bending the magnets in or out corrects this condition.

TROUBLE SYMPTOMS: LOW-VOLTAGE POWER SUPPLY

SYMPTOM: No raster—no sound (tubes do not light).

PROBABLE CAUSES: All functional sections are inoperative, and the common power supply is at fault. There may be a dead wall outlet, a disconnected plug, a break in the line cord, or a break in the filament supply.

SERVICING PROCEDURE:

Step 1. Disconnect the interlock, turn the set on, and check across the interlock terminals with an ohmmeter. This check is shown in Fig. 3-9. If the meter shows a reading, the filament supply line is probably good, and the trouble is in the wall outlet, line cord, or interlock. Go on to steps 2 and 3. If the meter shows open, go on to step 4.

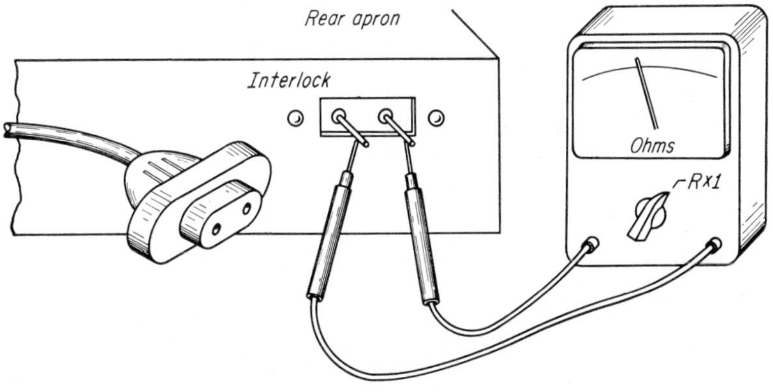

Fig. 3-9. Check line continuity of the power supply.

Step 2. Check the wall outlet by plugging in a lamp.

Step 3. Check the line cord by substituting a cheater cord and trying operation.

Step 4. If the receiver is of the series-heater type, check the tubes looking for an open heater, as described earlier in this chapter. If the receiver uses a power transformer for the filament supply, repairs are more conveniently made on the bench. Pull the chassis, and apply the procedure described in Chap. 4.

SYMPTOM: No raster—no sound—tubes light.

PROBABLE CAUSE: All functional sections are inoperative, pointing to the low-voltage power supply as the section at fault. Since the tubes are lighting, the outlet, line cord, and filament supply are working, and the trouble is in the rectifier or in some other component in the B power supply.

Replacing a rectifier is usually considered a field-servicing job even for the selenium type. Two field-servicing test procedures are given below, one for the transformer-tube type, and the other for the dry-plate selenium type.

SERVICING PROCEDURE: Tube type: Turn the set on, locate the rectifier tube, and look to see if the plates inside the tube are glowing red. If they are, there is a short or overload in the set. Do not put in a new tube. Check the set according to the bench-servicing procedure given in Chap. 4.

If the tube looks normal or is not lighting or is filled with a blue glow, put in a new tube, and watch it for a minute or two. If the plates become red, turn the set off. The short or overload is present, and continued use will ruin the new tube. If the tube looks normal, check to see if operation has been restored.

SERVICING PROCEDURE: Selenium type: Receivers which use selenium rectifiers incorporate a fusible resistor to protect the rectifiers against overload. This resistor is often equipped with plug pins for easy replacement. A typical mounting is shown in Fig. 3-10 for identification purposes.

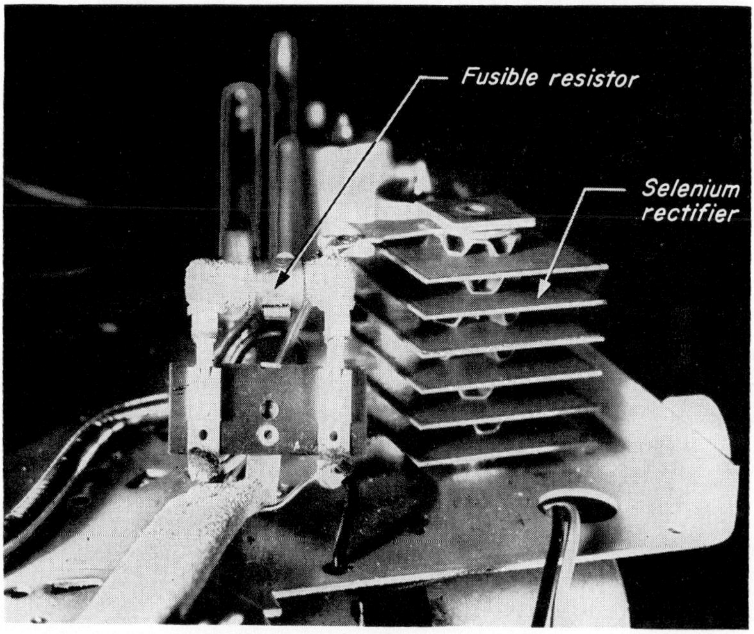

Fig. 3-10. The fusible resistor used in sets with selenium rectifiers.

Look for the fusing resistor, and examine it for signs of burning. If it is charred, there is a short or overload in the receiver. Apply the procedure described in Chap. 4.

The fusing resistor sometimes opens without being burned. This will also stop reception. Check the resistor with an ohmmeter. A good resistor will read 5 to 7 ohms. If it reads open, and there are no signs of overheating, replace the resistor. If it reads good, and there are no signs of overheating, the probability is that the selenium rectifiers are bad. Replace the rectifiers, being careful to use the correct size and to observe polarity. If reception is not restored, apply the bench-servicing procedure described in Chap. 4.

SYMPTOM: Small raster, small picture, sound normal (brightness below normal).

PROBABLE CAUSE: The symptom previously described as no raster or sound but with tubes lighting was diagnosed as being caused by a dead low-voltage B supply. The small-raster condition is generally caused by abnormally low output voltage at the B plus point.

The r-f amplifiers in the picture-sound chain operate at almost full efficiency even with a low plate supply; so that a picture is present. Lowered sound output is rarely noticed. The deflection circuits require full B voltage for proper operation, and the reduced deflection at lower plate supply accounts for the reduced height and width. In addition, reduced efficiency in the horizontal circuit lowers the high-voltage output, accounting for the reduced brightness. Advancing the brightness control to compensate causes the picture to bloom and lose focus. The usual cause for the drop in B voltage is a drop in line voltage or a weak rectifier.

Note that all four edges of the raster are visible, so that we do not have the off-center condition of Fig. 3-7*a*. Another possible cause of the small raster is that the deflection yoke is too far back, and part of the deflection becomes lost in neck shadow. The following field-servicing procedure investigates all the possibilities.

SERVICING PROCEDURE:

Step 1. Inspect the yoke and see that it is snugly fitted against the bell of the picture tube.

Step 2. Check the line with an a-c voltmeter. If it measures 105 volts or less, recommend the installation of a line-regulating transformer like one of those illustrated in Fig. 4-10. If it measures a normal 110 to 120 volts, replace the rectifier tube, or selenium rectifiers, as the case may be.

SYMPTOM: Dark band covering half the screen—sound normal.

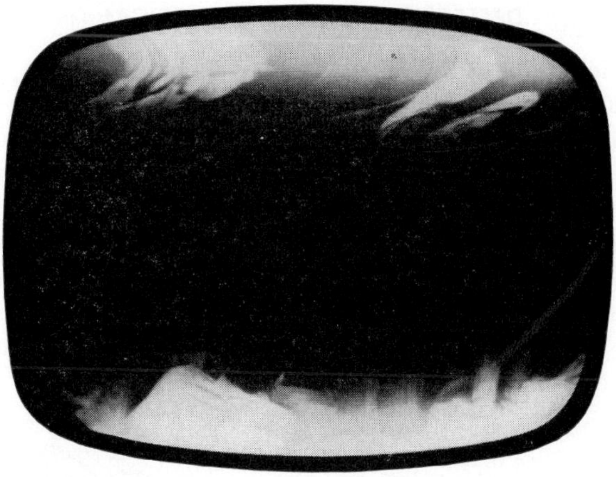

PROBABLE CAUSE: A dark band over half the screen is usually caused by leakage of 60-cycle line current into the picture-signal section. The most likely point of entry is cathode-heater leakage in one of the tubes in the signal chain.

This symptom is often called "hum in picture" as an outgrowth of radio terminology. In a-m radio receivers, leakage of line current into the signal chain causes a hum in the loudspeaker. The video equivalent is the dark bar on the screen of the picture tube. The symptom listed above states that sound is normal. If there is hum in the sound as well as the bar in the picture, the trouble is probably in the filter circuits of the low-voltage power supply.

SERVICING PROCEDURE: The picture-signal chain may contain eight or more tubes, any one of which may be causing the trouble. In order to limit testing to fewer tubes, determine if the dark bar is always present or only on those channels where a station is broadcasting. To do this turn the channel selector to an unused channel. If the dark bar disappears, the station signal is being modulated by leakage, and the trouble is before the detector, in one of the tuner or i-f tubes. These are changed one at a time until the trouble is cleared up.

If the dark bar persists on the raster, even on an unused channel, the defective tube is behind the video detector. The tubes involved are the detector, if it is of the vacuum-tube type, the video amplifiers, and the picture tube itself. Change the video-amplifier and detector tubes. Leakage troubles in the picture tube can be cleared up by the use of a booster of the isolation type.

DEFLECTION TROUBLES: VERTICAL

Field servicing in the vertical section includes setting the vertical service adjustments. The diagram of Fig. 3-11 illustrates all the vertical controls, including the hold, which is a front-panel-operating control. The vertical service, or rear-apron, controls are usually labeled VERT. SIZE or HEIGHT, and V. LINEARITY or VERT. LIN.

How to Set the Vertical Adjustments. Changing the setting of either of the vertical service adjustments has an interacting effect on the other. For this reason it is advisable to go through the entire adjustment procedure even though the symptom description seems to indicate that only one of the controls is misadjusted.

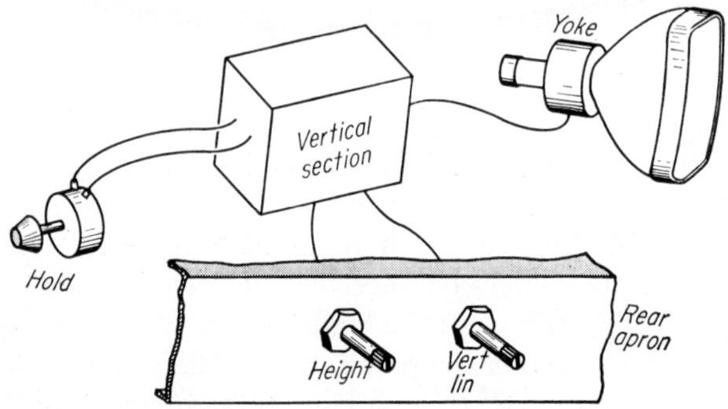

Fig. 3-11. Adjustments in the vertical section.

Step 1. Refer to the receiver service notes or examine the rear apron of the chassis and locate and identify the vertical adjustments.

Step 2. Check for proper centering. To do this, temporarily unseat the vertical oscillator tube, and see that the bright horizontal line that results is in the center of the screen. If the set is of the series-filament type, the line will fade and disappear as the tubes cool down, but it will remain long enough to check for centering. If the line is off center, adjust the centering tabs shown in Fig. 3-5.

Step 3. Turn both the height and linearity controls to the maximum counterclockwise position so that the picture is compressed into a narrow band in the center of the screen. The use of a mirror set up in front of the receiver makes it convenient to observe the screen while making adjustments at the rear of the cabinet. If the picture begins to roll while making this or any subsequent adjustment, retouch the hold knob on the front panel.

Step 4. Advance the height control slowly while observing the unlit portions of the screen above and below the picture. Both of these will narrow as the picture increases in size, but at different rates. As soon as the unlit portions are noticeably different in size, leave the height control, and advance the linearity control. This will cause the unlit portions to become more nearly equal in size as both narrow. Continue by slowly advancing first one control, and then the other, always trying to keep the unlit portions of the screen about equal in size, until finally both the top and bottom edges touch the mask opening simultaneously, and the picture fills the screen.

Step 5. Check for correct adjustment of the linearity control by misadjusting the vertical hold so that the picture rolls very slowly. The dark blanking bar between the frames of the picture can now be seen moving slowly up or down the screen. If the height of the blanking bar does not change as it moves from top to bottom, or vice versa, the vertical linearity is correctly adjusted.

Step 6. The correct setting of the vertical-hold control is obtained as follows: Swing the control through its full rotation. At one end, the picture is rolling up; at the other, it is rolling down. Midway between, there is a small area of rotation where the picture is stationary. The correct setting is the center of this stationary area of rotation.

SYMPTOM: Reduced height of raster—width normal—sound normal—figures on picture have a squat appearance.

PROBABLE CAUSE: It is important to note the full width. If the width is also reduced, an abnormally low voltage at the B plus point is indicated. Full width indicates normal operation of the power supply and horizontal sections. Reduced height indicates insufficient deflection in the vertical section. The obvious first choices for field servicing are weak tubes or misadjusted controls in the vertical section.

SERVICING PROCEDURE:

Step 1. Change the vertical oscillator and output tubes, one at a time. If replacing a tube causes the picture to roll, retouch the vertical-hold control.

Step 2. Reset the height and vertical linearity controls. This step should be made, even if replacement of one of the tubes restored the height, since the new tube will probably require readjustment of the settings.

SYMPTOM: Bright horizontal line across the center of the screen—sound normal.

PROBABLE CAUSE: Vertical deflection has been reduced to zero by complete inoperation of one of the stages in the vertical section. Since the beam is being deflected in the horizontal direction only, the result is the bright horizontal line.

SERVICING PROCEDURE:

Step 1. Change the tubes in the vertical section one at a time.

Step 2. If the height is restored, readjust the vertical controls. If the trouble persists, apply the bench-servicing procedures described in Chap. 5.

SYMPTOM: Improper vertical linearity.

Objects in the top half of the picture, like heads and necks, appear too long (eggheads) or too short (flatheads). Width is normal and sound is normal.

PROBABLE CAUSE: The vertical deflection generator is not sweeping evenly, and the raster lines are therefore farther apart (expanded) or closer together (compressed) at the top or bottom of the screen, as compared with the spacing in the middle. The picture therefore seems elongated or flattened. The vertical linearity circuit is out of adjustment. Although the symptom can happen on either the top or bottom of the screen, it is more apparent when at the top because abnormal heads are more noticeable than abnormally long or short legs.

SERVICING PROCEDURE: Go through the complete vertical-adjustment procedure, including the check for centering. The symptom seems to indicate that only one control is out of adjustment. However, because of interaction of the controls, it is better to go through the complete procedure.

SYMPTOM: Picture rolls vertically—sound is normal.

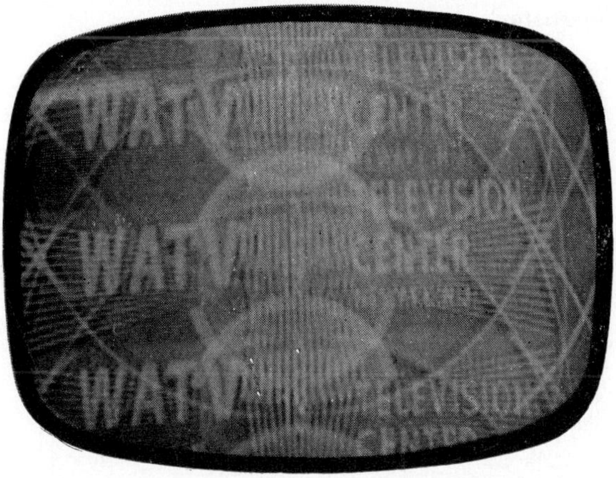

PROBABLE CAUSE: The deflection rate of the vertical oscillator in the receiver is faster or slower than the vertical deflection circuits at the transmitter.

SERVICING PROCEDURE:

Step 1. Locate and readjust the vertical-hold control. If this corrects the condition, and the picture "locks" vertically, instruct the consumer in the use of this control. In some receivers, the hold control is located on the back panel, and the consumer may not be aware of its existence. The term vertical lock implies that the picture remains stationary over a short range of the rotation of the hold control. Correct setting is the center of this lock-in range. If the lock occurs at either end of the hold control's range, aging has caused some component in the vertical circuit to change in value. Continued aging may place the lock outside the range of the control. The vertical circuits should be checked by the bench procedure described in Chap. 5.

Step 2. If manipulation of the hold control brings in one complete picture, but this does not lock and starts to roll again, the trouble is probably caused by failure of the transmitted vertical sync pulse to reach the vertical section. Change the sync tubes one at a time. If the trouble persists, apply the bench procedure described in Chap. 8 for this symptom.

If the hold control cannot bring in one complete picture, even momentarily, or reverse the direction of the rolling, its range is not great enough to bring the vertical section to its proper frequency. Apply the bench procedure described in Chap. 5.

TROUBLE SYMPTOMS: HORIZONTAL AND HIGH-VOLTAGE SECTION

SYMPTOM: No raster—sound normal

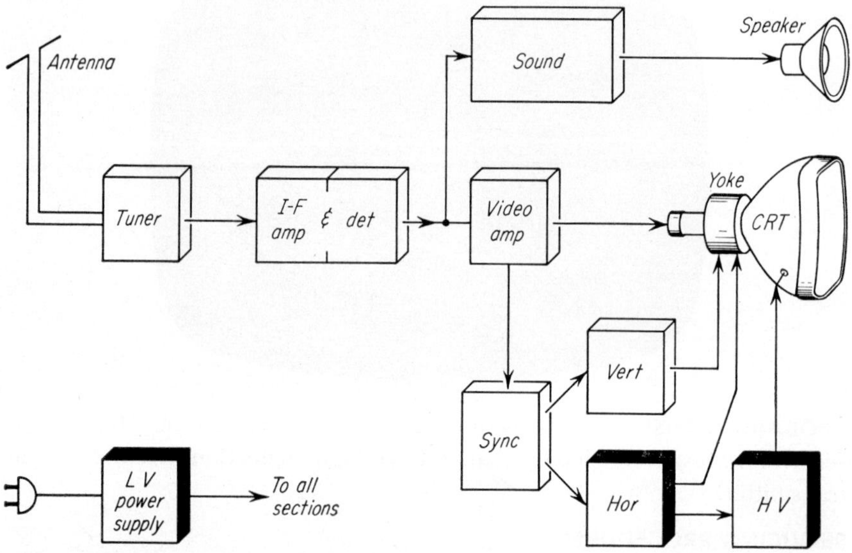

Fig. 3-12. Sections to investigate for loss of raster.

PROBABLE CAUSES: The section diagram of a receiver is repeated in Fig. 3-12 to help visualize the possible causes of the completely dark screen. The obvious first choices are a dead picture tube or the absence of high voltage at the second anode. Since the high-voltage stage is associated with the horizontal section, this also may be at fault. The presence of sound indicates that the signal chain and low-voltage power supply are working. However, as explained before, the power supply cannot be neglected because the deflection circuits require full B voltage, and the sound circuits may still be operating almost normally at a B voltage that reduces the horizontal section below the point where sufficient high voltage is developed. Another point which must not be forgotten is that a completely misadjusted ion trap may also darken the tube. To summarize, the symptom of a dark screen coupled with normal sound may be caused by a defective picture tube, a misplaced ion trap, or weakness in the horizontal system, the high-voltage power supply, or the low-voltage power supply. Some of these defective conditions may be caused by misadjusted controls or accessories. The controls and accessories relating to the horizontal/high-voltage system are illustrated in Fig 3-13.

SERVICING PROCEDURE:

Step 1. Look for a PHONO-TV switch, and make sure it is in the TV position. The switch blanks out the raster in the PHONO position. If this is the trouble, instruct the consumer in the use of the switch.

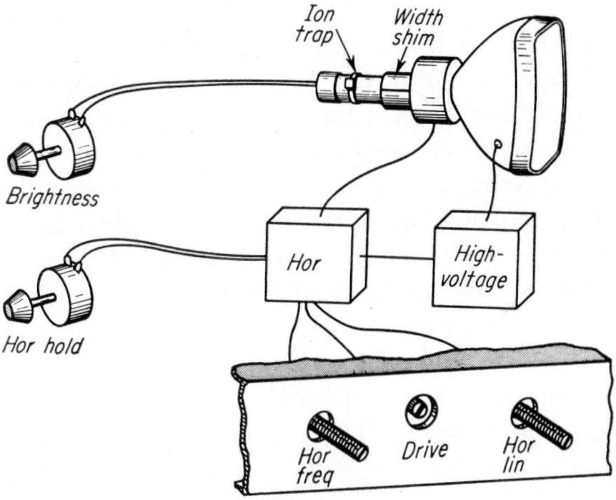

Fig. 3-13. Controls in the horizontal/high-voltage section.

Step 2. Check to see if the filament of the CRT lights. If it is out, resolder the filament pins (Nos. 1 and 12) on the tube. If the tube still fails to light, check for filament voltage. If this is present, replace the picture tube.

Step 3. Set the brightness control to maximum and swing the ion trap through its adjustment. If a glimmer of brightness comes on, adjust carefully for maximum brightness.

Step 4. In some receivers the horizontal-hold control on the front panel is of the permeability type which permits several full turns of rotation, rather than the three-quarter turn of the ordinary control. In receivers of this type, misadjustment of this control may be serious enough to stop the high voltage. Drive it through its range, and if brightness comes on, adjust for a normal picture. Leave the control at the center of its lock-in range, and instruct the consumer in its use. Omit this step with conventional three-quarter turn controls. These cannot misadjust the horizontal circuits sufficiently to stop the high voltage.

Step 5. Look for a width shim on the picture tube (see Fig. 3-13). This type of width control can darken the screen if pushed in too far. Pull it all the way out. If brightness comes on, push it back carefully for normal width. If the screen remains dark, leave it all the way out, and adjust it after the brightness has been restored. Other types of width control, if misadjusted, will not cause this symptom.

Step 6. Look for a horizontal drive control (see Fig. 3-13). If present, turn it fully counterclockwise, which is the position of maximum drive to the horizontal system. If this restores brightness and a normal picture, leave it in the full position. If a bright vertical line called a drive line is observed in the middle of the raster, reduce the drive until this line disappears.

Step 7. Check the brightness circuit. To do this, connect a voltmeter across the grid and cathode at the 2 and 11 contacts in the CRT socket, and vary the brightness control through its range. In a normal receiver, the voltage will vary from zero to 100 as the control is turned. If an abnormal reading is found, change the video-amplifier tube. If the abnormal condition remains, apply the checks described in Chap. 6.

Step 8. Bring the high-voltage lead close to the chassis. If you can draw a strong blue spark about ½ in. long, the high-voltage circuits are working, and the picture tube is bad. Replace it (see Chap. 10).

Step 9. If the previous check produces a weak spark, or no spark, remove the plate clip from the horizontal-output tube, and check for voltage at the clip. The normal voltage at this point is the B plus value of the receiver 250 to 300 volts depending on the type of low-voltage power supply. If the reading is approximately 170 volts, the rectifier is probably weak. Replace it. If the reading is zero, look for a high-voltage fuse. If this is blown, replace it. Also replace the damper tube as a preventive maintenance measure, since intermittent arcing in the damper is a common cause of the blown fuse. If the fuse is good, replace the damper tube. If the reading is normal, change the horizontal oscillator, afc phase detector, horizontal-output and high-voltage rectifier tubes, one at a time. If the trouble still persists, continue with the bench-servicing procedure described in Chap. 6.

These checks take longer to describe than to perform. An experienced field serviceman can do all of them in a matter of minutes. If the chassis is taken to the shop for bench servicing, make sure to take along the deflection yoke.

SYMPTOM: Blooming.

Normal pattern — Blooming

PROBABLE CAUSE: In this trouble, the customer complains that, when he increases the setting of the brightness control, the picture seems to grow in size, or "bloom," and become indistinct at the same time.

In the illustration above, a normal test pattern is shown on the left for comparison with the pattern obtained when blooming is present. Note the lack of brightness. Advancing the brightness control causes the over-all picture to expand so that the outer rings labeled in the normal pattern have moved out toward the mask opening. The fault is most often caused by a decrease in high voltage. This points to the horizontal/high-voltage system. It may also be due to weakness in the picture tube.

SERVICING PROCEDURE: Change the horizontal output, damper, and high-voltage rectifier tubes one at a time. If the symptom persists, try a booster on the CRT. If this improves the condition, the picture tube is weak; if it does not, apply the bench-servicing procedure given in Chap. 6 for this symptom.

SYMPTOM: Insufficient width.

PROBABLE CAUSE: The cause of this symptom is, of course, insufficient horizontal deflection, which points to weakness in the horizontal system, most likely in the output stage. However, there may be a misadjusted width control, and there is the additional possibility that the low-voltage power supply is weak. This usually causes a reduction in height also, but often enough the height is less affected than the width, or the height control has been readjusted.

SERVICING PROCEDURE:

Step 1. Look for a width shim or width control and adjust for normal width. If these controls are missing or ineffective, look for a drive control. This is not intended as a width control, but since it has some effect on the width, its adjustment should be checked.

Step 2. Replace the horizontal-output tube.

Step 3. Check the line voltage. If it is below 105 volts, recommend a line-regulating transformer. If the line measures a normal 110 to 120 volts, replace the low-voltage rectifier.

SYMPTOM: Bright vertical lines and foldover.

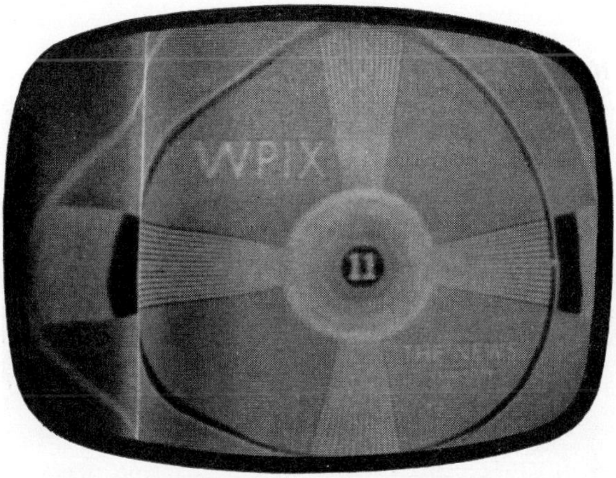

PROBABLE CAUSE: In this symptom the picture appears to be "pleated" with sections folded in at the center or folded over at the edges. It is caused by uneven horizontal deflection, with the trouble generally in the horizontal output or damper stage.

SERVICING PROCEDURE: If the receiver is equipped with a drive control, check its adjustment, since the symptom may be caused by overloading the horizontal output tube. If this does not clear up the picture, replace first the damper tube, and then the horizontal output tube. Handle these tubes carefully, since they normally run very hot. If the symptom persists, apply the bench-servicing procedure described in Chap. 6 for this symptom.

SYNC TROUBLES

SYMPTOM: "Marble cake."

PROBABLE CAUSE: In this symptom, the loss of synchronization between the deflection circuits at the transmitter and receiver causes the display on the screen to be a meaningless scramble of black and white areas similar to the illustration.

In sync troubles the fault is not necessarily only in the sync section. The sync pulses may never get to the sync section since they are transmitted signals and may get lost in the picture chain. Also, the sync section may be delivering normal sync pulses to the deflection circuits, but these are not operating anywhere near proper frequency, and the normal sync pulses cannot correct the situation. A servicing procedure for out-of-sync operation must investigate all these possibilities.

SERVICING PROCEDURE:

Step 1. Decouple the antenna by disconnecting the transmission line from the antenna connections and allowing it to lie near the antenna cable in the receiver. If it is now possible to tune in a normal picture, the deflection and sync circuits are operating normally, and there is an overload condition in the picture chain. Refer to the service notes for the receiver and look for an agc, noise-control, or local-distant switch. If the control is provided, readjust it in accordance with the manufacturer's instructions. If there is no such control, or if the readjustment does not effect a repair, apply the bench procedure given in Chap. 12 for this symptom.

Step 2. Decoupling the antenna made no improvement, and so the antenna is reconnected. The horizontal-hold control on the front panel, which is ineffective in clearing up the condition, is set to its mid-rotation position. The horizontal frequency control on the rear apron is adjusted, while watching the screen. A picture or a vertically rolling picture may come on. A touch of the vertical-hold control should take care of the rolling.

If the picture now locks in, the repair is completed. Check operation on all channels, and retouch the horizontal frequency control, if needed.

If the picture comes on but will not lock, the deflection circuits can be brought to proper frequency, but the absence of the sync pulse prevents a lock. Change the sync tubes, and if the symptom persists, apply the bench procedure given in Chap. 8 for this symptom.

Step 3. If adjusting the horizontal frequency control does not produce a picture, the horizontal oscillator cannot be brought to its proper frequency. Change the horizontal oscillator, afc, and phase-detector tubes, and try the horizontal frequency control again. If the symptom still persists, apply the bench procedure given in Chap. 8 for this symptom.

SYMPTOM: "Christmas treeing" or "gunboating."

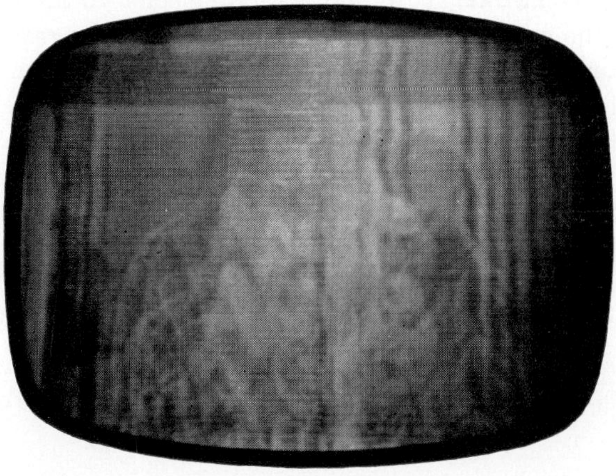

PROBABLE CAUSE: In this symptom, although the picture remains in sync, erratic operation of the horizontal oscillator causes the bright lines which give the effect called Christmas treeing. The name gunboating is also applied to this symptom because it is accompanied by a chirping sound from the high-voltage cage. The trouble occurs in receivers that use the synchroguide system.

SERVICING PROCEDURE: Readjust the horizontal frequency controls as described in Chap. 8 for a synchroguide system.

SYMPTOM: Pulling or bending.

PROBABLE CAUSE: Some of the horizontal lines are misplaced to either the left or the right, causing the picture to appear as though it were being viewed through a fun-house mirror. This is usually caused by 60-cycle or hum currents interfering with the sync pulses. The hum frequency is usually introduced by cathode-heater leakage in the sync or horizontal oscillator tubes, or indirectly, through any of the tubes in the picture chain.

SERVICING PROCEDURE: Turn the station selector to an unused channel; turn up the contrast and observe the raster. If the raster shows the dark hum bar, as seen in the right-hand illustration above, change the video amplifier and detector tubes one at a time. In the case of a crystal detector, do not disturb it, since this will not introduce hum. If the bar persists, try the use of an isolation-type booster on the picture tube.

If the bar is not seen on the unused channel, the following tubes are changed one at a time: horizontal oscillator, phase detector, sync tubes, i-f amplifier, and tuner. After each change check for curing of the symptom. If the tubes do not improve matters, apply the procedure given in Chap. 8 for the symptom.

SYMPTOM: Multiple pictures.

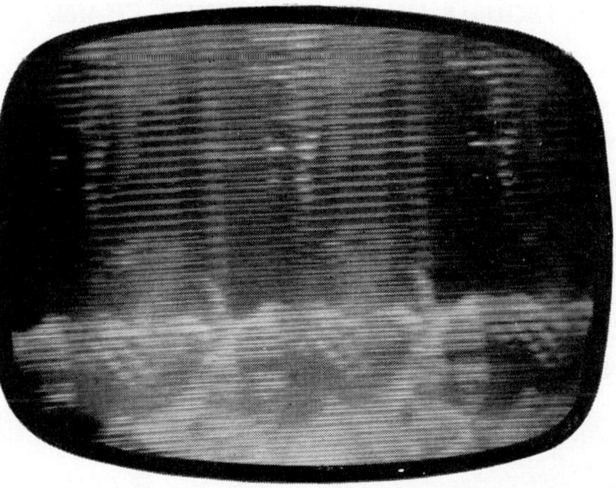

PROBABLE CAUSE: The symptom of multiple pictures as illustrated is caused by off-frequency operation of the horizontal oscillator. The condition can be caused by a defective oscillator-afc tube, an off-value component in the oscillator-afc circuit, or a misadjusted horizontal frequency control.

SERVICING PROCEDURE: Set the front-panel horizontal frequency control in its mid-rotation position, and adjust the rear-apron frequency control while watching the picture. If the condition is cleared up, check operation on all channels, and if needed, retouch the frequency control. If the condition remains unimproved, or a change in the number of multiple images is the best that can be accomplished, change the horizontal oscillator tube and try the frequency control again. If the symptom persists, apply the bench-servicing procedure given in Chap. 8 for off-frequency operation.

TROUBLE SYMPTOMS: PICTURE CHAIN

Breakdown of the horizontal/high-voltage section stops the raster, and, of course, the picture. We can also have the condition where the raster is present but there is still no picture. The normal raster indicates normal operation of the deflection circuits and low- and high-voltage power supplies. Lack of picture coupled with full raster places the trouble in the picture-sound signal chain. The presence or absence of sound is of minor importance in the servicing procedure, because the sound take-off point is so far back in the chain as to be of little diagnostic value.

SYMPTOM: No picture or weak picture—snow present.

PROBABLE CAUSE: These symptoms are grouped together because the same field-servicing procedure is applied to both. The symptom of no picture is caused by the complete failure of one of the stages in the picture chain; the weak picture is caused by partial failure or inefficiency in one of the stages. The presence of snow is very important in localizing the trouble.

Snow is the result of random electrical pulses on the screen of the picture tube. In a radio receiver, random electrical pulses cause crackling noises; in the television signal chain, similar pulses cause irregular flashes which give the screen a speckled appearance. The moving speckles are called snow.

Noise pulses may be picked up by the antenna or may be generated by any tube in the signal chain. However, tubes well along in the signal chain are followed by less amplification, and so any noise they generate will not affect the screen. If a break comes early in the chain, noises generated by the first tube after the break are followed by sufficient amplification to cause snow.

The presence of snow on the screen is therefore of considerable importance in servicing. It shows that the picture tube and most of the signal chain are working, limiting the trouble to the antenna or tuner tubes.

SERVICING PROCEDURE: Change the two tuner tubes one at a time. If this does not restore normal operation, disconnect the antenna, connect a portable indoor antenna, and try operation. If this produces better results, the regular antenna installation is at fault. Check the antenna as described in Chap. 16. If the indoor antenna does not improve matters, in an area where indoor antennas are known to work, the tuner is defective. Apply the bench-servicing procedures described in Chap. 14.

In fringe areas, where an indoor antenna cannot be expected to work, other methods must be used to determine whether the fault lies in the tuner or antenna. These are described in Chap. 16.

SYMPTOM: No picture—no snow.

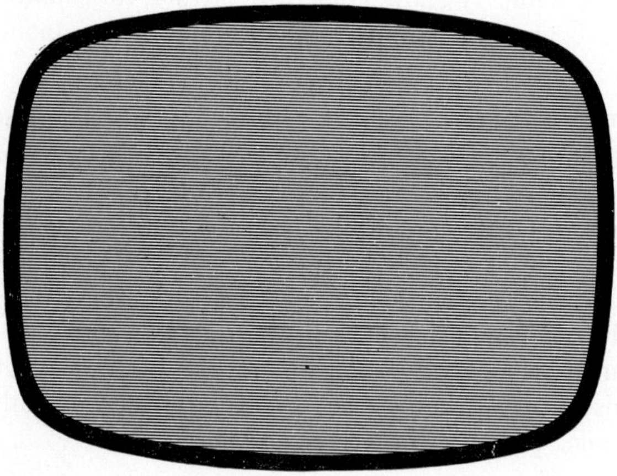

PROBABLE CAUSE: The presence of only the normal raster shows a break in the picture-signal chain. The absence of snow places the break behind the tuner.

SERVICING PROCEDURE:

Step 1. Decouple the antenna by disconnecting the transmission line and allowing it to lie near the antenna line in the receiver. Tune the receiver to all channels normally received, and if a picture or at least snow can now be seen, there is an overload condition in the picture chain. Look for an agc or noise control and adjust it in accordance with the manufacturer's service notes. If there is no such control, or if the adjustment does not effect a repair, apply the bench procedure given in Chap. 12 for this symptom.

Step 2. If decoupling the antenna has no effect, replace the tubes in the picture chain in the following order: video amplifier, detector (tube or crystal), video i-f tubes, and agc tube, if used. On the chance that the receiver uses a shelf-type power supply, try changing the sound-output tube. If this does not effect a repair, apply the bench-servicing procedure given in Chap. 9.

SYMPTOM: Weak picture—no snow.

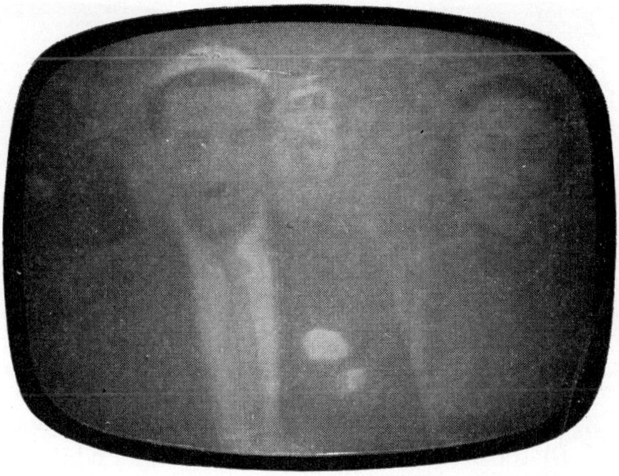

PROBABLE CAUSE: The weak picture is one that lacks blacks, even though the contrast control is advanced to the full on position. The loss of contrast is caused by weakness or inefficiency in one of the stages of the picture-signal chain. The absence of snow places the fault behind the tuner.

SERVICING PROCEDURE: Replace all tubes in the picture chain one at a time. Also change the sound-output tube, even though the sound seems normal, because of the possible use of a shelf-type power supply. If the symptom persists, apply the bench-servicing procedure given in Chap. 9.

SYMPTOM: Overloaded picture and negative picture.

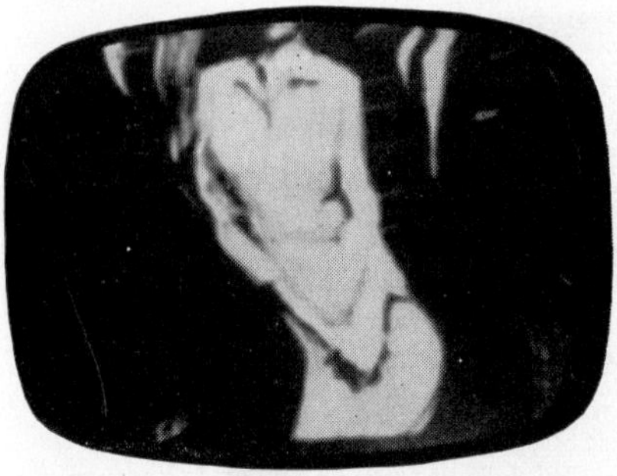

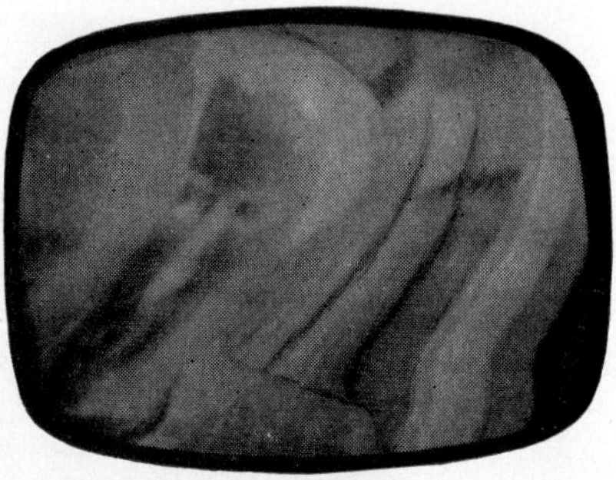

PROBABLE CAUSE: The overloaded picture is too rich in blacks, and reducing the contrast control has little effect. Synchronization is likely to be unstable. In the negative picture, blacks and whites are reversed, and the picture has the appearance of a photographic negative. The two symptoms are treated together, because the negative picture is caused by a slightly greater degree of overload than the overloaded picture. As a matter of fact, careful manipulation of the fine-tuning control can usually change one symptom to the other.

SERVICING PROCEDURE: Look for the manual control in the agc system, and adjust it in accordance with the manufacturer's instructions. If there is no such control, or if its adjustment does not clear up the complaint, change the first and second i-f tubes, and the detector, diode, or crystal. If the set uses an agc tube, change it, and also the video-amplifier tube. If the symptom persists, apply the bench-servicing procedure given in Chap. 12 for these symptoms.

SYMPTOM: Sound bars in the picture.

PROBABLE CAUSE: In this symptom, a series of dark bars stretch across the screen and come and go with variations of the sound. The condition may be caused by a microphonic tube in the picture-signal chain or by improper tuning of the receiver. A microphonic tube is one that reacts to sound vibration. In a properly tuned receiver, the responses to the picture and sound carriers are maintained at fixed levels. The ratio is somewhat controlled by the fine tuner. Sound bars often result when the receiver is tuned for too large a response to the sound carrier.

SERVICING PROCEDURE: Tune in a station where the symptom is noticeable, and reduce the setting of the volume control. If the sound bars disappear at a low volume level, there is likely to be a microphonic tube. Advance the volume control until just before the sound bars start, and gently tap each tube in the picture chain. When the microphonic tube is reached, flashes will appear in the picture. Replace this tube.

If the volume control has little effect on the sound bars, keep the volume at a low level, and adjust the fine tuner. This should affect the sound bars but will probably be unable to reach the point of best picture and best sound. Set the fine tuner to its mid-rotation position, remove the knobs, and look for the oscillator-adjustment slug. Refer to the manufacturer's service data for instructions on the adjustment of the oscillator slugs. Reset the adjustments on all channels normally received. Generalized adjustment procedures for oscillator-tuning slugs are given in Chap. 14.

Summary of Trouble Index

Field Servicing

Symptom	*Section*	*Tubes*	*Check or Adjust*	*For Further Information See Chapter*
Tilted picture 	Picture tube		Yoke-positioning screw	10
Insufficient brightness 	Horizontal/high voltage Picture tube	Video amplifier	Clean face of CRT Ion trap CRT booster	Check brightness circuit as described in Chap. 6
Neck shadow 	Picture tube		Centering device	10
Blurred picture 	Picture tube		Focus control	10 If focus is good, check for bandwidth as described in Chap. 9
No raster, no sound (dead set, tubes do not light)	Low-voltage power supply	All tubes in a series-filament type of receiver	Outlet Line cord Interlock Low-voltage fuse	4
No raster, no sound (tubes light)	Low-voltage power supply	Rectifier		4
Small raster 	Low-voltage power supply	Rectifier	Line voltage Yoke positioning	4

Summary of Trouble Index (Continued)

Symptom	*Section*	*Tubes*	*Check or Adjust*	*For Further Information See Chapter*
Hum bar 	Low-voltage power supply	R-f amplifier Mixer-oscillator I-f tubes Video detector Video amplifier	Filter condenser Try isolation type of CRT booster	8
Lack of height 	Vertical	Vertical oscillator Vertical amplifier	Height Vertical linearity	5
Horizontal line 	Vertical	Vertical oscillator Vertical amplifier		5
Vertical distortion 	Vertical		Centering Height Vertical linearity	5
Rolling 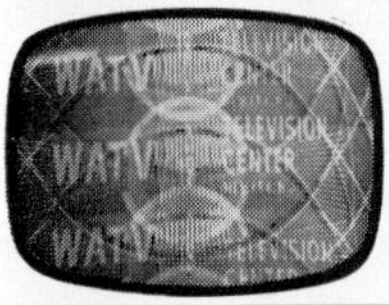	Vertical Sync	Vertical oscillator Sync tubes	Vertical hold Automatic gain control	5 and 8
No raster, sound normal	Horizontal/ high voltage	Horizontal oscillator/afc Phase detector Horizontal amplifier Damper High-voltage rectifier Video amplifier	Phono-TV switch CRT filament Ion trap Horizontal frequency (coarse) Width shim Horizontal drive High-voltage fuse	6

Summary of Trouble Index (Continued)

Symptom	*Section*	*Tubes*	*Check or Adjust*	*For Further Information See Chapter*
Blooming 	Horizontal/ high voltage	Horizontal oscillator/afc Horizontal amplifier Damper High-voltage rectifier		6
Lack of width 	Horizontal/ high voltage	Horizontal amplifier Low-voltage rectifier	Line voltage Width control or shim Horizontal drive	6
Foldover 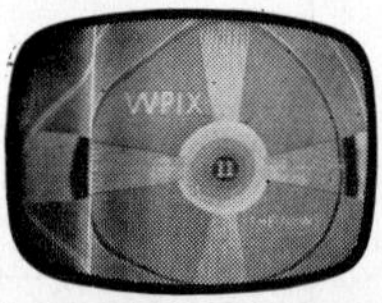	Horizontal/ high voltage	Damper Horizontal amplifier	Horizontal drive	6
Marble cake 	Sync	Sync tubes Horizontal oscillator/afc Phase detector	Agc Horizontal frequency Vertical hold Noise control	8
Pulling or bending 	Sync	Sync tubes Video detector Video amplifier Horizontal oscillator/afc Phase detector I-f amplifier tubes R-f mixer- oscillator	Try isolation booster on picture tube	8
Multiple pictures 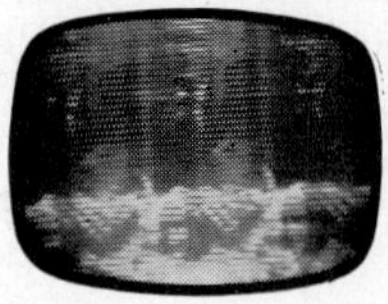	Sync	Horizontal oscillator/afc	Horizontal coarse- frequency control Horizontal- hold control	8

Summary of Trouble Index (Continued)

Symptom	*Section*	*Tubes*	*Check or Adjust*	*For Further Information See Chapter*
Christmas treeing 	Sync	Horizontal oscillator/afc	Horizontal frequency Horizontal stabilizer Horizontal lock	8
Snow 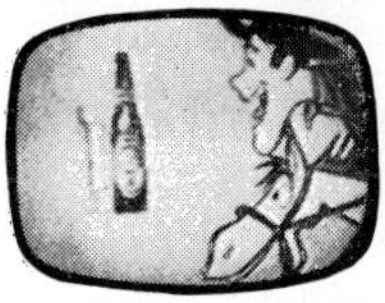	Picture chain	R-f amplifier Mixer-oscillator	Antenna	14
No picture, no snow 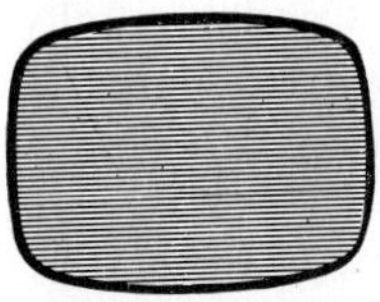	Picture chain	Video amplifier Video detector I-f tubes Agc tube A-f output	Agc	9
Weak picture, no snow 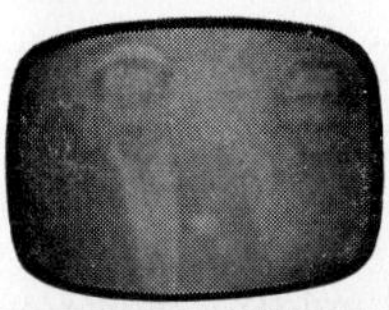	Picture chain	Video amplifier Video detector I-f tubes Audio output	Contrast	9
Sound bars 	Picture chain	R-f amplifier Mixer-oscillator I-f tubes Video detector Video amplifier	Fine tuner Oscillator frequency-adjustment slugs	14
No sound, picture normal	Sound	Sound i-f Sound detector Audio tubes		15
Intercarrier buzz	Sound		Fine tuner Buzz control	15

4

Servicing Low-voltage Power Supplies

The low-voltage power supply is so called to distinguish it from the high-voltage supply, which is part of the horizontal deflection system and furnishes 12,000 to 16,000 volts for the second anode of the picture tube. The low-voltage power supply furnishes filament and bias potentials to the picture tube, as well as filament, plate, screen, and bias potentials to all the other tubes in the television receiver. Fundamentally, there will be a transformer or series-filament arrangement to light the tubes from the 120-volt line, and a rectifier and filter system to change the 60-cycle line current to the direct current required by the plate circuits of the receiver tubes.

Indications of Trouble in the Power Supply. If the tubes in the receiver fail to light, or if turning on the set causes house fuses to blow or house lights to dim, the trouble is definitely in the low-voltage power supply. If the tubes light, but the set is completely dead, that is, no light on the screen and no sound from the loudspeaker, the trouble is most likely to be in the power supply. Finally, servicing procedures for other symptoms, such as lack of brightness or lack of height and width, call for a check of the low-voltage power supply as a possible cause of these troubles.

Quick Check for Trouble in the Power Supply. The power supply is functioning normally when all the tubes light and B plus measures 250 to 325 volts. An easily found point at which to measure B plus is the plate pin of the damper tube. This is the No. 5 pin for the 6AX4, 6AU4, and 6W4 types of damper tube. The actual B plus voltage measurement should be within 10 per cent of the normal reading as set down by the manufacturer in his service notes.

Types of Power Supplies. Low-voltage power supplies may be classified into three general groupings. One uses a power transformer and tube rectifier. Another uses a pair of selenium rectifiers in a voltage-doubler circuit and a filament transformer to light the tubes. The third is transformerless. The tube filaments are lined up in a series or series-parallel string and are lit from the 117-volt line, and the B plus supply is obtained from a half-wave voltage doubler. These groupings present different servicing problems and so will be studied separately.

TRANSFORMER-TUBE TYPE OF POWER SUPPLY

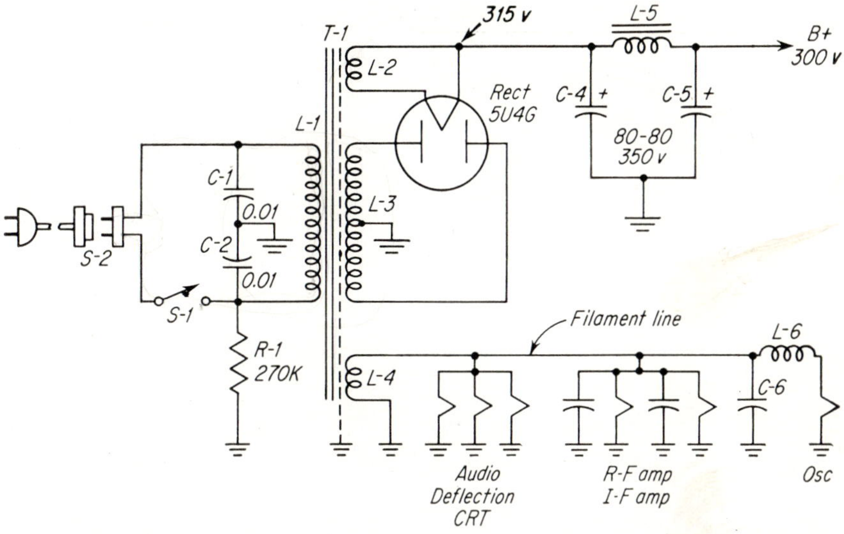

Fig. 4-1. Basic transformer-tube type of low-voltage power supply.

Basic Circuit. A basic circuit of the transformer-tube type of low-voltage power supply is shown in Fig. 4-1. The primary winding *L*-1 of power transformer *T*-1 is connected to the line through the power switch S-1. Switch S-1 is usually ganged with the audio volume control. Interlock S-2 is a safety device. It is simply a 120-volt connector in the line cord, the receptacle half of which is fastened to the rear panel of the cabinet. Removing the back of the cabinet or high-voltage cage therefore automatically disconnects the line. This ensures the consumer's safety, but since the serviceman must often work inside a live receiver, he carries a spare line cord with the proper connection for this purpose. The line cord–connector combination is called a "cheater" and is illustrated in Fig. 4-2*a*. Field servicemen often make up a special cheater cord like the one illustrated in Fig. 4-2*b*. This acts as an extension cord between the removed back of the cabinet and the receptacle plug in the receiver. It provides connectors for the two common types of interlocks, as well as outlets for soldering gun, trouble light, test equipment, etc.

Condensers *C*-1 and *C*-2, and resistor *R*-1 in the primary circuit of the power transformer act as a line filter. The diagram of Fig. 4-1 shows commonly used values. Individual manufacturers may omit any or all of these components. The line filter rarely gives any servicing difficulty.

Filament winding *L*-2 and plate winding *L*-3 are hooked up in a standard full-wave rectifier circuit. The tubes used are the 5U4-G, 5U4-GB, or 5AS4 types of rectifier. The 5U4-G and 5U4-GB are alike except for current rating. The 5U4-GB is rated for an output current of 275 ma, whereas the 5U4-G is rated at 225 ma.

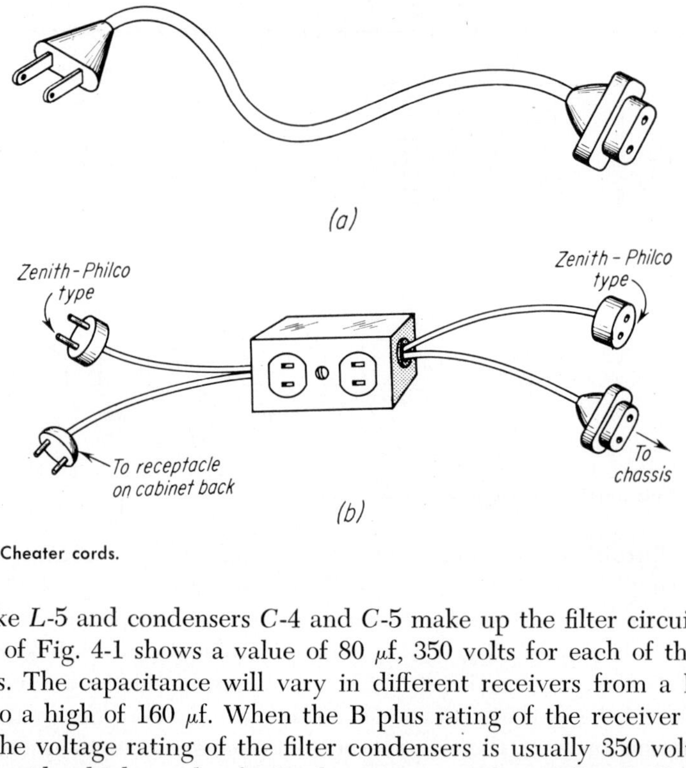

Fig. 4-2. Cheater cords.

Choke *L*-5 and condensers *C*-4 and *C*-5 make up the filter circuit. The circuit of Fig. 4-1 shows a value of 80 μf, 350 volts for each of the condensers. The capacitance will vary in different receivers from a low of 40 μf to a high of 160 μf. When the B plus rating of the receiver is 300 volts, the voltage rating of the filter condensers is usually 350 volts. Receivers with a higher value for B plus may use 400- or 450-volt filter condensers. Both filters are usually contained in one dual unit. Choke *L*-5 is sometimes the field coil of a dynamic speaker, but most often the speaker is of the p-m type, and choke *L*-5 is a standard filter choke. A representative filter-choke rating is 1 henry, 250 ma, and 50 ohms.

Winding *L*-4 on the power transformer furnishes 6.3 volts at about 8 amp for the filament circuit. The voltage is applied directly to the filaments of the picture tube and the tubes in the audio and deflection circuits. To prevent possible interaction between the tubes in the high-frequency circuits, r-f filters are often incorporated in the filament line. Condenser *C*-6 and r-f choke *L*-6 in the feed line to the filament of the oscillator tube are representative of this type of filter. Additional filters are often found across the filaments of the r-f and i-f amplifier tubes. From the serviceman's point of view, these filters need no further discussion, since they rarely present servicing problems.

Variations in the Transformer-Tube Type of Low-voltage Power Supply. The basic tube-transformer power-supply circuit of Fig. 4-1 is used with very little variation for a large number of television receivers containing up to about 20 tubes. About 225 ma of B current is required to supply the plate current for this number of tubes, and this is the full-load current for a 5U4-G rectifier tube. In larger sets where more tubes are used, the B voltage requirement is the same, but more current is needed for both the filament and plate circuits.

A circuit representative of one of these heavier power supplies is shown in Fig. 4-3. The plates of the 5U4-G tubes are connected in parallel, and

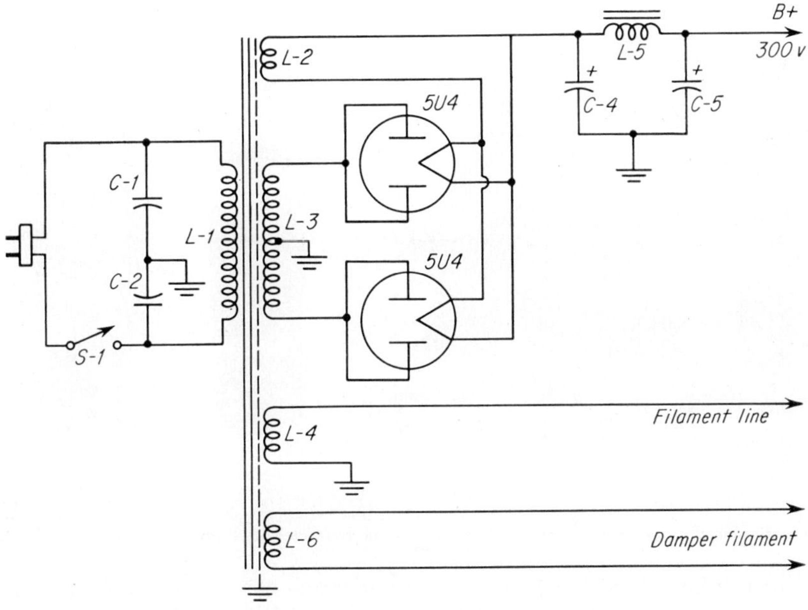

Fig. 4-3. Power supply designed to deliver 350 ma or more of B plus current.

they are connected as two half-wave rectifiers to make up the same full-wave rectifier circuit. Now each tube carries half the load, and the maximum current rating of the rectifier is therefore 450 ma. In some receivers, the circuit of Fig. 4-1 is left untouched, the second rectifier tube being simply connected in parallel. Operation is the same in either case.

The power transformer and choke are wound with heavier wire to handle the extra current and therefore are bulkier than those used in the basic circuit. A typical rating for filter choke *L*-5 is 0.8 henry, 375 ma, and 25 ohms. In the power transformer, the 5-volt winding is rated at 6 amp to light the two rectifier tubes, and the 6.3-volt winding may be rated at 10 amp to light the extra tube filaments.

Winding *L*-6 on the power transformer is rated at 6.3 volts and 1.2 amp. This winding is designed to feed the damper tube. This is not necessarily an adjunct of the larger sets. The separate damper winding may be used in smaller sets also. It is used where one of the damper-filament leads is connected to its cathode, or a tap on the horizontal-output transformer. In these circuits, the damper-filament winding and the connecting leads are well insulated, since they must withstand the 4,000-volt pulses which may be present. Some receivers also use a separate 6-volt winding for the picture tube.

How to Identify the Transformer-Tube Rectifier Type of Power Supply. A typical receiver is shown in Fig. 4-4 with the power transformer and

Fig. 4-4. Identification of the transformer-tube rectifier type of power supply.

rectifier tube labeled. The two common types of rectifier tube are also shown in Fig. 4-4 for identification purposes. The tubes are easily recognized since they are the largest in the receiver. Another pointer is that transformer-type receiver chassis are considerably heavier than those which do not use this component.

Servicing Charts for the Transformer-Tube Rectifier Power Supply. Most of the troubles relating to the power supply can be found by voltage, resistance, and substitution checks. Charts showing normal voltage and resistance checks are added to the basic diagram in Fig. 4-5 as an aid in service work.

Before resistance checks are made, the receiver plug is disconnected, and the filter condensers are discharged by shorting them. The resistance chart shows a variance in values. As a general rule, the lower ohmic values will be found in receivers with more tubes.

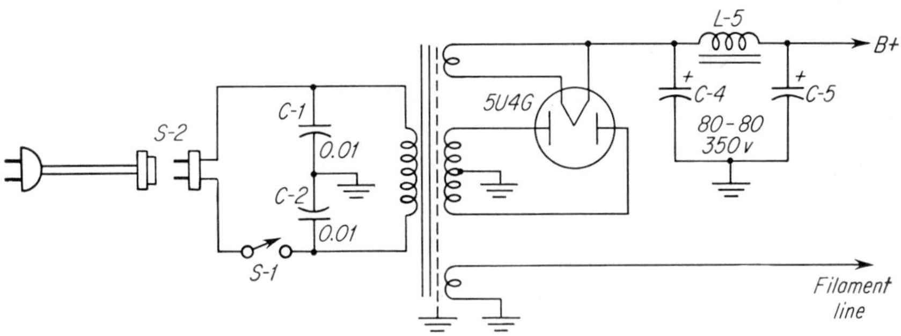

Fig. 4-5. Basic transformer-tube rectifier power supply with voltage and resistance charts.

Normal voltage readings:

Line	117 volts a-c
Across filament line	6.3 volts a-c
Chassis to rectifier plates	250–380 volts a-c
Chassis to rectifier filament	240–340 volts d-c
Chassis to B plus	220–325 volts d-c
Across filter choke *L-5*	15–20 volts d-c

Normal resistance readings:

Plug, prong to prong	1–2 ohms
Chassis to rectifier plates	15–30 ohms
Across filter choke *L-5*	25–60 ohms
Chassis to rectifier filament	15,000–100,000 ohms

The wide divergence shown for the rectifier cathode-to-chassis reading is caused by differences in voltage-divider or bleeder systems. Some receivers have no bleeder, or a light bleeder. These will have a high ohmic value for this reading. In all cases the true reading will be masked by the charging current and leakage resistance of the electolytic filter condensers. Reverse test prods, and take the higher reading when making this test.

Servicing Procedure When Tubes Do Not Light. When the tubes in a receiver using a power transformer do not light, the trouble is usually in the line cord-to-transformer circuit. This circuit is shown in Fig. 4-6. Some sets use a fuse in this line; so a fuse is included. The trouble will be an open circuit in the line cord, interlock, fuse, switch, or transformer primary winding.

A neon-lamp checker like the one shown in Fig. 4-6 can be used for most of the checking in this circuit. If the lamp lights when its two test prods are inserted in the receptacle half of interlock S-2, the line cord is good, and it is connected to a live outlet.

Next the two halves of the interlock are pressed together and the neon-lamp checker is connected across the terminals on the receiver side of the interlock. If the interlock is making contact, the neon checker will light. If it does not, wiggle the interlock plug. Intermittent contact will give intermittent lighting of the neon lamp. With no light on the neon checker, or intermittent light, replace the interlock.

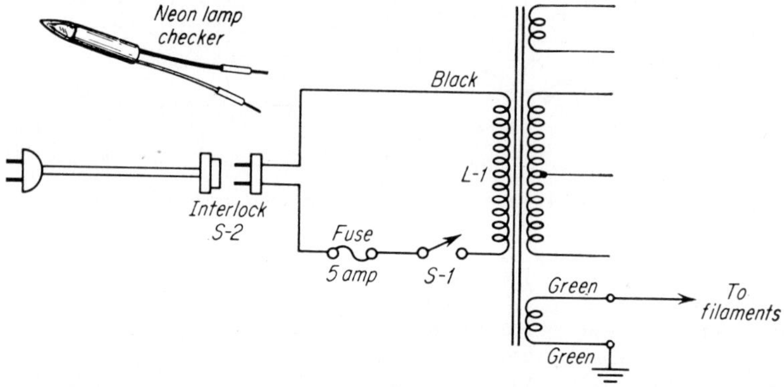

Fig. 4-6. Primary circuit of the power transformer.

If the neon lamp shows a steady glow on the receiver side of the interlock, the trouble is confined to an open circuit in the fuse, switch, or primary winding of the power transformer. Disconnect the line cord, turn the switch to the on position, and check the winding, switch, and fuse with an ohmmeter. The switch will be found on the back of the audio volume control, and should give a reading of zero ohms. The transformer primary leads may be found by tracing the wiring. They are usually color-coded black, and the normal reading is between 1 and 2 ohms. The fuse, if present, is usually located on the rear apron of the chassis near the interlock, or near the power transformer. The normal reading for the fuse is zero ohms. An open switch or transformer calls for replacement. If the fuse is open, replace it, but do not connect to current until making the checks for shorts described in the next section.

If line cord, interlock, fuse, switch, and transformer all check good, the serviceman must look for an unusual cause for the trouble. This might be a broken lead or possibly a poor contact or rosin joint at the filament leads of the power transformer. The latter is sure to be the case if only the rectifier tube lights. The 6.3-volt filament leads are usually color-coded green. Trace them to their connection points; inspect the joints and resolder them with a hot iron.

In some receivers, the filament line is split and powered from two 6-volt windings. This circuit is shown in Fig. 4-8. Here, a rosin joint in the filament line will cause half the tubes to go dark.

Servicing Procedure When the Set Blows Fuses. When the complaint states that turning the set on causes power-line fuses to blow or lights to dim, the receiver is drawing too much current because of a short. A receiver which has its own fuse and continues to blow that fuse is in the same category. The procedure to be given will prevent blowout.

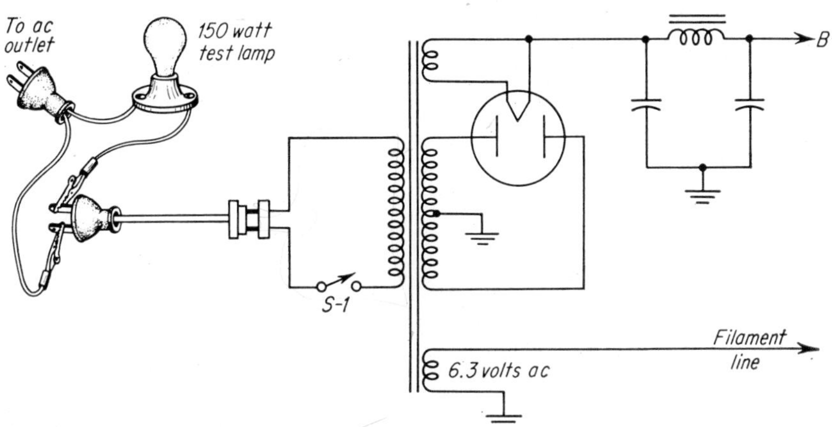

Fig. 4-7. Using the lamp short checker for finding shorts in the power supply.

Step 1. Disconnect the line cord and replace all fuses.

Step 2. Use a 150-watt lamp in the lamp-short checker, and connect it as shown in Fig. 4-7. This places the 150-watt lamp in series with the receiver. When the set is turned on, the lamp will light brightly because of the short, but no fuses will blow since the lamp limits the current.

Step 3. Remove the rectifier tube from its socket, and watch the effect on the brightness of the lamp in the line. If the short is in the rectifier tube or the B plus line, removing the tube opens the short, and the lamp in the line dims to a dull glow. If this happens, look for the short as described in the next section. If the lamp in the line remains bright with the rectifier tube out, the short is still in the transformer. Leave the rectifier tube out, and proceed to step 4.

Step 4. Open the hot end of the 6-volt filament line. If the 150-watt lamp in the line goes out entirely, the short has been removed from the transformer. Check along the hot, or ungrounded, filament line from tube to tube looking for a bit of wire or solder lodged between a tube lug and chassis or a frayed lead that is shorting to the chassis. If the set has a dial lamp, do not neglect to look for the short in the dial lamp or its wiring. When the short is found and cleared, reconnect the hot 6-volt lead from the power transformer, and check again with the test lamp in the power line. If the short is cleared, the lamp will now just glow, and the receiver tubes will light.

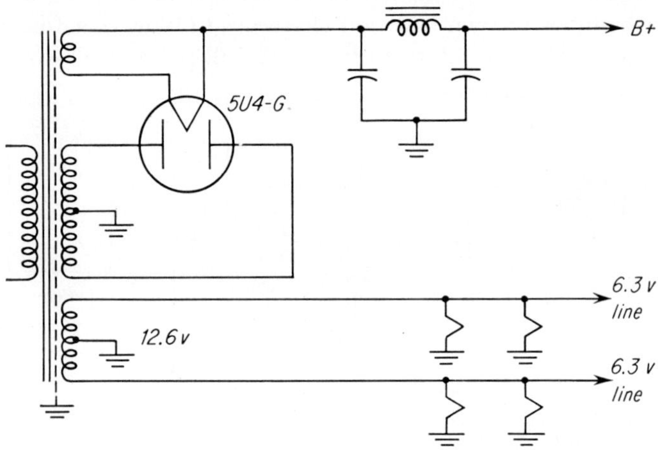

Fig. 4-8. Power supply using the 6-volt lines.

If the 150-watt lamp in the line remains bright with the rectifier out and the 6-volt filament line open, the short is still present. Refer to the schematic diagram of the receiver, and open all transformer windings except the primary, one at a time. As each winding is opened, observe the test lamp. When the short is taken off, the lamp will go out. Check this circuit for the short. If the lamp remains bright after all circuits are open, the short is an internal one in the power transformer. It must be replaced.

How to Find a Short in the Rectifier Circuit. When the checks just given place the short in the rectifier or B plus line, the common faults are a shorted or gassy rectifier, a shorted filter condenser, or a short or overload on the B plus line. These defects are sometimes obvious in that the defective rectifier usually glows with a purple light, and a short on the rectifier output causes the plates in the rectifier tube to glow at red heat. However, these indications may not be noticeable if fuses are blowing, and the set is operating at the reduced current allowed by the 150-watt lamp in series with the line.

The best check, therefore, is to replace the rectifier tube while the lamp is still in the line. If the rectifier were at fault, the lamp which had been bright with the old rectifier would now dim to about half brightness. The set is not likely to operate because of the reduced current, but normalcy can be established by checking voltage at the rectifier cathode.

With a 150-watt lamp in series with the line, in a good set, the reading at the rectifier cathode would be about 150 volts, or half normal.

If the lamp remains bright with a new rectifier tube installed, the rectifier output is probably shorted. Disconnect the interlock, and look for the short with an ohmmeter. The ohmmeter reading from rectifier cathode to ground will probably be SHORT or a low ohmic value. Then open the filter condensers and the B plus line at the choke until the short is found.

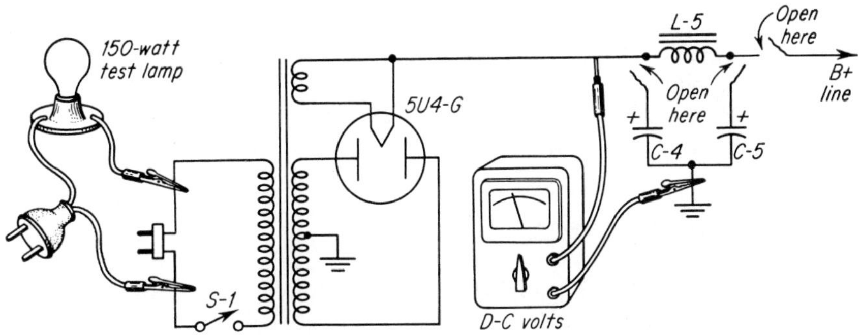

Fig. 4-9. How to check for shorts in the B power supply.

It may be that the short does not manifest itself under the low-voltage conditions of the ohmmeter check. In this case, use the same procedure of opening possible shorting branches one at a time, with a voltmeter connected to the rectifier cathode to tell when the shorting branch is opened. This procedure is illustrated in Fig. 4-9. If the short is not cleared until the B plus line is opened at the choke, the power supply is good. The short is now difficult to find since B plus branches all over the receiver. A detailed procedure for finding this type of short is given later in this chapter.

Servicing Procedure for Low Voltage at B Plus. When the voltage reading at the B plus point is low, coupled with a symptom of insufficient height and width, the likely causes are a weak rectifier, input filter condenser, or both. Replace the rectifier tube, and if this does not bring up the voltage, bridge the input filter condenser with a test condenser. If this brings the voltage at B plus to normal, the input filter condenser is replaced.

It may be that the low voltage at B plus is caused by a low line voltage. This is primarily a field-servicing problem, since line-voltage conditions at the bench are known. When the complaint is lack of width, the serviceman checks the line voltage. If this checks 105 volts or lower, the condition is caused by the low line voltage. A tapped autoformer, like the one shown in Fig. 4-10*a*, will step up the line voltage and restore normal operation. It may be that the symptom is of an intermittent nature in that the line voltage fluctuates depending on load conditions at different times of the day. In this case, a metered variable unit like the one shown in Fig. 4-10*b* is installed, and the consumer is instructed in its use. An

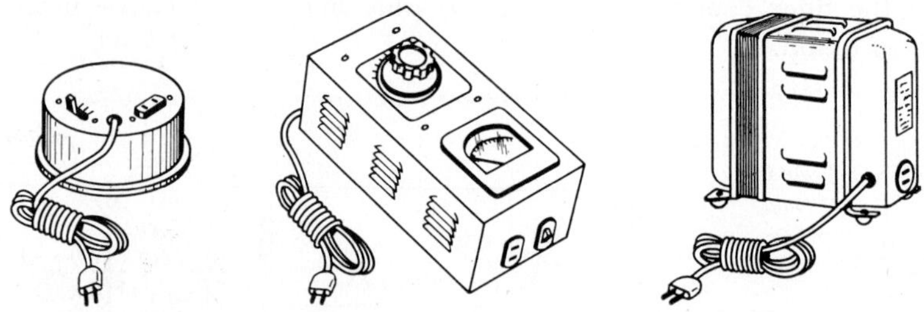

(a) (b) (c)

Fig. 4-10. Line-adjustment transformers.

automatically regulated transformer which delivers a constant voltage over a wide fluctuation in line voltage is also available. This transformer is illustrated in Fig. 4-10*c*. When installing any of these units, make sure to choose one with a rating high enough to handle the receiver. A rating of about 300 watts will be sufficient for most television receivers.

Replacement Notes on Power Transformers. When the power transformer in a receiver is defective, reference is made to replacement catalogues for a proper replacement for the particular receiver. If the receiver is not listed, the choice of a replacement to be used is governed by the following factors. The total number of secondary windings must be the same as the original. The filament windings should have ample current ratings, that is, a 5-volt 3-amp winding is needed for one rectifier tube, and a 5-volt 6-amp winding is needed for two rectifiers. The 6.3-volt filament winding should also be heavy enough to carry the filament load. The rectifier plate winding should have about the same voltage rating as the normal B plus value of the receiver. For example, if service notes show a normal B plus of 300 volts, the transformer plate winding should have a rating of 300-0-300 or 320-0-320. The current rating of the plate winding can be judged from the rectifier tube or tubes. A current rating of 200 to 225 ma would be right for a single 5U4-G, 225 to 250 ma for a single 5U4-GB, and 250 to 300 ma for a pair of 5U4-G tubes in parallel. If the receiver uses two 6BQ6-G or equivalent horizontal amplifier tubes, the power-transformer rectifier winding should be rated at 300 to 350 ma.

Many power transformers used in television receivers have a copper band like the one illustrated in Fig. 4-11. This reduces the external field of the power transformer so as to keep it from radiating to other parts of the receiver, particularly the vertical deflection circuit. Choose a replacement transformer equipped with a shield of this type, even if the original transformer did not use one. This is of particular importance when transmitter and receiver are not powered from the same line, since it prevents the power transformer from triggering the vertical deflection circuits. It is likewise of importance when receiving color-television programs on a black-and-white receiver. Here the transmitted vertical sync pulse is slightly off 60 cps, and of course, the vertical deflection circuit should be in step. When the vertical deflection circuits are also in the field of a 60-cycle power transformer, a dark horizontal bar will be noted periodically moving up the screen.

Fig. 4-11. Copper shield used on power transformer to reduce 60-cycle radiation.

Power-transformer Color Code. Most transformer manufacturers color their leads in accordance with the Electronics Industries Association (EIA) color code. This can be used to identify leads in replacement work, and is given in Fig. 4-12.

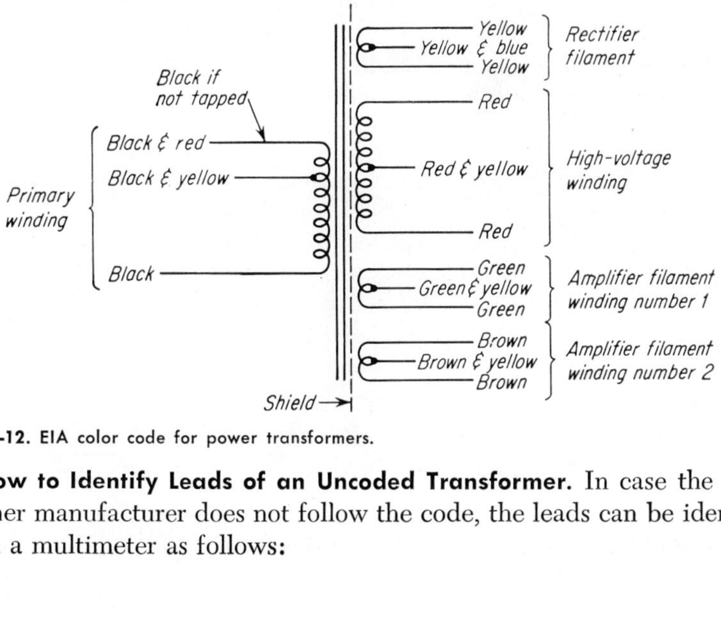

Fig. 4-12. EIA color code for power transformers.

How to Identify Leads of an Uncoded Transformer. In case the transformer manufacturer does not follow the code, the leads can be identified with a multimeter as follows:

1. Pair up the winding leads by means of an ohmmeter.

a. First connect the ohmmeter to any lead, and check for continuity with all the other leads, as shown in Fig. 4-13*a*. The lead that shows continuity is the other end of that winding, or a tap. In the case of a tapped winding, three leads will show continuity.

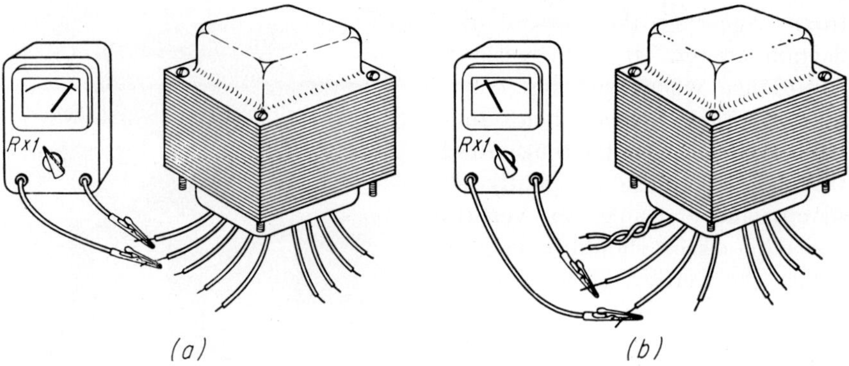

Fig. 4-13. Pairing leads.

b. Separate these two or three leads, as the case may be, and repeat to find the other windings as shown in Fig. 4-13*b*.

2. Read the resistance of each winding for identification purposes.

a. The primary will show a resistance of 1 to 2 ohms.

b. The rectifier plate windings will show a resistance of 15 to 30 ohms for each section. This identifies the primary and rectifier plate windings.

3. Separate the leads so as to avoid shorts, and connect the primary to the a-c line. Then measure each of the secondary windings with an a-c voltmeter, as shown in Fig. 4-14.

a. The rectifier filament winding will read 5 to 6 volts.

b. Each 6-volt filament winding will read 6 to 7 volts.

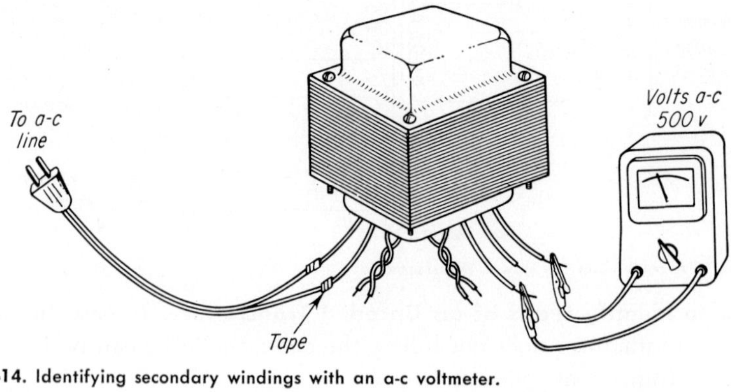

Fig. 4-14. Identifying secondary windings with an a-c voltmeter.

c. Each half of the rectifier plate winding will read 300 to 380 volts.

The readings will be somewhat higher than the rating because the transformer is being tested under no-load conditions. After making allowance for the no-load voltage of the rectifier plate winding, compare this voltage with the B plus specification of the receiver to confirm the choice of the transformer as a replacement. The correct a-c voltage should be approximately the same as the B plus rating.

Replacement Notes on Filter Chokes. An exact replacement for a filter choke is desirable but not essential. The important consideration in choosing a replacement choke is the current rating. Here again, if data are not available, choose the choke on the basis of the rectifier tubes in the set. Table 4-1 lists appropriate current ratings to choose for various rectifier-tube combinations.

Table 4-1

Rectifiers in Receiver	*Current Rating for Filter Choke, ma*
Single 5U4-G	200–225
Single 5U4-GB	225–250
Two 5U4-G	250–300
Two 5U4-G and two 6BQ6	300–350

The catalogue may list two chokes at the same current rating. For example, a 300-ma choke is decided upon, and the catalogue lists a 1-henry choke and a 3-henry choke with this rating. The 3-henry choke is, of course, much larger than the 1-henry unit. Choose the replacement which comes closest to the original choke in physical size.

In some receivers, the field winding of an electrodynamic speaker is used as the filter choke. If trouble develops in this field winding, the entire speaker must be replaced. The replacement speaker must be either an exact replacement of the original, or one that is very close to it. The important characteristics for matching are diameter of the cone, impedance of the voice coil, and inductance, resistance, and current rating of the field coil. If a close match cannot be found, the best procedure is to use a p-m speaker of the correct diameter as a replacement for the speaker, and a standard filter choke as a replacement for the field coil. This procedure is described in more detail in Chap. 15.

As a final note on replacement filter chokes, if it is desired to keep on hand a stock of this item, a 1-henry 300-ma choke with an ohmic resistance of about 50 ohms will work satisfactorily in most television receivers.

Replacement Notes on Tube Rectifiers. When the rectifier tube is replaced, be sure to use the type called for in the service notes. A 5U4-G may be badly overloaded if used in a set designed to operate with a 5U4-GB. In the case of receivers that use two rectifiers, if only one is found to be weak, check to see if both are the original tubes or are of the same age. If this is the case, it may be assumed that, although one tube still checks good, it has performed its normal expectation and will soon weaken in turn. It is considered good practice to replace both tubes to prevent a service recall for the same defect.

TRANSFORMER-SELENIUM TYPE OF POWER SUPPLY

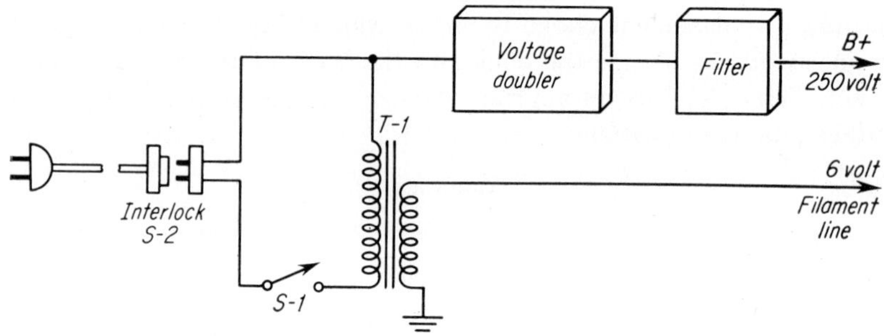

Fig. 4-15. Basic transformer-selenium type of power supply.

A basic view of the transformer-selenium type of power supply is shown in block-diagram form in Fig. 4-15. A heater transformer is used to step the line voltage down to 6 volts for the filament circuit. A pair of selenium rectifiers in a voltage-doubler circuit provides a d-c supply of 250 volts for the plate circuits of the tubes in the receiver. These receivers are also provided with a safety interlock switch indicated by S-2 in the diagram.

Transformer *T*-1, feeding the filament circuit, is considerably smaller than the power transformer in the transformer-tube type of power sup-

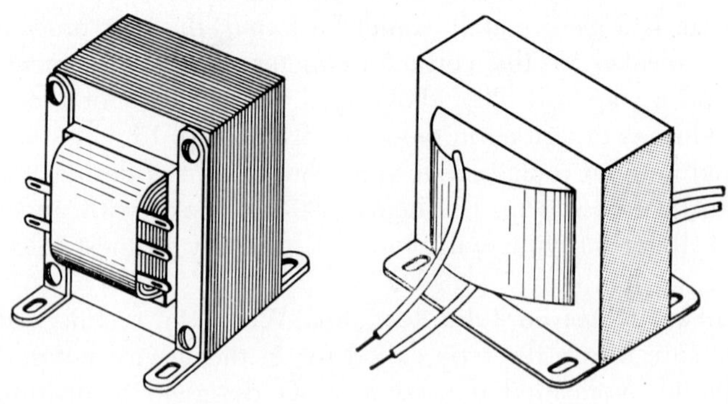

Fig. 4-16. Types of heater transformers used in the transformer-selenium type of power supply.

ply, since it provides filament power only. Some representative heater transformers are shown in Fig. 4-16 for identification purposes. In all other respects, the filament circuit is similar to the filament circuit in the transformer-tube rectifier type of power supply. Sometimes an extra 6-volt winding is provided for the damper tube. Service notes previously given for the transformer and filament circuit will apply here. This section is mainly devoted to the selenium rectifier and voltage-doubler circuit.

How the Voltage Doubler Works. A commonly used voltage-doubler circuit operates on the basis of charging two condensers, connecting them in series, and using the double voltage that results. Consider Fig. 4-17*a*. Condensers *C*-1 and *C*-2 are connected in series, and each one is charged to a potential of 125 volts, with the polarities as shown. This is the equivalent of two 125-volt batteries connected in series, giving an available potential of 250 volts across the pair.

The condensers are charged by the a-c line through rectifiers as follows: In Fig. 4-17*b*, with the line polarity as shown, condenser *C*-1 is charged to the peak of the line voltage through RECT. 1. On the reversal of line polarity, the rectifier will not conduct, and condenser *C*-1 maintains its charge. At the same time, the line polarity is correct to charge con-

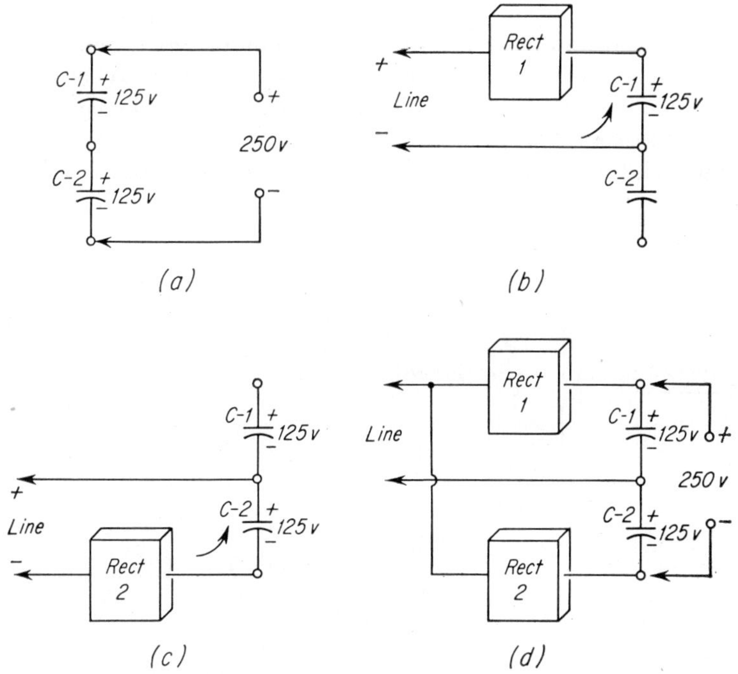

Fig. 4-17. How the voltage doubler works.

denser *C*-2 through RECT. 2, as shown in Fig. 4-17*c*. The combined circuit is shown in Fig. 4-17*d*. The total voltage across the two condensers is now available for the B circuit.

Since each condenser is charged alternately on each half cycle of the input line voltage, this type of circuit is called a full-wave voltage doubler. Each condenser will be charged to the peak of the line voltage, about 160 volts, if there is no load on the power supply. When the load is applied, the voltage drops, depending on the size of the load and the capacitance of the condensers. For the usual values in a television receiver, the total output of the voltage doubler is 250 to 270 volts.

Circuit of Typical Transformer-selenium-voltage-doubler Power Supply. A complete circuit of a typical power supply using a filament transformer and two selenium rectifiers in a full-wave voltage doubler is given in Fig. 4-18. Transformer *T*-1 supplies the filament line. Condensers *C*-1 and *C*-2 are the two condensers which are charged singly by rectifiers *SR*-1 and *SR*-2, on each phase of the applied line voltage. The condensers discharge in series to provide the double voltage into the load which is the B circuit of the receiver. Since this load is a heavy one, the 200 to 300 ma of the usual television receiver, these condensers must have a large capacitance to maintain voltage. The values shown in the diagram are usual. Choke *L*-5 and condenser *C*-5 form a conventional filter circuit.

The rectifiers used in practically all the voltage-doubler circuits in television receivers are of the dry-plate selenium type pictured in Fig. 4-18. Two rectifiers are used, and the common rating is 130 volts, 350 ma. Some sets may use slightly different current ratings, but the voltage rating is always the same.

Resistor *R*-2 is in series with both rectifiers. Its purpose is to limit the

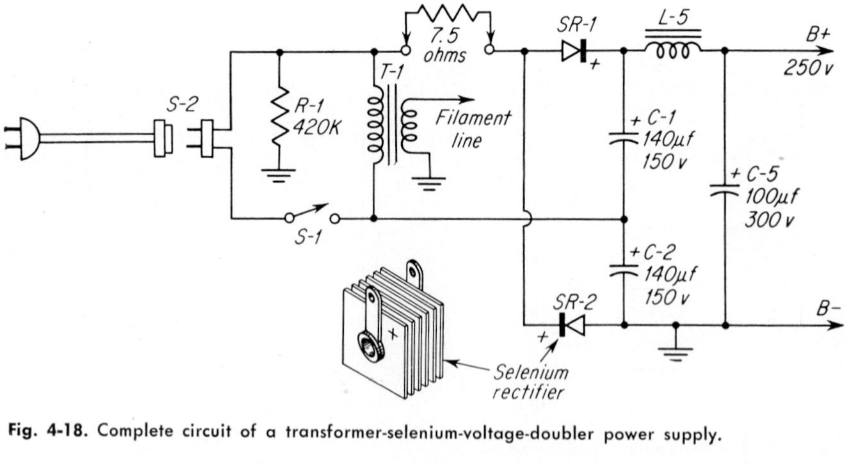

Fig. 4-18. Complete circuit of a transformer-selenium-voltage-doubler power supply.

charging current of the large-capacitance filter condensers which might overload the rectifiers on initial charge. It also has a second function. In case of a short it passes a heavy current and will usually burn out before the rectifiers are harmed. For this reason, it is known as a fusible resistor. It is usually a 5-watt wire-wound resistor mounted on a strip with pin leads for easy replacement. A fusible resistor is pictured in Fig. 3-10 for identification.

Note that the line is connected to the center point of series condensers *C*-1 and *C*-2. This places this line connection at 125 volts positive with respect to B minus. The B minus point is connected to the chassis, and so there is a considerable potential between chassis and the line. This is a shock and short hazard.

In most receivers, the hazard is avoided by insulating, or "floating," the chassis in the cabinet so that the consumer has no contact with it. Those parts of the receiver which come through the cabinet are the mounting screws, the control shafts, and the antenna. These are insulated as indicated in Fig. 4-19. The mounting legs are insulated from the chassis by being attached to an insulating strip mounted in a cutout section of the chassis. A common mounting of this type is shown in Fig. 4-19. The con-

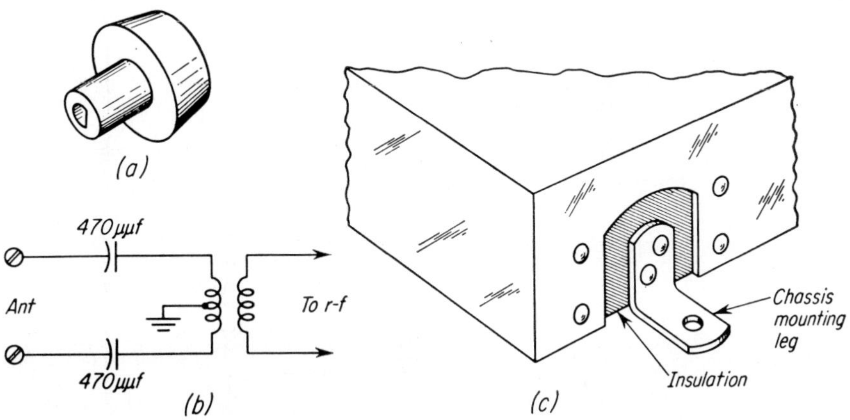

Fig. 4-19. Provisions for insulating a "hot" chassis. (a) Long-shank knob. (b) Antenna-insulating capacitors. (c) Insulated chassis mount.

trol shafts are recessed from the front panel and operated by means of long-shank insulating knobs which go through the front panel. The antenna is insulated by means of 470-$\mu\mu$f condensers in series with each lead. A condenser of this capacitance is a good conducting path at very high frequency and practically an open circuit at line frequency.

This insulating protection is for the consumer. When the serviceman is working on such a receiver, he has no such protection and must recognize that he is working on a "hot" chassis. He must be careful to avoid grounding either the chassis or himself. Another nuisance is to remember that a mounting leg which is usually a convenient connecting point for a test clip cannot be used as a common ground connection on this type of receiver. Also, spurious effects are sometimes noted when line-operated equipment is connected for test purposes. Perhaps the best way to work on these sets is to connect them to the line through an isolation transformer which insulates the chassis from the line. A tapped isolation transformer rated at about 350 watts is a useful piece of bench equipment for this purpose. The RCA TV isotap WP25A is suitable.

Receivers Using a Power Transformer and Selenium Rectifiers. Some receiver manufacturers avoid the hot chassis by using a winding on the power transformer rather than the raw line to feed the voltage doubler. A receiver of this type is the RCA KCS66 chassis. The schematic diagram of the power supply in this receiver is shown in Fig. 4-20. Note that the

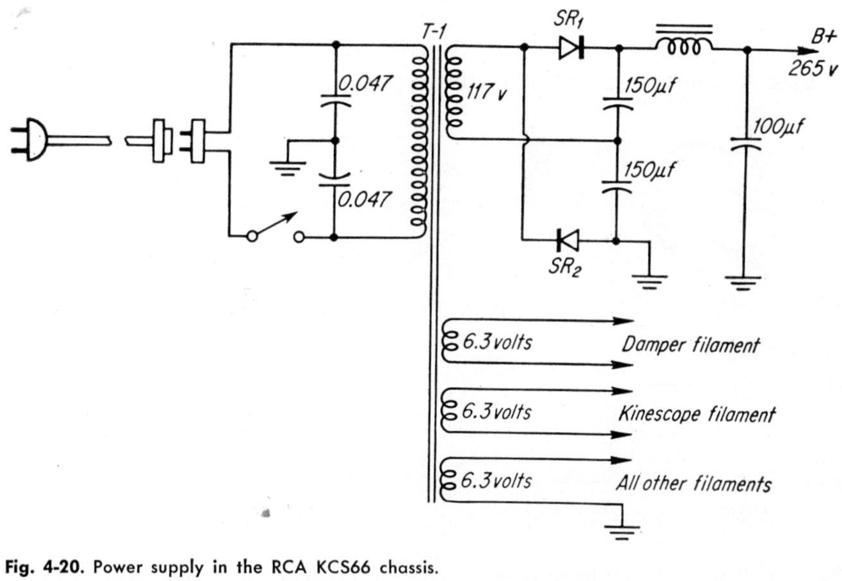

Fig. 4-20. Power supply in the RCA KCS66 chassis.

power transformer has a 117-volt secondary to supply the B power. This winding feeds two selenium rectifiers in a full-wave voltage-doubler circuit. The negative B point is connected directly to the chassis, and because of the isolation characteristic of the transformer, the chassis is insulated from the line. The transformer also has three separate 6.3-volt windings, so as to provide separate filament feeds to the damper and picture tubes. All other tube filaments are connected to the third winding. Because of the power required by the B circuit, this transformer will be considerably larger than the filament-only type, being comparable in size with the transformer in the tube-rectifier types of power supply.

The Half-wave Voltage-doubler Circuit. The full-wave voltage-doubler circuit of Fig. 4-18 cannot be used with a series filament line in a transformerless circuit. This is because the filaments are on the line, and the negative B point cannot be connected to the filament circuit. The full-wave voltage-doubler circuit, therefore, is always used with a filament transformer. This insulates the filament circuit so that it can be connected to the negative B point.

There is another voltage-doubler circuit where the negative B point is connected to the line, thereby permitting operation with a series filament circuit. Consider the circuit of Fig. 4-21*a*. Here, rectifier *SR*-1 is used to charge condenser *C*-1 from the 117-volt 60-cycle line. Note the polarity of the charged condenser, and the line.

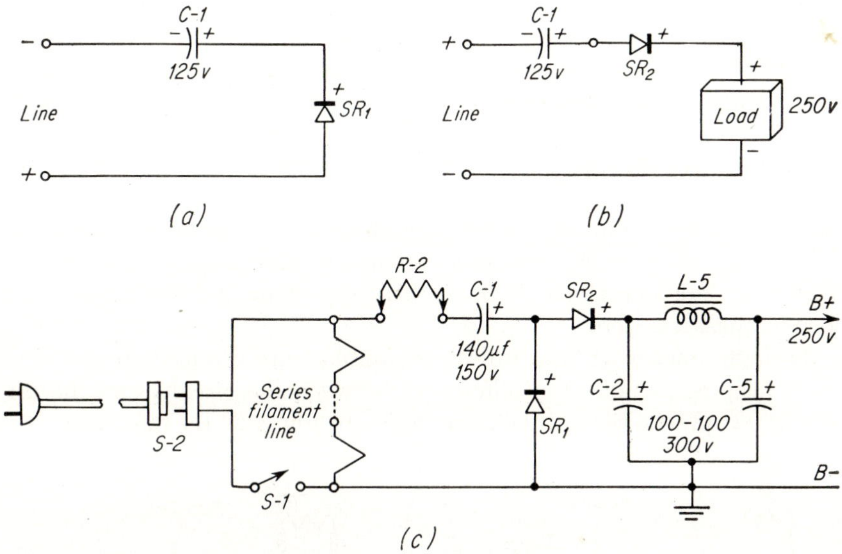

Fig. 4-21. A typical half-wave voltage doubler and how it works.

When the line polarity changes we have the condition shown in Fig. 4-21*b*. Now condenser *C*-1 and the line are in series with the proper polarity to apply a double voltage to rectifier *SR*-2. This double voltage is the source of supply to the B load. Note that the negative B point is one side of the line. Now we have a circuit which can be used with a series filament line, as well as with a filament transformer. The circuit is called a half-wave voltage doubler because the B load is fed on only one half cycle of the line supply.

A typical half-wave voltage-doubler circuit is shown in Fig. 4-21*c*. Switches S-1 and S-2 are the line switch and interlock, respectively, serving their usual functions. The position of the series filament circuit is shown directly across the line. Resistor *R*-2 is a fusible resistor performing its same function in protecting the rectifiers from overload. Condenser *C*-1 is charged by rectifier *SR*-1. Its charge is added to the line voltage when the line polarity changes. Its usual size is 140 μf, 150 volts. Rectifier *SR*-2 acts on the doubled voltage to feed the B load through the conventional filter circuit made up of choke *L*-5 and condensers *C*-2 and *C*-5. These are 100-μf 300-volt filter condensers. The rectifiers are the same type used in the full-wave voltage-doubler circuit.

Although the chassis and one side of the line are at the same potential, the chassis is still hot depending on how the set is plugged into the line receptacle. Since one side of the line is earth-grounded, the position of the line plug in the receptacle may place the chassis on the hot line wire. Here again, we have a short and shock hazard. This chassis, therefore, is also floated by insulating the mounting legs, etc., as was previously described for the full-wave-rectifier voltage-doubler circuit.

Servicing Voltage Doublers. The voltage-doubler type of power supply is subject to the same ills that happen to any B power supply. The selenium rectifiers weaken with age, the electrolytic condensers dry up and lose capacitance, and sometimes there is an overload in the B circuit. In all cases the B plus output drops from its normal reading of 250 volts. The voltage doubler has one other common trouble peculiar to itself. The fusible resistor sometimes opens. If this happens the a-c line is open, and B plus measures zero.

Both the half-wave and full-wave doublers are serviced in the same way, keeping in mind the differences between the circuits and the different test points. Both circuits are drawn in Fig. 4-22 for reference.

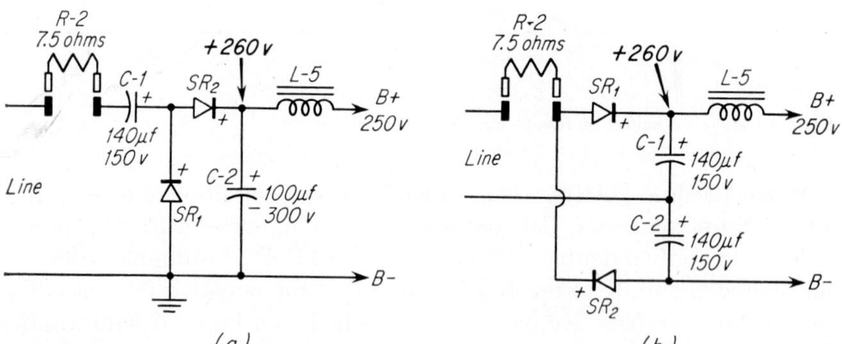

Fig. 4-22. Servicing voltage-doubler circuits. (*a*) Half-wave voltage doubler. (*b*) Full-wave voltage doubler.

If the tubes light and the voltage at B plus measures zero, the most probable trouble is an open fusible resistor *R*-2. This can be confirmed with an ohmmeter check. If the resistor proves to be open, before replacing it look for a short which is still present and may have been the cause of the open resistor. This is most easily done with an ohmmeter check from B plus to chassis. The normal resistance reading from B plus to ground represents the charge and leakage readings of all the condensers in the B plus line, as well as the resistance across the two backed-up rectifiers. If this reading is 10,000 ohms or more, assume no shorts, and replace the resistor. If the check shows a low ohmic value from chassis to B plus, the short must be found and cleared before the resistor is replaced.

If the voltage at B plus measures considerably below its normal 250 volts, look for the filter choke which is labeled *L*-5 in both diagrams and measure the voltage across it. The diagram shows normal readings of 260 volts on the hot side of the choke, and 250 volts on the B plus side. This, of course, is measured from chassis, making a normal drop across the choke of 10 volts. In the case of a reduced voltage at B plus, both these readings will probably be low, but the drop across the choke is significant. If it is considerably higher than 10 volts, the B plus line is drawing too much current. Service analysis of the B plus line is taken up later in the chapter. If the drop across the choke is 10 volts or less, the cause of low voltage is in the power supply--a weak rectifier, filter condenser, or both.

It is easier to check filter condensers; so these are checked first. When working on these receivers for an open condenser do not make the mistake of connecting the negative end of the test condenser to chassis and then probing for various B plus points with the positive condenser terminal. First, examine the schematic diagram of the receiver being tested, and trace the leads from the rectifiers to identify the filter condensers which are the equivalent of *C*-1 and *C*-2 in Fig. 4-22. These are not likely to be in the same can, since the filter condensers do not have a common negative terminal in either the half-wave or the full-wave doubler circuits. Leave the voltmeter connected to B plus to observe the result on the output voltage. Being careful to observe polarity, connect a test condenser across first one and then the other filter condenser. If the output voltage comes up to normal when either condenser is checked, that condenser is open and must be replaced. If there is no change or little change when the condensers are checked, they are probably good, and the rectifiers are defective.

When a defective filter condenser is found it must be replaced. If it is part of a multiple condenser in a can, it is not necessary to replace the entire can. Unsolder the lead or leads from the defective section and connect them to the positive terminal of the replacement unit. Then connect the negative terminal of the replacement condenser to the common terminal of the multiple unit. Make it a practice to follow this procedure in all cases, even when the common terminal of the multiple condenser does connect to ground. This will avoid mistakes.

Selenium rectifiers have a plus sign stamped on or near the connection lug which is wired to the positive terminal of its associated filter condenser. Schematic diagrams usually indicate polarity on the condensers and rectifiers also. To check a selenium rectifier, unsolder one lead from the one in the set, so as to disconnect the original. Provide yourself with a replacement rectifier equipped with clips for test purposes. The setup is shown in Fig. 4-23. Clip in the test rectifier, being careful to observe

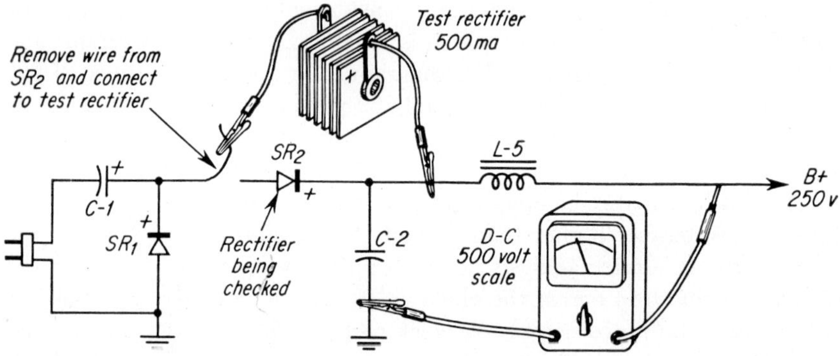

Fig. 4-23. How to check a selenium rectifier by substitution.

polarity. Connect the voltmeter to measure the output B voltage and turn on the set. If the output voltage is restored to normal, the old rectifier is bad and needs to be replaced. If there is no change, the second rectifier in the receiver is checked in the same way.

If either rectifier is defective, and both appear to be the same age, it is usually considered good practice to replace both. Selenium rectifiers rated at 500 ma, 130 volts will usually work satisfactorily in any receiver. If they cannot be mounted because of a space requirement, an exact replacement should be obtained.

Series Filament Circuits. Quite a number of television receivers use a series or series-parallel arrangement of tube heaters lit directly from the line, thereby avoiding the expense of a power transformer. The early transformerless receivers were designed around existing tubes. It must be remembered that there is one value of current in a series setup. All picture tubes have a 6-volt 0.6-amp filament. This means that the filament string must be built up to 0.6 amp. Some other tubes like the 6SN7-GT also have a 0.6 amp filament, and so these are connected in series with the picture tube. Then there are any number of tubes with 0.3-amp filaments. Two lines of these connected in parallel require 0.6 amp, and so these paralleled lines are also connected into the series circuit. Resistors piece out the circuit to fit the line.

As transformerless receivers gained in popularity tube manufacturers developed a line of tubes with 0.6-amp filaments. These are more commonly called 600-ma tubes, and only one series line is needed. Both types of filament circuit are found in receivers, and so a representative circuit of each will be described together with applicable servicing hints.

A Typical Series-Parallel Heater Chain. The low-voltage power supply in the Motorola TS-314 chassis is shown in Fig. 4-24 as a typical example of a receiver using both 0.6- and 0.3-amp tubes. As a matter of fact, this receiver was chosen because it also uses some 0.45-amp tubes, thereby making the circuit even more representative for study purposes.

First note that the B power supply in the upper portion of the diagram is a standard half-wave voltage doubler providing 250 volts to the

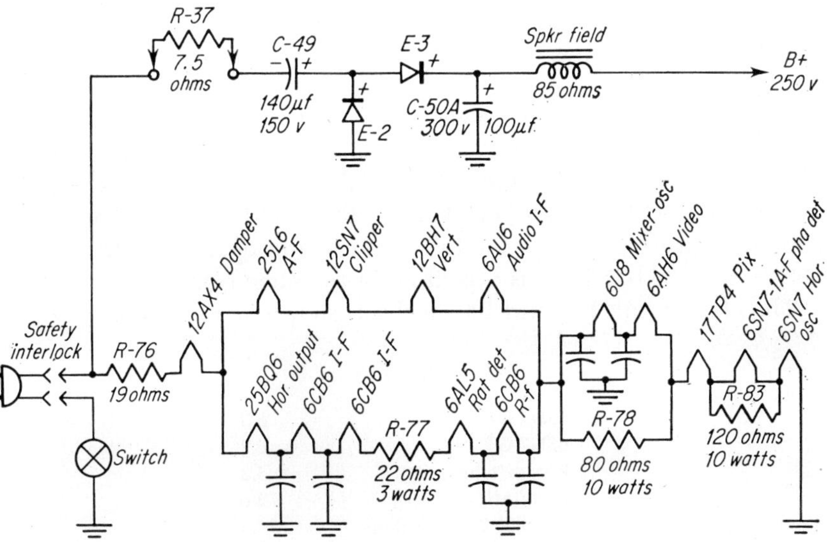

Fig. 4-24. Low-voltage power supply in Motorola TS-314 chassis showing a filament string using 0.6- and 0.3-amp tubes.

receiver. In the filament line, the 12AX4 damper at the beginning of the line and the picture tube and two 6SN7-GT tubes at the end of the line are the 0.6-amp tubes. Resistor *R*-83 across the audio 6SN7-GT reduces the current in this tube very slightly and may be neglected. The first audio, picture, and horizontal oscillator tubes are most sensitive to hum effects, and these will usually be found near the grounded end of the line. The damper, power audio, and horizontal output tubes are least sensitive to hum, and so these will usually be found near the hot end of the line.

The two paralleled 0.3-amp groups have four tubes in the upper branch beginning with the 25L6-GT power audio, and five tubes in the lower branch beginning with the 25BQ6 horizontal output tube. The two parallel branches must have the same voltage across them. The upper four tubes require more voltage than the lower five tubes, and so resistor *R*-77 is added to the lower branch to take up the excess, as seen in Table 4-2.

Table 4–2

	Voltage Required		*Voltage Required*
25L6-GT	25.0	25BQ6-GT	25.0
12SN7-GT	12.6	6CB6	6.3
12BH7	12.6	6CB6	6.3
6AU6	6.3	Dropped in *R*-77	6.3
		6AL5	6.3
		6CB6	6.3
Total	56.5		56.5

The 6U8 and 6AH6 tubes which come next in the chain are 0.45-amp. tubes. Resistor *R*-78 is shunted across the pair to pass the extra 0.15 amp in the 0.6-amp line. This resistor normally dissipates about 2 watts. An oversize 10-watt resistor is used here, because the full 0.6 amp of current would flow through it if either of the two tubes should open. Note that resistor *R*-83 across the 6SN7-GT is oversized for the same reason.

The filament line requires 100.6 volts as shown in Table 4-3.

Table 4–3

6SN7-GT	6.3
6SN7-GT	6.3
Picture tube	6.3
6U8-6AH6	12.6
Parallel 0.3-amp line	56.5
12AX4-GT	12.6
	100.6

The extra 17 volts on a normal 117-volt line is taken up in line-dropping resistor *R*-76. This is a resistor of special construction, sometimes called a globar. Its resistance is high when cold, approximately 250 ohms, and decreases to 19 ohms as the resistor warms up. The tube filaments have the reverse characteristic, low resistance when cold and high resistance when hot. Without the globar, the tubes themselves would take a very heavy current when the set is first turned on. The initial heavy current taken by the cold filaments might burn one of them out. With the globar in the circuit, the initial current is low because of the high resistance of the cold globar. Then, as the tubes and globar warm up, the current reaches its normal 0.6 amp.

Note the decoupling filter condensers around the r-f and i-f tube filaments. These are usually Ceramicon condensers with a capacitance of approximately 1,000 $\mu\mu$f. It was previously mentioned that filament decoupling filters give very little service trouble. Although this is true for a 6-volt filament line, trouble in the series chain, while still remote, will happen more frequently because of the higher voltages in the series filament circuit. If one of these condensers were to short, the trouble would be found by a thorough routine check of the filament circuit.

Servicing a Filament String Using 0.3- and 0.6-amp Tubes. The common trouble in the series or series-parallel filament circuit is that one of the tube filaments opens. Then, since effects are different in a single series line and the series-parallel line, it might be wise first to determine the type of circuit. This can be done by referring to the manufacturer's service notes or by examining the tubes. The horizontal output tube is easily identified. If it is a 25BQ6-GT, there are 0.3-amp tubes in the chain, and it is a series-parallel combination. For confirmation, look for 6CB6 tubes in the i-f strip. These are also 0.3-amp tubes. If the horizontal output tube is a 12CU6, look for 3-volt tubes like the 3CB6 in the i-f strip. If these are present, it is pretty sure to be a single 600-ma filament line.

If the set has a series-parallel filament line, and one of the 0.6-amp tubes or the globar opens, all tubes go out, and we have a dead set. If one of the tubes in either of the 0.3-amp lines should open, that line goes dead, but the other is still working and passing current through the 0.6-amp tubes. None of the tubes will be at correct temperature, but it will be observed that some tubes light and some do not.

When working on a completely dead set, it might be wise to check the line cord, interlock, and on-off switch first. If these are good, check the globar. A defective globar shows signs of overheating. When checking with an ohmmeter, remember that the globar will normally check about 250 ohms—its resistance when cold. If the globar is good, the 0.6-amp tubes are checked next. Check the 12AX4-GT damper tube first, since experience shows that this tube is often the cause of the trouble. An easy way to check the picture tube is by plugging in a CRT booster. If the picture tube is open, the booster completes the circuit, and all other tubes light. Checking the other 0.6-amp tubes is routine.

If some of the tubes light, one of the 0.3-amp chains is probably open. Check the audio and 6AL5 tubes first, since these are common offenders. If these are good, all tubes in the dead chain must be checked. A fast way to check tubes for this purpose is to use a filament continuity checker like the one shown in Fig. 4-25. Replacing miniature tubes in blind corners is a considerable nuisance, often causing bent pins and frayed nerves. An appliance called a tube pilot disk makes this job somewhat easier. A tube pilot disk is also illustrated in Fig. 4-25.

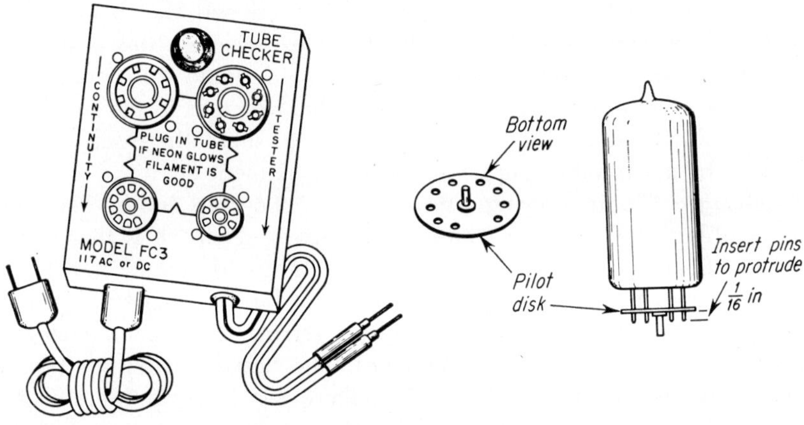

Fig. 4-25. Filament continuity checker and tube pilot disk.

There are some less common causes of the condition where some tubes light and some do not. One of these is the shorted filament bypass condenser mentioned previously. This would ground out the tubes below it in the chain. This possibility must be kept in mind when the checks do not produce an open tube filament. It then becomes necessary to refer to the schematic diagram of the receiver and check these condensers with an ohmmeter.

A final cause for some of the tubes to fail to light is a cathode-heater short in a tube whose cathode is wired to ground. This short will again ground out filaments below it in the series chain. A short of this type may not show up on an ohmmeter check, since it may be present only when the filament is heated and expands, touching the cathode. This must also be kept in mind when other checks fail. This type of short is difficult to find. The best way is to replace all tubes, one at a time, and observe results. Begin with those where the cathode is at ground potential. These are usually the vertical oscillator, sync tubes, r-f mixer-oscillator, horizontal output, r-f and i-f tubes. Check the schematic diagram of the receiver to see which tube cathodes are at ground potential.

In printed-circuit receivers, the filament wiring and connections are fragile and often break in handling. In conventionally wired receivers, filament wiring is more stable and rarely requires checking. The possibility of breaks in the filament wiring on printed boards must be kept in mind when working on this type of receiver. Servicing hints for printed wiring are given later in this chapter.

The 600-ma Series Filament String. The series-parallel filament circuit has certain disadvantages. One of these is that the tubes do not heat up at the same rate. The tubes that heat up fastest increase in resistance, take a disproportionate share of the voltage, and slow up the slower ones even more. This sometimes causes a long wait for the set to warm up, and also makes for short life of some tubes. The 600-ma tubes overcome most of these disadvantages in that they were designed for a controlled warm-up time. They all heat at about the same rate, so that the set warms up faster and no one tube is overloaded during the warm-up period. Because of these advantages the 600-ma tubes are replacing the older types in television receivers.

The low-voltage power supply of the Crosley 431-3 television chassis is given in Fig. 4-26 as an example of a receiver with a single 600-ma fila-

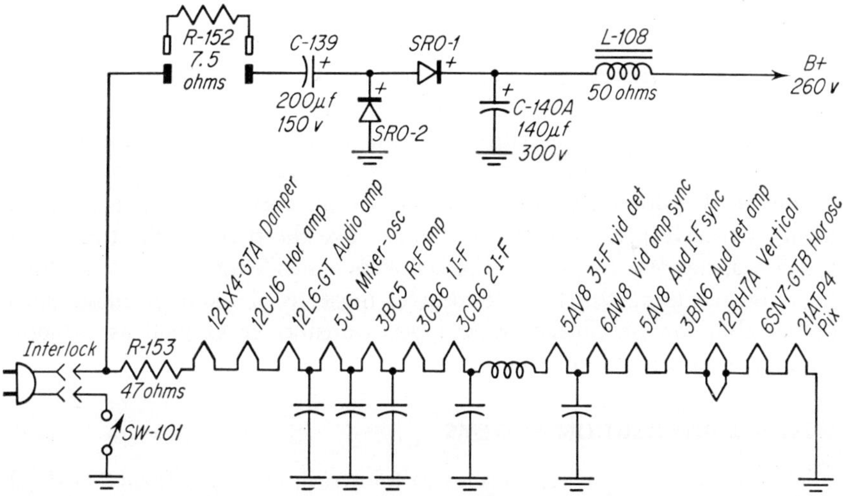

Fig. 4-26. Low-voltage power supply of the Crosley 431-3 chassis.

ment string. Again, note the standard half-wave voltage doubler which is always used in receivers with the series filament string. Note the tubes near the grounded end of the filament string are the 3BN6 first audio amplifier, the 12BH7A vertical oscillator, the 6SN7-GTB horizontal oscillator, and the picture tube. The 12BH7A has two filaments and may be connected as a 12-volt 0.3-amp tube with its filaments in series, or as a 6-volt 0.6-amp tube with its filaments in parallel. As the diagram shows, the parallel connection is used in the 600-ma heater chain. The 6SN7-GTB is the controlled warm-up version of the standard 6SN7-GT twin triode.

At the high end of the filament line are the tubes used in circuits less sensitive to hum. These are the 12AX4-GTA damper, the 12CU6 horizontal oscillator, and the 12L6-GT audio output tube. Note also the decoupling filters around the r-f and i-f tubes. Line-dropping resistor *R*-153 limits the current so as to maintain correct voltages on the tubes in the string. Because of the controlled warm-up time of the tube, a globar is not necessary, and resistor *R*-153 is an ordinary wire-wound heavy-duty resistor.

Servicing a 600-ma Filament String. The common trouble in the series 600-ma filament circuit is again that one or another of the tube filaments opens. This time, all tubes go out, and we have a dead set. The dead tube may be any one of the 14 or more tubes in the chain and calls for a lot of checking. The tubes that seem to burn out most often are the 3AL5, if used, the 12AX4-GTA damper, and the 12L6 audio output; so these should be checked first. If these are good, check all others one at a time in any order. The filament continuity checker of Fig. 4-25 is an aid in fast checking.

Shorts in the filament bypass condensers, or cathode-heater shorts may also develop in the 600-ma filament string. This would cause those tubes below the short to go dead. This symptom would be serviced by the same procedures just described for similar shorts in the series-parallel chain.

One final point must be mentioned. Some very few receivers use a filament transformer to light some tubes, like the tubes in the tuner, and a series string for the rest of the set. When servicing a set where some tubes do not light, this type of receiver must also be kept in mind, since the trouble can be caused by an open transformer as well as cathode-filament shorts in one of the tubes.

VOLTAGE-DISTRIBUTION SYSTEMS

The low-voltage power supplies in television receivers furnish a d-c B voltage of 220 to 350 volts to loads of 200 to 300 ma. In voltage-doubler systems, the B voltage is usually 250 volts. In transformer-tube rectifier systems, the voltage at B plus depends on the design of the power transformer. The load varies with the number of tubes in the receiver, larger receivers usually requiring more current. It will also vary with the design of the B voltage-distribution system.

Basic B Voltage-distribution Systems. Let us first examine a basic voltage-distribution system where all the tubes are fed in parallel from the B plus output of the rectifier-filter circuit. This setup is shown in Fig. 4-27.

As will be explained later in Chap. 6, the action of the horizontal output transformer, yoke winding, and damper tube produces a voltage which is added to B plus. This is called the boost voltage and is represented in the top portion of Fig. 4-27 by the line labeled BOOST LINE 500 v. The actual boost voltage varies considerably in different receivers, but for purposes of discussion, 500 volts may be taken as an average figure. The boost line feeds the horizontal output tube, an average current drain for which is 100 ma.

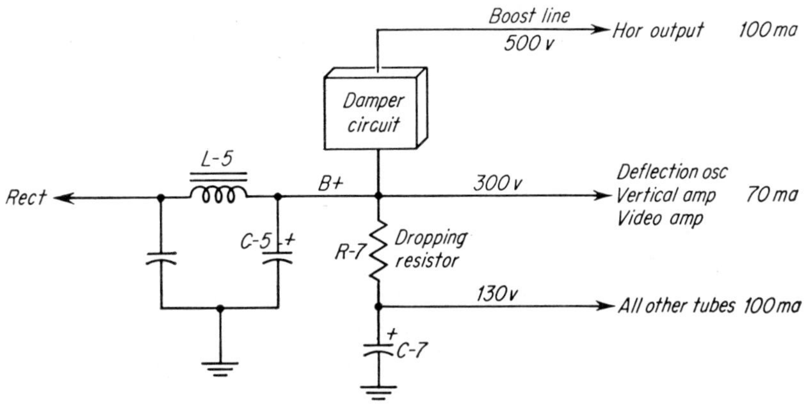

Fig. 4-27. Basic voltage-distribution system in B circuits of a television receiver.

Next comes a brace of tubes requiring normal B plus voltage. These include the deflection oscillators and the vertical and video-amplifier tubes. This line draws roughly 70 ma from the power supply.

The r-f pentodes used in the i-f amplifier stages work best at a plate and screen potential of about 130 volts. Tubes in the sync circuits also require low voltages. These tubes are fed through dropping resistor *R*-7. Condenser *C*-7 bypasses the 130-volt line.

There is considerable variation in television receivers as to the supply voltage fed to the audio-amplifier tubes. In some receivers, especially where the loudspeaker is small, the supply voltage to the audio-amplifier tubes is kept low. This is the case in the basic circuit of Fig. 4-27 where the audio tubes are fed from the 130-volt line. This puts a heavy drain on the 130-volt line. The diagram shows an average value of 100 ma for such a circuit, making a total of 270 ma for the entire set.

In some receivers, the audio amplifiers are removed from the 130-volt line and fed a higher voltage through a separate dropping resistor. This allows a better audio output but makes an even greater drain on the B supply system. A circuit of this type will now be examined in somewhat greater detail so as to provide a basis for service work in voltage-distribution systems.

Analysis of a Voltage-distribution System Where All Tubes Are Fed in a Parallel Arrangement. A detailed circuit where the B circuits of all tubes are fed in a parallel arrangement is given in Fig. 4-28. Examine the boost line first at the top of the diagram. Note that B plus is fed to the damper plate through the high-voltage fuse. The damper tube is easily identified in all sets, and since the plate is always the No. 5 pin, we have an easily located point to check the voltage at B plus. The damper cathode at the No. 3 pin is the source of the boost voltage. Note how the horizontal linearity coil and condenser are connected as a filter on the boost voltage. In addition to feeding the horizontal amplifier, the boost voltage is often used to feed other circuits in the receiver. There is considerable variation in the circuits fed, but in almost all cases, the boost line is used to feed the vertical oscillator and the No. 2 accelerator grid at the No. 10 pin of the kinescope. The boost voltage will vary with the efficiency of the horizontal output circuit, and therefore with the width. Note that the height control is in the plate circuit of the vertical oscillator. Using the boost line to feed this circuit tends to keep the aspect ratio between height and width constant for small changes in width. The high voltage is also obtained from the horizontal output circuit, and so the boost voltage will vary with the high voltage. The No. 2 accelerator grid of the picture tube is usually connected to the boost voltage. Since a change in high voltage will change focus, the focusing electrode is also often connected to the boost line. Thus, by connecting picture tube electrodes to the boost line, automatic regulation is obtained for small changes in high voltage.

The boost voltage cannot be measured at the plate of the horizontal output tube because of the presence of the horizontal pulse which may reach peaks of several thousand volts. If desired, the boost voltage can be measured at the No. 3 cathode pin of the damper tube or at the No. 10 pin of the picture tube. Linearity coil terminals should not be used to measure boost unless the diagram of the receiver is checked. In some circuits this coil is in the plate lead of the damper tube, where the potential is B plus, not boost.

The next line in Fig. 4-28 is the B plus line which feeds the horizontal oscillator and vertical and video-amplifier tubes. Note the vertical linearity and contrast controls in the cathode circuits of their respective tubes. This line also feeds the two halves of the r-f amplifier tube connected in series. This places approximately half the B plus voltage on each tube in the cascode amplifier and is a very commonly used circuit. Resistor *R*-11 and condenser *C*-11 form a decoupling filter for this stage. The B plus line also goes on to feed the screen of the horizontal amplifier tube through dropping resistor *R*-12.

The next line down is at approximately 200 volts to feed the audio-

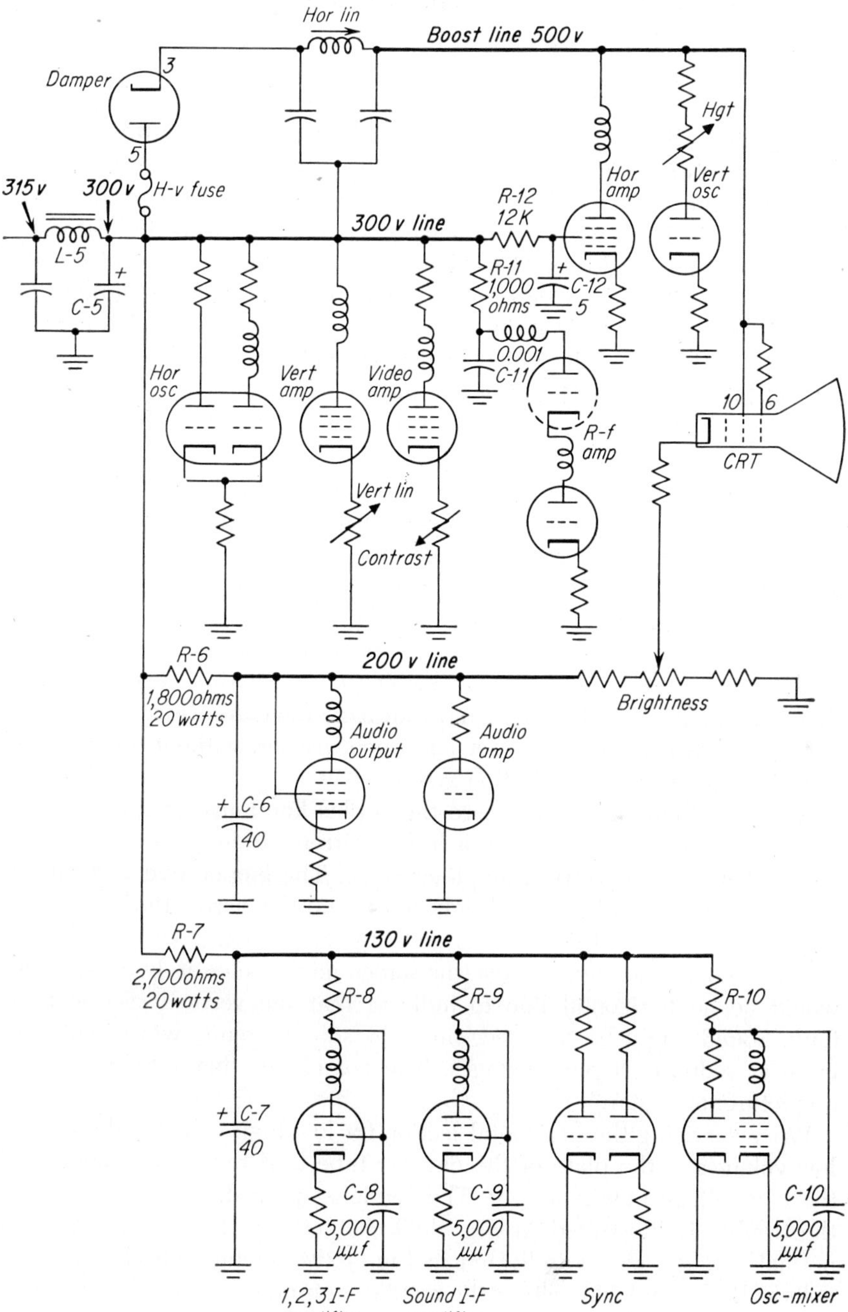

Fig. 4-28. Voltage distribution where all B circuits are fed in a parallel arrangement.

amplifier stages. Resistor *R*-6 drops the voltage to this line, and condenser *C*-6 serves as its bypass condenser. In this receiver, the 200-volt line goes on to feed cathode voltage to the picture tube through the brightness control.

The bottom line is the 130-volt line feeding all the other tubes in the receiver. Resistor *R*-7 drops B plus to 130 volts, and condenser *C*-7 bypasses the line. For simplicity, the three i-f amplifier tubes are indicated by one circuit. Each will have its own decoupling network indicated by resistor *R*-8 and condenser *C*-8. Next comes the sound i-f amplifier with the same circuit as the video i-f amplifiers. As seen in the diagram, the 130-volt line goes on to feed the sync tubes and the mixer-oscillator in the tuner. Note the decoupling filter in the latter circuit.

Servicing the B Plus Line. For the most part, defective components in the voltage-distribution system affect only the stage involved, and there is no need of checking the entire B plus line. For example, suppose decoupling filter *C*-8 in the i-f section of Fig. 4-28 were to become shorted. It would cause heavy current through resistor *R*-8 and burn it open. The stage would become inoperative and stop the signal at that point. All other stages would receive proper B voltages. The deflection and high-voltage circuits would be working and producing a normal raster. The complaint of "no sound, no picture, normal raster" leads to investigation of the stages before the sound take-off point. Routine stage checking shows the inoperative i-f stage, and routine voltage and resistance checking shows the shorted condenser and open resistor, without any need of checking the voltage-distribution system.

However, there are cases where the B plus line must be checked, and a procedure is needed. Assume a heavy current drain in one of the circuits, let us say the vertical amplifier stage. The inoperative vertical deflection circuit should cause a bright horizontal line across the screen, but if at the same time, the overload were to cause enough drop in B plus voltage to drop the high voltage, the screen might be entirely dark. There would be no horizontal line to indicate that the vertical section is at fault. Sound might be received, since the sound circuits will usually operate at reduced B plus voltages. The complaint, therefore, is "sound, no picture, no raster."

When servicing the horizontal section for low high voltage, abnormally low voltages at the plate of the damper tube lead to an investigation of the low-voltage power supply. The servicing procedure which localizes the trouble is illustrated in Fig. 4-29. Let us say that the circuit and normal voltages are as indicated in Fig. 4-29*a*. Voltage checks give 200 volts at B plus and 240 volts at the rectifier output, as shown in Fig. 4-29*b*. If the reading at the rectifier side of the choke were 210 volts, the trouble would probably be a weak rectifier or input filter condenser. But since it is 240 volts, showing a 40-volt drop across the filter choke, instead of its normal 15-volt drop, we have evidence of an overload.

The overload might be in any one of the B plus lines. The only way of finding it is to open all leads at the B plus end of the choke and clip the voltmeter to the choke as shown in Fig. 4-29*c*. The meter will now

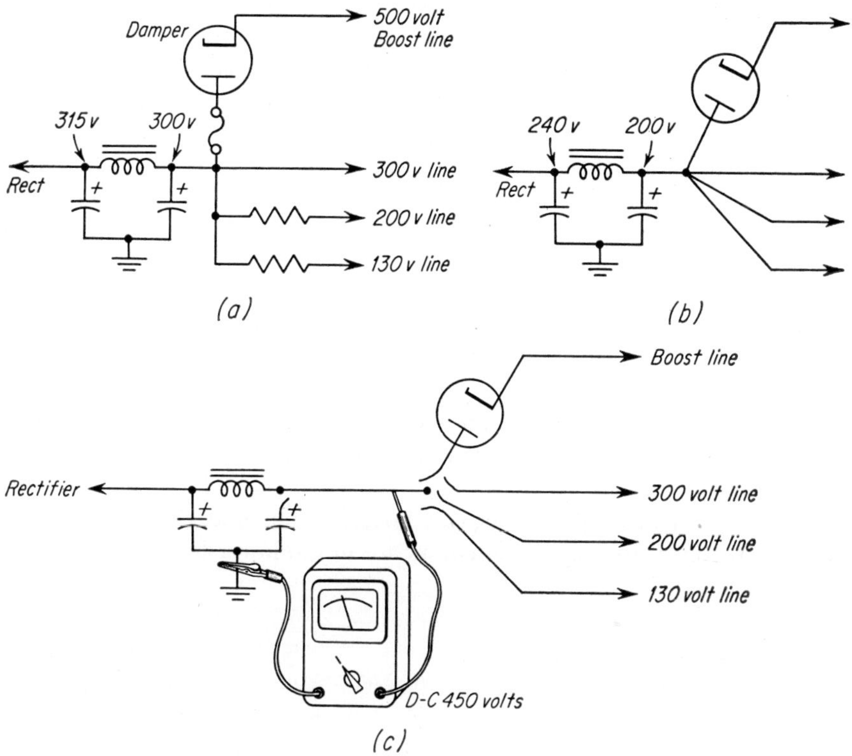

Fig. 4-29. Servicing the B plus line for an overload.

read approximately 450 volts, the peak voltage of the unloaded line. Now restore the leads one at a time. As each lead is returned, the voltage reading will drop closer to its normal 300 volts. When the line containing the overload is put back, the voltage reading will drop to the 200-volt reading. Then, it is necessary to go along the overloaded line, opening all circuits in the same way until the circuit containing the overload is found. Look for an overheating voltage-dropping resistor which will point the way to the overloaded circuit.

The actual receiver being serviced may not be wired exactly as shown in the diagram of Fig. 4-29. All the branches may not be so conveniently assembled at the filter choke. Some of the connections may be made farther down on the B plus line. However, this does not alter the servicing procedure. The filter choke is the beginning of the B plus line, and all wires are removed and checked individually for the short or overload by observing the effect on voltage as each wire is reconnected. One will contain the short. This wire is traced to its next connecting point where again all wires are removed and checked individually. This process is continued until the trouble is narrowed down to the one circuit which causes it.

When this one circuit is found, the actual trouble should be easily ascertained. It may turn out to be a defective tube which is drawing heavy current. It may be that a leaky coupling condenser is placing a positive bias on a grid with heavy current drain as a result. It may also be a type of short which appears only under operating conditions. For example, the high pulse voltage present in the vertical output transformer may break down the insulation between the winding and core, shorting the B plus line. A short of this type may not be apparent under the conditions of an ohmmeter check, but the effect on the B plus voltage will be immediately noted when the wire feeding the vertical output transformer is removed and reconnected.

Looking for an Overload on the B Plus Line in a Printed-circuit Receiver. The B plus wiring of a receiver is shown in Fig. 4-30 as an example of how leads are opened in a search for a short or overload. The receiver chosen for study uses three printed-circuit boards. Note that the power transformer, rectifier, and filter condensers are in the lower left corner but the

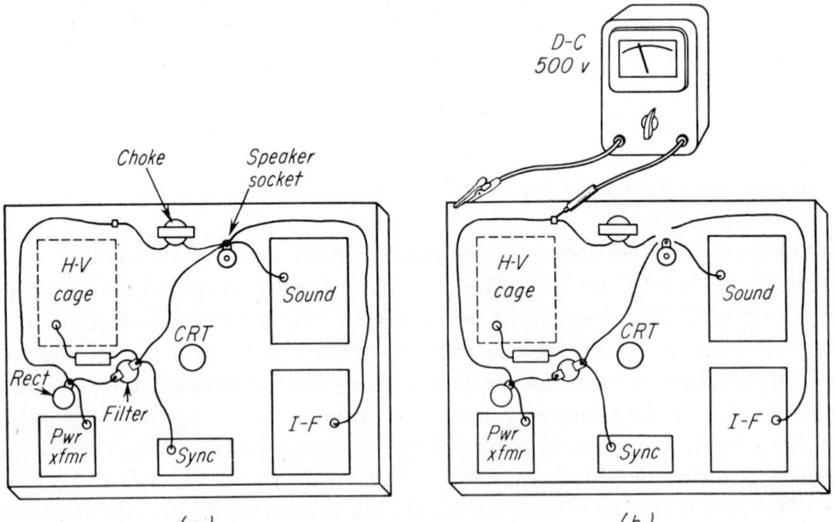

Fig. 4-30. Opening B plus leads on a printed-circuit receiver.

choke is mounted in the center of the top near the speaker socket. In this receiver, most of the B plus wiring is concentrated at the B plus terminal on the speaker socket, where the choke lead and B plus feeds to the sound and i-f boards are located. The B plus feeds to the high-voltage cage and sync board are connected at the filter condenser.

If this receiver were being worked on for an overloaded B plus line, it would be logical to open all B plus connections at the speaker socket, as shown in Fig. 4-30*b*. The voltmeter is left connected to the choke as the indicating device. Then each wire would be checked separately for the short. If the short proved to be in the wire going to the filter condenser, the B plus wires would be removed from the filter condenser, and the procedure would be repeated until the short is found.

This is a good point at which to insert some servicing hints for working on printed-circuit receivers. Printed-circuit chassis do not handle easily. Probably the best method of working on them is to remove the picture tube and clamp the chassis in a heavy-duty stand of the type made for working on phonograph turntables. Soldering on a printed-circuit board is also difficult. Too hot an iron may burn the thin conductors or lift them from the boards. A soldering gun, therefore, is impractical. Best results are obtained by using a small soldering iron of about 60 watts and working carefully so as to avoid rosin joints. Also recommended is a special low-melting solder, which flows easily at comparatively low temperatures. Another hint along these lines is to crush out or clip away defective components, leaving as much of the original leads as possible. The replacement component is then soldered to the leftover leads rather than to the embedded conductor on the board. Still another hint which may be of some value is to work with a strong light so placed as to shine through the printed-circuit board. This enables the serviceman to locate components on the back side of the board, since they can be seen through the translucent material.

Bus connections on printed-circuit boards usually employ a solderless wire-wrap method. These connections consist of several turns of wire tightly wrapped around a special square stud. The turns are made by a machine designed for the purpose. This type of connection is equivalent to a soldered joint for conductivity and permanence, but if the connection is unwound for test purposes or replacement of a part, the new connection must be soldered. The solderless wire-wrap connections are also found on modular units and controls.

Sometimes, a break is found in a connection which connects a lug on the top of the board to one on the bottom. The best way of repairing such a break is to drill a small hole through the board at the connection. Use a hand drill and a drill size of about No. 30. Then thread a piece of hookup wire through the hole, and solder to the proper lugs. In all cases, lug-to-lug soldering with a piece of wire is the preferred method of repairing a break. Run the wire over the connector it is repairing, and if there is a bend in the original, hold the wire in place at the bend with a gob of sealing wax.

Series-Parallel Voltage-distribution System. In the voltage-distribution system just discussed, all the tubes are fed in parallel from the B plus line, and dropping resistors are used to provide lower voltages where needed. The basic form of this circuit is shown in Fig. 4-31*a*. There is

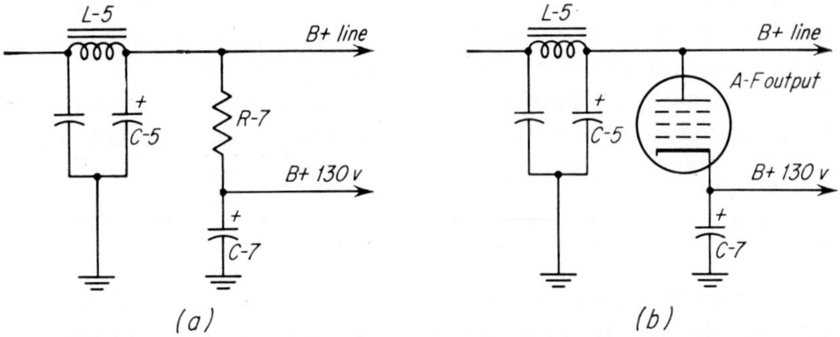

Fig. 4-31. How the audio-output tube is used to replace the dropping resistor to make a series-parallel, or shelf, voltage-distribution system.

another widely used circuit where the dropping resistor is replaced by the audio power tube. This type of circuit is sometimes called a "shelf," or a "stacked" power supply. The basic diagram is shown in Fig. 4-31*b*. The circuit is practical in that the audio output tube takes about the same current as the rest of the tubes on the 130-volt line. When the series-parallel circuit of Fig. 4-31*b* is used, the drain on the rectifier and filter system is about 50 ma less than in Fig. 4-31*a* where the audio output tube is fed from the same 130-volt line. Series-parallel voltage-distribution systems are widely used with both transformer-tube and selenium voltage-doubler rectifier circuits.

A skeleton B diagram showing the voltage-distribution system of the Dumont RA-164 television chassis is given in Fig. 4-32 as a typical example of a series-parallel, or shelf, B system. The audio output tube is a 6W6-GT, the cathode of which is the source of the B plus 135-volt line. This line feeds all the i-f tubes and the two sync tubes, each with its own decoupling filter. Note the heavy bypass condenser on the second sync clipper which must have a steady voltage. The same point feeds the screen of the video-amplifier tube. Condenser *C*-292B is the bypass for the entire 135-volt line. The mixer-oscillator tube also operates at a low B voltage, but if it were fed from the 135-volt line, it would require more cur-

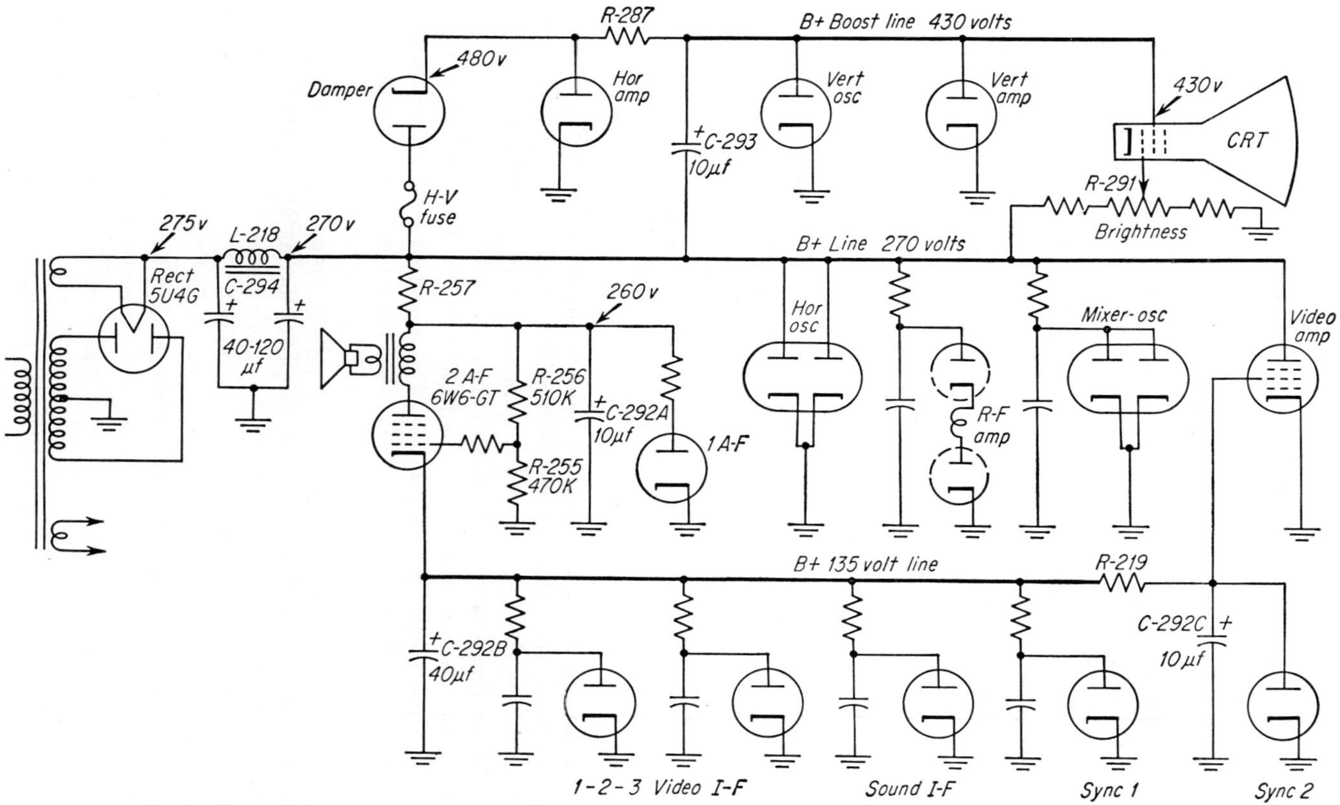

Fig. 4-32. Shelf-type voltage-distribution system in the Dumont RA-164 chassis.

rent than the 6W6-GT audio tube could handle. So this tube is fed from the B plus line through its own dropping resistor.

Since the cathode of the 6W6-GT audio power tube is at a positive voltage of 135, some provision must be made for a correct bias for this tube. Resistor *R*-257 and condenser *C*-292A make up a decoupling filter providing 260 volts to the audio-amplifier section. Resistors *R*-255 and *R*-256 make up a voltage divider across this 260 volts, and the grid leak of the 6W6-GT returns to the center of the voltage divider. This puts the grid of the 6W6-GT tube at a positive voltage with respect to ground. If the grid is made +125 volts by the voltage divider, and the cathode is at +135 volts, both with respect to ground, the grid is 10 volts more negative than the cathode, and the bias on the tube is therefore 10 volts.

The potential at the damper cathode is given as 480 volts. This is applied to the horizontal amplifier tube. Before going on to feed the vertical tubes and the first anode of the picture tube, the boost line is dropped to 430 volts by resistor *R*-287. Condenser *C*-293 is the bypass for the boost line. Note that the brightness control is in the control-grid circuit of the picture tube. In this receiver, the cathode is at a positive voltage, since it is fed by direct coupling from the plate of the video-amplifier tube.

Variations of the Series B Line. In the series-parallel, or shelf, voltage-distribution system, it is essential that the tubes on the 135-volt line draw the same current as the second audio output tube. In order to bring about this balance, the tubes on the 135-volt line are shifted about as each set is designed. The 135-volt tubes that are left over are then operated from the B plus line through a separate voltage-dropping system. As a result, there is considerable variation in this regard.

One variation that seems to be used quite often is to string the leftover tubes in series on the B plus 250 line. This variation is illustrated in Fig. 4-33. The second audio tube is in series with a batch of 135-volt tubes in the usual shelf connection. Two leftover i-f tubes are then run in series from the B plus line, each acting as a voltage-dropping resistor for the other. Condenser *C*-6 bypasses the B supply of the audio output tube, and condenser *C*-7 bypasses the 135-volt line. Both condensers in series act as the output filter condenser of the power supply. This also is a very commonly used circuit.

Service Notes on Series-Parallel Voltage-distribution Systems. When making voltage checks in a receiver with a series-parallel voltage-distribution system, it is confusing to find high positive voltages on cathodes and control grids. Such readings are an indication of trouble in ordinary sets where B plus is fed in parallel to all tubes. Here such readings are normal, and the serviceman must be alert to read diagrams and voltage charts correctly.

One other point must be kept in mind. A symptom may often be due to

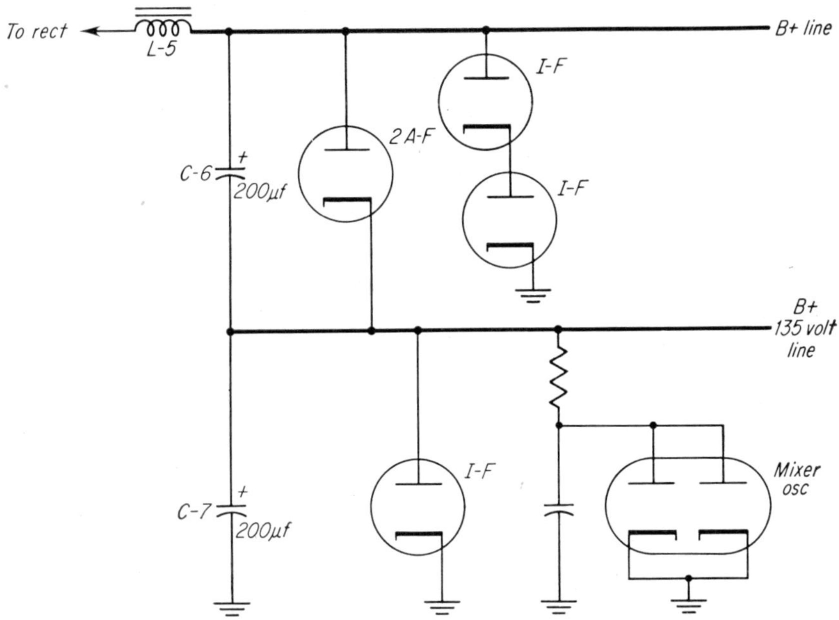

Fig. 4-33. A series-parallel, or shelf, voltage-distribution system where i-f tubes are operated in series from the B plus line.

a seemingly farfetched cause. For example, assume that an audio output tube loses emission and becomes weak. One would expect the sound output to become weak, as it would where tubes are fed in parallel from the B plus line. But in a series-parallel arrangement, the weak power tube (audio output) would act like a dropping resistor that increased in ohmic value. The voltage across the dropping resistor would increase, and the voltage left for the line would decrease. As a result, the audio output at the increased voltage might be close to normal, while the tubes on the 135-volt line would be operating at low voltage and causing the trouble. The symptom in the receiver will vary depending on which tubes are on the 135-volt line. Assume it is the video i-f tubes, and the sound i-f and first audio tubes are fed from B plus. The loss of video gain would bring this set in for loss of contrast, and the sound might be considered normal.

When working on this receiver, all checks in the seemingly defective video i-f amplifier would show normal results, until a voltage check is made. The low B voltage is noted, and leads to a check of the power supply, where it is discovered that the B plus output is normal. The serviceman must then be aware of the possibility of a series-parallel distribution system, and the effects of weakness in the tube used as a voltage-dropping resistor. Many manufacturers include a skeleton voltage-distribution diagram in their service notes as an aid in work of this kind.

SUMMARY

When tubes do not light, or fuses blow, or service checks for other symptoms show a below-normal voltage at B plus, the power supply is at fault.

Typical Power Supplies. Basic diagrams for three types of power supply are given in Figs. 4-34 to 4-36.

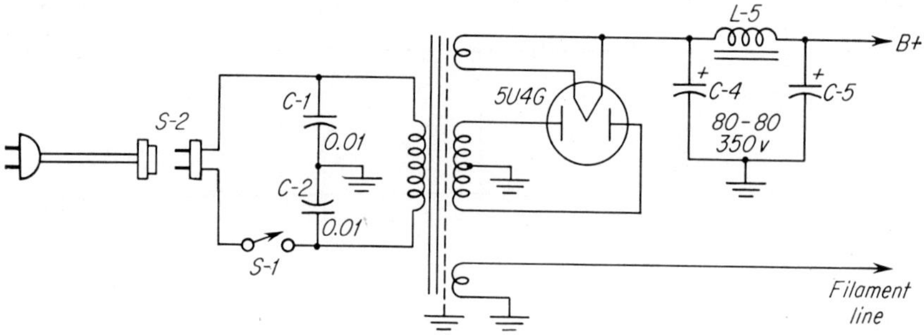

Fig. 4-34. Basic transformer-tube rectifier power supply with normal voltage and resistance measurements.

Normal voltage readings:

Line	117 volts a-c
Across filament line	6.3 volts a-c
Chassis to rectifier plates	250–380 volts a-c
Chassis to rectifier filament	240–340 volts d-c
Chassis to B plus	220–325 volts d-c
Across filter choke *L*-5	15–20 volts d-c

Normal resistance readings:

Plug, prong to prong	1–2 ohms
Chassis to rectifier plates	15–30 ohms
Across filter choke *L*-5	25–60 ohms
Chassis to rectifier filament	15,000–100,000 ohms

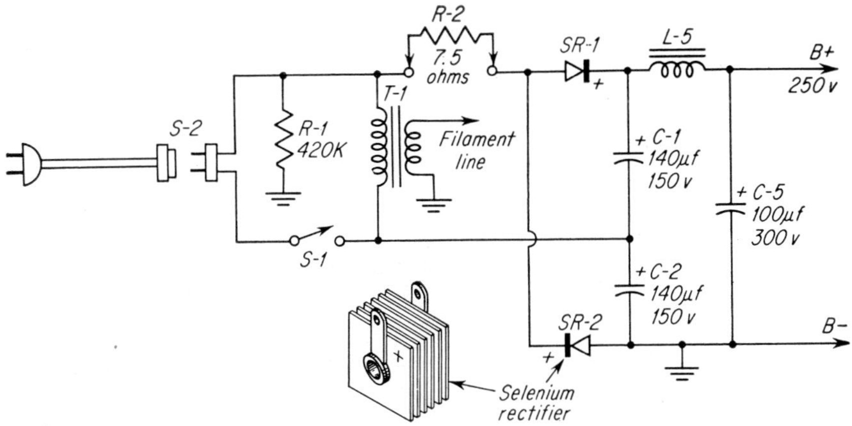

Fig. 4-35. Basic circuit of a full-wave voltage doubler with filament transformer.

Normal voltage readings:

Line	117 volts a-c
Chassis to B plus	250 volts d-c
Across filter choke	10 volts

Normal resistance readings:

Across line plug	15–25 ohms
Chassis to B plus	2,000 ohms
Across fusible resistor	5–7 ohms

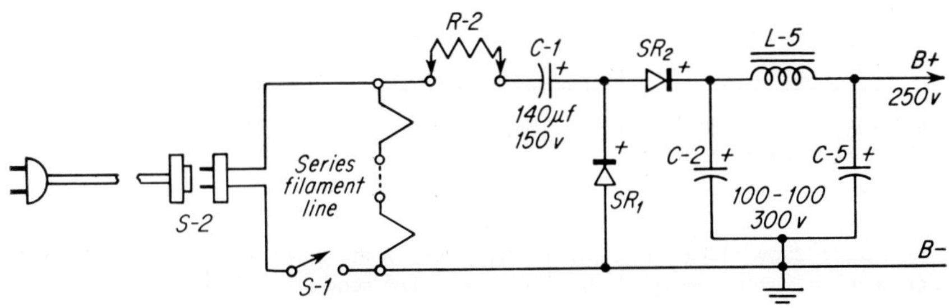

Fig. 4-36. Basic circuit of a half-wave voltage doubler used with tube heaters connected in series.

Normal voltage readings:

Line	117 volts a-c
Chassis to B plus	250 volts d-c
Across filter choke	10 volts d-c

Normal resistance readings:

Across line plug	80–100 ohms
Chassis to B plus	60,000 ohms

Service Data Chart

Symptom	*Abnormal Reading*	*Look for*
Set blows fuses, all types of power supply		Defective rectifier Shorted filter condensers Short on the B plus line Short on a filament line Short in the power transformer
Tubes do not light: All types		Defective line cord No contact in interlock Defective switch
Transformer type		Open primary of power transformer Rosin joint in the filament line
Series filament type		Open globar or line-dropping resistor Open heater in one of the tubes
Some tubes light, series or series-parallel heater types		Cathode-heater short in one of the tubes Shorted filament decoupling filter condenser Open 0.3-amp filament in series-parallel heater string
Dead receiver—rectifier tube shows purplish glow		Gassy rectifier tube
Dead receiver—rectifier tube plates show red glow or selenium rectifiers overheat	B plus voltage zero or low. Ohmmeter check on the B plus line shows low reading	Shorted filter condenser Short in the B plus line
	Ohmmeter check on the B plus line shows normal reading	Overload on the B plus line Short or overload which appears under voltage conditions. Check for leaky coupling condensers causing heavy drain, or shorts which appear under operating conditions (see Fig. 4-29)

Service Data Chart (Continued)

Symptom	*Abnormal Reading*	*Look for*
Dead receiver	B plus voltage checks zero	Dead rectifier, open fusible resistor, open filter choke (look for short or overload on the B plus line before replacing). In receivers using the circuit of Fig. 4-36, look for an open in condenser C-1.
Insufficient height and width, or no raster	B plus voltage checks below normal	Weak rectifier, input filter condenser or both (in field servicing, check for low line voltage)
Washed-out or dim picture	B plus voltage checks normal Low B voltage on the i-f tubes	Incorrect voltage-dropping resistor Weak audio output tube used in a shelf-type power supply

QUESTIONS

1. A television receiver of the transformer-tube rectifier type is equipped with a 6-amp line fuse which is blown. List the possible causes for the blown fuse. Describe the series lamp check for finding the trouble.

2. When using the series lamp check on a receiver which blows fuses, what inference can be made if the lamp remains at full brightness after the rectifier tube has been removed? What if it dims when the rectifier tube is removed?

3. What is the significance of an overheating rectifier?

4. A television receiver used two 5U4-G rectifier tubes and two 6BQ6 horizontal output tubes. What current rating should you look for in a replacement filter choke?

5. Describe how to identify leads in an uncoded replacement power transformer.

6. A television receiver is working normally except for reduced height and width. Service notes show a normal B plus reading of 250 volts. You measure it and find 210 volts. List the likely causes of trouble, and how to check for each.

7. What is the purpose of the copper band around the core of a power transformer in a television receiver?

8. What is the purpose of the fusible resistor generally found in selenium voltage-doubler circuits?

9. What methods are used to float the chassis when the rectifier system operates directly from the line in a transformerless circuit? What precautions should be observed when working on one of these receivers?

10. In a voltage-doubler circuit, how do you check the filter condensers? the rectifiers?

11. In any type of low-voltage power supply, when checking for below-normal output voltage, if the voltage drop across the filter choke is also below normal, what is likely to be at fault? If the drop across the choke is abnormally high, what is likely to be at fault?

12. The tubes in a set with a 600-ma heater string do not light. List the possible causes. How would you check for each?

13. Some tubes in a set with a 600-ma heater string do not light. List the possible causes. How would you check for each?

14. The tubes in a set with a series-parallel heater string do not light. Describe a servicing procedure to find the trouble.

15. When servicing a dead receiver, the first thing you notice is that the tubes in the i-f strip are dark but that others in the set are lit. You remove one of the i-f tubes and check its filament. It is a 6CB6, and its filament shows continuity. Go on with your test procedure from there.

16. Describe the servicing procedure to find a short or overload on the B plus line.

17. What is the effect of a weak audio output tube in a receiver with a series-parallel, or shelf, voltage-distribution system?

5

Servicing the Deflection Section: Vertical Sweep Circuits

As was previously stated, the key to successful servicing of a television receiver is to make use of the visual and aural effects inherent in the television system to isolate the trouble to one section of the receiver. Then a series of tests places the trouble in one stage within that section, and finally a voltmeter or ohmmeter shows the defective component in that stage.

Let us first review how the visual and aural clues bring us to investigate the vertical sweep section. A receiver is brought in for repairs. Sound is normal, and there is a bright horizontal line across the face of the picture tube. Refer to the section diagram of Fig. 5-1, and you will see

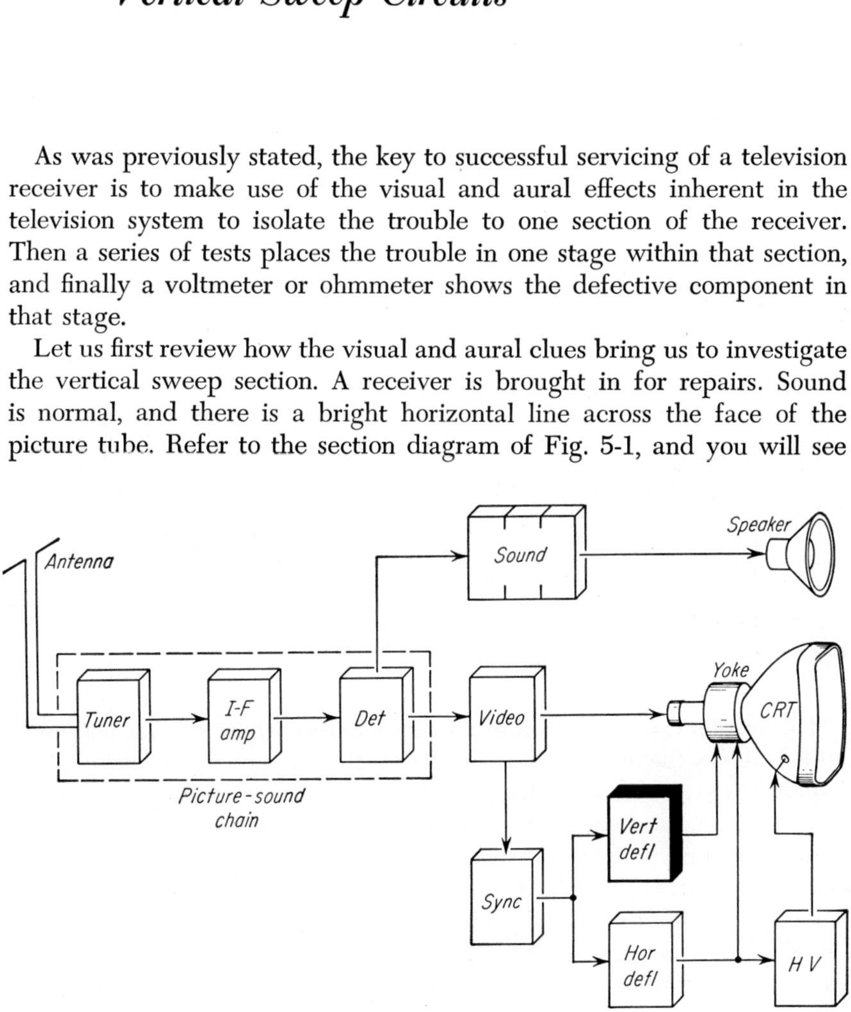

Fig. 5-1. The position of the vertical sweep (deflection) section in the section diagram of a television receiver.

that normal sound indicates that the tuner, i-f, and sound sections are functioning. The presence of any brightness on the screen proves that there is high voltage, and so the horizontal sweep and high-voltage section is in operation. This leaves the video-amplifier, synchronizing, and vertical sweep sections. The function of the vertical sweep section is to deflect the beam from the gun in the picture tube up and down the face of the tube. The bright line shows the presence of the beam, but since it is only a line and not a complete raster, there is no vertical deflection, and this is the defective section. The video-amplifier and synchronizing sections may or may not be working, but the vertical section is definitely not working, and this is our starting point.

The Stages in the Vertical Section. The vertical section consists of an oscillator stage, an amplifier stage, and a deflection yoke. These are shown in block-diagram form in Fig. 5-2. The oscillator generates voltage at 60 cycles—the vertical deflection rate. The amplifier builds this up so that it is strong enough to drive the yoke. It is usually called the vertical output stage. The vertical deflection winding of the yoke is so placed around the neck of the picture tube that its field causes the electron beam in the tube to be deflected up and down.

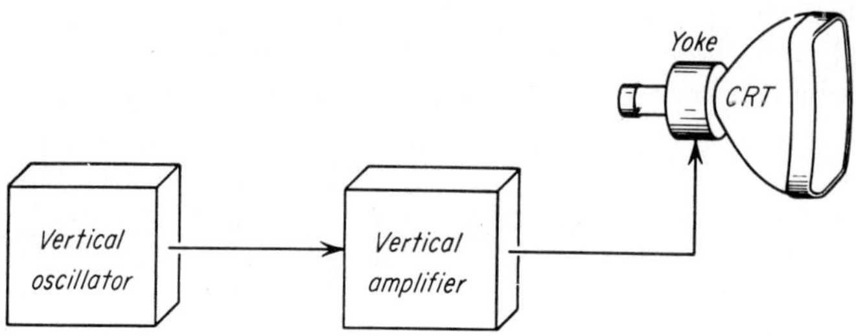

Fig. 5-2. Block-diagram representation of the vertical deflection section in a television receiver.

The section can be likened to an audio system where the vertical oscillator becomes the audio generator, that is, the microphone or pickup. The amplifier serves the same function in both systems, and the deflection yoke replaces the loudspeaker. Considering that the vertical section operates at 60 cps, an audio frequency, the comparison is heightened, since components will be similar to familiar audio-frequency circuits.

Controls in the Vertical Section. A simplified diagram of a vertical section is shown in Fig. 5-3. The oscillator circuit is a standard feedback oscillator and transformer *T*-1 furnishes the feedback between the plate and grid circuits. Condenser *C*-1 and variable resistor *R*-2 form the oscillator grid leak and condenser. The values are chosen so that the oscillator blocks periodically, giving a signal waveform which can be shaped to the pulses needed to operate the deflection yoke. Adjusting grid leak *R*-2 changes the frequency of the oscillator so that it can be brought in step with the transmitted vertical synchronizing pulses. This makes it act as the vertical-hold control.

The generated signal is coupled through condenser *C*-4 to the input of the amplifier tube. Potentiometer *R*-4 controls the amount of signal fed into the grid circuit. It is similar to a volume control in audio circuits. In the vertical section, it controls the amount of the beam deflection, and

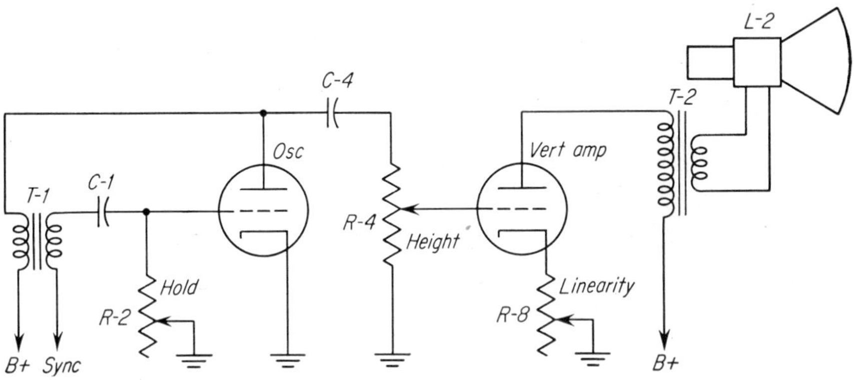

Fig. 5-3. Simplified vertical section.

so acts as a height control. Transformer *T*-2 is the vertical output transformer coupling the energy from the power tube to the deflection yoke. Variable resistor *R*-8 is in the cathode of the vertical output tube. It controls the bias in this amplifier stage, thereby giving a control over the operating point of the tube. This in turn allows for linear amplification of the input signal. The control is known as the vertical linearity control.

Location and Appearance of the Vertical Section. When a tube-function and -location chart is not available, the vertical section may be identified by tracing the leads from the deflection yoke. Trace one pair of leads to the vertical output transformer. The other pair of leads is for the horizontal yoke winding and goes to the high-voltage cage. These connections are shown in Fig. 5-4. Now trace the blue lead from the transformer to the vertical output tube. The vertical oscillator tube and blocking oscillator transformer are usually nearby.

Sometimes, the vertical output and oscillator tubes are the two sections of a twin triode like the 6SN7 type. In this case, there is only one tube socket in the vertical section. The location diagram of Fig. 5-4 includes the audio output transformer and filter choke to show comparative sizes. The filter choke is usually about the same physical size as the vertical output transformer but has two leads, whereas the transformer has three or four.

Quick Check for Finding a Defective Stage in the Vertical Section. Assume a receiver with a defective vertical section as evidenced by a bright horizontal line across the picture tube. The vertical section can be serviced just like an audio amplifier by a signal-substitution procedure.

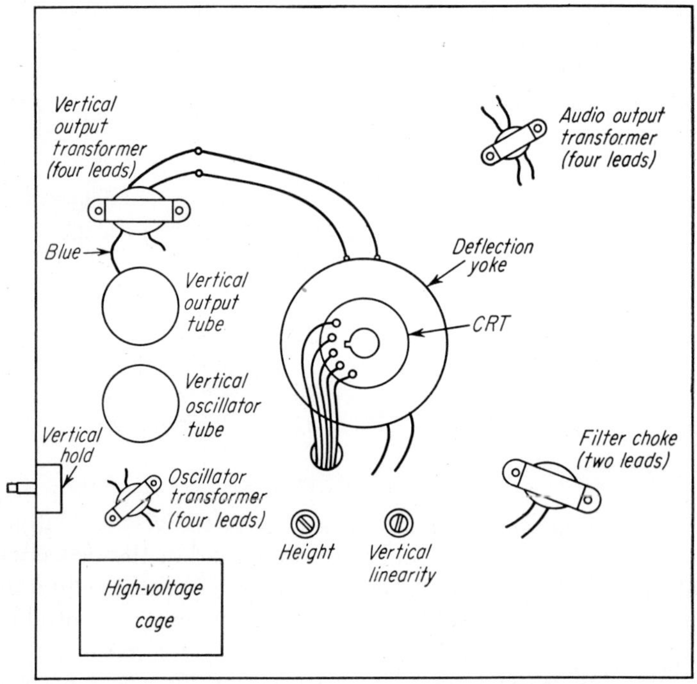

Fig. 5-4. Location and identification of the vertical sweep section on a vertical chassis.

The normal signal at the plate of the output tube is a 60-cycle sawtooth wave at a potential of several hundred volts. It is not necessary to reproduce this signal. Ordinary 60-cycle sine-wave current at a potential of about 100 volts will serve. If the vertical output transformer and yoke are in working order, applying a 60-cycle test signal of about 100 volts will cause a definite vertical deflection on the screen of the picture tube.

Do not use the a-c line for this test. In many transformerless receivers, one side of the line is connected to common negative in the low-voltage power supply. Using the line as a test-voltage source may have disastrous results in such a receiver. An isolation transformer can be used, but the simplest device would be to equip a small speaker-output transformer with test clips. Simply clip the voice-coil winding across any 6-volt filament in the receiver, and a 60-cycle test signal of 100 to 200 volts is available across the plate winding of the audio transformer. The setup is shown in Fig. 5-5. The 0.05-μf condenser keeps the transformer winding from interfering with any B plus potential at the test point. The test procedure follows.

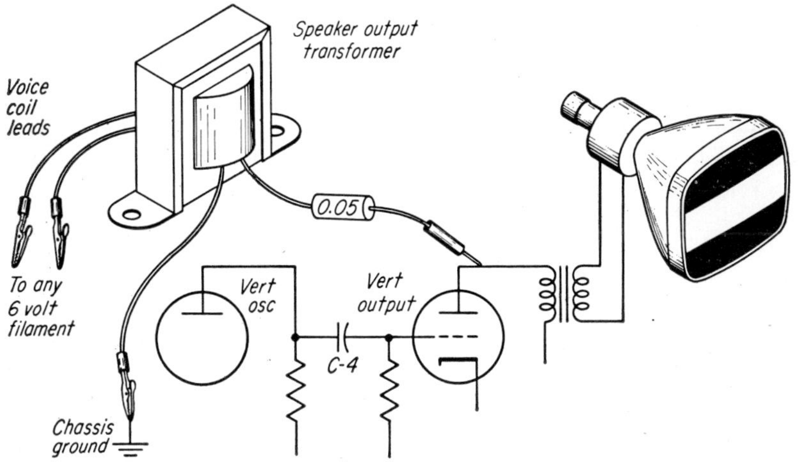

Fig. 5-5. How to find a defective stage in the vertical section.

Set the channel control to an unused channel—one which normally gives no reception in your area. Clip the test transformer voice-coil leads to the filament pins of any convenient socket containing a tube with a 6-volt filament. This is now the primary winding of the test transformer. Clip one of the secondary leads to the receiver chassis. Then, as shown in Fig. 5-5, touch the test prod to the plate pin of the vertical output tube while watching the bright horizontal line across the picture tube. If the bright line spreads into a raster about 3 in. high, the output transformer and yoke are good. If the line remains unaltered, either the transformer or yoke is defective, and an ohmmeter will show which one is at fault.

The amount of the deflection depends on the transformer chosen to produce the test signal. It would be good practice to use the same transformer for all testing to enable you to get a good idea of how much deflection to expect under normal conditions.

If the first test shows deflection, the trouble is before the plate of the output tube in the oscillator or amplifier stages. Now shift the test prod to the grid pin of the vertical amplifier tube, while once again watching the screen of the picture tube. If the bright line spreads to a full raster, the output stage is working. This time, because of the gain of the output stage, the normal result, with the same test signal, is a full raster. Actually, the raster will overspread the bottom edge of the tube. A full raster places the defect in the oscillator stage. If the bright line remains unaltered, the defect is in the vertical output stage.

One further test can be made with the same setup. If the vertical output stage is good, shift the test prod to the plate pin of the oscillator tube. If a full raster results, coupling condenser *C*-4 in Fig. 5-5 is good. If the bright line remains unaltered, the coupling condenser is open.

These tests locate the defective stage in the vertical section. The defective component is then found by voltage, resistance, and substitution checks.

Standard Circuit. A circuit representative of common practices used in vertical deflection is given in Fig. 5-6. This circuit will be used as a basis for servicing analysis.

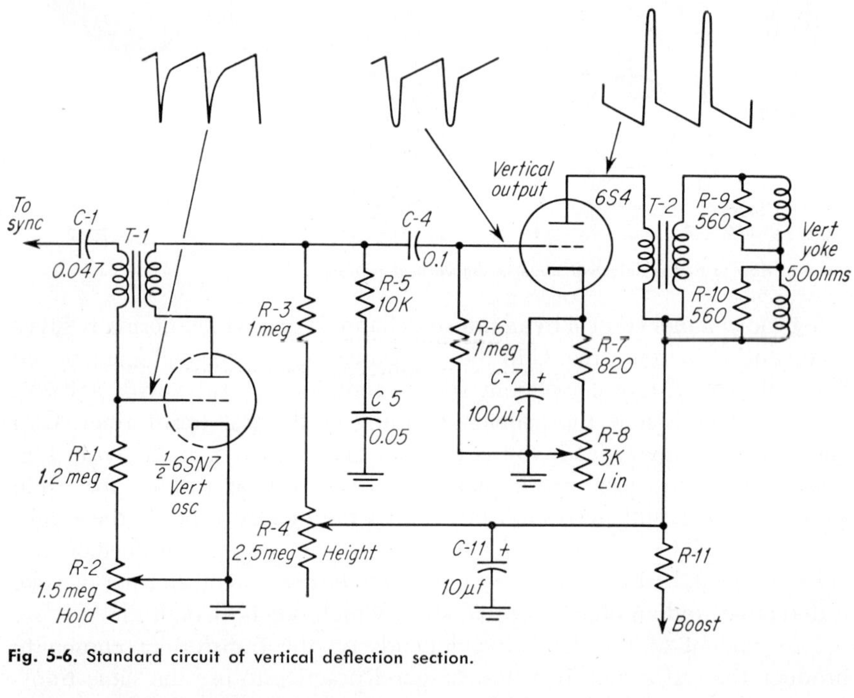

Fig. 5-6. Standard circuit of vertical deflection section.

Functions and Values of Components. The standard circuit shows one half of a 6SN7 twin triode being used as the vertical oscillator tube. The other half is used elsewhere in the receiver. The miniature base 12AU7 is often used in the same way. Sometimes, the circuit calls for a single triode like the 6C4. The 6S4, shown as the vertical output tube, has been specially designed for this purpose and is found most often. Older receivers use 6K6 or 6AQ5 beam pentodes in this position. These tubes are usually triode connected by a jumper between the screen and plate pins. Occasionally, a receiver is found in which both vertical tubes are contained in one envelope, the tube used being the twin triode 6SN7 or 12BH7.

As previously mentioned, variable resistor *R*-2 in the grid of the vertical oscillator tube acts as the hold control. Resistor *R*-1 in series with it makes for a minimum grid leak of 1.2 meg when the hold control is in the minimum-resistance position. Often both resistors are combined in one unit which is a 3-meg variable, with a stop at 1 meg. A similar condition exists for height control *R*-4 and its minimum-position resistor *R*-3. Note that these resistors are connected as a variable supply of B plus to the oscillator tube. This controls the output voltage of the oscillator, and therefore the height. In the simplified diagram of Fig. 5-3, the height control was shown as an audio volume control so that its function could be more easily understood.

Resistor *R*-5 and condenser *C*-5 in the plate circuit of the oscillator tube act to shape the sawtooth wave generated in the oscillator to the pulse needed to operate the deflection yoke. Note the waveshapes indicated at the test points in the diagram of Fig. 5-6. When an oscilloscope is connected to the grid of the vertical oscillator, the trace will also show the synchronizing pulse fed in from the synchronizing circuit. If the latter is deactivated, the trace shown in the diagram will appear. The waveform at the grid of the output tube shows the shaping effect of the *R*-5, *C*-5 circuit. Note the sharp pulse which is the output of the 6S4 tube. Do not connect your oscilloscope to this point unless it is equipped to handle high voltages, since the pulse at this point may be close to 1,000 volts.

Condenser *C*-4 and resistor *R*-6 in the grid circuit of the output tube serve their conventional functions of coupling condenser and grid load, respectively. Likewise, condenser *C*-7 in the cathode circuit acts as the cathode bypass condenser. The values shown in the diagram are usual. Resistor *R*-7 in the cathode circuit is a minimum-bias resistor; *R*-8 is the adjustable bias control and acts as the vertical linearity control. In some circuits, resistor *R*-7 may be as low as 330 ohms and *R*-8 as high as 5,000 ohms. In all cases, these resistors dissipate heat and should be of the 2-watt size.

Vertical output transformer *T*-2 couples the output of the tube to the vertical deflection winding in the yoke. In some receivers, the B plus lead is connected to a secondary lead internally, making this a three-lead transformer or autoformer. The usual ratio is 10:1.

Oscillatory currents may be set up between the transformer and yoke windings in the interval between pulses. These oscillations are damped by resistors *R*-9 and *R*-10 across each half of the yoke winding. The vertical windings of the deflection yoke usually have an ohmic resistance of 50 ohms.

Resistor *R*-11 and condenser *C*-11 make up a decoupling filter in the

B plus feed to the section. The condenser is usually a 10- or 20-μf 450-volt electrolytic unit. Resistor *R*-11 is a 2-watt carbon type, with a value between 470 and 5,600 ohms depending on the receiver.

Normal Voltage Readings in Vertical Section. The voltages in Table 5-1 refer to the standardized circuit of Fig. 5-6.

Table 5-1

Test Point	*Normal Voltage*
Oscillator control grid	−20 to −30
Oscillator plate	90 to 150
Amplifier cathode	20 to 30
Amplifier plate	400 to 500

The voltages given show a spread representing differences between receivers. The serviceman should also remember that the oscillator plate voltage reading will vary with the setting of the height control. Similarly, the vertical linearity control affects the voltage at the amplifier cathode.

The negative voltage at the oscillator grid is an important test voltage, since it shows that the oscillator is working. In most receivers, the plate supply for the vertical section (Table 5-2) is obtained from the boost voltage in the horizontal output circuit. Boost voltage is 400 to 500 volts in all receivers. In those receivers where the amplifier is fed from B plus, the plate voltage will measure approximately 250 volts.

Table 5-2. Resistance Data for Components in the Vertical Section

Component	*Normal Resistance*
Blocking-oscillator transformer:	
Grid coil	800–1,000
Feedback coil	150–200
Oscillator grid resistor *R*-1	½–1 meg*
Hold control *R*-2	1–3 meg
Oscillator plate resistor *R*-3	½–1½ meg*
Vertical size control *R*-4	2–5 meg
Amplifier cathode resistor *R*-7	300–800
Vertical output transformer:	
Primary	400–1,300†
Secondary	7
Vertical linearity control *R*-8	1,500–5,000
Yoke shunting resistor *R*-9	560
Yoke shunting resistor *R*-10	560
Vertical winding in yoke	50
Decoupling filter resistor *R*-11	470–5,600

* Resistor *R*-1 may be included in hold control *R*-2 as a mechanical stop on its rotation. Resistor *R*-3 may also be incorporated in vertical size control *R*-4.

† This check should be made against specific data for the receiver being serviced. The variation in resistance for the primary winding found in different receivers is very large. The data in the table are useful for finding an open winding.

OTHER TROUBLES IN THE VERTICAL SECTION

There are other visual indications of trouble in the vertical section besides the bright horizontal line. These are lack of height, poor vertical linearity, rolling, and foldover. The troubles are illustrated in Fig. 5-7.

Lack of height, illustrated in Fig. 5-7*a*, is caused by insufficient vertical deflection. Note that the width is normal. This is important since it definitely places the trouble in the vertical section. If both height and width were deficient, the indication would be trouble in a circuit common to both the horizontal and vertical deflection systems. As a matter of fact, in all symptoms pointing to the vertical section, full horizontal deflection is assumed.

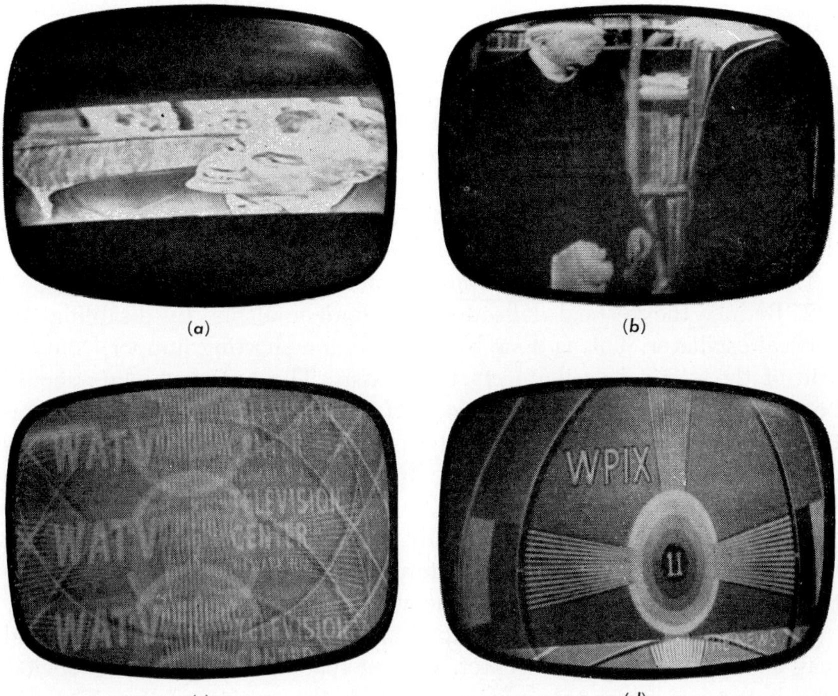

(*a*) (*b*) (*c*) (*d*)

Fig. 5-7. Trouble in the vertical section. (*a*) Lack of height. (*b*) Vertical distortion. (*c*) Rolling. (*d*) Foldover.

Another trouble is poor vertical linearity, illustrated in Fig. 5-7*b*. In this case, either the top or bottom end of the picture is expanded or contracted. This makes for distorted pictures where heads are elongated or flattened as the case may be. Since insufficient height and poor linearity often occur together and may be due to similar causes, a servicing procedure will be presented for either or both troubles.

Rolling, illustrated in Fig. 5-7*c*, is a lack of vertical hold. This happens when the operating frequency of the vertical oscillator changes, and manipulation of the hold control will not keep a steady picture.

These troubles may be caused by the vertical controls being out of adjustment. A first step in the servicing procedure, therefore, would be to check the control settings. In addition, abnormal control action may give a clue to the trouble. The settings of the height and linearity controls are interdependent in that adjusting the linearity control will affect the height and vice versa. Because of this, the controls should be adjusted in accordance with a fixed procedure. Test patterns which would make this work easier are not always available. The procedure to be presented has been designed to produce a full raster with good linearity, without the aid of a test pattern.

Foldover, illustrated in Fig. 5-7*d*, is a bright band on the bottom of the picture. The symptom is sometimes called curtain raising, since the bright band gives an appearance of the raising of a translucent curtain on a stage. Foldover is a special case of poor vertical linearity and is serviced by the same procedure.

How to Adjust the Vertical Controls. The following procedure is for the adjustment of the height and vertical linearity controls:

1. Tune the receiver for best picture on a strong channel.

2. Reduce the vertical deflection to a horizontal line by disabling the vertical oscillator. This is done by connecting a shorting jumper from the grid of the vertical oscillator to the chassis. The horizontal line should be exactly in the center of the screen. If it is a little off, as in the illustration of Fig. 5-8*a*, readjust the centering controls. Centering adjustments are described in Chap. 10. After the line is properly centered, remove the shorting jumper.

3. Compress the picture vertically by turning both the height and vertical linearity controls fully counterclockwise. If this causes the picture to roll, steady it by readjusting the vertical-hold control.

The picture will now have the squat appearance shown in Fig. 5-8*b* and may also appear to be off center. Do not readjust the centering device. The beam has been properly centered in the previous step. The off-center appearance is caused by uneven or nonlinear amplification at this setting of the controls.

4. Advance the brightness control. Bright lines will appear above the picture, as shown in Fig. 5-8*b*. These lines represent the vertical blanking interval between frames and can be observed at advanced positions of the brightness control. Since the lines will interfere with the adjustment, reduce the setting of the brightness control until they blank out. Readjust

the contrast control for rich blacks in the picture, which should now appear as shown in Fig. 5-8*c*.

5. Note the unlit portions of the screen above and below the picture tube. Their respective heights are indicated by the arrows in Fig. 5-8*c*. The height and linearity controls are at their minimum positions. Advancing either one will expand the picture and reduce the dark areas.

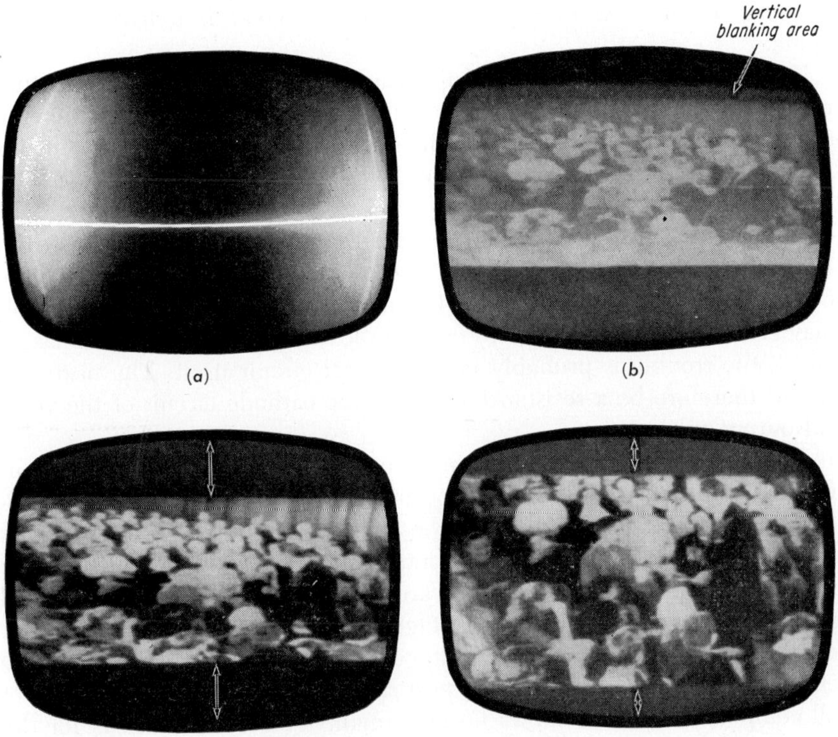

(*a*) (*b*) (*c*) (*d*)

Fig. 5-8. Steps in vertical adjustment. (*a*) Center. (*b*) Reduce height and linearity. (*c*) Reduce brightness. (*d*) Expand height.

However, one will be more effective on the top, and the other on the bottom. Advance whichever of the controls tends to equalize the dark areas. Continue to advance the control slowly while observing the unlit portions of the screen above and below the picture. As soon as the unlit portions are again different in size, leave this control, and advance the other. This will again cause the dark areas to become more equal in size, as both narrow. Continue in this way, advancing first one and then the other control, always in small steps, and always keeping the dark areas about equal in size, until finally both disappear entirely when the picture touches the top and bottom edges of the face of the tube. The photograph of Fig. 5-8*d* shows an intermediate step in the adjustment. Note that the dark areas are about equal in height. If the picture should start to roll during the adjustment, retouch the hold control.

In the case of round picture tubes, if the cabinet is not available to the benchman, he cannot be sure of the final height. These receivers are adjusted by the same procedure, approximating the limits of the mask opening. Final touchup is done when the receiver is delivered to the home.

To check for correctness, misadjust the hold control so that the picture rolls very slowly. The vertical blanking bar can now be seen moving slowly up or down the screen. If the height of the blanking bar remains unchanged as it moves from one edge to the other, the controls have been correctly set.

Servicing the Vertical Section for Poor Height, Linearity, and Foldover. When it is not possible to obtain full height with good linearity, there is trouble in the vertical section. The adjustment of the vertical controls may give a clue to the trouble. For example, if the linearity control has no effect, the trouble is probably in the control circuit itself. The next step would therefore be a resistance check of the cathode circuit of the vertical output tube.

Lacking such clues, the servicing procedure becomes a routine. Insufficient height is caused by a lack of gain in the output stage, or the generation of insufficient signal in the oscillator stage. Poor linearity is caused by distortion either in the output stage or in the coupling device feeding it. Defects of this nature are often accompanied by abnormal voltage readings; so a routine voltage check becomes the next step. A detailed analysis of abnormal voltage readings follows. If the voltage check shows normal readings, the next step is a careful resistance check of all components. If this shows no abnormality, the final check is for defects in components which will not affect voltage or resistance readings.

The checks which are to be described will also point out the defective component which causes a complete lack of height, that is, the bright horizontal line. Of course, in this case, the signal checks have pointed out the defective stage, and only that stage need be checked.

Voltage Analysis of the Vertical Section. The voltage check should be made against normal readings given by the receiver manufacturer in his service notes. If these data are not available, the average normal voltage readings given in this chapter should be helpful. For convenience, the standardized circuit of the vertical section is repeated in Fig. 5-9, with normal voltage readings as measured from chassis or common negative lettered in at the test points.

Assume a condition of no vertical deflection. Voltage check shows normal readings for the vertical output tube and readings of zero volts at the oscillator plate and grid pins. Referring to the diagram, there must be an open from the plate pin to the B plus boost point. The components involved are the plate winding of the oscillator coil, plate resistor *R*-3, and height control *R*-4. If there is a voltage reading at the plate side of resistor *R*-3, the transformer is open. If there is no voltage at this point either, move the test prod to the junction of resistor *R*-3 and the height control. Voltage at this point proves that the height control is conducting, and resistor *R*-3 is open. No voltage indicates an open in the height control. The condition is confirmed with an ohmmeter.

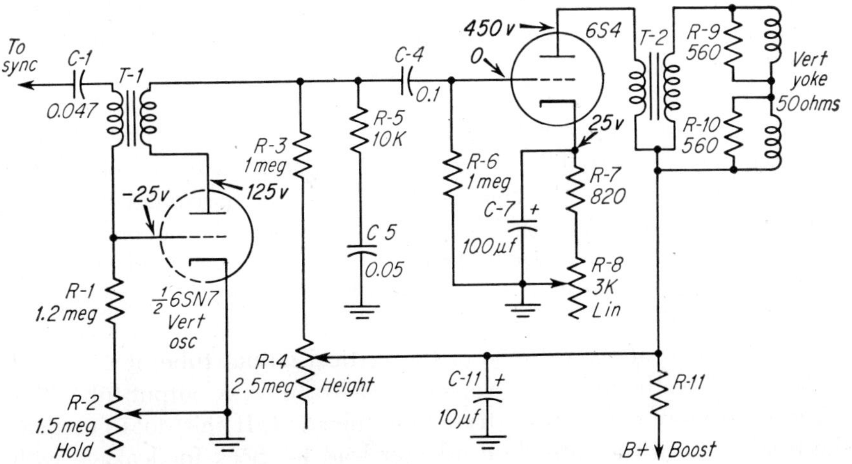

Fig. 5-9. Normal voltage readings in the vertical section.

If the voltage check shows a higher-than-normal reading at the oscillator plate, and zero at the oscillator grid, the fault can be a dead oscillator tube. The tube is checked by substituting one known to be good.

Should the voltage check at the oscillator plate show a lower-than-normal reading, the tube is drawing heavy current, and the oscillator is probably inoperative. A check of the grid voltage helps place the cause. If the grid reading shows a positive indication, there is leakage from B plus into the grid circuit through either condenser *C*-1 or transformer *T*-1. Open the connection between condenser *C*-1 and the transformer. Check the two opened ends with a voltmeter for a positive indication. Whichever component shows positive voltage is leaking and must be replaced. If the grid shows a below-normal negative reading, the same factors may be at work to a lesser degree. The same procedure will show which component is leaking. If the grid shows a zero reading, the grid winding of the transformer is probably open. This is checked with an ohmmeter.

Assume insufficient height, and a voltage check shows normal readings for the vertical output tube, a low reading of −10 volts at the oscillator grid, and a high reading of 200 volts at the oscillator plate. Such a combination of readings indicates a low plate current in the oscillator tube, probably caused by low emission. Try another tube.

Similarly, leakage across coupling condenser *C*-4 between the two vertical tubes will place a positive potential on the grid of the vertical output tube, with adverse effects on height and linearity. Again, the key to the trouble is the positive voltage reading on the grid. In this case, the positive grid reading may come from leakage in coupling condenser *C*-4 or also from grid emission in the output tube. The latter is easily checked. Keep the meter clipped to the grid pin, and pull out the output tube. If the positive voltage disappears immediately, the tube is at fault. Replace it. If the positive voltage remains, the coupling condenser is leaky.

If the voltage check shows no reading at the plate of the output tube, B plus or boost supply being normal, the trouble is an open in the plate winding of output transformer *T*-2. Confirm with an ohmmeter check. Here, of course, the complaint is the bright horizontal line indicating no vertical deflection. This would have been found on the first signal check, when a test signal at the vertical plate caused no spreading of the raster. It is repeated here as a confirmation check.

In checking for poor height and linearity, when an abnormally low reading is found at the cathode of the vertical output tube, it is caused by leakage in cathode bypass condenser *C*-7 or a weak output tube. The latter is the more common; so try a new tube first. If this does not clear the trouble, open the cathode condenser lead to check for leakage with an ohmmeter.

If the voltage at the cathode of the output tube checks zero, plate voltage remaining the same, the vertical output tube is dead. The same condition may be caused by a short in cathode bypass condenser *C*-7, but this is a rare occurrence.

Another abnormal voltage reading is an unusually high voltage at the cathode of the output tube. This is caused by an open in cathode resistor *R*-7 or linearity control *R*-8. An ohmmeter check tells which one is at fault.

Finally, there may be the case of no voltage at all at any of the tube check points in the vertical section. This indicates a lack of supply voltage to the section and places the trouble in the decoupling network consisting of *R*-11 and *C*-11 in the feed line. Condenser *C*-11 may be shorted, or resistor *R*-11 may be open. An ohmmeter will check both possibilities. If the trouble proves to be a short in the condenser, after replacing it check the resistor carefully before calling the repair completed. This is necessary because the resistor may have been harmed by feeding heavy current to the shorted condenser.

Servicing the Vertical Section for Rolling. As explained previously, rolling is caused by the inability of the vertical oscillator to operate at its normal frequency of 60 cps. A similar symptom may result when the synchronizing pulse is missing, since this pulse has the function of holding the vertical oscillator at 60 cps.

In servicing for rolling, therefore, it is first necessary to determine whether the fault is in the oscillator or the synchronizing system. To do this, manipulate the vertical-hold control slowly throughout its full range. If a point is found where one complete picture can be viewed, the oscillator is operating at its proper frequency of 60 cps. If the picture does not lock in and stay but slowly drifts up or down and starts to roll, the

(a) (b) (c)

Fig. 5-10. Effects of an off-frequency vertical oscillator. (*a*) Vertical oscillator off 60 cps. (*b*) Two pictures; oscillator locks at 30 cps. (*c*) Half of picture; oscillator locks at 120 cps.

synchronizing pulse is absent. Servicing the synchronizing stage is taken up in a later chapter. It may be that the set has a noise-gate control. A simple readjustment of this control may restore the sync signal. Refer to the manufacturer's service notes for the presence of the control and its adjustment instructions.

If it is not possible to get one complete picture by adjusting the hold control, the trouble is in the vertical oscillator. The oscillator is working since there is vertical deflection, but it is not working at 60 cps. If the picture is of insufficient height as well as rolling, service for the lack of height first by the procedure given previously.

Two other symptoms are coupled with rolling. If the vertical oscillator is operating below 60 cps, a lock may be achieved at 30 cps. Two pictures will appear on the screen, as shown in Fig. 5-10*b*. Similarly, if the vertical oscillator is operating near 120 cps, a lock occurs at this multiple of 60 cps. The result is that the top half of the picture spreads over the screen with the bottom half making a sort of double-exposure appearance in the background. A picture locked at 120 cps is illustrated in Fig. 5-10*c*.

Having established that the rolling is caused by an off-frequency vertical oscillator, the first step in finding this fault is to try operation with another vertical oscillator tube. It is best to take the vertical oscillator tube from another set which is working normally. If this clears up the trouble, replace the tube with a new one. Incidentally, this is a good general rule to follow when checking a tube which is suspect. The expression "replace with a known good tube" can best be met by taking a similar tube from a similar circuit in a set which is operating normally. Then, if the known good tube clears the trouble, return it to its set, and put a new tube in the set being worked on. If the new one repeats the trouble try another new one. Finally, when a good tube has been found, allow the receiver to operate for at least a half hour to make sure that the new tube does not develop an off-frequency characteristic after it is thoroughly heated.

If the known good oscillator tube does not stop the rolling, the vertical oscillator stage is checked for the fault. To do this, it may be best to remove the vertical sync signal, thereby allowing the stage to operate as a free-running oscillator. Break the sync feed line at the point indicated in Fig. 5-11, and now there is simply an off-frequency oscillator to deal with.

The main frequency-determining components in the oscillator are *C*-1, *R*-1, and hold control *R*-2 in the oscillator grid circuit. The integrator net work, consisting of resistors *R*-12 and *R*-13 and condensers *C*-12, *C*-13, and *C*-14, will also have an effect on the operating frequency. Any one of these components may have changed in value, causing the off-frequency operation of the oscillator.

Resistors *R*-1 and *R*-2 are checked with an ohmmeter, and the readings are compared with the values printed on the service notes of the receiver being worked on. The resistors and condensers in the integrator network may be separate units, or they may be combined in a single printed-circuit component called a couplate. A typical couplate is illustrated in Fig. 5-11.

If a couplate is being used, the individual resistors cannot be checked. It will then be necessary to check across points 2 and 3 in the couplate where the normal reading is 16,000 ohms. Sometimes, resistor *R*-14 is included in the couplate. In this case, the normal reading is 38,000 ohms. When any reading varies more than 10 per cent from the rated value, the component should be replaced.

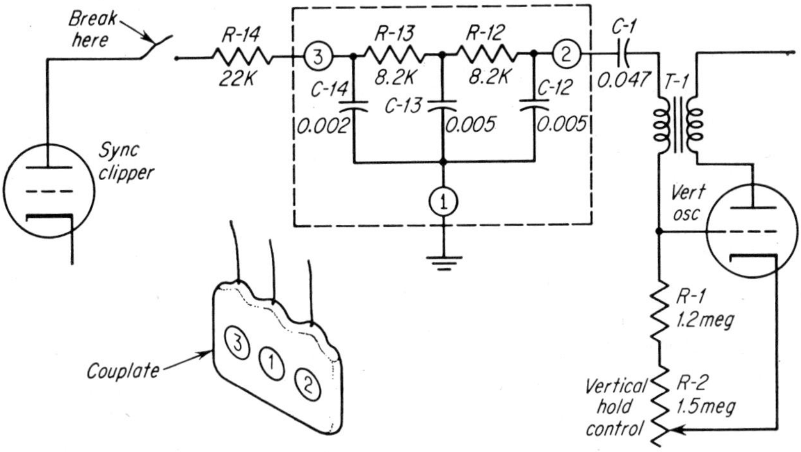

Fig. 5-11. A vertical integrator couplate and its position in the circuit.

The condensers in the couplate are checked for leakage by taking an ohmmeter reading from point 2 or 3 to ground. The correct reading is open circuit. Any reading at all indicates leakage, and the couplate should be replaced. It is not necessary to check coupling condenser *C*-1 for leakage, since this would place a positive voltage on the oscillator grid and stop the vertical sweep entirely. But the condenser may have changed in capacitance, causing the rolling. This is checked for by disconnecting the condenser and trying operation with another of the correct capacitance. Similarly, the condensers in the integrator network may have changed in value. This is checked by substituting another couplate for the suspected unit.

When the off-value component has been found and replaced, the oscillator will operate at 60 cps, as evidenced by the ability to obtain one complete picture by manipulation of the hold control. The integrator network is then reconnected to the plate of the synchronizing tube, and operation is tried again. When checking for proper operation, adjust the hold control so that the picture rolls slowly toward the top of the screen. As it approaches the correct frequency, the vertical blanking bar will bounce into position when the sync pulse takes hold.

Servicing for Vertical "Jittering." Another trouble akin to rolling is the condition where the picture, or a part of it, has a small rapid bounce in the vertical plane. This symptom is called jittering. This may be due to off-frequency operation of the vertical oscillator. The picture starts to roll, but before it gets very far, it is snapped back by the vertical sync pulse. Continued repetition produces the bounce.

Jittering, therefore, may be considered as a mild form of rolling in that the oscillator drifts just a little too far to be steadily held by a normal sync pulse, or the sync pulse is just a little inadequate to hold a normally drifting oscillator. However, from the serviceman's point of view, this small trouble is more difficult to handle. The defective component may be so close to normal as to be considered normal.

The complaint is serviced in a manner similar to the procedure just given for rolling. First the sync feed line is broken at the input to the integrator network as shown in Fig. 5-11. This time, it would be best to operate with a substitute sync signal. This is obtained by feeding a 6-volt point on the filament line to the integrator network through an isolating condenser. In localities where the transmitter uses the same power line, the power frequency makes a good synchronizing source. If the picture now locks and holds without jitter, the trouble is in the sync line. Servicing procedures for the sync signal are given in Chap. 8. If the jittering continues, the frequency-determining components are checked in the same manner as for rolling. This time the checks must be made with extra care, since a rough estimate of tolerance may pass the defective component. Make sure that components are within 10 per cent of rated values.

The tendency to stray off frequency may also be due to a periodic change in value of one of the components under operating conditions. This may not be found with an ohmmeter check. A good way to find such a faulty component is to operate the receiver and to apply heat from a soldering iron or gun to each suspected component in turn, while observing results on the screen. The breakdown will be more apparent under the added heat.

Another possible cause for the bounce may be that random noise pulses generated in the receiver are triggering the oscillator. This is especially true if the bounce is confined to the top half of the picture. One source of these noise pulses may be that previous service work or adjustments or dirt may have broken the grounding contact on the aquadag coating of the picture tube. The aquadag coating is one plate of a filter condenser in the high-voltage supply. The condenser discharges intermittently through leakage paths to ground. If the intermittent pulses are near the sync frequency, they will trigger the vertical oscillator, causing the bounce. To correct this condition, clean and respring the grounding contacts. If the aquadag coating is worn or peeled, it can be renewed by painting or spraying with a conductive coating like Tube coat.

Another source of noise pulses within the receiver is corona discharge in the high-voltage supply. Repair work in the r-f cage may have disturbed the lead dress. Or there may be resoldered connections left with sharp points. The misplaced leads or sharp points carrying high voltages at radio frequency are prone to corona discharge which may cause the vertical bounce.

It may be possible to find this trouble by inspection. The characteristic purple glow of the corona discharge will point out the location. Darkening the room helps. If the glow cannot be found, and corona discharge is suspected, go over the entire r-f cage, resoldering any connections which appear pointed, and wiping the joint with the iron, so as to leave a rounded connection. The high-voltage rectifier plate lead is the worst offender for corona effects. Make sure this lead is dressed away from all components, including chassis or r-f cage parts. A preparation known as corona dope has been found helpful when painted over connections. Another aid of this type, marketed under the name of Glyptol, is a plastic which is sprayed inside the r-f cage. Finally, check to see that all holding screws for the r-f cage are present, and tight.

REPLACEMENT PROBLEMS

It is not always possible to obtain an exact duplicate when replacing a defective component. This is of no particular importance when replacing resistors and condensers. For these, the products of any reliable parts manufacturer will serve, as long as due care is exercised in selecting components of the proper wattage, voltage rating, etc. However, many specialized components have additional important specifications which must be met to ensure proper operation of the receiver. This section will outline the test which confirms the suspicion of defects in components and will give pointers on replacement problems such as selecting, wiring, and "making do" with available units when an exact duplicate is unobtainable.

Troubles Common to the Vertical Output Transformer. The output transformer is a common cause of trouble. High pulse surges cause voltage breakdown, and heavy primary plate currents result in open windings.

If the transformer plate winding opens, plate voltage is cut off from the output tube, and the vertical section is completely inoperative. The bright horizontal line will be the symptom. Feeding test signal to the plate of the output tube finds the fault quickly, since no spread brings suspicion to the output transformer or yoke. The ohmmeter then shows the open winding.

If the transformer develops shorted turns, there will still be deflection, but not enough, depending on the severity of the short. There will also be distortion of the waveshape of the output signal. This will result in poor linearity and possibly foldover.

The combination of these symptoms may be caused by a number of conditions, as explained in a previous section. When tube substitution makes no improvement, and voltage and resistance checks appear normal, the serviceman suspects the presence of shorted turns in the transformer.

This may not be revealed by the resistance check since a few shorted turns will have very little effect on the total resistance of the winding. The suspicion can be confirmed as follows.

How to Check the Vertical Output Transformer for Shorts. First unplug the receiver line cord and open one of the vertical yoke winding leads from its connection to the vertical output transformer. Then, using a lamp short checker like the one illustrated in Fig. 5-12, apply the 120-

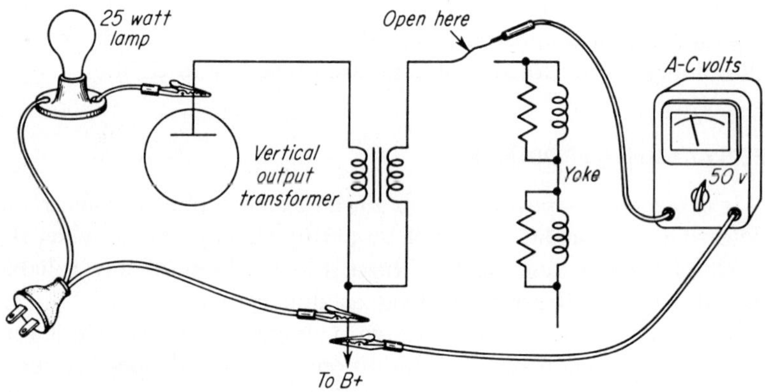

Fig. 5-12. How to check a vertical output transformer.

volt line to the primary of the output transformer. Measure the voltage developed across the secondary with an a-c voltmeter. A good transformer will show a reading of about 12 volts since most vertical output transformers have a stepdown ratio of 9:1 or 10:1. If the transformer is shorted, the reading across the secondary winding will be considerably less than 12 volts depending on the severity of the short.

The above test is valid for most receivers except those where the vertical output transformer has a stepdown ratio of 44:1. This ratio will cause a normal reading of about 3 volts for a good transformer on the same test. The reading for a shorted 44:1 transformer will be less than 3 volts. In case of doubt, measure the resistance of the vertical winding in the yoke.

The more common receivers use a yoke winding with a resistance of 50 ohms. These take the 10:1 ratio output transformer. Some receivers use a vertical yoke winding with a resistance of 3 or 4 ohms. These require a 44:1 ratio transformer for proper matching.

This check can also be used for determining the presence of shorts in any low-frequency transformer where the turns ratio is known. For example, the blocking-oscillator transformer with a turns ratio of 4 in the grid winding to 1 in the feedback winding will produce a voltage of 120 divided by 4, or 30, volts. Similarly, an audio output with a ratio of 20:1 will produce 6 volts.

How to Select a Replacement Vertical Output Transformer. When replacing a defective transformer, it is advisable to obtain an original manufacturer's replacement or one made for replacement purposes. Manufacturers of replacement components distribute catalogues which list components by make and model numbers of receivers. Part of a page of the Stancor Replacement Guide is shown in Fig. 5-13 as a typical example. The manufacturer's part numbers are listed first, with the Stancor replacement number below it.

The use of these catalogues is strongly recommended for choosing correct replacement components. This is especially true when the specifications for the part are critical. The vertical output transformer is such a part. It must make the correct match between the vertical output tube and the vertical winding in the deflection yoke.

MODEL NUMBER	FLYBACK	DEFLECTION YOKE	POWER	CHOKE	VERTICAL OUTPUT	VERTICAL BLOCK.-OSC.	C
ADMIRAL							
(1) Chassis 19A1	-- --	-- --	80B13 --	74A13 C-2303	79B70 --	69B70 --	79 A
(2) Chassis 19A2/A/AZ/Z, 19B1/C, 19B1A/AZ/Z, 19C1, 19D2/A, 19E1, 19E2/A/, 19F1/A/B/C, 19F2Z/AZ, 19G1/A, 19H1, 19J1/A, 19K1/K2Z/AZ, 19L1/2, 19M2, 19N1, 19N2Z, 19P1, 19S1, 19S2, 19T1C, 19T2A, 19W1/A/B/C, 19Y1A, 20A2/Z, 20D2, 20L2	79D41-1/2 A-8135	94C51-1 DY-11A	80C-35-1/-2 P-8164	74B18-2 C-2327[a]	74B40-2 A-8143	79A18-4 A-8125	A-3
(3) Chassis 20A1, 20B1	79C19-2[2] A-8127	94B2-2 DY-1A	80B17 P-8160	74A12[1] C-2325	79B20 A-8116	79A18-1 A-8125	79A A-38
(4) Chassis 20AX5/A/B/CZ/D/EZ/F1, 20X5/A/B/CZ/EZ, 20XP5/A	76C60-1 A-8254	-- --	-- --	-- --	-- --	-- --	- -
(5) Chassis 20T1, 20V1	79C32-1[4] A-8130	A-3222 DY-2A	80C26-2 P-8164	-- --	79A29-2 A-8123	79A18-3,-4 A-8125	79B3 A-38
(6) Chassis 20X1, 20Y1, 20Z1	79C23-1[6] A-8127	79B2-1[7] DY-1A	80B19 P-8171[15]	74A12 C-2325[a]	79B24-1 A-8112	79A18-1 A-8125	79A A-
(7) Chassis 21A1	79C17-2 A-8119	94B2-2 DY-1A	80B17 P-8160	74A12[1] C-2325	79B20 A-8116	79A18-1 A-8125	79A A-3
(8) Chassis 21B1, C1, F1, G1, H1, J1	79C30-2,-4	A-3222[9]	80C26-1	74B18-4	79B29-1	79A18-2	

Fig. 5-13. Part of the Stancor Replacement Guide catalogue.

If the receiver being serviced is not listed in any catalogue, the choice of a replacement becomes difficult. First look to see if the receiver is listed under a different name. Very often, manufacturers make receivers which are sold under various trade names, and not all trade names are listed. A good cross index for this purpose is obtained by referring to the index of Rider's "TV Manual."

If a listing cannot be found by this method either, a replacement must be selected from the available information. Examine Fig. 5-14, which reproduces another section of the Stancor Replacement Guide listing output transformers. Note the column showing the d-c resistance of secondary windings. Most of these vary from 10 to 15 ohms, but one reading is 0.3 and another 4.5. The secondary winding of these transformers consists of comparatively few turns of heavy wire. Since it is unlikely that this winding was harmed in the original transformer its d-c resistance can be measured. If the original winding checks less than 1 ohm, the transformer with a ratio of 44:1 and a d-c secondary winding of 0.3 ohm is the best choice. If the original winding checks about 5 ohms, the 18:1 transformer is the likeliest one. Note also the high impedance and low current rating of the primary winding of this transformer. These make it suitable for use with one triode of a 6SN7, 12AU7, or 12BH7 type of output tube. The other half of the tube is the vertical oscillator. For confirmation check the type of tube used in the receiver.

Vertical Deflection Output Transformers

Part No.	Turns Ratio Pri./Sec.	Primary Impedance#	D.C. Res. In Ohms		Height Overall	Base Area	Mtg. Ctrs.	Mtg. Type	Sh
			Pri.	Sec.					
A-8112	10:1	18,000 Ω @ 12 DCMA	1300	10	2	1¾ x 3¼	2 13/16	A	
A-8113	8.8:1	16,500 Ω @ 10 DCMA	700	12	2	1¾ x 3¼	2 13/16	A	
A-8115	10:1	19,000 Ω @ 13 DCMA	600	7	3 1/16	2½ x 2½	1 19/32 x 2	NV	
A-8116	10:1	18,000 Ω @ 10 DCMA	525	7	3 1/16	2¼ x 2½	1 19/32 x 1¾	NV	
A-8123§	11.4:1	17,000 Ω @ 20 DCMA	1200	11	2	1¾ x 3¼	2 13/16	A	
A-8140	44:1	11,000 Ω @ 20 DCMA	400	0.3	3 1/16	2½ x 2½	1 19/32 x 2	NV	
A-8141§	18:1	30,000 Ω @ 10 DCMA	1650	4.5	2¼	3¾ x 2⅛	3⅛	A	
A-8142	8:1	19,000 Ω @ 13 DCMA	540	10.5	2⅝	2¼ x 3	2⅜ x 1⅝	NH	
A-8143	10:1	14,000 Ω @ 15 DCMA	625	14	2¼	2¼ x 3⅝	3⅛	A	
A-8144	9:1	9,500 Ω @ 30 DCMA	540	15	2¼	1¾ x 3⅝	3⅛	A	
A-8145	9:1	11,000 Ω @ 19 DCMA	540	14	2⅝	2¼ x 2½	—	N1	
A-8146§	6.9:1	6,000 Ω @ 15 DCMA	375	10	2	1⅞ x 3¼	2 13/16	A	
		4,700 Ω @ 50 DCMA							
A-8147§	6:1	3,200 Ω @ 40 DCMA	300	9	2	1¾ x 3¼	2 13/16	A	
A-8148§	8:1	6,000 Ω @ 15 DCMA	375	6.5	2	1⅞ x 3¼	2 13/16	A	
		4,700 Ω @ 50 DCMA							
A-8149§	6 9:1	11,500 Ω @ 20 DCMA	330	8.5	2	1⅝ x 3¼	2 13/16	A	

Fig. 5-14. Catalogue listing of vertical output transformers. (*From the Stancor Replacement Guide.*)

If the d-c resistance of the original transformer is 10 to 15 ohms, too many of the listed replacement transformers would fit, and the selection must be based on other factors. If the output tube is a 6K6, 6V6, or other high-current tube, the best choice would be the transformer rated at 40 ma for the primary winding. When the output tube is a 6S4, or the two halves of a 6SN7 connected in parallel, the transformer primary should be able to handle 20 to 30 ma. In any case, the transformer chosen will have a ratio of 9:1 or 10:1. Your choice may not match the original exactly, but the mismatch may be compensated for by manipulation of the height and linearity controls. After the transformer has been replaced, and the controls adjusted, if you think you have chosen the wrong replacement unit, try operation with another output tube of the same type. Also, some other choice of output tube may make for a better match. For example, substitute a 6V6 for a 6K6 or vice versa, or a 6BL7 for a 6SN7.

Wiring in a Replacement Vertical Output Transformer. When wiring the replacement transformer into the circuit, there may be some confusion because of the lack of standardization in the color coding of the leads. However, the plate lead is universally blue, and this leaves only the secondary for trial connections. Wire in the secondary with temporary connections. If the resulting picture is upside down, reverse the secondary leads, wiring them in permanently.

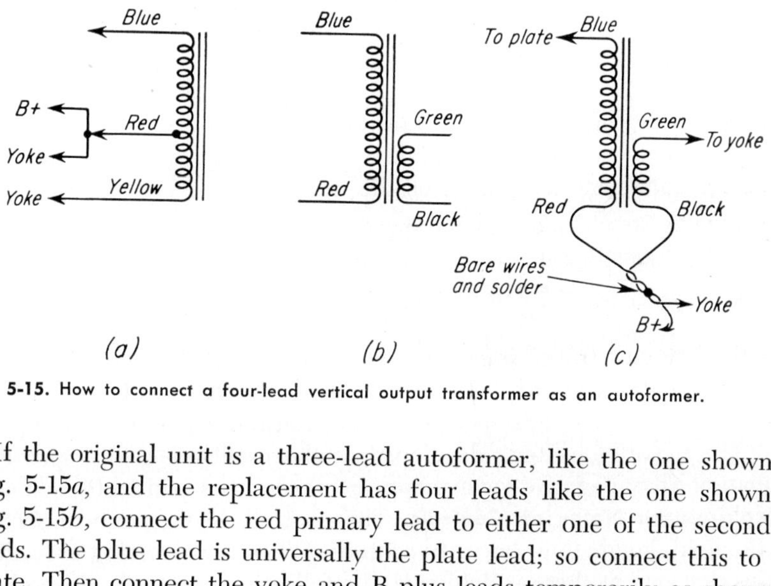

Fig. 5-15. How to connect a four-lead vertical output transformer as an autoformer.

If the original unit is a three-lead autoformer, like the one shown in Fig. 5-15*a*, and the replacement has four leads like the one shown in Fig. 5-15*b*, connect the red primary lead to either one of the secondary leads. The blue lead is universally the plate lead; so connect this to the plate. Then connect the yoke and B plus leads temporarily as shown in Fig. 5-15*c*. Then, if the picture is upside down, reverse the secondary leads.

Replacement Notes on Blocking-oscillator Transformers. The oscillator transformer commonly develops an open winding. This stops oscillation, resulting in no deflection and a bright horizontal line. Signal and voltage checks bring the serviceman to suspect the transformer, and the open winding is found with an ohmmeter.

When replacing a defective oscillator transformer, the use of an exact duplicate of the original is desirable but not necessary. If the set is listed in a replacement catalogue, the recommended replacement will require, at most, the redrilling of mounting holes. If the set is not listed, rely on the fact that most receivers use a vertical blocking-oscillator transformer with a turns ratio of 1:4.2. The only exception is that some Philco receivers use a transformer with a ratio of 1:1.5. The listed transformers are alike except for mounting details. The types in most common use are illustrated in Fig. 5-16. Try to use the mounting type which matches the original.

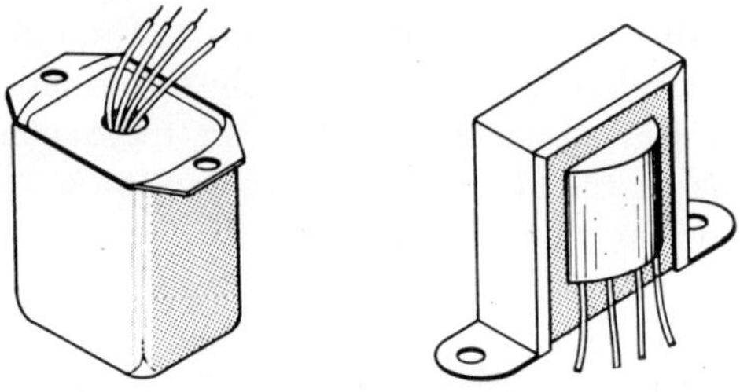

Fig. 5-16. Typical vertical blocking-oscillator transformers.

When wiring the replacement transformer, make sure to wire the leads correctly, since the oscillator will not operate with a reversed winding. The color code commonly used on the transformers is as follows:

Blue	Plate
Red	B plus
Green	Grid
Yellow	Synchronization circuit

If the original transformer or replacement uses different colors, an ohmmeter will distinguish between the plate and grid windings since the latter has the higher resistance. Wire the transformer, and try its operation. The negative voltage developed at the oscillator grid is proof of oscillation. If no negative voltage appears, reverse the grid and sync leads.

Replacement Notes on Controls. The controls in the vertical section are the hold, height, and linearity controls. When the servicing procedure proves that one of these is defective, the following points should be kept in mind when making the repair.

As a general rule, do not clean the control; replace it. The cleaning job may not be too lasting and will result in a service recall. If the defective control is the hold, this is usually mounted on the front panel, and the replacement must match the original in shaft length and knob-fastening arrangement as well as in electrical characteristics. If the control is part of a dual unit with concentric shafts, the other section being the horizontal-hold control, the entire unit must be replaced. It is necessary to obtain a duplicate replacement made by the receiver manufacturer, or one designed for that receiver. In the case of the height and linearity controls, the electrical specifications are all that need be met, since these are usually mounted on the rear apron of the chassis. Generally, these controls have knurled or slotted shafts for ease in making adjustments. Remember also that the linearity control should be a 2-watt size.

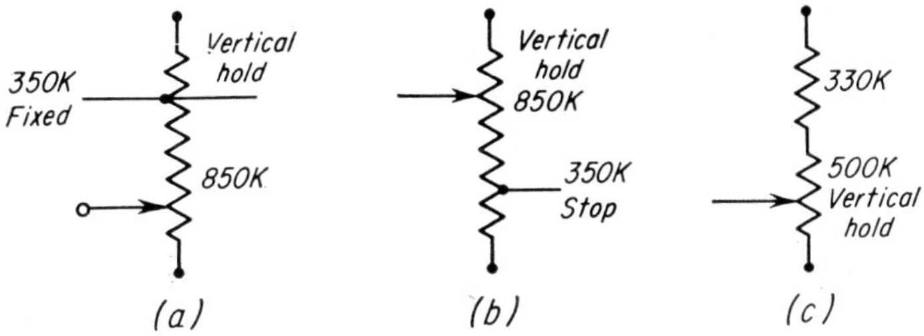

Fig. 5-17. Use of fixed and variable resistors to replace a vertical-hold control that incorporates a stop.

When the original control is of the type which incorporates a stop for the minimum position, it is not necessary to obtain an exact duplicate. The stop can be taken care of with an external fixed resistor. Refer to the schematic diagram of the receiver, which should list the ohmic position of the stop. Two conventional methods of indicating the stop are shown in Fig. 5-17*a* and *b*. A replacement for either of these controls could utilize a 330,000-ohm ½-watt resistor in series with a 500,000-ohm variable as indicated in Fig. 5-17*c*.

When information as to the ohmic size of a control is not available, a replacement value may be chosen in the following manner. Make up a test potentiometer like the one illustrated in Fig. 5-18. A 5-meg potentiometer is a good size to use for the vertical hold or height control. Remove the wires from the defective control, and clip on the test potentiometer. If it is the hold control which is being replaced, adjust the test unit until the picture locks in. Without changing the setting of the test potentiometer, turn the set off, and use an ohmmeter to measure the resistance which gives correct operation. Then, since the normal correct setting of the control should be approximately in the center, multiply the ohmmeter reading by 2 (TV controls being linear) for the proper value of replacement control to choose. Do not use too large a value since it may allow the oscillator to operate at 30 cps in its extreme position.

The same test potentiometer may be used for many similar operations

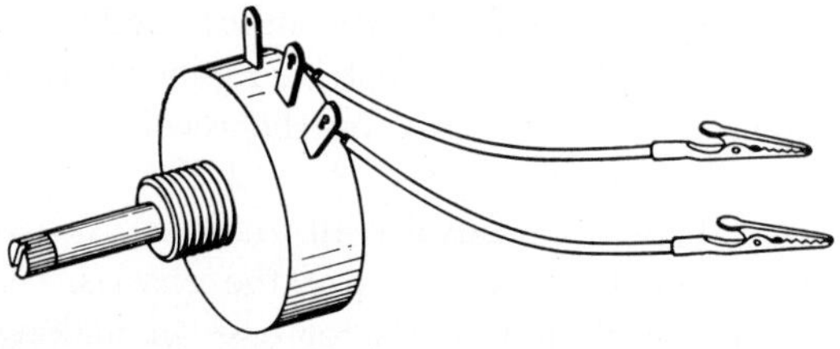

Fig. 5-18. Test potentiometer for finding correct value of replacement resistor.

in finding correct replacement values of resistors. For example, if no information is available for a replacement of the stop resistor in the hold circuit, clip in the test potentiometer as the replacement. Adjust the hold control to the center of its rotation, and adjust the test potentiometer for a lock. Its resistance value at this point will make a good replacement for the stop resistor.

Troubles Common to the Deflection Yoke Assembly. The damping resistors across each half of the vertical winding are in the yoke assembly. Occasionally, one of the resistors may open. It will be remembered that the function of the resistors is to damp any oscillation that may occur between the output transformer and yoke winding. If a resistor opens, the damping action is affected, and the oscillation may be set up. The oscillatory current results in vertical stripes on the raster (Fig. 5-19).

The trouble is found from the symptom. If the complaint is about the vertical bars, the serviceman should suspect the damping resistors and remove the yoke from the picture tube for inspection. Remember to observe the tube-handling safety precautions described in Chap 10. When

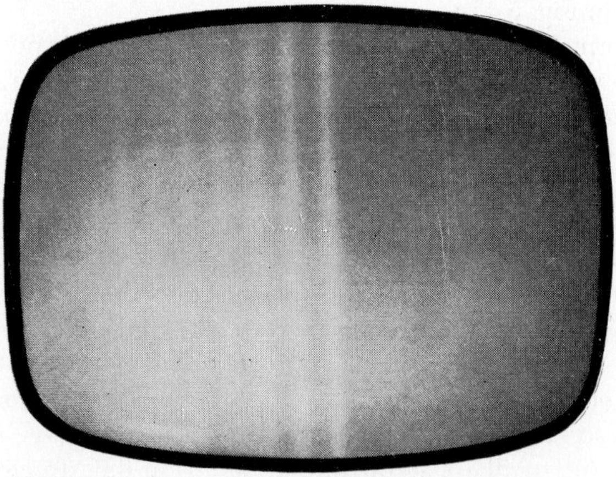

Fig. 5-19. Vertical stripes caused by an open damping resistor.

checking the resistor, remember that the yoke winding is in parallel, and the connection must be opened. Replace a defective resistor with one of the same value in ohms and watts. The usual value is a 560-ohm 1-watt resistor.

As for the yoke itself, the common troubles are that sometimes the winding opens, and sometimes it develops shorted turns. Should a winding open, together with its damping resistor, there is no vertical deflection, and we have the bright horizontal line. This is quickly found by resistance checks, after the signal check places the trouble in the transformer or yoke.

If the winding becomes shorted or partially shorted, there will be greatly reduced vertical deflection in the section of the winding that contains the short. The reduction in height depends on the severity of the short, but it is always greater on the half of the winding containing the defect. The raster appears as shown in Fig. 5-20. This keystone or trapezoidal shape is characteristic of a short in the yoke. It could also be caused by an open in either vertical winding, deflection current being applied to the other winding through the damping resistor.

Fig. 5-20. Trapezoidal raster resulting from a short in one half of the vertical winding of the deflection yoke.

There is one other possibility which may cause a similar effect. This is a magnetized spot in a metal picture tube causing an irregularity in beam deflection. To see if this is the case, rotate the picture tube. If the keystone effect remains unchanged the trouble is in the yoke. If the keystone moves with the tube, the tube is at fault. A demagnetizing or degaussing procedure for metal picture tubes is given in Chap. 10.

If the trouble is in the yoke, one end of each damping resistor is opened. Then each resistor and each half of the yoke winding may be checked separately with an ohmmeter.

The deflection yoke assembly also includes the horizontal winding. The windings cannot be separated, and a defect in either winding necessitates replacement of the entire yoke assembly.

How to Remove the Deflection Yoke Assembly. The deflection yoke is mounted on a frame. When it is desired to remove the yoke for inspection or test, it is usual practice first to loosen the tube-holding straps and remove the tube socket and ion trap. Then the tube is drawn out of the yoke assembly. It will sometimes be found that the tube does not move. This is because the horizontal winding which is closest to the tube has welded itself to the glass. Force should not be applied, since there is danger of tube breakage.

The bond can be loosened by heating the entire yoke assembly. Clamp an infrared lamp in such a way as to expose the yoke to the heat waves. In about 10 minutes, the bond will be sufficiently softened to free the neck of the picture tube so that it can be moved in the yoke. Apply Lubriplate to the bakelite base to ease the removal. Holding the tube with both hands, apply a slight back-and-forth rotation while withdrawing the tube from the yoke. Allow the neck of the tube to cool before washing off the gooey mess left by the bond. Carbon tetrachloride makes a good cleaning agent for this purpose.

How to Select a Replacement for a Deflection Yoke. The vertical winding in the deflection yoke must match the existing vertical output transformer. At the same time, the horizontal winding must match the horizontal output transformer. This makes it imperative to obtain either the original manufacturer's part or one made for the receiver by a reputable maker of replacement components.

If you can find no listing for an odd or unusual receiver, and a repair must be attempted, the choice of a proper replacement is very difficult.

Deflection Yokes

Part No.	Coil Inductance in MH		Coil Resistance in Ohms		Max. Scan.	Case Diam.	Unit Length
	Horizontal	Vertical	Horizontal	Vertical			
DY-1A	8.3	50	13.5	66	53°	3⅛	3
DY-2A	10.3	50	14.5	66	70°	3⅛	2¾
DY-8A	8.5	50	14.5	52.5	70°	3⅛	2¾
DY-9A	13.5	50	17.5	53.5	70°	3⅛	2¾
DY-10A	30.0	3.5	45	3.5	70°	3⅛	2¾
DY-11A	20.0	50	31	50	70°	3⅛	2¾
DY-12A	30.0	50	41	50	70°	3⅛	2¾
•DY-13A	12.5	48	15	52	90°	3⅛	3
•DY-14A	30.0	3.5	45	3.5	70°	2⅛	2¾
•DY-15A	25.0	50.0	35	50	70°	3⅛	2¾
•DY-16A	20	44.0	26	45	90°	3⅛	3⅛

Fig. 5-21. Catalogue listing of replacement deflection yoke.

Yokes are manufactured to meet a specified inductance for each of the windings. A typical listing of replacement yokes is given in Fig. 5-21.

The listing of inductance is of little value in this regard, since few service shops are equipped to measure inductance. The other columns, however, are all helpful. The RCA Tube Manual gives the recommended deflection angle for the picture tube. The table of resistance values is of great aid. Even if the windings of the original yoke are open, it is unlikely that both halves of each winding have been harmed. To measure the windings, first identify the vertical winding by the presence of the two damping resistors (see Fig. 5-22). Since the damping resistors form a parallel circuit, one end, preferably the lead end of each resistor, is opened, and then the resistance is measured directly at the coil terminals. Similarly, the horizontal winding is identified by the shunt balancing condenser. Again, this condenser lead is opened to measure the d-c resistance of the winding. Then, if half of the winding is intact, the resistance of the entire winding is obtained by multiplying by 2.

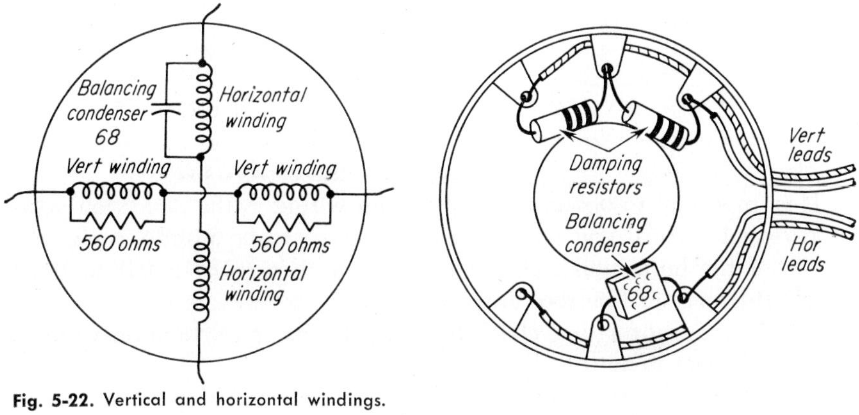

Fig. 5-22. Vertical and horizontal windings.

The horizontal windings in deflection yokes differ more than do the vertical windings. In addition, the match in the horizontal circuit is far more critical than the match in the vertical. For these reasons, after a certain type of yoke has been decided upon, it is wired in temporarily, and tried. If there is mismatch in the horizontal circuits, the probability is that the high voltage will be reduced and the picture screen will not light. Before trying a different yoke, check to see that the shunting network in the replacement is the same as in the original. If not, change to the original network.

When the raster is obtained, check for centering of the raster. If the original yoke cover had a centering device attached to it, make sure to cap the replacement yoke with this cover. Then adjust the vertical height and linearity controls. Variance in the deflection angle of the replacement yoke may cause a change in width. Readjust the width control for compensation. If this does not have sufficient range, a condenser of approximately 0.01 μf across the width coil will increase deflection, and a width shim (see Fig. 3-8) will reduce it. If vertical bars are produced on the right side of the picture, this is due to a slight mismatch in the horizontal winding. The bars can be eliminated by connecting a resistor in series with the balancing condenser which is across half of the horizontal yoke winding. Temporarily, connect a 5,000-ohm variable resistor in series with the condenser as shown in Fig. 5-23, and adjust it until the bars disappear. Then, turn off the set, and without touching the adjustment of the variable resistor, measure its resistance with an ohmmeter. Solder a resistor of this size permanently into the circuit. If the network already has a resistor in this position, and the vertical bars are present, it is the wrong size. Remove it, and use the 5,000-ohm variable resistor as before to determine the correct size.

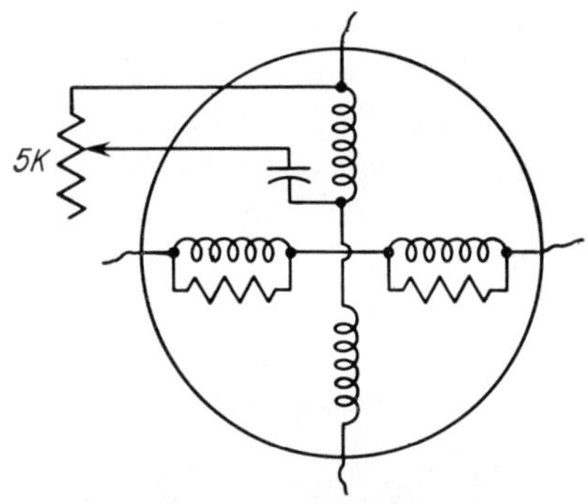

Fig. 5-23. Balancing a replacement yoke.

It is important to observe polarity when wiring in the deflection yoke. A reversed vertical coil will cause the picture to be upside down. Reversal of the horizontal coil will reverse the picture from left to right. The latter may not be too obvious until writing appears on the screen. To avoid waiting for this, check for such details as the side on which a man's suit coat is buttoned.

Replacing the yoke may also necessitate readjustment of pincushioning corrector magnets, which are provided on some receivers, particularly those using cylindrical face tubes like the 21EP4. The corrector magnets are mounted on metal strips which are attached to the sides of the yoke mounting bracket and extend out near the flange of the picture tube. The location of the magnets is shown in Fig. 5-24. To check if readjustment is needed, reduce the picture width so that the edges of the raster may be seen. If either edge is bowed (pincushion effect), the corrector magnet on that side is adjusted by repositioning the metal strip until the edge is straightened. The diagram of Fig. 5-24 shows a curved raster edge caused by a misadjusted corrector magnet. Do not look for pincushion corrector magnets on sets using metal picture tubes.

Replacement Notes on Vertical Integrator Couplates. The integrator network is sometimes composed of separate resistors and condensers. If one of these is found to be faulty, only the defective unit need be replaced. When the integrator network is a combined couplate, and a defect is found in it, the entire couplate is replaced. There are only a few types of integrator networks in use, and it is only necessary to obtain the correct replacement part.

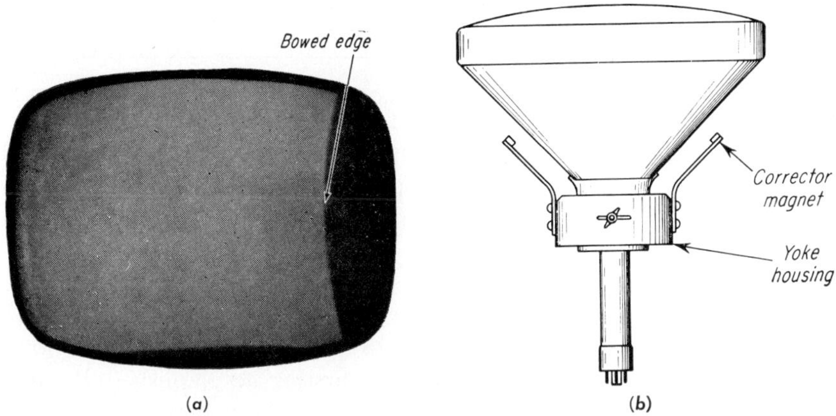

Fig. 5-24. (*a*) Pincushion effect. (*b*) Location of corrector magnets.

Variations in the Vertical Section. The output stage in the vertical section varies very little from set to set. Different tubes are employed in different receivers, requiring different output transformers for matching purposes, but the circuit remains substantially the same. Servicing problems caused by the use of different output tubes have already been covered.

The oscillator stage shows more variance. The basic vertical section of Fig. 5-6 uses a type of circuit known as the blocking oscillator, where feedback is obtained through a transformer and frequency is established by a condenser and variable resistor in the grid circuit. This is probably the most common type of vertical oscillator, but others are used and they may call for some changes in the servicing procedure.

The Plate-coupled Multivibrator. An oscillator circuit often used as a deflection generator is a multivibrator. A basic circuit, known as a plate-coupled multivibrator, is shown in Fig. 5-25. Here, we have a two-stage resistance-coupled amplifier. Feedback which creates the oscillation is obtained when the output of the second stage is coupled back to the input circuit of the first stage through coupling condenser *C*-2. The circuit forms a free-running oscillator, the frequency of which is determined mainly by the values of coupling condenser *C*-1 and grid leak *R*-3.

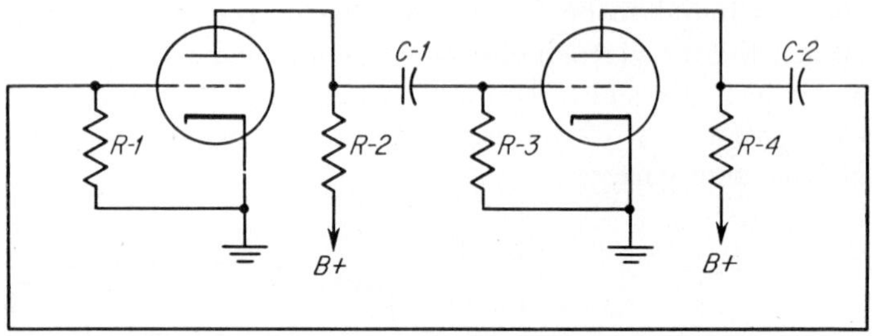

Fig. 5-25. A basic plate-coupled multivibrator.

The schematic diagram of the vertical section of the Magnavox 300 TV chassis is given in Fig. 5-26 as a typical example of the use of a plate-coupled multivibrator to generate the vertical deflection current. The 6SN7 dual triode is used for both tubes in the oscillator. The sync pulse is applied through a conventional integrator couplate to the grid of the first triode. Coupling condenser *C*-306 and resistors *R*-314 and *R*-315 control the frequency of oscillation, *R*-315 being the vertical-hold control. Feedback from the second tube to the first is accomplished through condenser *C*-305. Note that the height control is potentiometer *R*-318 feeding a variable amount of the generated voltage to the output tube.

When this type of vertical section is serviced, the dynamic checks of applying a 60-cycle test signal to the grid and plate of the output tube will still isolate the fault to the inoperative stage. If the oscillator stage

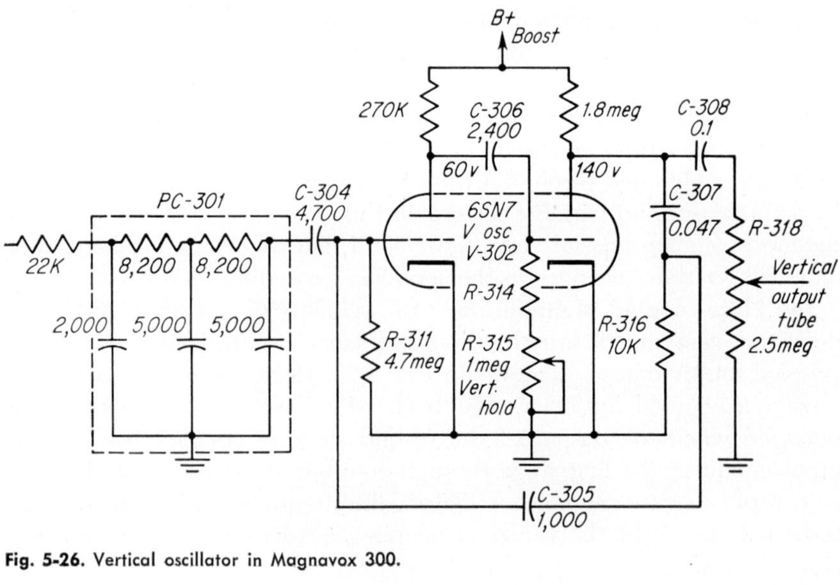

Fig. 5-26. Vertical oscillator in Magnavox 300.

is defective, it is serviced by voltage, tube substitution, and resistance checks. When voltage checks are made on this oscillator, two plate and two grid voltages are to be measured. Normal voltage values are found on the diagram, and any deviation is investigated by basic voltage-analysis principles. If the complaint is vertical rolling, the symptom is first isolated to an off-frequency oscillator, and then the frequency-determining components are checked, just as was done for the blocking-oscillator circuit. In the circuit of Fig. 5-26, these are condenser *C*-306 and resistors *R*-314 and *R*-315. However, if these prove good, in addition to checking the integrator network it will also be necessary to check condensers *C*-307 and *C*-305 and resistor *R*-316 in the feedback chain.

Cathode-coupled Multivibrator. Another type of deflection oscillator, known as a cathode-coupled multivibrator, is shown in the vertical section of the Hoffman Model 24M725 receiver of Fig. 5-27. Here again,

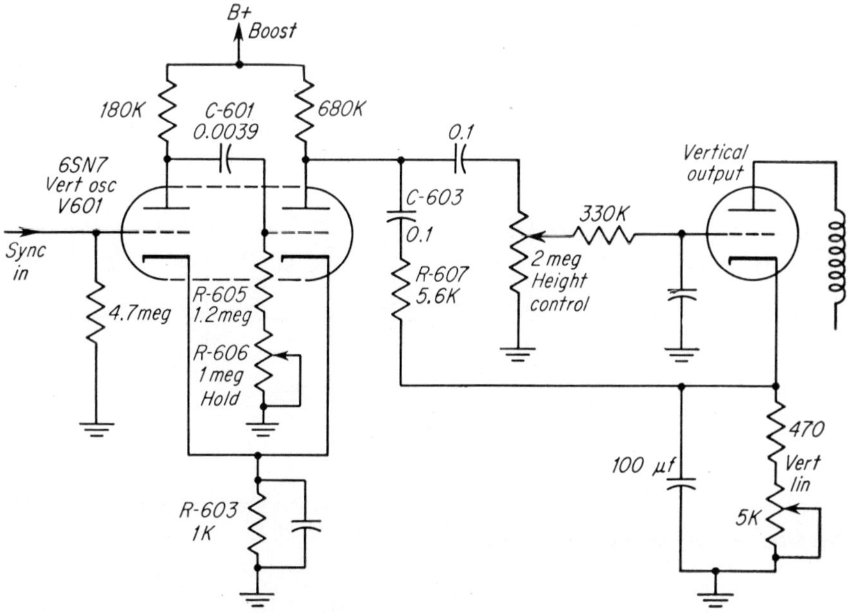

Fig. 5-27. Vertical oscillator in Hoffman chassis 403-24.

a 6SN7 twin triode is used for the oscillator. This time, plate-circuit variations of the second tube are fed back to the first through the common cathode resistor *R*-603, giving the circuit its name. The coupling condenser and grid leak between the tubes control the frequency of oscillation, grid leak *R*-606 being the vertical-hold control. Note that the waveshaping circuit consisting of condenser *C*-603 and resistor *R*-607 returns to the cathode of the output tube rather than ground. This connection is often found but introduces no new servicing problems.

This type of oscillator is also serviced by voltage and resistance analysis and tube substitution. However, if servicing for off-frequency operation, concentrate on coupling condenser *C*-601 and resistors *R*-605 and *R*-606. The integrator network will have less effect on the free-running frequency of the oscillator, since its function here is to feed a pulse to a triggering grid.

The Output Tube as the Second Tube in the Multivibrator. There is another type of plate-coupled multivibrator where the output tube does double duty in that it is the second amplifier tube in the multivibrator circuit. An example of this type is found in the vertical section of the Teleking TVJ chassis, illustrated in Fig. 5-28. The first tube in the multi-

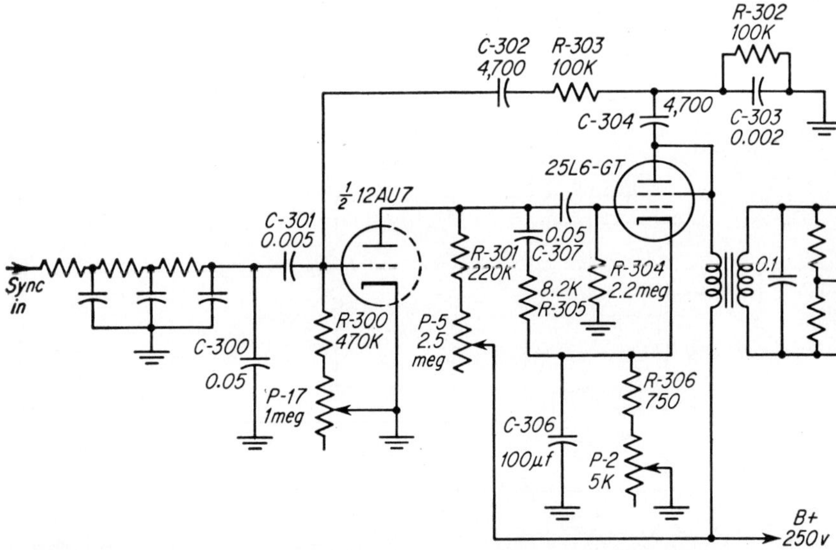

Fig. 5-28. Vertical section of the Teleking TVJ chassis.

vibrator is one triode section of a 12AU7 tube. This is coupled in a conventional circuit to the 25L6-GT output tube. Condensers *C*-304 and *C*-303 form a voltage divider for the output signal in the plate circuit of the 25L6-GT. The tap between the condensers feeds signal back to the grid of the 12AU7 through resistor *R*-303 and condenser *C*-302, thereby completing the multivibrator.

This circuit introduces two new servicing problems. One has to do with the high pulse voltages normally present in the plate circuit of the output tube, and the strain this places on condensers. In the circuit of Fig.

5-28, if condensers *C*-302, *C*-303, or *C*-304 should become defective, the replacement unit should have a voltage rating of at least 1,000 volts.

The other problem has to do with the more critical nature of the control adjustments. Since the amplifier is the second tube in the multivibrator, the height and linearity controls are in the feedback chain. Adjustment of these controls will therefore affect the oscillator frequency and cause rolling. The generalized procedure for vertical adjustments is applicable, but more emphasis must be placed on the direction to retouch the hold control if the picture starts to roll during the adjustment.

Receivers with Vertical Retrace Eliminators. In some receivers, a blanking pulse is fed from the vertical section to the picture tube to eliminate vertical retrace lines. A typical example of this type of circuit is given in Fig. 5-29.

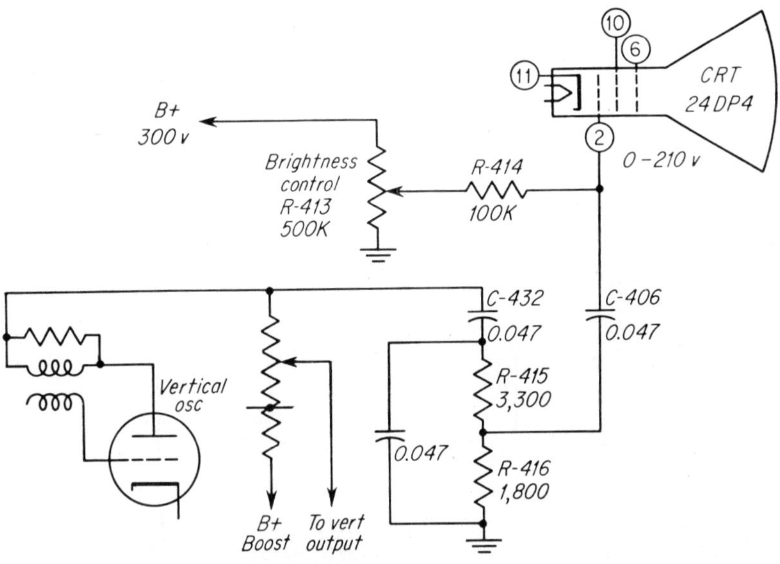

Fig. 5-29. Raytheon 24T3 vertical retrace eliminator.

The waveshaping circuit of the vertical oscillator consists of condenser *C*-432 and resistors *R*-415 and *R*-416. The junction of the two resistors is the source of the vertical retrace blanking pulse. This is fed through condenser *C*-406 to the control grid of the picture tube. The control grid is also fed its normal potential from the brightness control.

If condenser *C*-406 should short or develop leakage, the control-grid voltage would be greatly reduced through resistor *R*-416 in the vertical oscillator circuit. As a result, the screen would go dark. One of the checks when servicing for no raster is the brightness circuit. Voltage check at the control-grid pin of the picture tube would show 0 to 10 volts instead of the normal 0 to 210 volts indicated on the diagram. Resistance checks would then reveal the short in condenser *C*-406.

SUMMARY

Symptoms. A bright horizontal line across the face of the picture tube indicates a lack of vertical sweep and trouble in the vertical section. Lack of height, poor vertical linearity, and vertical rolling are other troubles in the vertical section.

Standard Circuit. The schematic diagram of a basic vertical section is given in Fig. 5-30.

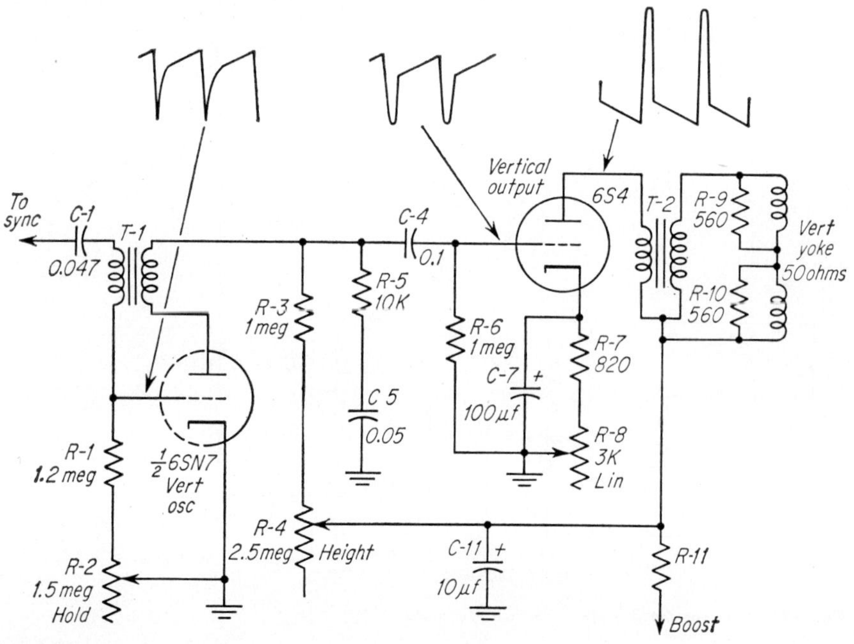

Fig. 5-30. Standard circuit of vertical deflection section.

Service Data Chart for the Vertical Section

Trouble. No vertical sweep as indicated by a bright horizontal line across the face of the picture tube.

Procedure	*Result*	*Trouble and Subsequent Check*
Apply 60-cycle test signal to the plate of the output tube	Bright line remains unaltered	Output transformer or yoke is defective Make ohmmeter check
	Bright line spreads	Proceed to next step
Apply 60-cycle test signal to grid of output tube	Bright line remains unaltered	The trouble is the output tube or its supply voltages
	Bright line spreads	Proceed to next step
Apply 60-cycle test signal to plate side of coupling condenser C-4	Bright line remains unaltered	Coupling condenser C-4 is open
	Bright line spreads	The trouble is in the oscillator stage. Replace tube. Make voltage and resistance checks

Service Data Chart for Other Troubles in Vertical Section

Symptom	*Illustration*	*Look for*
Insufficient height, poor linearity, and foldover		Weak tubes Misadjusted controls Off-value resistors and condensers Shorted output transformer Low supply voltage
Vertical bars on raster	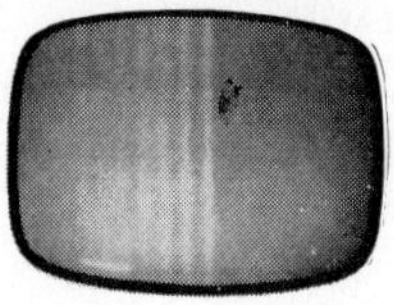	Open damping resistors *R*-9 and *R*-10 in vertical deflection winding of yoke
Rolling (sync normal)		Defective oscillator tube Off-value hold control *R*-2 or stop resistor *R*-1 Off-value input condenser C-1 Defective integrator network

QUESTIONS

1. A receiver shows a bright horizontal line. Describe a servicing procedure which would isolate the trouble to either the output stage and yoke, or the vertical oscillator.
2. In checking the yoke, how may the vertical windings be quickly identified?
3. In measuring the d-c resistance of the yoke, what caution must be observed?
4. How would you identify a 10:1 vertical output transformer without referring to a catalogue? How would this check help in servicing when a transformer is suspected of having shorted turns?
5. Approximately what d-c resistance will the vertical section of the yoke have in a receiver which uses a 44:1 vertical output transformer?
6. Upon checking a receiver for no vertical sweep, B plus voltage at the plate of the vertical blocking oscillator is absent. Describe the voltage checks which should be made. How would you confirm your suspicion of an open vertical blocking-oscillator transformer?
7. Describe a procedure for setting up the vertical controls.
8. Describe a procedure for determining if vertical rolling is due to absence of vertical synchronizing pulse.

6

Servicing the Deflection Section: Horizontal Output, Damper, and High-voltage Stages

The primary function of the horizontal sweep section is to deflect the electron beam in the picture tube back and forth across the face of the tube in much the same manner as the vertical section deflects it up and down. Loss of vertical sweep is indicated by a bright horizontal line across the face of the picture tube. Similarly, loss of horizontal sweep might be expected to give a bright vertical line. This is not so because, in almost all cases, another function of the horizontal section is to energize the high-voltage circuits which produce brightness on the screen. Thus, without high voltage, the vertical line cannot be seen. In those few receivers where the high voltage for the picture tube is obtained from a separate source, the bright vertical line will be observed when the horizontal section is inoperative. For most sets, therefore, loss of raster is the indication of trouble in the horizontal and high-voltage section. Sound is present, indicating that the low-voltage power supply, tuner, and sound sections in the receiver are operative.

Stages in the Horizontal and High-voltage Section. Like the vertical deflection system, the horizontal section begins with a deflection generator. This is labeled HORIZONTAL OSCILLATOR in Fig. 6-1. This oscillator

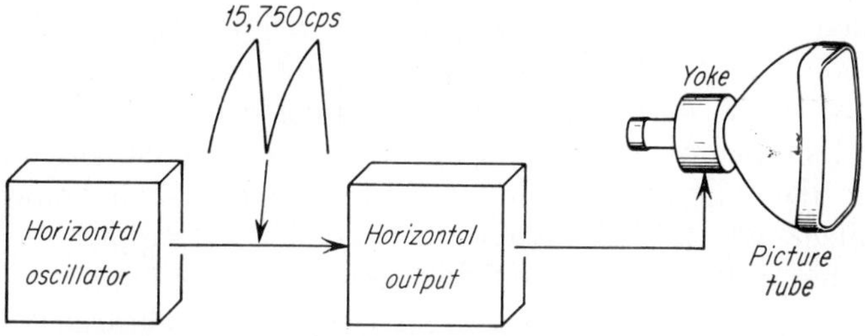

Fig. 6-1. Basic stages in the horizontal deflection section.

is followed by an output amplifier stage which in turn feeds the horizontal winding of the deflection yoke.

The oscillator stage generates a signal at 15,750 cps, the rate of horizontal deflection. Waveshaping circuits alter this to a sawtooth form which is fed to the control grid of the horizontal output tube. This waveshape is indicated in Fig. 6-1 and is an important oscilloscope check for the horizontal section.

The frequency of the vertical oscillator is controlled by a sync pulse directly, since noise pulses have very little effect at 60 cps, the vertical rate. At the horizontal rate, however, noise pulses may also trigger the oscillator, causing off-frequency operation. This causes "tearing" or "pulling" of the picture. To prevent this, a more elaborate control system is used. An extra stage called automatic frequency control (afc) is inserted to control the frequency of the horizontal oscillator. As shown in Fig. 6-2,

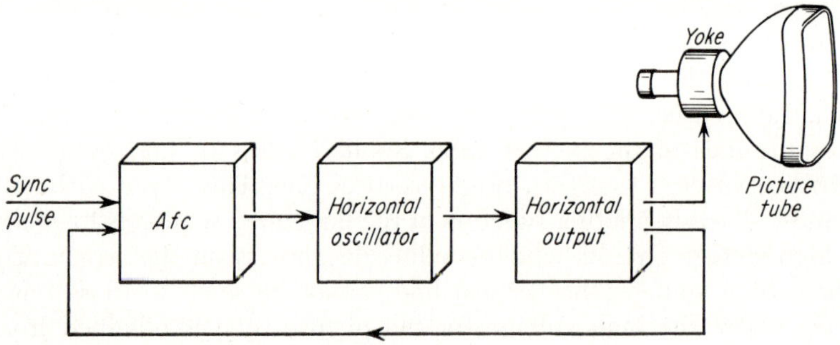

Fig. 6-2. Position of the afc stage in the horizontal deflection chain.

the afc stage is fed two signals: the sync pulse, and the output of the oscillator. Together, these produce a control voltage for the oscillator.

The horizontal output stage is a power amplifier which drives the horizontal winding of the deflection yoke. At the same time, the output transformer between the tube and yoke has an extra winding which supplies input voltage to a high-voltage power supply. Here it is rectified and filtered to furnish the high d-c voltage needed for the second anode of the picture tube. This is shown in block-diagram form in Fig. 6-3.

In the vertical winding of the yoke, damping resistors serve to suppress oscillation. In the horizontal circuit, damping energy, instead of being dissipated in damping resistors, as in the vertical circuit, is utilized through a damping tube to obtain a d-c voltage. This is known as the boost voltage, which is additive to the low-voltage B supply.

Regardless of the type of oscillator circuit used, the input signal to the amplifier stage is the characteristic sawtooth wave illustrated again in Fig. 6-3. Different types of oscillators require different servicing pro-

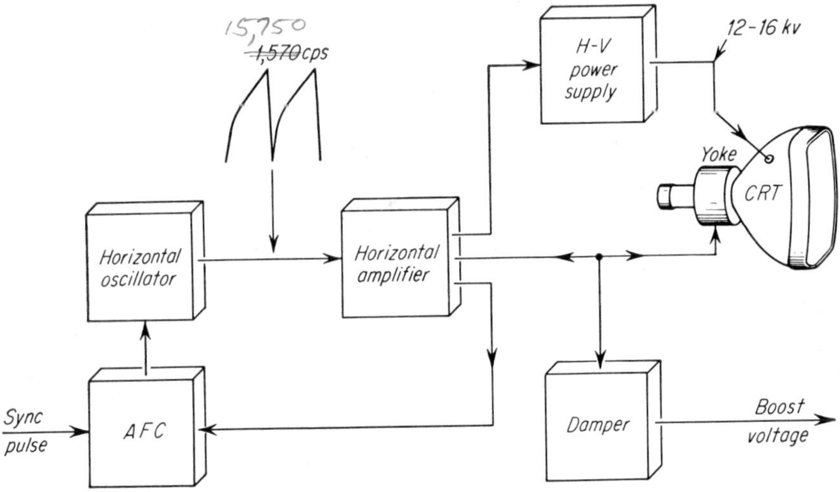

Fig. 6-3. Block diagram of the horizontal deflection and high-voltage system.

cedures. These will be dealt with in the next chapter. The main emphasis in this chapter will be placed on the output stage, together with the high-voltage and damper stages.

Basic Circuit Diagram of the Horizontal Output, High-voltage, and Damper Stages. Refer to Fig. 6-4. The tube in most common use for the horizontal output stage is the 6BQ6/6CU6. This tube is adequate to deflect a picture tube fully up to 21 in. Larger sets use a 6CD6 output tube. In transformerless sets with a series filament string, the tube most commonly found is a 25BQ6 or its 600-ma equivalent, the 12CU6.

The horizontal output tube is used in a class C amplifier circuit where bias is obtained by the drive voltage. The drive voltage is, of course, obtained from the horizontal oscillator stage. The peak-to-peak amplitude of the drive voltage is controlled by condenser *C*-1, which is called the horizontal drive control. Grid condenser *C*-2 and grid leak *R*-1 determine the bias voltage by grid-leak action. If the horizontal oscillator should stop, there would be no drive voltage and therefore no bias voltage. This would cause the 6BQ6 tube to draw a very heavy plate current. This is somewhat offset by resistor *R*-5 in the cathode circuit which provides a minimum bias. Even with this minimum bias, the plate of the amplifier tube often glows at red heat from excessive plate current when the drive voltage fails.

Resistors *R*-2 and *R*-3 in the control-grid and screen-grid circuits serve to stop parasitic oscillations which may be generated in these circuits. Condenser *C*-3 is the screen bypass condenser and resistor *R*-4 is the screen dropping resistor.

The normal drive voltage is large enough to bias the 6BQ6 beyond cutoff. Plate current, therefore, flows on the positive peaks of the input signal. These 15,750-cycle pulses flow through winding *W*-1 of the horizontal output transformer. By transformer action, voltage is induced in winding *W*-4, which feeds the horizontal winding *L*-1, *L*-2 in the yoke through blocking condenser *C*-4. Between pulses the horizontal yoke winding and winding *W*-4 set up an oscillatory circuit which feeds energy back into the transformer. On the first half cycle of this oscillatory current, some of this energy shows up at point 2, which connects to the plate of the horizontal output tube, as a 4,000- to 6,000-volt pulse. It also appears

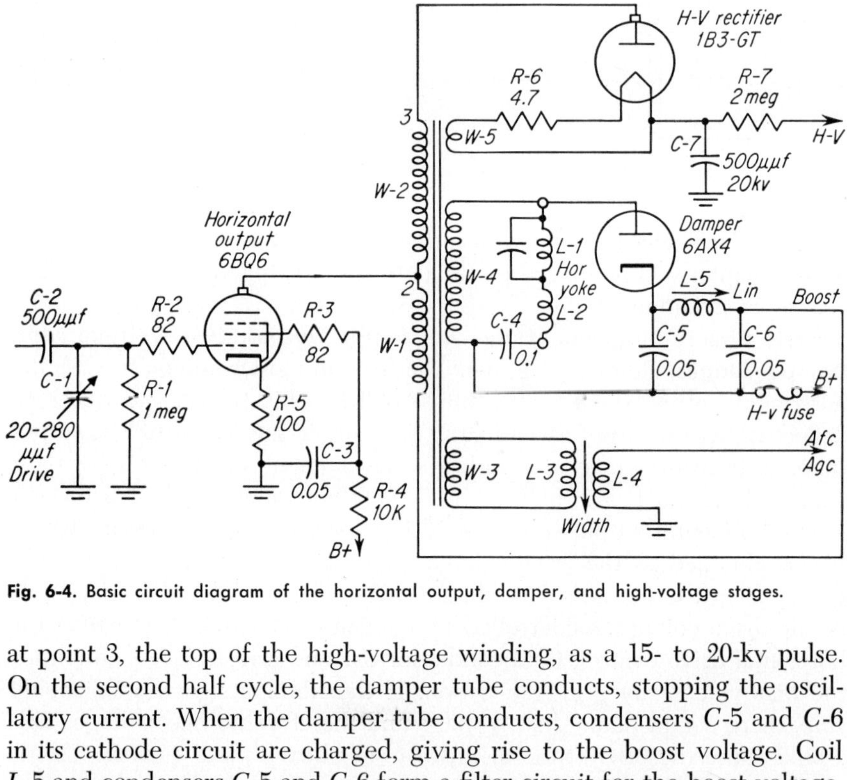

Fig. 6-4. Basic circuit diagram of the horizontal output, damper, and high-voltage stages.

at point 3, the top of the high-voltage winding, as a 15- to 20-kv pulse. On the second half cycle, the damper tube conducts, stopping the oscillatory current. When the damper tube conducts, condensers *C*-5 and *C*-6 in its cathode circuit are charged, giving rise to the boost voltage. Coil *L*-5 and condensers *C*-5 and *C*-6 form a filter circuit for the boost voltage. Coil *L*-5 is a permeability-tuned inductance. Adjusting the tuning slug governs the conduction time of the damper tube, thereby speeding or slowing the start of the horizontal trace, and acting as a control of the horizontal linearity.

Winding *W*-5 of the output transformer is a single turn of well-insulated wire placed around the ferrite core of the transformer. It develops the heater voltage (1.25 volts) for the 1B3-GT high-voltage rectifier tube.

This is fed to the filament through series resistor *R*-6. In some circuits the series resistor is omitted. Resistor *R*-7 and condenser *C*-7 form the filter for the high-voltage system. In some receivers, this filter circuit is omitted, and the only filter used is the condenser formed by the inner and outer coatings of the picture tube, the glass acting as the dielectric. A high-voltage circuit of this type is illustrated in Fig. 6-5. A grounding spring touching the outer aquadag coating is essential to complete this filter circuit.

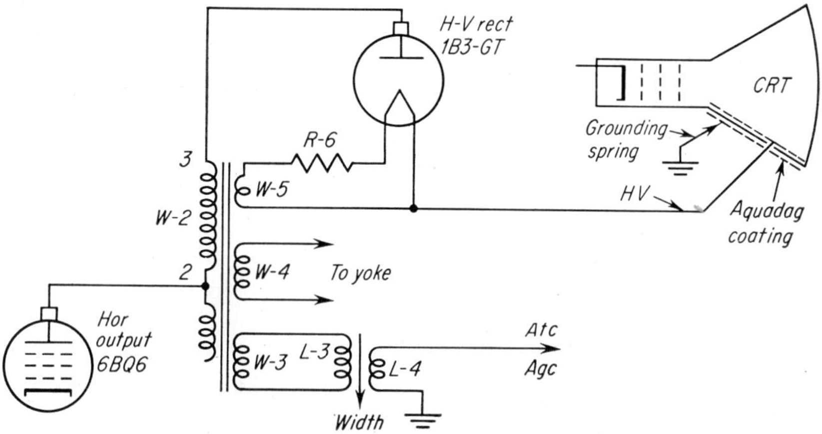

Fig. 6-5. Circuit where the aquadag coating on the picture tube is used as the filter in the high-voltage supply.

Winding *W*-3 of the output transformer is connected to *L*-3, a permeability-tuned coil which is called the width coil. Adjusting the tuning slug in this coil tunes the entire output transformer, thereby controlling the deflection current in the yoke winding. In this way, it gives a control of the width of horizontal sweep. An extra winding *L*-4 on the width coil furnishes a pulse voltage which is fed back to the afc circuit controlling the horizontal oscillator. In those receivers which use a keyed agc system, this point is also the source of the keying pulse for the keyed agc stage. Sometimes the two feedback voltages are produced in separate windings, and sometimes pulse coil *L*-4 is omitted entirely and the feedback pulses are obtained from separate taps or windings on the output transformer.

In some receivers, the width is controlled by a metallic shim which is inserted between the neck of the picture tube and the deflection yoke. The shim is a cylinder of thin aluminum or brass which slides along the neck of the tube into and out of the yoke. Inserting the shim causes the picture to become narrower; removing it causes more deflection and a wider picture. A width shim is shown in Fig. 6-6*a*.

Location and Appearance of the Horizontal Section. The horizontal deflection and high-voltage section contains a high-frequency generator and power amplifier which can radiate considerable energy. To cut down the radiation effects, the section is enclosed in an r-f cage for shielding purposes. The r-f cage is easily found and identifies the section. Sometimes it is a small cage like the upper illustration in Fig. 6-6. Here, only the flyback output transformer and high-voltage rectifier, which will radiate the greatest amount of r-f energy, are enclosed in the cage. The output tube, identified by its top plate lead, the damper, and horizontal oscillator tubes are generally close to the cage. In Fig. 6-6*b*, the r-f cage encloses most of the section.

The r-f cage is generally to the rear of the chassis for short leads to the horizontal control adjustments on the rear apron. The names of the controls are usually stamped on the chassis.

Test Procedure for Locating the Cause of No Raster. Throughout this chapter and the next, we are working on a condition of normal sound and no raster—a completely dark screen. This symptom can be caused by the failure of any one of a large number of components in the four stages of the horizontal and high-voltage section, and some few outside the section. The procedure to be presented will trace the fault to the defective component whether within or outside the horizontal section.

How to Check the High Voltage. When the screen of the picture tube is dark, the obvious first step is to see if high voltage is being applied to the second anode of the picture tube. Some servicemen do this by removing the high-voltage lead, holding it close to the chassis, and observing the spark. But, in addition to the harm that this may do to components in the high-voltage section, it is difficult to judge the amount of voltage by the intensity of the spark. The best way is to use a voltmeter equipped with a high-voltage probe. If one of these is not available, a voltage divider like the one shown in Fig. 6-7 is easily made up.

The largest stock size of resistor is 22 meg. Four of these and one 12-meg resistor make up the 100-meg section, and the 1.2-meg unit completes the divider. With the values chosen, the high-voltage system is loaded to less than a quarter of a milliampere, and the voltage to be measured across the 1.2-meg resistor is one one-hundredth of the total. This can then be read on the 250- or 300-volt scale of a VTVM or 20,000-ohms-per-volt meter, and the reading multiplied by 100 to give a maximum range of 25 or 30 kv. This device is also useful in color-service work where the high voltage must be adjusted to manufacturer's specifications. One- or two-watt carbon resistors with a tolerance of 10 per cent may be used since no great accuracy is required. However, the entire divider should be enclosed in a plastic tube or have some other means of dust

(a)

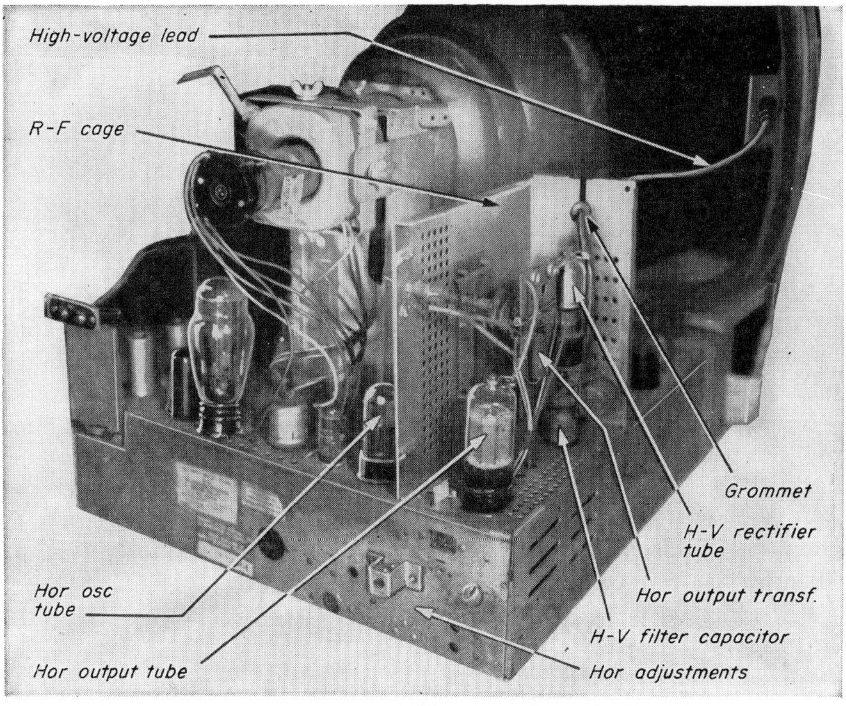

(b)

Fig. 6-6. Identification and location of components in the horizontal high-voltage section.

protection, since a layer of dust over a resistor of the order of 100 meg may make a considerable difference.

The method of measuring the high voltage is shown in Fig. 6-7. Observe normal precautions when doing this. Turn the set off. Discharge the high-voltage lead, and connect the divider as shown. Make sure the high-voltage lead is not close to the chassis, turn the set on, and read the meter.

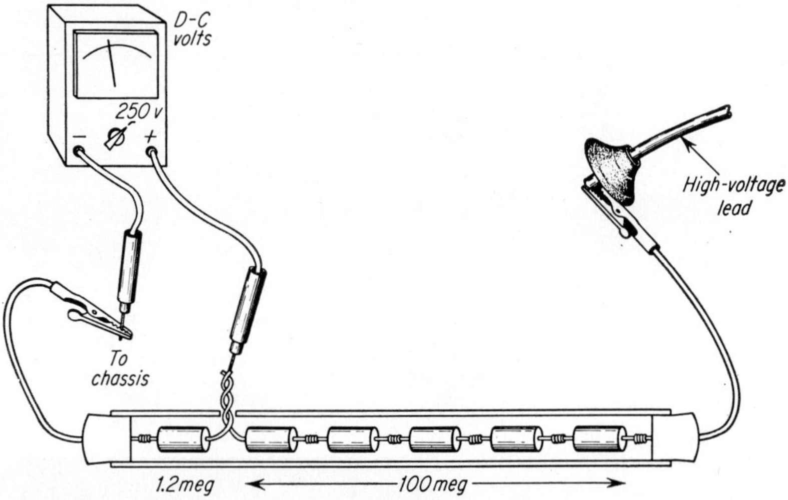

Fig. 6-7. Voltage divider for checking potential at the high-voltage lead.

The normal voltage for sets with picture tubes up to 17 in. is 12 to 16 kv. When the picture tube is 21 in. or larger, the high voltage ranges from 16 to 18 kv. A reading as low as 8 kv may put some brightness on the screen but indicates weakness; readings from 0 to 2 kv indicate trouble in the horizontal section.

Servicing Procedure for No Raster Coupled with Normal High Voltage. Let us first assume the condition of a normal high-voltage output of 12 to 18 kv, and no raster. This obviously places the trouble in the picture tube or its associated circuits.

When handling the picture tube, remember to observe the safety precautions described in Chap. 10. First check to see that the filament of the picture tube lights. If light cannot be seen, before discarding the tube resolder the connections on the filament pins with a hot iron. Quite often, the heater connections lose contact in the tube pins, and it is possible to effect a repair by a simple resoldering process.

If the filament lights, the trouble can be a defective brightness-control circuit, a misadjusted ion trap, or a defective picture tube. The possibilities must be investigated. The easiest of these to check is the ion-trap adjustment; so this is tried first. A procedure for setting up the ion trap now follows.

How to Adjust the Ion Trap. The ion-trap magnet should be placed on the neck of the picture tube so that the magnet is over the space between grids *G*-1 and *G*-2, and parallel with the glass gun supports (see Fig. 6-8). Turn the brightness control to the halfway point. With the set now

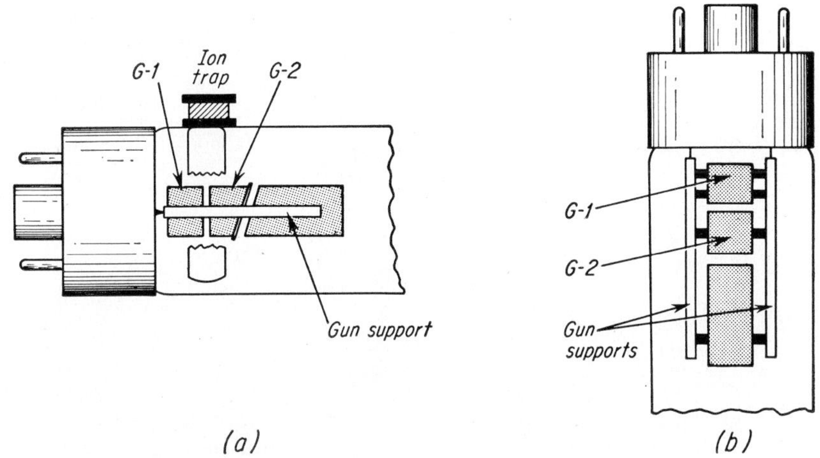

Fig. 6-8. Approximate setting of ion trap. (*a*) Side view of gun. (*b*) Top view of gun.

turned on, rotate the ion trap through a full turn while moving it slightly forward and backward until maximum brightness is obtained. This adjustment should be made rapidly, as possible damage to the gun may result. Readjust the brightness control for correct picture brightness. It may be necessary to retouch the focus adjustment after positioning the ion trap.

Checking the Picture Tube. If adjustment of the ion trap does not restore the raster, the next step is to try operation with a test picture tube. The 8XP4 is a small picture tube especially designed for test purposes. It does not use an ion trap, and focusing is unimportant for this check.

Turn the set off, remove the high-voltage lead from the picture tube, and discharge it by shorting it to the chassis. Then remove the socket from the original tube, and plug it on the base of the 8XP4, as shown in Fig. 6-9. Connect the high-voltage lead to the 8XP4, and turn the brightness control all the way off. Then turn the set on, and after allowing time for the high voltage to come on, advance the brightness control gradually.

If a spot of light comes on in the center of the test 8XP4 and the brightness of the spot can be adjusted by manipulation of the brightness control, the brightness circuit is working, and the picture tube in the set is defective. Replacement notes on picture tubes are given in Chap. 10. When making the replacement, check particularly for the presence of the aquadag coating. Sometimes a replacement picture tube with the correct type number may not have the aquadag coating that is present on the original tube, and this absence will affect the high-voltage output.

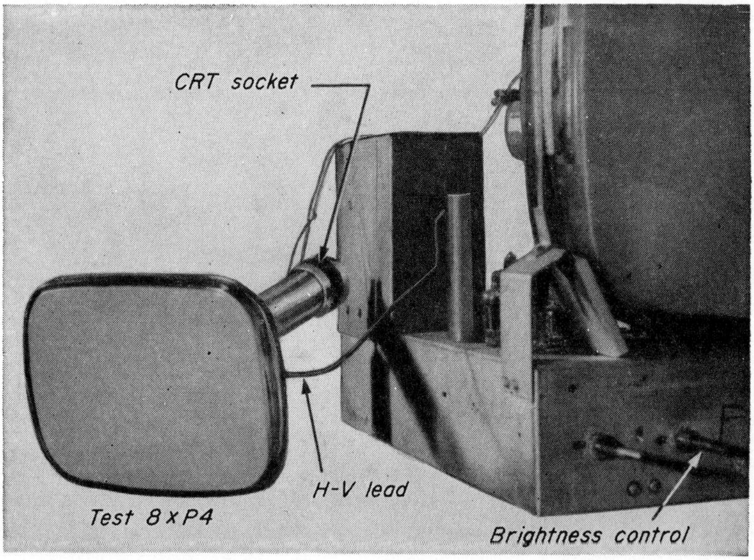

Fig. 6-9. Using a test 8XP4 to check operation of the original picture tube.

If the spot does not come on, even after the brightness control is adjusted to the full on position, the picture tube is probably good, and there is trouble in the brightness-control circuit.

Do not replace the picture tube until you have made a careful recheck of the positioning of the ion trap.

How to Check the Brightness Circuit. Before checking the brightness circuit, it might be a good idea to describe how this circuit operates. To do so, consider the picture tube as though it were an ordinary amplifier tube as shown in Fig. 6-10. The cathode emits electrons which are attracted by the high positive voltage at the second anode. The electrons strike phosphors on the face of the tube and fluorescence occurs, emitting light. The control grid is very near the cathode and controls this electron stream. As it is made more negative than the cathode, it repels some electrons, cutting down the electron flow to the picture screen, making it less bright. If it is made even more negative, it will cut off the electron stream entirely, making the face of the picture tube dark. A negative potential of 33 to 75 volts, depending on the type of picture tube, will cut the tube off. Voltages below this value give varying degrees of brightness on the screen. The brightness control varies the negative potential on the grid with respect to the cathode.

Representative brightness circuits are shown in Fig. 6-11. In the circuit of Fig. 6-11*a*, the cathode is connected to a voltage divider across the low-voltage B plus line. The cathode potential can be adjusted between zero and about 100 volts by the brightness control. The video signal is fed to the control grid at pin No. 2 of the picture tube. When the cathode is at a low potential, the tube produces a bright picture. As the control is turned counterclockwise, the negative cathode-grid voltage relationship increases, dimming the tube. At the cutoff point and beyond, the screen on the picture tube remains dark.

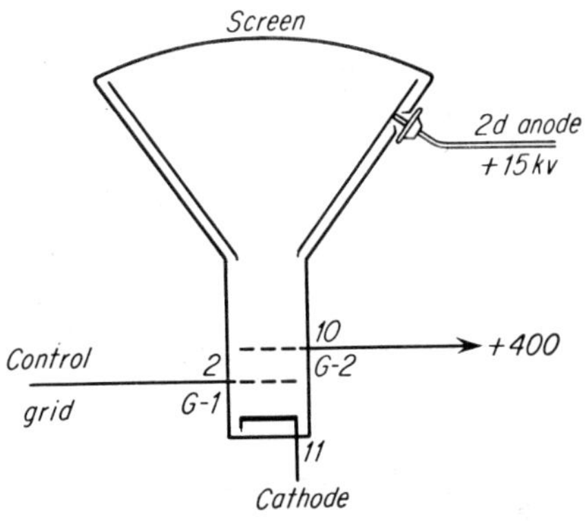

Fig. 6-10. Electrodes in a picture tube.

Some receivers include provision for blanking out the tube during vertical retrace time. This is accomplished in the circuit of Fig. 6-11*a* by feeding a pulse from the vertical deflection circuit to the cathode of the picture tube through condenser *C*-1. The pulse momentarily cuts off the tube.

Another brightness-control circuit is shown in Fig. 6-11*b*. Here the video signal is fed to the cathode of the picture tube, and the brightness-control voltage is applied to the control grid, at pin No. 2. The cathode is direct-coupled to the plate of the video-amplifier tube and is at a high d-c potential.

In the circuit of Fig. 6-11*c*, both the brightness-control voltage and the video signal are brought to the cathode, and the control grid is at ground potential.

In all cases, the brightness control is a method of varying the grid-bias voltage of the picture tube, thereby controlling the electron flow, and so the brightness of the image on the screen.

To check for normal operation of the brightness circuit, it is not necessary to determine the type of circuit employed. Simply check for the relative voltage existing across the grid and cathode pins directly at the picture-tube socket as shown in Fig. 6-12. The socket is removed from the tube for this test. Color coding on the CRT socket leads is fairly well established and may be used as an aid in getting to the correct pins. The positive prod of the voltmeter is connected to the No. 11 pin of the picture-tube socket, and it will be noted that the yellow wire in the

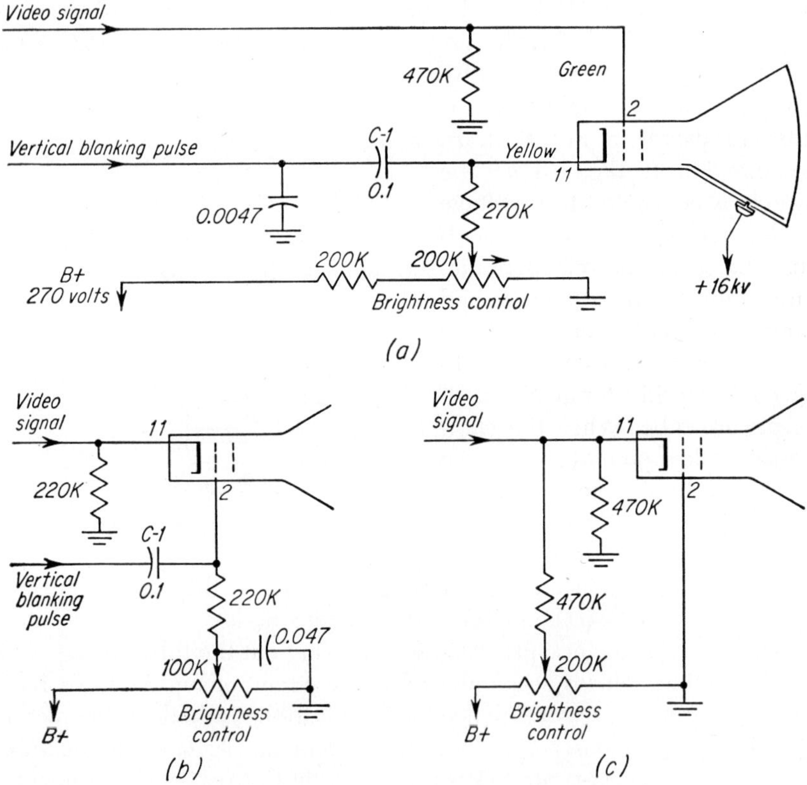

Fig. 6-11. Methods of controlling brightness in television receivers.

CRT cable goes to the same pin. The negative prod is pushed into the No. 2 pin, which has the green cable wire attached to it.

The voltage readings are taken directly at the socket pins of the picture tube because these will always show the true voltage difference. If the voltage at these pins were to be measured from the chassis, different brightness circuits would produce a wide difference in readings for the two test points. Very often both electrodes are positive with respect to chassis. However, in all cases, the circuits are so designed that the grid is more negative than the cathode, the differential being varied by the brightness control.

The brightness control is varied through its range, and the meter reading is taken. When the brightness circuit is operating normally, a reading varying from zero to approximately 150 volts will be observed. If this voltage swing is not obtained, while the circuit is varied, the brightness circuit must be checked. Refer to the manufacturer's service notes for the circuit. Voltage and resistance checks will then locate the defective

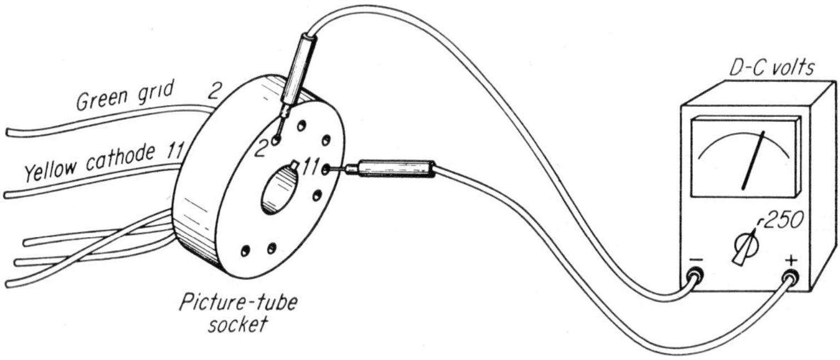

Fig. 6-12. How to check for normal operation of the brightness circuit.

component. Do not neglect the components in the vertical blanking-pulse circuit since breakdown here will disturb the cathode-grid voltage of the picture tube. Also, especially in circuits where the video signal is fed to the cathode of the picture tube, the voltage at the cathode is affected by the plate voltage of the video-amplifier tube. Therefore, defects in the video output tube and its associated circuits can result in a case of no brightness.

At this point, we must mention that receivers often incorporate switches for phonograph operation. The switch is connected so as to darken the picture tube in the phonograph position. Therefore, when looking for a cause of no brightness, look for a phonograph switch and make sure it is in the TV position.

CHECK FOR ISOLATING TROUBLE TO THE OSCILLATOR, AFC OR OUTPUT, DAMPER, AND HIGH-VOLTAGE STAGES

In the previous pages, we were checking for the cause of no raster when the high voltage showed a normal reading of 12 to 18 kv. Now let us consider the condition where the voltage check at the second anode of the picture tube shows no voltage, placing the trouble in the horizontal section. The stages in the horizontal section are shown in block-diagram form in Fig. 6-13. A logical test point is between the oscillator and output stages, and the next check to be described will determine whether the trouble is located before this test point, in the oscillator and afc stages, or after it, in the output, damper, and high-voltage stages.

The normal input signal to the grid of the horizontal output tube is a sawtooth wave at 15,750 cps, the horizontal deflection rate. This characteristic waveform is shown in Fig. 6-13 between the oscillator and output stages. If this signal is present at correct frequency, waveshape, and amplitude, the oscillator and afc stages are working, and the trouble is

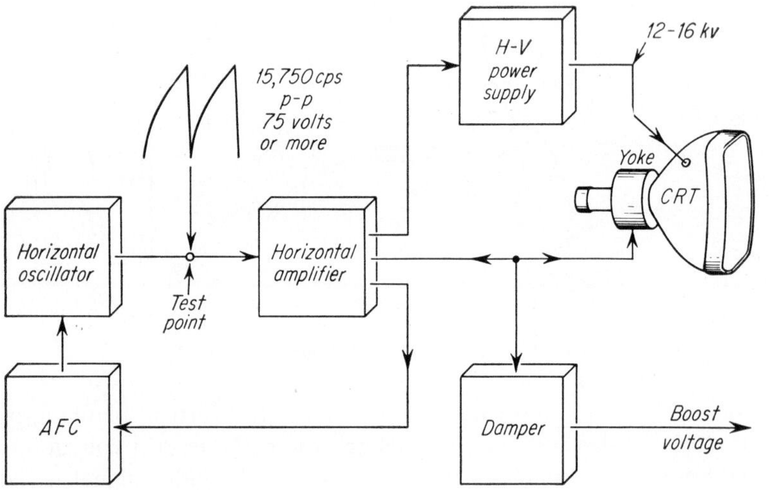

Fig. 6-13. The first check point to localize trouble in the horizontal sweep section is between the oscillator and amplifier stages.

therefore in the output, damper, or high-voltage stages. Conversely, if the signal is absent or incorrect, the oscillator and afc stages must be checked.

The best way to determine the presence and correctness of this signal is by observing its trace on a cathode-ray oscilloscope which is specially calibrated for this purpose. No particular instrument is needed; any ordinary type may be calibrated by the following procedure.

Calibration of the Oscilloscope. A 21-in. television receiver producing a normal good picture is used as the standard for calibrating the oscilloscope. When the instrument is connected to the input of the horizontal output tube, a continued normal good picture ensures the accuracy of the signal as regards waveshape, frequency, and amplitude. The setup is shown in Fig. 6-14.

The vertical input terminals of the oscilloscope are connected through a 200-$\mu\mu$f isolating condenser to the grid circuit of the horizontal output tube. The test point is easily identified as the ungrounded terminal of the horizontal drive trimmer. If the receiver does not use this type of drive control, locate the grid pin of the output tube (pin No. 5 for the 6BQ6, 6BG6, and 6CD6 types), and clip the test lead to the more accessible end of the grid suppressor *R*-2.

Working at a reduced setting of the sync control on the oscilloscope, its two frequency controls are adjusted to give a 2-cycle trace on the oscilloscope screen. The 2-cycle trace is obtained by setting the oscilloscope frequency to 15,750 divided by 2, or 7,875 cps. The coarse control is set for 2 to 10 kc, and the fine control is adjusted to bring in an almost steady 2-cycle trace. The sync control is then advanced just far enough to keep this trace steady. If the oscilloscope is not equipped with a reference scale, one may be contrived by the use of two thin strips of masking tape, as shown in Fig. 6-14. Then the vertical gain and positioning controls on the oscilloscope are adjusted so that the trace is just touching the two reference lines.

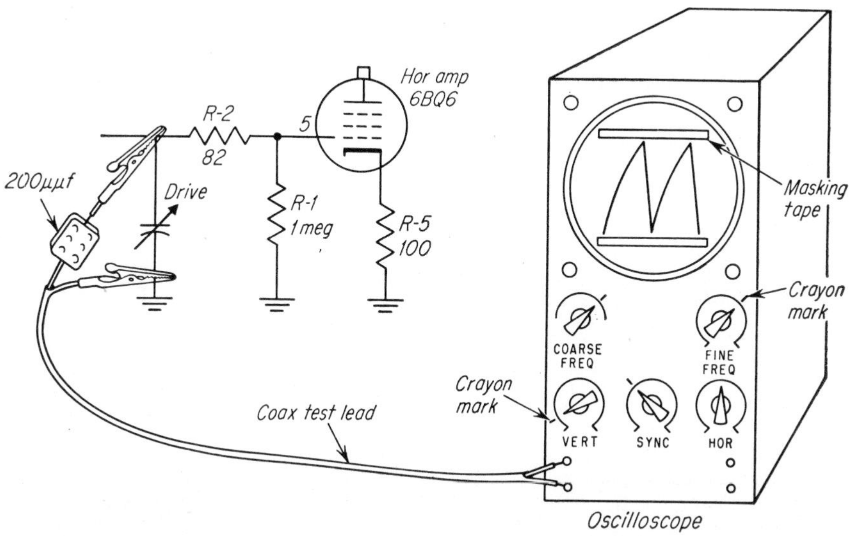

Fig. 6-14. Oscilloscope setup for checking the input signal to the grid of the horizontal output tube.

The oscilloscope is now showing a 2-cycle trace of the grid signal of a horizontal output tube which is at the correct frequency, waveshape, and amplitude, as evidenced by a normal good picture on the screen of the test set. The positioning of the oscilloscope controls may now be marked with a wax crayon, and the instrument is calibrated. The frequency controls are calibrated for any set, since the horizontal deflection rate is always 15,750 cps. The waveshape varies very little from set to set and is easily remembered. Besides, waveshape information is often included in manufacturers' service notes. The calibration of the vertical gain control is also set for the receiver under test, but here again there is little variation from set to set. A good working minimum for most receivers is a 75-volt peak-to-peak signal at the grid of the horizontal output. This will hold for most receivers using 6BQ6 or 25BQ6 output tubes, and a picture tube up to 21 in. Larger picture tubes using a 6CD6 output tube are generally driven by a 90- to 125-volt peak-to-peak signal.

Checking the Input Signal to the Horizontal Amplifier. The calibrated oscilloscope is now used to check the receiver being serviced. The controls are set to the crayon marks, and the input lead is connected to the set as shown in Fig. 6-14. If a normal 2-cycle trace appears, the oscillator and afc stages are working, and the trouble must be in the output, damper, or high-voltage circuits.

We cannot jump immediately to the converse statement that the trouble must be in the oscillator-afc stage if the correct trace does not appear. The correct trace, together with abnormal traces produced by commonly found troubles, is shown in Fig. 6-15. These will be analyzed and discussed with reference to the next step in the servicing procedure.

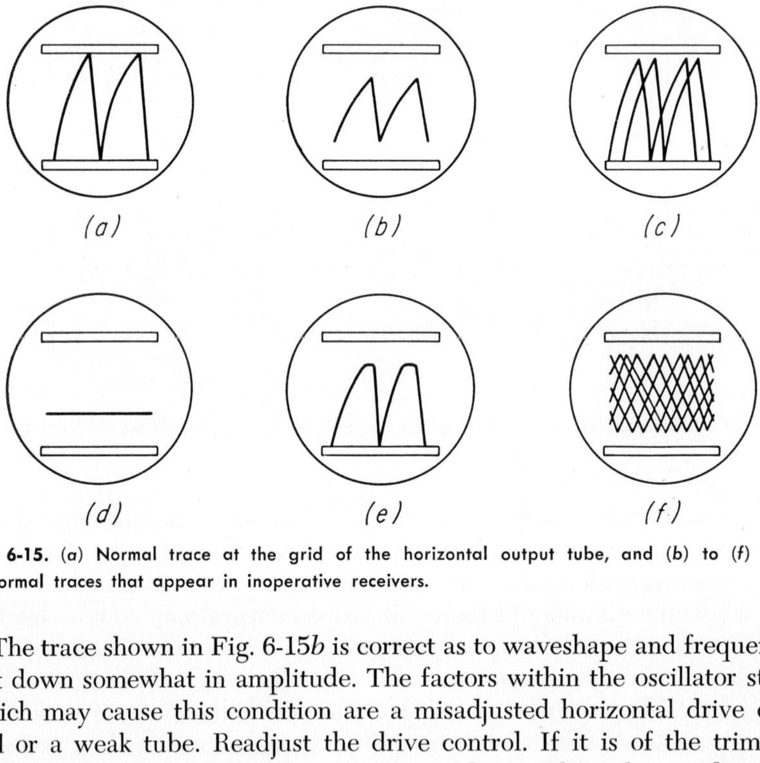

Fig. 6-15. (a) Normal trace at the grid of the horizontal output tube, and (b) to (f) some abnormal traces that appear in inoperative receivers.

The trace shown in Fig. 6-15*b* is correct as to waveshape and frequency, but down somewhat in amplitude. The factors within the oscillator stage which may cause this condition are a misadjusted horizontal drive control or a weak tube. Readjust the drive control. If it is of the trimmer type, minimum capacity gives maximum drive. The tube condition is checked by trying operation with a known good oscillator tube. Another factor which may cause the trace at Fig. 6-15*b* is reduced B supply voltage to the oscillator stage. Refer to manufacturers' service notes for the type of B supply circuit. If it is obtained from the low-voltage supply, a voltage reading more than 10 per cent below normal calls for an investigation of the low-voltage power supply. If plate voltage for the oscillator is obtained from the boost circuit, call the trace normal, since the reduced-boost-voltage condition will be rectified when the trouble is found.

The trace shown in Fig. 6-15*c* represents a slight off-frequency condition which is not likely to stop the high voltage. If this trace can be changed to the normal trace by a slight adjustment of the frequency control on the oscilloscope, call the trace normal, because a slight adjustment of the horizontal frequency controls in the receiver will correct it. Procedure for adjusting these controls is given in Chap. 8.

The trace shown in Fig. 6-15*d* shows a condition of no drive signal being fed to the output stage. This may be caused by a short at the test point or a completely inoperative oscillator stage. If an ohmmeter check does not reveal the short, the trouble is in the oscillator-afc stages, which are then serviced as outlined in Chap. 7.

The trace shown in Fig. 6-15*e* is correct as to frequency, but off somewhat in amplitude and waveshape. The peak of the normal trace has been flattened out. In analyzing this condition, we must remember that the output stage is of a type which loads its input signal. Abnormal loading may cause the abnormal trace. To check for this possibility, remove the output tube and observe the trace. If the abnormal trace remains, the fault is in the oscillator-afc stages. If the normal trace, possibly reduced in amplitude because of a boost-voltage feed, appears when the output tube is removed, the oscillator stage is working, and the output stage is then checked.

The trace shown in Fig. 6-15*f* shows an off-frequency hash which cannot be resolved by a wide swing of the fine-frequency control of the oscilloscope. The oscillator is therefore considerably off in frequency. Here again, we have a condition which may be brought about by interaction between the oscillator and output circuits. The same procedure will again isolate the trouble to its proper area. Remove the output tube. If the hash remains on the screen of the oscilloscope, the oscillator stage is at fault. If the trace reverts to normal with the output tube removed, a factor common to both stages is involved.

In the above checks, the horizontal output tube is removed. If the receiver being serviced uses a 25BQ6 or similar tube in a series filament string, its removal will stop operation. When working on a set of this type, it is suggested that you have a prepared check tube which will accomplish the purpose without interrupting the series filament string. Take a weak or defective 25BQ6 which has a good heater, and clip off all but the No. 2 and No. 7 heater pins. Save the dummy heater for this and similar operations. When removing a 3- or 6-volt tube for test purposes, it is not necessary to prepare a dummy heater for use in series filament sets. Simply short the heater prongs.

Servicing Procedure to Be Followed When Removal of the Output Tube Restores a Normal Trace at Its Grid. When incorrect traces like those shown in Fig. 6-15*e* and *f* are restored to normal by the removal of the output tube, the trouble is improper loading or a common factor between the stages in the horizontal section. Proceed as follows:

First replace the output tube with one known to be good since grid emission or a gassy tube can cause the improper loading. Next, make voltage and resistance checks of the cathode, control-grid, and screen circuits of the output tube since incorrect supply voltages can also cause this trouble. Representative values for the basic circuit are given in Fig. 6-16,

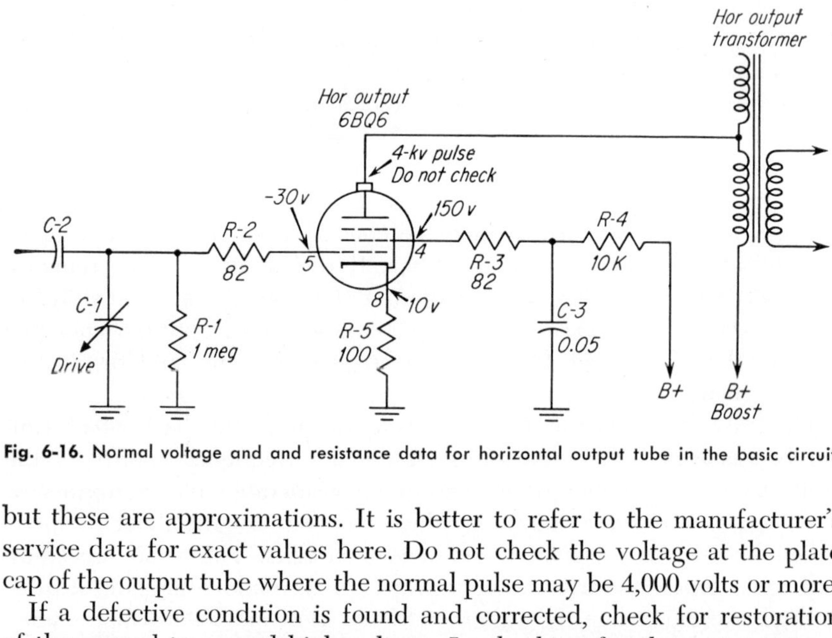

Fig. 6-16. Normal voltage and and resistance data for horizontal output tube in the basic circuit.

but these are approximations. It is better to refer to the manufacturer's service data for exact values here. Do not check the voltage at the plate cap of the output tube where the normal pulse may be 4,000 volts or more.

If a defective condition is found and corrected, check for restoration of the normal trace and high voltage. In checking for the restoration of high voltage, use a high-voltage probe at the second anode of the picture tube. Do not depend on lighting up the screen of the picture tube for this purpose, because of the possibility of a misadjusted ion trap.

If no fault can be found in the supply circuits by the above checks, look for a coupling between the output and oscillator stages which is broken when the output tube is removed. A coupling of this type is the afc feedback line from the horizontal output transformer back to the frequency-control tube of the oscillator. Check the diagram of the receiver to see if it has such a circuit. A typical circuit of this type is shown in Fig. 6-17. Faults in this circuit affect the operation of the oscillator. Then removal of the output tube deactivates the feedback line, restoring the oscillator to normal. To check if this is the trouble, break the circuit at the horizontal output transformer as indicated in the diagram. If this restores the high voltage, check along the feedback chain, looking for defects in the resistors and condensers.

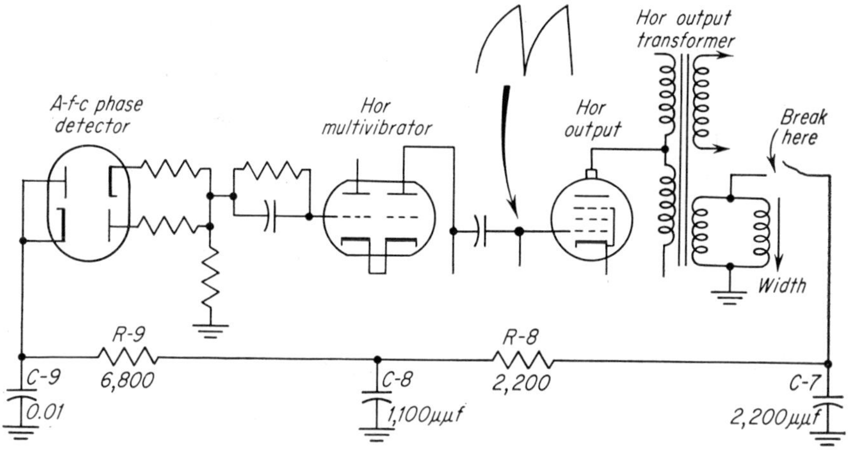

Fig. 6-17. Typical afc feedback circuit.

Another type of coupling between the output and oscillator stages may be present when the plate supply for the oscillator is obtained from the B plus boost line. The boost line is often used for plate supply to both vertical and horizontal oscillators for stabilizing purposes.

Examine Fig. 6-18. The boost line feeds the plate circuit of the output tube and also of the oscillator tube through the decoupling filter *R*-8, *C*-8. If condenser *C*-8 should open, the two circuits are coupled through resistor *R*-8. The result is an off-frequency oscillation which stops the high voltage. When the output tube is removed, only the oscillator is working, and the normal trace is restored.

Therefore, if the oscilloscope check at the grid of the output tube shows an off-frequency trace which becomes normal when the tube is removed, this possibility must be investigated also. Disregard a loss of amplitude, since the oscillator is working at reduced B voltage. Examine the schematic diagram of the receiver. If it shows a boost feed to the horizontal oscillator, look for a condenser which is the equivalent of *C*-8 in Fig. 6-18, and monitor the high-voltage output while trying operation with a test condenser connected across the suspected one. If this restores the high voltage, the original condenser is open and must be replaced. Use a condenser with a high voltage rating for this purpose, since it is connected to the boost voltage, which is higher than B plus. A 20-μf condenser rated at 475 volts or higher should be satisfactory.

Servicing Procedure for the Output, Damper, and High-voltage Stages. When the oscilloscope check at the grid of the output tube shows the normal trace of Fig. 6-15*a*, the oscillator stage is operating and the fault causing the loss of high voltage is in the output, damper, or high-voltage stages. In the procedure to be presented, a check is first made of the supply voltages to the output stage, and then, having ascertained that the output tube is operative, functional checks are made of the damper

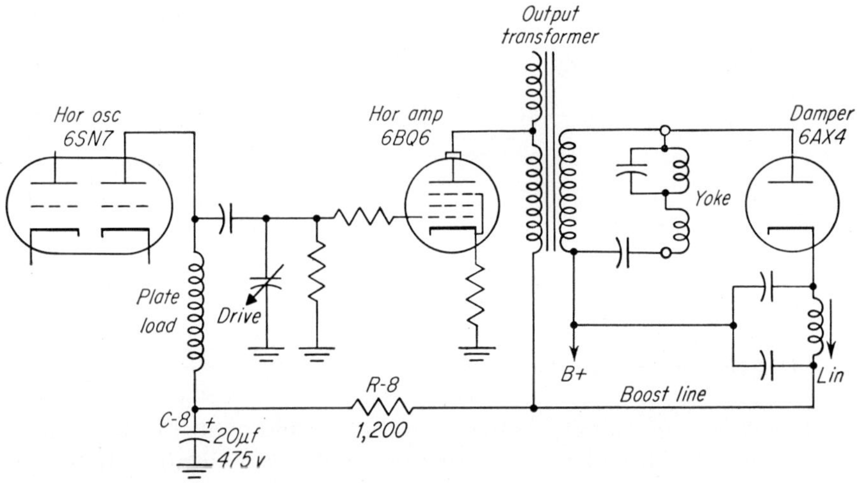

Fig. 6-18. Decoupling filter *R*-8, C-8 in the boost-line feed to the horizontal oscillator tube.

and high-voltage circuits. In checking supply voltages remember to check filament potential on the horizontal oscillator, output, and damper tubes in series string receivers. Irregularities in the filament supply line may place a below-normal voltage on one of these tubes, reducing its emission, and consequently the high voltage.

Checking the Horizontal Output Tube and Circuit. The first step in checking the horizontal output stage is to try operation with another tube known to be good. If this does not restore the high voltage, the next step is to check the voltages and resistances in the cathode, control-grid, and screen-grid circuit. Typical values are given in Fig. 6-16. If these supply circuits appear to be normal, the plate cap connection is removed, and a measurement is made of the voltage appearing at the plate lead. The cap is removed so as to stop the high pulse voltage which is normal at this point. Taken in this way, the tube is inoperative, the boost voltage is absent, and the normal reading is the low-voltage B plus value of the receiver. The actual reading on the defective receiver will be the B plus value or zero.

In order to trace the cause of a zero reading, the plate supply line of the basic circuit has been redrawn in Fig. 6-19. Coil *W*-4 is the yoke winding of the horizontal output transformer, and *W*-1 is the plate winding of the same transformer. The zero reading can be caused by a completely dead damper tube, an open fuse, an open winding in the output transformer or linearity coil, or a short in linearity condenser *C*-5. The latter condition is most likely to cause the fuse to blow.

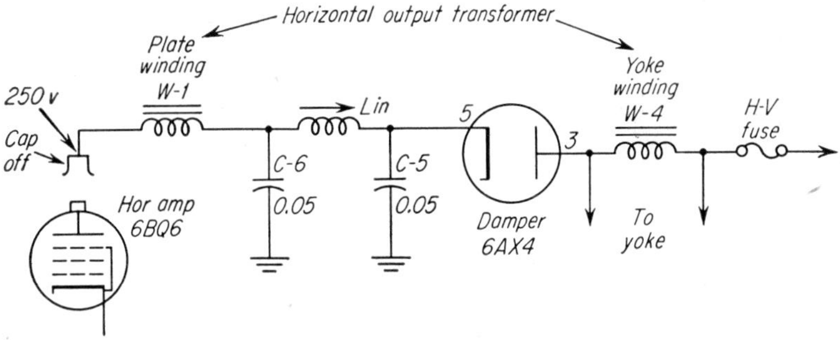

Fig. 6-19. **Plate supply line of the horizontal output tube.**

In checking which component is at fault, move the voltmeter test prod along the supply line. Terminals on the fuse, damper tube, and linearity coil are readily accessible. From the plate cap, the next check point would be both terminals of the linearity coil. If voltage appears at both terminals, the winding *W*-1 in the output transformer is open. If it appears at only one terminal, the linearity coil is open. If the zero reading remains at both terminals, the next check point is the plate pin of the damper tube. If voltage appears here, the damper tube is dead. If the voltage does not appear until the B plus side of the fuse is reached, the fuse is open. Check for a short from the damper cathode to ground before replacing the fuse.

The diagram of Fig. 6-19 shows linearity condenser *C*-5 connected to ground, and the above checks will show a condition of a shorted condenser. However, the linearity condensers are often connected back to the B plus line rather than to ground, in which case the check will not reveal a short in the condenser. This condition, if present, will be found on the next check.

A reading of the normal B plus value at the disconnected plate lead to the output tube does not necessarily indicate correct operation of the components shown in Fig. 6-19. It simply is a continuity check showing no opens in the plate supply line. To make further checks in this circuit, proceed as follows.

Replace the lead on the plate cap of the output tube. This allows the tube to operate and brings the damper and boost circuits into play. It might also be a good idea to remove the fuse and replace it with an inexpensive ammeter having a full scale of 300 to 500 ma. The circuit is shown in Fig. 6-20. Abnormalities in the plate circuit may draw excessive currents which may harm components. If the meter shows a heavy plate current, the serviceman should work quickly, turning the set off whenever possible, so as to avoid further damage to the receiver.

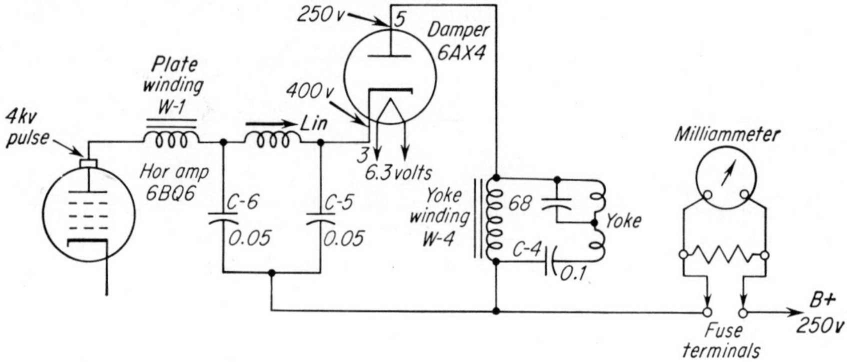

Fig. 6-20. Voltage distribution in the plate circuit of the horizontal output tube.

In a normal receiver, oscillatory currents between the horizontal winding of the yoke and the yoke winding W-4 of the output transformer give rise to the high pulse voltage at the plate of the output tube. The damper circuit, in partially suppressing these oscillations, produces the boost voltage which is applied to the tube through winding W-1 of the output transformer. This time, linearity condensers C-5 and C-6 are connected to the B plus line, where a short in either condenser would not be revealed by the previous check. Normal representative voltages are indicated at check points in the diagram.

With the plate cap reconnected, check the voltage at the cathode (No. 3) and plate (No. 5) pins of the damper tube. Check these against the manufacturer's service notes since there is considerable variation in these voltages in different receivers. If the voltage checks are normal, the circuit is probably working correctly, and the trouble is in the high-voltage stage.

In sets using an autoformer or direct-drive horizontal output system, there is another possibility in addition to the defective high-voltage stage. In these sets, boost voltage, and even sometimes high voltage, may be obtained without the oscillatory currents between the yoke and autoformer. Even though the yoke is open, shock excitation in the high-efficiency coil may produce an almost normal boost voltage. If the high voltage is not sufficient, the screen remains dark. In some cases the high voltage produces brightness but the open yoke does not allow deflection, and a vertical line is seen. When servicing sets of this type, the vertical line indicates an open yoke winding. When servicing for no raster, the presence of normal boost voltage tells the serviceman to check the yoke winding. If this is not open, the trouble is most likely to be in the high-voltage stage.

Procedure for Servicing Abnormal Voltage Readings at the Damper. The voltage at the plate pin of the damper tube is the normal B plus voltage of the receiver. Assuming this to be 250 volts, let us first consider a pair of abnormal readings like 200 volts at the plate and 240 volts at the cathode of the damper tube. A reading of this type indicates trouble in the low-voltage power supply of the receiver. Voltage at B plus is down 20 per cent. This reduces efficiency in all circuits, but the reduction might allow the sound circuits to operate and appear normal. In the horizontal output stage there would likewise be a reduced signal in the plate circuit. This would in turn cause a reduction in the oscillatory current between the yoke and output transformer. There is therefore less current for the damper to turn into boost voltage, which furnishes the strong operating potential for the output tube. A small drop in the B plus voltage, therefore, which has little effect on the sound circuits, may prevent the boost (sometimes called "bootstrap") voltage "from getting off the ground." This reduces the high voltage below the point where the screen of the picture tube lights up. A condition of low voltages at plate and cathode of the damper tube calls for an investigation of the low-voltage power supply. There may be nothing wrong in the horizontal circuit.

Let us now consider a condition of normal voltage at the plate of the damper, and zero at the cathode. The meter in the feed line will show a high current, and the damper plate may glow. This is obviously a short in the damper cathode circuit, but one that was not present previously, when the circuit was checked with the plate lead removed from the output tube. This short is therefore not present under normal B plus voltage but appears when the high pulse voltage is generated in the plate circuit of the output tube.

A likely spot for this breakdown is between the cathode and heater of the damper tube. Try operation with another tube known to be good. If this does not restore the high voltage, the trouble may be in the damper filament winding of the power transformer. In some receivers, the damper heater is connected to its cathode or to a tap on the flyback, and a separate heater winding is provided in the power transformer for the damper heater. This places the pulse voltage on the filament winding and leads, and they must be insulated accordingly.

since the short will appear only under high pulse-voltage conditions. To check for it, first move the heater wiring around with a bakelite rod. If the insulation is defective, there will be sparking through the insulation to the chassis. Replace the leads with varnished high-voltage wire. If the wiring appears to be good, the breakdown is probably in the power transformer itself. To check if this is the case, remove the filament leads from the damper tube, and temporarily connect the heater pins to a 6-volt filament transformer. Keep this test transformer away from the receiver chassis, and try operation. If the high voltage is restored, the pulse voltage is shorting in the power transformer. It is not necessary to replace the entire transformer. Special filament transformers made for this purpose are a standard replacement item listed in catalogues. The primary winding is 6.3 or 117 volts; the secondary winding is 6.3 volts with 5,000-volt insulation.

Going back to the voltage readings at the damper tube, let us now consider a condition of normal voltage at the plate, and about the same voltage at the cathode. This indicates a lack of boost voltage and may be due to a number of causes. These include a weak damper tube, defects in the flyback transformer and yoke, a short in one of the condensers in the linearity circuit, or leakage or a short on the B plus boost line. The tube is checked by replacement with one known to be good. A procedure for localizing troubles to the output transformer or yoke is given in the next section. A shorted linearity condenser is checked for as follows.

Remove the damper tube, and turn on the set. If the high voltage comes on, check the condensers in the linearity circuit, since one of them is likely to be shorted. To see how removal of the damper tube restores high voltage in this case, examine Fig. 6-21.

If either of the condensers in the linearity circuit is shorted, the yoke winding becomes shorted through the damper tube, stopping the oscillation between the yoke and transformer. When the damper tube is re-

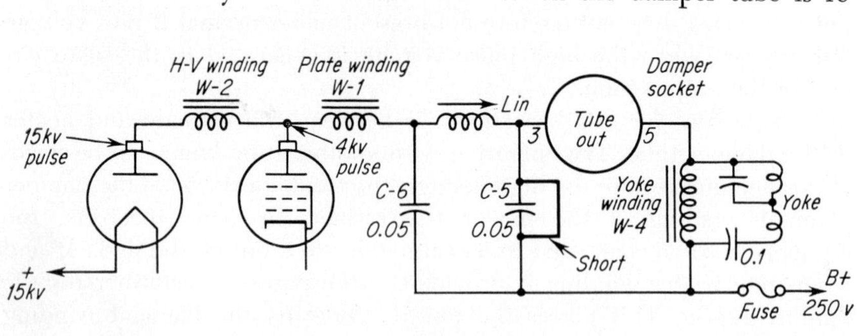

Fig. 6-21. If condenser C-5 or C-6 is shorted, removal of the damper tube removes the short across the yoke winding and allows the circuit to operate, restoring high voltage.

moved, it removes the short, and the oscillations are set up again. The shorted condenser carries B plus to the plate of the output tube, and transformer action between the windings restores the high-voltage pulses.

An ohmmeter check shows which condenser is shorted. When replacing it, use the same capacitance as the original, and at least the same voltage rating. A higher voltage rating is to be preferred. Condensers rated at 1,600 volts are good.

If removal of the damper tube does not restore the high voltage, a check is made of the boost line for leakage to ground. The boost voltage is fed through decoupling filters to the vertical section, and to the accelerating or focusing anodes at the No. 10 and No. 6 pins of the picture tube. Shorts in the decoupling condensers load the boost line through the decoupling resistors, thereby lowering the boost voltage. An ohmmeter check of these feed lines locates the short.

Procedure for Finding Trouble in the Flyback-yoke Circuit. So far, we have a receiver with insufficient or no high voltage, the drive signal to the grid of the output tube is correct, a known good output tube is in the socket, its cathode and screen voltages appear correct, its plate line shows continuity to B plus, but the tube is not operating since the B plus boost voltage is not being applied to the plate. The oscillatory current between the output transformer and horizontal winding in the yoke either fails to start or is damped out by some form of short in the system. To proceed from this point, a second working receiver is used in a signal-tracing type of servicing procedure. Make sure that at least one of the two receivers is insulated from the line by a transformer.

First make the setup shown in Fig. 6-22. The two chassis are bonded together, and the plate caps are removed from both horizontal output tubes. It is not necessary for both output tubes to be of the same types. Next, the plate cap of the tube in the dead set is connected to the output transformer lead of the test set. Both sets are then turned on. A raster should appear on the test set, and we now know that the horizontal output tube in the dead set is capable of driving a good flyback-yoke setup. If the raster does not appear, check for the high voltage in the test set. It may be that the test set requires a considerable amount of voltage to produce a raster, and the mismatch of operating the two systems together will not generate a sufficiently high voltage. If the high voltage measures 6 to 10 kv, call it the equivalent of a raster. If the voltage measured is 0 to 2 kv, something was overlooked when checking the supply circuits of the horizontal output tube in the dead set. Recheck the input signal drive to the tube. Try another known good tube. Recheck the cathode and screen voltages against the service data. Check for open cathode and screen bypass condensers.

Once the raster on the test set is obtained, both sets are turned off, and

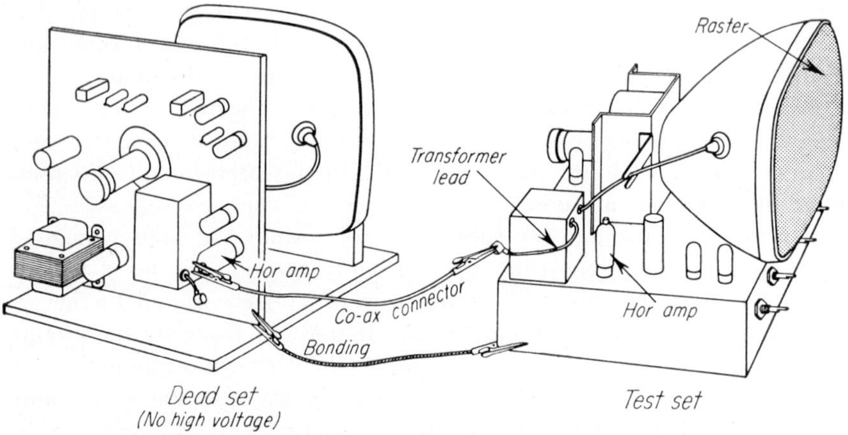

Fig. 6-22. Setup to determine that the horizontal output tube is working.

the caps are restored to their respective plates. Then both high-voltage leads are removed from their connection points on the picture tubes, and the high-voltage lead from the test set is brought over and connected to the second anode button in the picture tube of the dead set. The setup is shown in Fig. 6-23*d*.

Both sets are turned on. The ion trap, which may have been previously misadjusted, is rotated for light on the screen. The display on the screen of the dead set, operating with the borrowed high voltage, will indicate the type of trouble. A bright vertical line indicates an open horizontal winding in the yoke. A keystone display like the one shown in Fig. 6-23*a* indicates unequal horizontal deflection caused by a short in one section of the yoke. The narrow raster of Fig. 6-23*b* indicates insufficient deflection. This time a short is indicated, and is probably between turns of the flyback transformer or width coil. The foldover in Fig. 6-23*c* indicates trouble in the damper circuit.

If the displays on the screen lack full height, disregard this symptom for the present. The vertical deflection circuits may be powered from the boost circuit, which is still inoperative.

The procedure of bringing an external source of high voltage to a dead set could also be used in receivers where there is some boost voltage, but not enough. In this case, the short is less severe, although severe enough to stop the high voltage. The same displays, possibly less marked, would result. If a normal raster appears, it indicates full deflection, and the trouble is in the high voltage stage of the dead set.

Servicing for a Defect in the Deflection Yoke. When the check just made produces a bright vertical line, the suspicion that the yoke is open should be confirmed. An ohmmeter check will take care of the yoke

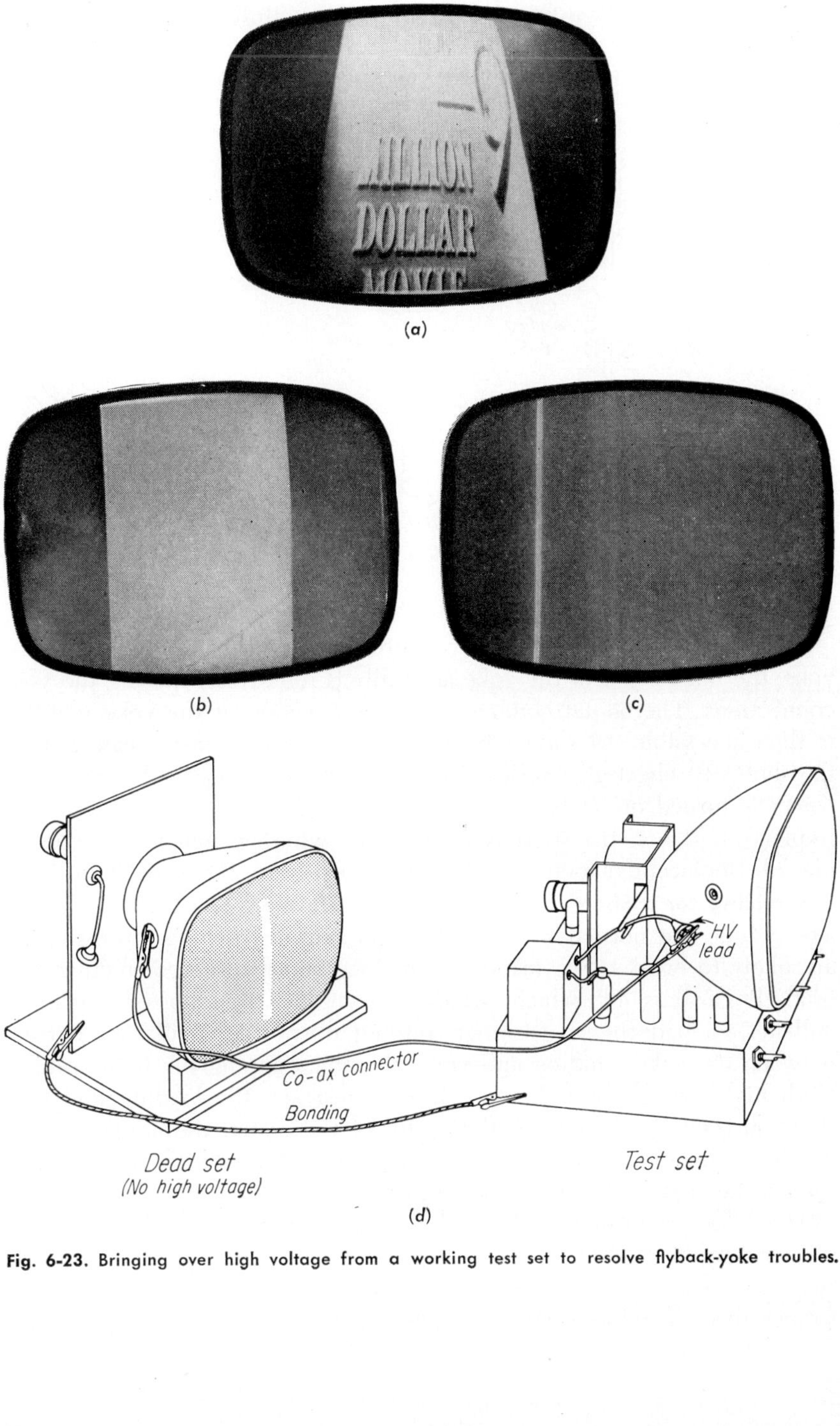

Fig. 6-23. Bringing over high voltage from a working test set to resolve flyback-yoke troubles.

winding and leads. One other circumstance can give the same indication. See Fig. 6-24. Condenser *C*-4 is in series with the yoke winding, and if it is open, it will produce the same result as the open winding. This condition is checked by bridging the suspected condenser with a test condenser.

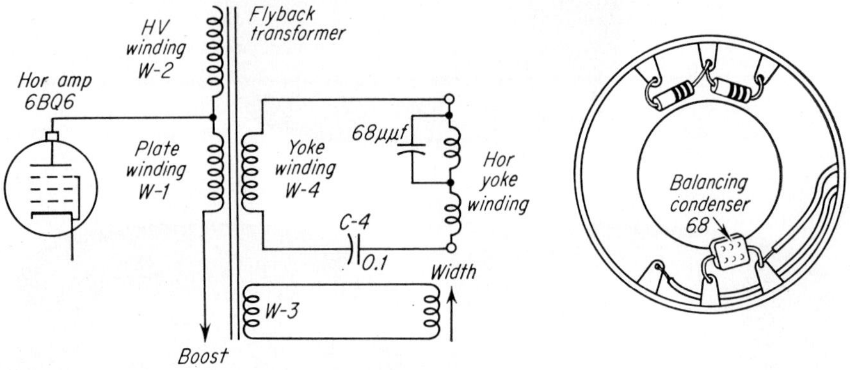

Fig. 6-24. Position of the horizontal yoke winding in the circuit. It may be identified in the yoke assembly by the 68-μμf balancing condenser.

When a keystone-shaped raster is produced, indicating a short in the yoke, the back cover of the yoke assembly is removed, exposing the yoke connections. The 68-μμf condenser across a section of the yoke winding is then accessible, as shown in Fig. 6-24. This condenser may contain the short. To check it, one lead of the condenser is opened and the receiver is turned on. If the keystone is gone, the condenser is defective. Replace it. If not, the short is in the yoke winding, which must be replaced. Replacement notes on deflection yokes are given in Chap. 5.

Servicing for a Short in the Flyback Transformer. The high pulse voltages which are normal in the horizontal output (flyback) transformer are likely to break down insulation, making shorts quite common in this unit. Regardless of which winding contains the shorted turns, it will act to load the transformer, pulling down the oscillatory current between the yoke winding and yoke, thereby stopping the high voltage. With high voltage brought over from a test set, the loading due to a short on the flyback reduces the width of the raster, the amount of reduction depending on the severity of the short. The narrow raster, therefore, is the sign of the short on the transformer.

The defect is not necessarily in the output transformer itself. The transformer may be loaded by a shorted component across it. For example, such a condition in the width coil in Fig. 6-24 would load the transformer through winding W-3. Another possible short may be reflected from the yoke by a width shim which has been positioned too far into the yoke.

There is considerable variation in output transformer circuits, and it would be better to work with the diagram of the receiver being serviced, when tracing such a short. The procedure is to remove these component leads from the output transformer one at a time. The removal of the short will be recognized by the spreading of the raster in the setup of Fig. 6-23, or by watching for the restoration of high voltage.

As a typical example of how this is checked, the horizontal output circuit of the Wells-Gardner 321MS31 receiver is shown in Fig. 6-25.

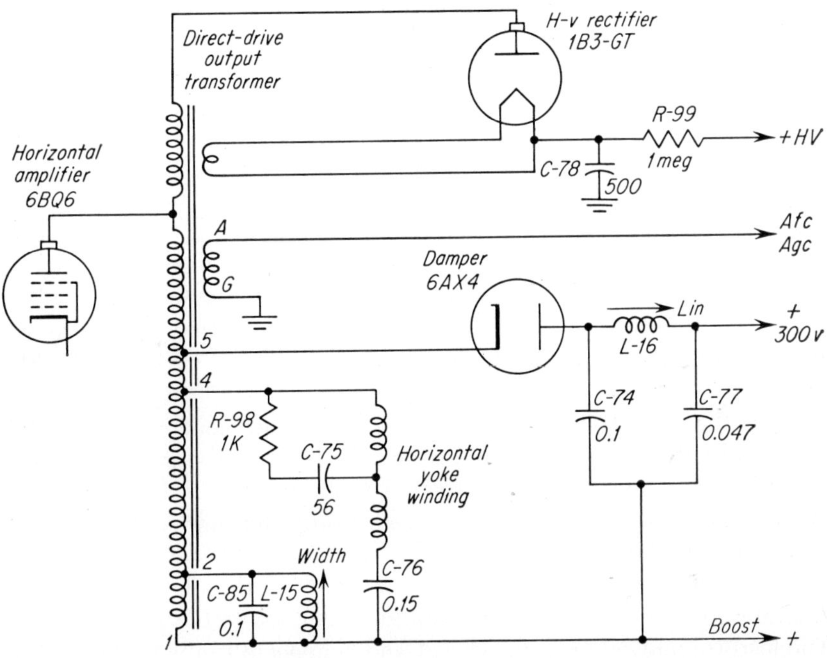

Fig. 6-25. Horizontal output circuit of the Wells-Gardner 321MS31 receiver.

The horizontal yoke winding is fed from a tap on the transformer rather than from a separate winding. This is called a direct-drive flyback transformer circuit. All servicing checks previously described may be applied. Note condenser *C*-85 and width coil *L*-15 across terminals 1 and 2 of the output transformer. A short in the condenser or shorted turns in the width coil would load the transformer. There are other connections at terminal 1; so the wire to terminal 2 is lifted. If the high voltage comes on, the short is in the coil or condenser. Terminal 4 goes to the yoke. A short in this circuit would already have been discovered by the keystone raster. Similarly, terminal 5 goes to the damper, which has also been checked for shorts. This wiring may therefore be left undisturbed. Terminal *A* goes to the afc-agc circuits which may contain a short that will load the transformer. The wire to terminal *A* is therefore lifted to see if the short is located here. Finally, condenser *C*-78 in the high-voltage circuit may be shorting and loading the transformer through the 1B3-GT high-voltage rectifier. This circuit may be checked by opening the grounded end of condenser *C*-78. If the raster remains narrow after all circuits have been opened, the short is in the windings of the transformer, which is then replaced.

Replacement Notes on Flyback Transformers and Width Coils. The output transformer must match the output tube and deflection yoke. In addition, there must be correct windings and taps for the width coil and afc-agc circuits. For these reasons, it is imperative that an exact replacement part be used.

If a width coil is found to be defective, its specifications are not so critical. Components listed in a replacement catalogue are generally satisfactory.

When replacing the output transformer or width coil, if an agc or afc winding is included, care must be taken to polarize this winding correctly. Incorrect polarity of the agc pulse will upset the agc system, resulting in loss of picture and sound. Incorrect polarity of the afc pulse will interfere with the operation of the phase detector and may stop the high voltage.

Since the new transformer is an exact replacement, it is wired in exactly like the old one. If there is any doubt as to the markings or operation of the agc or afc winding, correct polarity is found with an oscilloscope adjusted to the crayon markings for horizontal measurements. The probe is connected to the hot lead of the winding. The hot lead of the agc winding is the one connecting to the plate of the keyed agc tube; the hot lead of the afc winding connects to the phase detector. The lead is removed from the circuit and connected to the probe. The set is then turned on, and its horizontal frequency control adjusted, if necessary, for a steady trace. The vertical gain control on the oscilloscope may have to be reduced from the crayon marker because the pulse at this test point is usually of high amplitude. If the winding is polarized correctly, the trace shown in Fig. 6-26*a* will be observed on the oscilloscope. Incorrect polarity gives the trace shown in Fig. 6-26*b*. If the incorrect trace appears, reverse the winding leads.

Singing Horizontal Output Transformers. When a replacement horizontal output transformer is installed, it sometimes exhibits a trouble known as "singing." This is caused by the core or frame vibrating at the horizontal frequency. The condition is usually cleared up by tightening

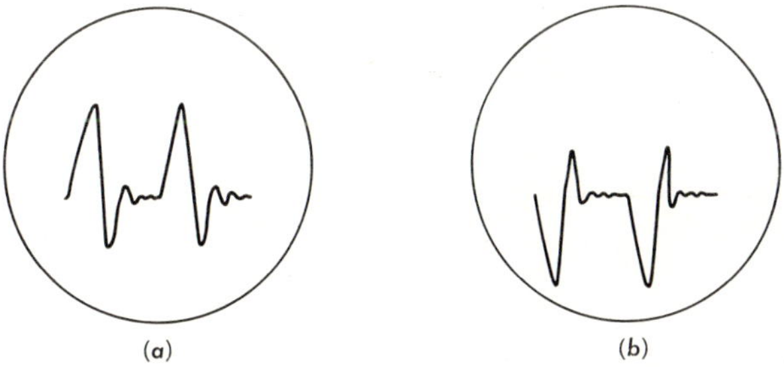

Fig. 6-26. How to check polarity of the agc-afc winding.

down the screws holding the core to the frame. In some cases, the entire transformer vibrates, and the vibration is not cut down when the mounting screws are tightened. When this happens, loosen the mounting screws and slip a piece of felt or other damping material between the frame and chassis. This usually stops the singing. A singing output transformer is treated as a replacement problem because, although any set is subject to this vibration, the consumer is not likely to make a service call for this trouble by itself.

Servicing for Defects in the Damper Circuit. When high voltage is brought from a test set to light the picture-tube screen, and the resulting display shows the foldover of Fig. 6-23*c*, damper trouble is indicated. Previous checks will have found a defective damper tube, a shorted linearity condenser, or an open linearity coil. A short in the linearity coil need not be considered, since this will not stop high voltage. The only check to be made at this time is for an open linearity condenser. This is done in the usual way by bridging a test condenser across each of the suspected condensers in turn while watching for the removal of the foldover line.

Troubles in the High-voltage Power Supply. Previous steps in the procedure for finding the cause of a lack of high voltage pointed to the high-voltage power-supply stage. For example, a normal reading at the damper cathode indicated that all is well in the output and damper stages, and the trouble is likely to be in the high-voltage stage. Similarly, when the dead receiver is operated with a high-voltage supply brought over from another receiver, and a normal raster results, this indication again points to the high-voltage stage.

For a quick review of the functional position of the high-voltage stage in the horizontal section, the block diagram is repeated in Fig. 6-27. Oscillations set up between the yoke and output transformer produce an r-f pulse of approximately 4,000 volts across the plate winding of the output transformer. This is the input pulse to the high-voltage power supply.

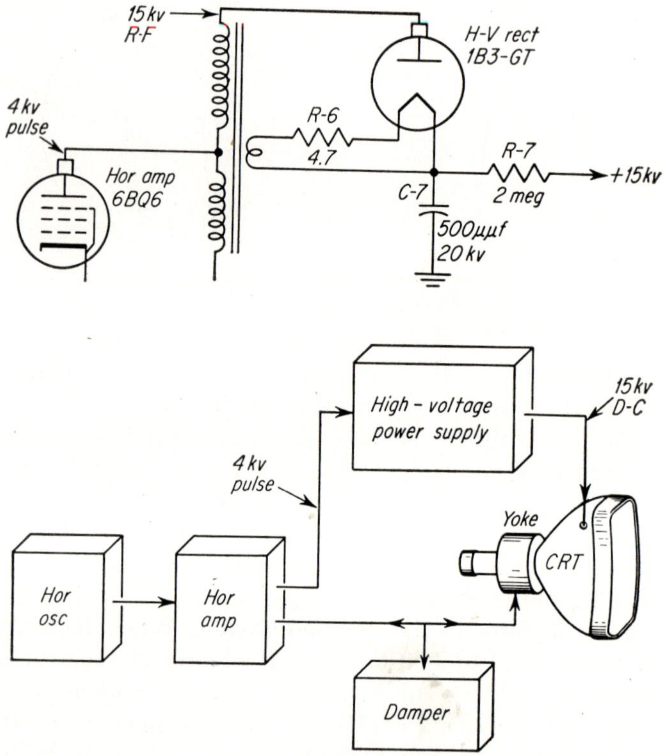

Fig. 6-27. The circuit of a basic high-voltage power supply and its position in the horizontal section.

In the power supply, the high-voltage secondary winding steps this pulse up to approximately a 15,000-volt pulse. This is then rectified to produce the 12- to 15-kv d-c output for the second anode of the picture tube.

The high-voltage stage is subject to the same troubles that are common to all power supplies. The transformer secondary may not deliver voltage, the rectifier tube may be inoperative, filter resistor *R*-7 may be open, or filter condenser *C*-7 may be shorted. A procedure for finding which of these is at fault now follows.

Finding the Defective Component in the High-voltage Power Supply. The high-voltage r-f pulses should produce a corona arc to a nearby piece of metal, and this is the basis of a quick check for its presence. The setup is shown in Fig. 6-28. Use a screwdriver with a well-insulated handle, since the shock from the r-f pulse can be dangerous. Turn the set on, and draw the tip of the screwdriver along the plate lead to the rectifier plate cap. If the high-voltage pulses are present, the screwdriver

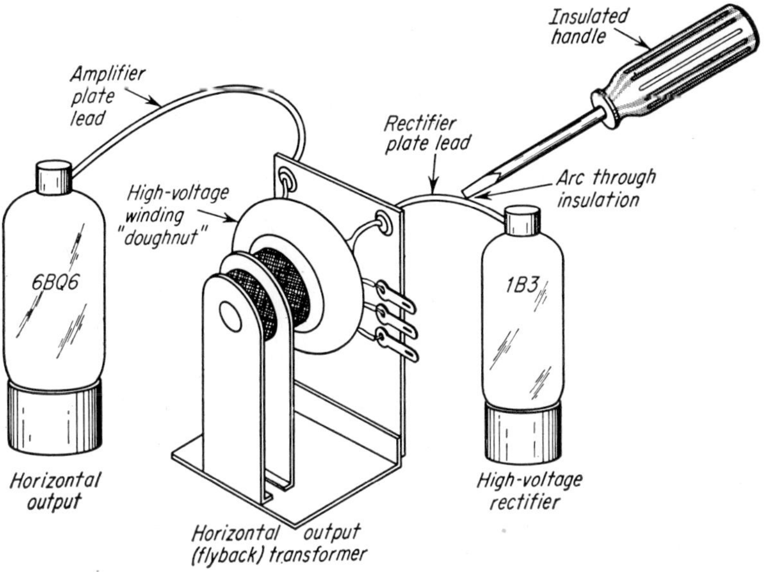

Fig. 6-28. Quick check for input voltage to the high-voltage power-supply stage.

will draw a corona arc through the insulation. At the plate cap itself, a large purple arc will be drawn.

A small arc or no arc does not necessarily mean that the transformer is defective. The r-f pulses may be loaded down by a shorted filter condenser. Turn the set off, discharge the plate cap by grounding it, lift the cap lead, and repeat the arc test. If the arc is still absent, high-voltage winding *W*-2 of the flyback transformer is probably open; if the normal arc now appears, the filter condenser is probably shorting.

Check the transformer with an ohmmeter, but before doing so, remember the high-voltage precautions. Turn the set off, and discharge both plate leads to ground. Then measure from the plate lead of the horizontal output tube to the plate lead of the rectifier tube with an ohmmeter. A good winding will show a reading of 300 to 500 ohms. If the winding is defective, the output transformer must be replaced.

A short in filter condenser *C*-7 may not show up on an ohmmeter check. This is because of the thick dielectric needed for high-voltage condensers. When the condenser becomes defective, current sparks across the broken dielectric, giving the shorting effect, but the plates need not touch. The short, therefore, may not be evident to the low voltage present in an ohmmeter. To check for a suspected short in the condenser, listen for the spark discharge. If this cannot be heard, open either of the condenser leads, and try operation with the condenser open. If this brings on the high voltage, the condenser is defective and must be replaced. A common location for the high-voltage condenser, often called a "cartwheel," is below the rectifier socket, as shown in Fig. 6-29.

If the high r-f voltage is present at the plate of the rectifier tube, the trouble may be an open in filament resistor *R*-6, or a defective rectifier tube. See Fig. 6-29 for the location of these components in the circuit.

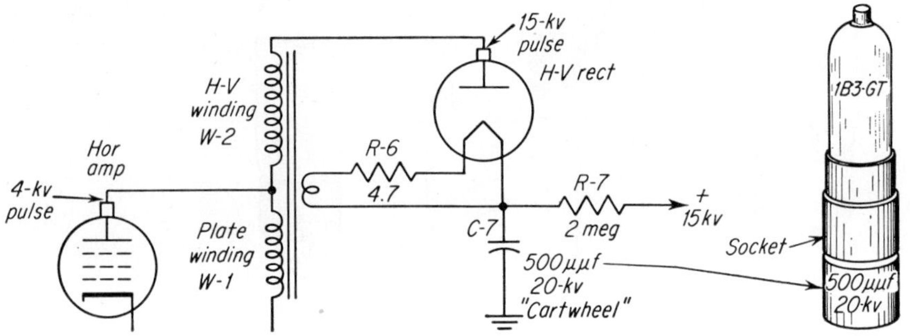

Filament pins of common high-voltage rectifier tubes	
1B3-GT	2 and 7
IX2A-1X2B	2, 5, 8 and 1, 4, 6, 9

Fig. 6-29. The "cartwheel" high-voltage condenser and its position in the circuit.

The filament circuit is checked by removing the rectifier tube and inserting the ohmmeter prods directly across the filament pins. The filament pin numbers for the common rectifiers are included in Fig. 6-29. Remember that you are counting pin numbers from the top of the socket, and work counterclockwise. Resistor *R*-6 varies in different receivers from 6.8 ohms down to no resistor at all. An ohmmeter reading, therefore, from 0 to 6.8 ohms is the normal reading for this check. A higher resistance reading or an open circuit shows a defective resistor. The check, of course, takes in winding *W*-5 of the output transformer, but this is a single loop of heavy well-insulated wire around the core of the transformer and leading directly to the rectifier socket. This winding does not open.

If the filament circuit proves good, the rectifier is checked next. Do not look for this tube to light. Its normal filament operating temperature is a rosy glow, the reflection of which can just barely be seen on the internal shield if the room is darkened. For check purposes, try operation with a tube known to be good. If this brings on the high voltage, the original tube was defective.

The last component, filter resistor *R*-7, is checked with an ohmmeter.

If defective, the resistor will probably be charred, as well as give a wrong reading. The usual value for this resistor is 1 or 2 meg.

In some receivers, the entire filter circuit is simply the capacitance existing between the inside and outside aquadag coatings on the picture tube. In this case, the servicing procedure is simply to check the rectifier tube by substitution, and a filament resistor if used. The condenser formed by the aquadag coatings need not be checked for leakage or shorting, but the outer aquadag coating may be worn off, or the grounding spring may not be making contact, opening the filter. To check for this, try operation with a 20-kv cartwheel, temporarily connected from chassis to the high-voltage lead. If this clears up the trouble, check for the coating and grounding contact. The coating may be repaired by a paint job with a product known as Tube coat. The spring is cleaned and resprung.

Another source of trouble in the high-voltage circuit is the tendency for arcing, sparking, or corona discharge from the high-voltage leads to the chassis or other components. This is accompanied by cracking noises and the characteristic smell of ozone. If the point where the discharge occurs cannot be readily found, darken the room. The area of corona discharge will then appear as a purple glow, and a spark location will easily be seen. There may be frayed insulation, but more often the fault is due to lead dress, or accumulation of conductive dust. The remedy is obvious. The discharge area is noted in the darkened room, and then examined. Any lead that is too close is moved. A dusty area is cleaned with carbon tetrachloride. If insulation is frayed, or there is a splice in the lead, the entire lead is replaced. Wire with good high-voltage insulation must be used for this purpose. If none is available, use a piece of RG59U coaxial cable. First, cut it to length. Then slice through the outer vinyl-plastic covering with a razor blade. Remove the covering, push the ends of the braiding in to loosen it, remove the braiding, and you have the inner conductor with its polyethylene-plastic insulation. Should the filament lead to the high-voltage rectifier need replacing, the same wire is satisfactory.

The corona discharge may also be coming from a sharp point protruding from a connection. If this proves to be the case, snip off the point and resolder the connection, wiping it to leave a rounded surface.

Replacement Notes on Components in the High-voltage Power Supply. When replacing a filament resistor on the rectifier socket, make sure to use the same tie points for the replacement as had been used for the original. This is because both filament legs on the rectifier tube are brought out to several pins so that sharp bends in wiring are avoided. Only two pins are left unconnected for use as tie points. Making connections to the original pins automatically gives the best wire dress as well as a correct tie-point pin.

When rewiring components in this circuit, remember the precautions pertaining to corona discharge. Wipe solder joints to leave rounded surfaces. If there are any spliced leads, replace the entire lead, using wire with high-voltage insulation. Check wire dress. Leads should not be too close to each other, the chassis, or components. Check particularly for grommets and sleeving where the leads go through the r-f cage.

Fig. 6-30. Replacement for a filter resistor. (*a*) Original resistor, 2-meg 4-watt high-voltage type. (*b*) Three 680-kc 1-watt resistors in series in a replacement.

When a burned-out filter resistor is replaced, if the original 4-watt long-type resistor is not available, a series string of ordinary resistors may be used as shown in Fig. 6-30. The series connection gives the needed high-voltage characteristics to the replacement.

As for the high-voltage filter condenser, there is only one type in common use: a 500-$\mu\mu$f ceramic condenser rated at 20,000 volts. However, different mounting arrangements are found. The replacement condenser illustrated in Fig. 6-31 is universally adaptable in this regard through the use of interchangeable terminals.

The basic circuit of Fig. 6-29 shows this condenser connected from the high-voltage point to chassis. In some receivers, the connection is from the high-voltage point to B plus or B plus boost. In making the replacement, wire in accordance with the original manufacturer's circuit.

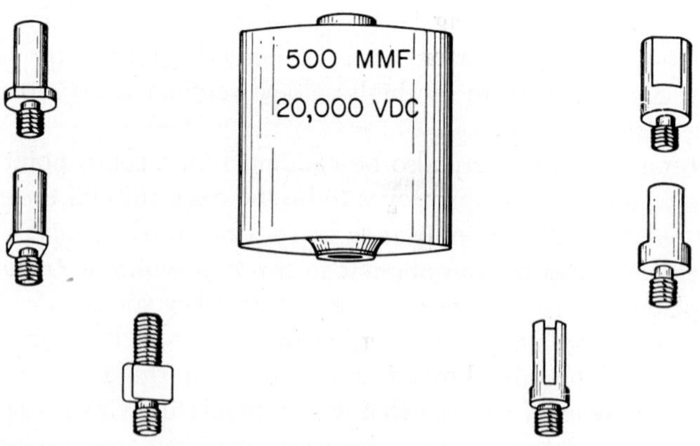

Fig. 6-31. Replacment high-voltage filter condenser.

OTHER TROUBLES IN THE HORIZONTAL SWEEP SECTION

There are other troubles in the horizontal section besides the loss of high voltage. In these cases there is enough high voltage to produce a raster, but the raster has a visible trouble pointing to the horizontal section. These include a narrow raster, poor horizontal linearity, foldover, and ringing. They are illustrated in Fig. 6-32.

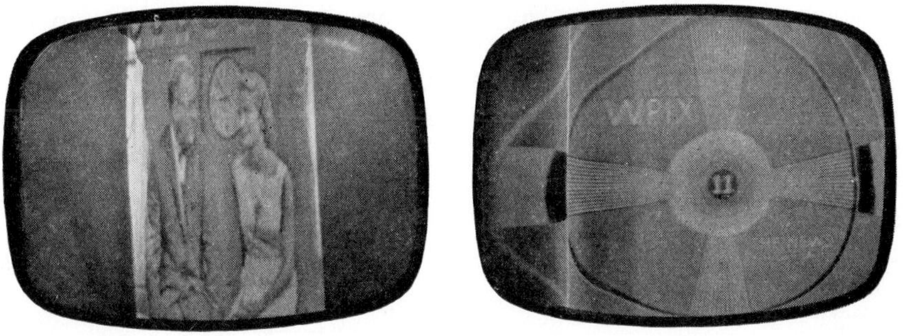

(a) (b)

Fig. 6-32. Other symptoms pointing to trouble in the horizontal section. (*a*) Narrow raster. (*b*) Poor linearity, foldover, and ringing.

An example of a narrow raster is shown in Fig. 6-32*a*. In this case, the linearity is normal, and the lack of width is due to insufficient horizontal deflection. Poor linearity is illustrated in Fig. 6-32*b*. Here, because of unequal speeds as the beam moves across the face of the picture tube, objects on the left side appear wider than objects on the right. Pincushioning may be mistaken for loss of linearity since this may also cause a crowding on one side of the picture. If readjustment of the pincushion corrector magnets relieves the crowding effect, the linearity may then be considered satisfactory. The corrector adjustment is described in Chap. 5 and illustrated in Fig. 5-24.

Poor linearity can occur together with incorrect width, since the width and linearity controls interact. Correct adjustment for either control requires a check of the other.

Figure 6-32*b* also illustrates foldover and ringing. In foldover, there are bright vertical bars with or without picture information. Ringing causes dark bars to appear on the raster.

Another complaint associated with the high-voltage and horizontal section is called "blooming." This name is applied to a condition where a change in the setting of the brightness control seems to cause objects on the screen to increase in size or bloom like a flower. A low output of the high-voltage power supply is a common cause of this symptom, because the beam is slowed down in its progress through the tube by the lowered anode potential. This causes the spot on the screen to be enlarged, giving the blooming effect.

The other symptoms just mentioned may also be accompanied by a drop in high voltage. The first step, therefore, is to check the potential at the second anode of the picture tube. If this reads below normal, correct this condition first by following the procedure given for low or no high voltage. When the high voltage is restored to normal, the accompanying complaint will most likely have been corrected. Merely touch up the drive, width, and linearity controls.

Controls in the Horizontal Deflection System. The horizontal drive, width, and linearity controls were described earlier in this chapter but will be quickly reviewed now before their adjustment is described.

The horizontal drive control allows for a variable input of the sawtooth signal feeding the amplifier tube. The control is usually a trimmer condenser, like condenser *C*-1 in the basic circuit of Fig. 6-33. Sometimes

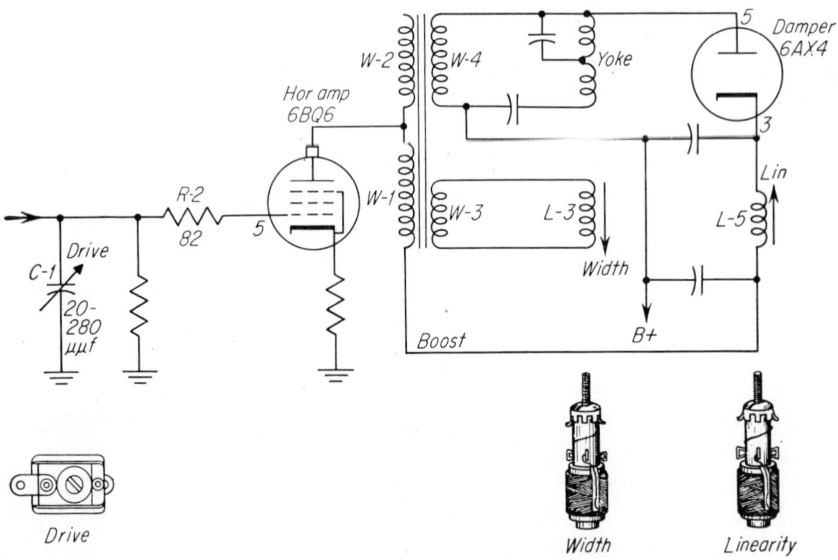

Fig. 6-33. Basic horizontal amplifier showing the drive, width, and linearity controls.

the trimmer condenser is connected in series with the amplifier grid. In some cases, a potentiometer is used in a "volume-control" type of feed circuit. Finally, the receiver may not be equipped with a drive control.

Most receivers that include a width control use a permeability-tuned coil across a tapped section of the horizontal output transformer. Sometimes the width coil is connected across a separate winding of the transformer. This is the method used in the basic circuit, where width coil *L*-3 is connected across transformer winding *W*-3. Tuning this coil changes the inductance of the transformer. This controls the deflection current into the yoke, and so determines the width. The minimum inductance of this coil is obtained with the tuning slug out of the coil, and gives the narrowest picture.

The linearity adjustment, when provided in the receiver, is a slug-tuned coil in the cathode of the damper tube. This is coil *L*-5 in Fig. 6-33. Adjusting the inductance in the cathode of the damper tube controls the starting time of damping action through the tube. This controls the beam speed on the left side of the picture-tube screen, and so acts as a linearity control.

All controls are usually on the rear apron of the chassis near the horizontal section. They are usually labeled DRIVE, HOR.LIN., and WIDTH or HOR.SIZE. If not marked, they are identified by tracing leads. The linearity and width coils look pretty much alike, but the linearity-coil leads go to the damper tube, while the width-coil leads go to the output transformer. The drive-control lead is traced to the grid pin of the horizontal amplifier tube, or its grid resistor *R*-2. When the receiver uses a width shim, it will be found on the neck of the picture tube.

How to Adjust the Horizontal Drive, Width, and Linearity Controls. Tune the receiver to a channel which usually gives a good picture and adjust the front-panel controls for good contrast, brightness, focus, etc. The picture should be steady, and normal except for size or horizontal linearity.

Turn the set off and remove the high-voltage fuse from its clips. If the fuse is of the type which is soldered in, unsolder one end. Then connect a milliammeter across the fuse leads, with the positive lead toward the B plus feed line. If the receiver does not have a fuse, unsolder the lead from the plate pin of the damper tube, and connect the meter in series with the damper plate. Adjust the multimeter to a range of approximately 200 ma. If your multimeter does not include milliammeter ranges, use a 27- or 33-ohm resistor instead of the meter. Then connect a low-scale voltmeter across the resistor. The meter connections are shown in Fig. 6-20. Since the meter is going to be used as a null indicator, the actual current flow and accuracy of the meter are unimportant.

Turn the set on again, see that the picture comes on, and note the meter reading. Adjust the multimeter to a range that keeps the reading at about center scale.

Now adjust the width coil so that the tuning slug is at about the halfway position in the coil form, disregarding the effect on the screen.

Next decrease the horizontal drive. As it decreases, note that the ammeter reading goes up and the width is reduced. It may happen that adjustment of the drive control throws the picture out of sync. This is because the drive control affects the boost voltage, and if the horizontal oscillator is fed from boost, its frequency may change with change in voltage. If the picture goes out of sync when the drive control is adjusted, bring it back by retouching the horizontal frequency control. Reduce the drive until both edges of the raster can be seen on the screen. The picture may lose some brightness. Disregard this.

The next step is to center the picture. Most sets have a mechanical adjustment near the yoke for this purpose; some use horizontal and vertical centering controls. Centering adjustments are described in Chap. 10. In either case, center the picture so that the dark portions on either side of the picture are equal. Readjust the drive control so that the picture is about 1 in. in from each end of the screen.

Now, with a narrow picture, properly centered on the screen, adjust the linearity control while watching the meter in the B plus line. There will be little change in current until a point is reached where the current begins to dip. Continued turning in the same direction causes the current to drop still further, and then to increase again. The correct adjustment of the linearity control is at the point which shows the lowest reading. This is called tuning for a "dip," or "null." Finally, check the dark portions on each side of the raster. If they are unequal, readjust the centering control.

Now readjust the drive control so as to widen the picture. Note that the raster is slightly wider than the picture. Continue to widen the picture until the edge of the picture touches the edge of the tube. Both edges should touch at once. If they do not, readjust the linearity control for the dip in the meter, and retouch the centering control, if needed. Then continue to widen the picture with the drive control until the picture is just over the edges of the tube.

Check the high voltage, and if it is normal, the adjustment is completed. If it is below normal, adjust the width control so that the picture edge is brought back on the screen. Then widen again with the drive control to bring the picture just over the edge of the tube. This should bring the high voltage back to normal. Thereafter, the width control can be used as a vernier adjustment to compensate for slight changes in the home.

The situation may arise where the drive control brings the picture correctly just beyond the edge of the tube but the high voltage is well below normal. A midway compromise should be acceptable. Note the high-voltage reading at correct width, advance the drive to the maximum high-voltage position, and then reduce it to give a high voltage halfway between the two readings.

Some manufacturers' lines include economy models which do not have drive, linearity, or width controls. These receivers are designed for a slightly overdriven horizontal deflection. Then, as components age, or in areas of low line voltage, the picture narrows to approximately the correct width.

A common complaint heard in connection with these receivers is that the consumer cannot read first and last words in titles, since they are cut off by the mask, or the ends of the picture tube. This is especially true in areas of high line voltage and is, of course, due to the overdrive.

Another possible cause of the overextended width is that the deflection yoke has loosened and slipped back from its correct position against the flange of the picture tube. This is easy to check. Loosen the yoke-holding bolts, check to see that it is snugly up against the outflaring sides of the picture tube, and finally tighten the bolts.

If the yoke is correctly placed, the overextended width may be remedied by installing a trimmer-type drive control. Check the schematic diagram of the receiver, looking for a fixed condenser in the grid circuit of the horizontal amplifier tube, in the position of drive condenser *C*-1 in the basic circuit of Fig. 6-33. Note the capacitance of this fixed condenser, and replace it with a trimmer whose maximum capacitance is somewhat larger than the original condenser. Then use the procedure given previously to adjust the newly installed drive control. Omit the steps applying to the linearity-control adjustments. This time, if a picture of the correct width results in a slightly lowered high-voltage output, leave the picture at the correct width setting of the drive control, since this was the original complaint. Another expedient is to install a brass or aluminum width shim if there is sufficient clearance between the yoke and the neck of the picture tube (see Fig. 3-8).

Servicing Procedure for Insufficient Width. About the first effect of a drop in line voltage would be reduced width. The line voltage, therefore, should be the first check when working on this complaint. If this is normal, the low-voltage power supply should next be checked, since here again the first effect to be noticed would be reduced width.

If there is normal output from the low-voltage power supply, the horizontal section comes up next for inspection. It may not be necessary to go through the entire procedure since the presence of the picture shows that most of these circuits are working. First try change of tubes and adjustment of controls. Do not neglect the width shim. If this does not correct the condition, go through the procedure as given for no high voltage.

Servicing Procedure for Poor Linearity. Poor horizontal linearity is most likely to be caused by misadjustment of the horizontal controls. There is interaction in the settings of the horizontal controls just as there is in the vertical controls. A condition of insufficient width caused by, let us say, a weak horizontal amplifier tube may have been compensated for by turning the linearity adjustment. This fills out the width but leaves poor linearity. The controls should be set in accordance with the procedure previously given. Then, if this leaves a narrow picture, the horizontal section is serviced for insufficient width.

Servicing Procedure for Blooming. It was previously mentioned that blooming was most often caused by a reduced high-voltage output from the horizontal section. The servicing procedure, therefore, is to check the high voltage, and correct it if it measures below normal. A weak high-voltage rectifier is a common cause of this trouble.

However, the high voltage may give a normal reading. In this case, the less common causes of blooming must be investigated. These are incorrect voltages on the cathode and grid of the picture tube, and a gassy picture tube. The cathode-grid voltage is checked as shown in Fig. 6-12. If the voltage is incorrect, the brightness circuit is investigated as previously explained. If this voltage reading is normal, the CRT is probably gassy. Try operation with another tube.

Servicing Procedure for Foldover and Ringing. Foldover and ringing, illustrated in Fig. 6-32*b*, are extreme cases of poor horizontal linearity. In foldover, the bright vertical sections are caused when the beam is slowed down for that section of the trace. In ringing, the alternate dark and bright bars are caused by an oscillatory condition. In both cases, of course, the deflection current is nonlinear.

These symptoms may or may not be accompanied by reduced high voltage, but a routine servicing procedure for low high voltage will find the trouble. However, it may be timesaving to go directly to likely causes for these symptoms. If the symptom appears on the left side of the screen, the damper circuit is at fault. The damper tube and the condensers around the linearity coil should be checked by substitution. If the symptom appears on the right side of the picture, the waveshape of the signal in the amplifier tube is probably at fault. In this case, an os-

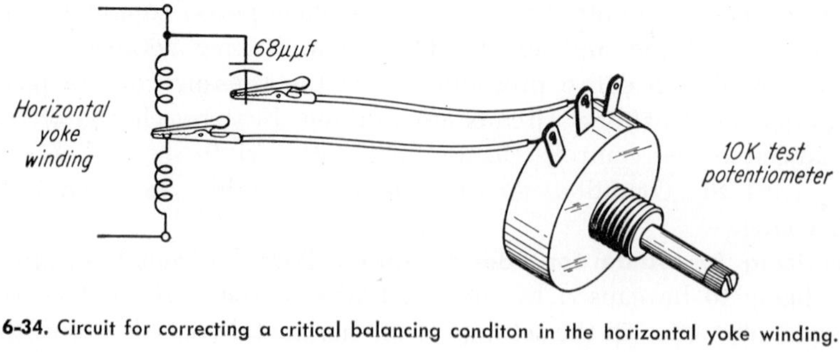

Fig. 6-34. Circuit for correcting a critical balancing conditon in the horizontal yoke winding.

cilloscope check is made at the grid of the output tube, to observe the waveshape. If incorrect (tube in or out), the trouble is probably defective components in the waveshaping circuits. If the wave shape is corrected by removal of the output tube, supply voltages to the output tube, and components in the feedback afc circuits are to be investigated.

One final cause of ringing is an open or wrong-value balancing condenser in the horizontal deflection yoke winding. If changing this condenser reduces the ringing but does not eliminate it, the following procedure should accomplish complete elimination. Clip a 10,000-ohm test potentiometer in series with the condenser, as shown in Fig. 6-34. Adjust the potentiometer until the ringing bars disappear. Use an ohmmeter to measure the resistance, and permanently solder in a 1-watt resistor of this value.

Servicing Procedure for Barkhausen Oscillations. Another oscillatory condition which cause a thin dark vertical bar or bars on the left side of the picture is called Barkhausen oscillations or "snivets." This symptom is illustrated in Fig. 6-35.

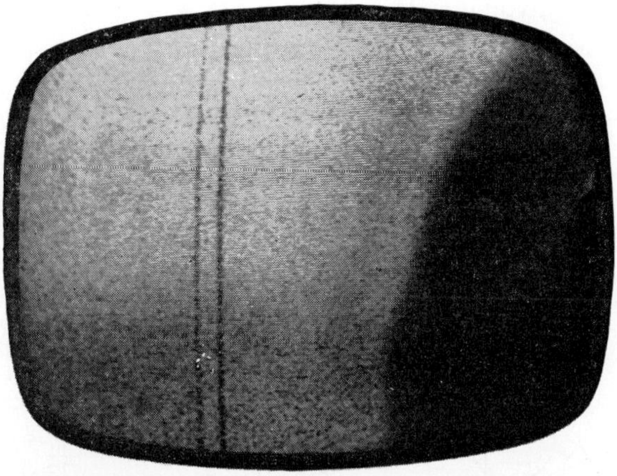

Fig. 6-35. Barkhausen oscillation.

It is a high-frequency oscillation generated in the horizontal output tube and radiated from the plate lead. If the radiated energy gets into the r-f portions of the receiver, the bar results. It is especially noticeable on weak stations.

The following suggestions have proved helpful in clearing up this complaint.

1. Try several different horizontal output tubes.
2. Dress the antenna leads away from the horizontal section.
3. Try a Barkhausen eliminator. This is an ion trap specially made to fit

over the output tube. If it will fit, an ordinary picture-tube single ion trap may be used. It is adjusted by rotation until the bars disappear.

4. Try a suppressor resistor in the plate lead of the horizontal output tube. Use a 1-watt resistor of approximately 100 ohms connected at the plate clip of the tube. Shorten the original lead for good wire dress.

5. If the bars are noticeable only on weak stations, inspect the antenna installation to see if it is possible to get more signal strength on these stations. If conditions warrant it, a signal booster may be recommended.

SUMMARY

Symptom. No light on the screen of the picture tube.

Procedural checks quickly isolate the trouble to the output, damper, and high-voltage stages, or to the oscillator-afc stages. A procedure for servicing the oscillator-afc stages is given in the next chapter. The servicing procedure for no high voltage is also used for conditions of low-high voltage.

Standard Circuit. The schematic diagram of basic horizontal output, damper, and high-voltage stages is given in Fig. 6-36.

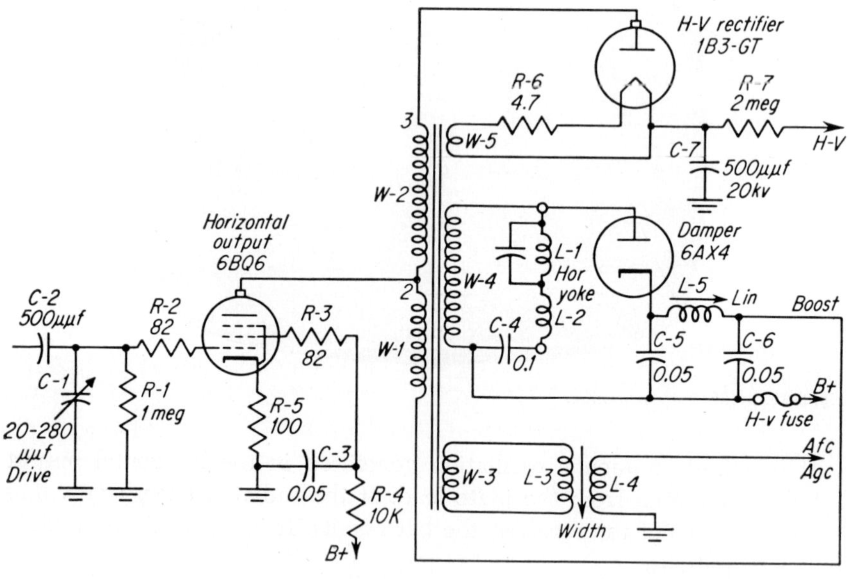

Fig. 6-36. Basic circuit diagram of the horizontal output, damper, and high-voltage stages.

Service Data Chart for the Horizontal Section

Trouble: No light on the screen of the picture tube.

Procedure	*Result*	*Trouble and Subsequent Check*
1. Check voltage at second anode of the picture tube	12–18 kv	Check brightness circuit Check ion-trap adjustment Check picture tube Check position of width shim
	0–2 kv	Leave the meter connected as an indicator, and proceed to step 2
2. Check trace at grid of output tube	Incorrect trace	For low amplitude check setting of drive control. Then proceed to step 3
	Normal trace	Proceed to step 4
3. Remove output tube and recheck trace at grid of output tube	Trace remains incorrect	Trouble is in the oscillator-afc stages (see Chap. 7)
	Trace is restored to normal	Check output tube by substitution Make voltage and resistance checks in output tube cathode, grid, and screen circuits Check cathode and screen bypass condensers Check for trouble in flyback coupling to afc stage Check for trouble in decoupling filter in boost line
4. Check output tube by substitution	High voltage comes on	Defective output tube
	No change	Proceed to step 5
5. Make voltage and resistance checks in output tube cathode and screen circuits	Abnormal condition	Correct condition and check for high voltage
	Normal condition	Proceed to step 6

Service Data Chart for the Horizontal Section (Continued)

Procedure	Result	Trouble and Subsequent Check
6. Remove plate cap from output tube and check voltage at plate lead	Plate volts = zero	Check for: open high-voltage fuse, output transformer, linearity coil; linearity condenser shorted to ground; dead damper tube
	Plate volts = B plus value	Proceed to step 7
7. Replace plate cap and check damper plate and cathode voltages	Damper plate = normal (250 V) Damper cathode = normal (480 V)	Check high-voltage stage (step 13) In direct-drive flyback-yoke system, check for open yoke
	Damper plate = low Damper cathode = low	Check low-voltage power supply (see Chap. 4)
	Damper plate = normal Damper cathode = low	Check for K-H short in damper tube Check for insulation breakdown in separate damper-heater winding in power transformer
	Damper plate = normal Damper cathode = approximately same voltage as plate	Proceed to step 8
8. Remove damper tube	High voltage comes on	Check for a shorted linearity condenser
	No change	Proceed to step 9
9. Replace damper with known good tube	High voltage comes on	Damper tube is weak
	No change	Proceed to step 10
10. Make ohmmeter check of the boost line	Boost line to ground shows a reading of 30,000 ohms or less	Check the boost feed lines for shorts or leakage
	Boost line to ground shows a normal reading of a condenser charging in a high-resistance circuit	Proceed to step 11

Service Data Chart for the Horizontal Section (Continued)

Procedure	*Result*	*Trouble and Subsequent Check*
11. Connect plate of output tube to flyback lead in a working set. See Fig. 6-21	Output tube will not produce high voltage in the test set	Recheck components in cathode and screen circuits of output tube
	High voltage appears in test set	Proceed to step 12
12. Connect high voltage from test set to second anode in dead set. Reset ion trap	Normal raster	Check high-voltage stage (step 13)
	Vertical line	Check for open yoke winding
	Keystone raster	Check for short in yoke winding
	Foldover	Check damper circuit especially for open linearity condenser
	Narrow raster (may be accompanied by foldover)	Flyback transformer loaded. Check for: misplaced width shim, short in width coil, short in afc or agc circuit If there is no external loading replace the flyback transformer
13. Check for arc at plate of high-voltage rectifier	Normal arc	Check high-voltage rectifier Check filter resistor for open Check filament resistor
	No arc	Proceed to step 14
14. Remove plate lead from high-voltage rectifier and recheck for arc	No arc	Check high-voltage winding on flyback for open
	Normal arc	Check for short in high-voltage filter condenser

Service Data Chart for Other Troubles in the Output, Damper, and High-voltage Stages

Symptom	*Illustration*	*Look for*
Insufficient width		Low line voltage Weak low-voltage power supply Inefficiency in the horizontal system. Check as for dead horizontal system. Look for lower-than-normal readings at all check points in the previous procedure for no raster
Excessive width, drive line may be present		Misadjusted horizontal controls Deflection yoke back from the funnel of the picture tube Install drive control if set does not have one
Poor horizontal linearity	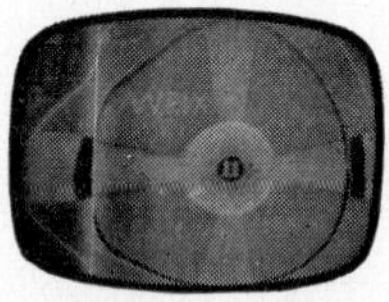	Misadjusted horizontal controls (correct adjustment of the linearity control may reduce width. Then service for insufficient width)
Blooming		Low high voltage Incorrect cathode or grid voltage for picture tube Gassy picture tube
Foldover and ringing		Defective damper tube and linearity condensers if on left side Incorrect waveshape if on right side Critical balancing circuit in horizontal yoke winding
Barkhausen oscillation	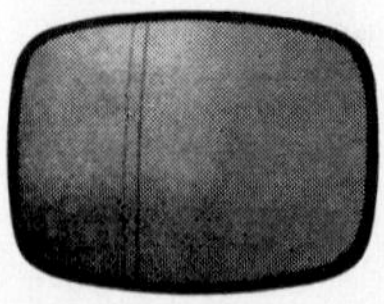	Defective output tube Lead dress on plate lead of output tube, and on antenna leads Try an ion trap on output tube

7

Servicing the Deflection Section: Horizontal Oscillator and AFC Stages

In the previous chapter, a procedure was presented for tracing the cause of a lack of high voltage to a defective component in the output, damper, or high-voltage stages in the horizontal section. It is the purpose of this chapter to find the cause of the same complaint if it is located in the horizontal oscillator or afc stages.

The line of demarcation between these two subsections is the grid of the output tube. If the waveform shown in Fig. 7-1 is present at this point, correct as to waveshape, frequency, and amplitude, the horizontal

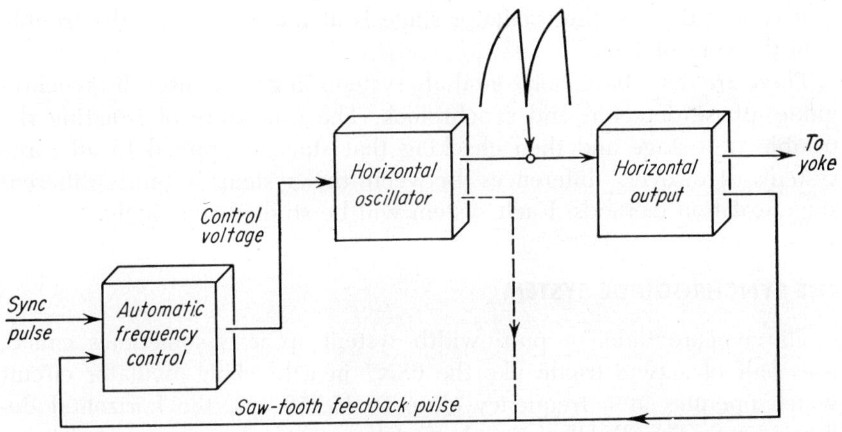

Fig. 7-1. The position of the oscillator and afc stages in the horizontal deflection system.

oscillator and afc stages are working, and the procedures of Chap. 6 were applied to find the trouble in the output, damper, and high-voltage subsection. Now, we do not find this correct trace and therefore must investigate the oscillator-afc stages for the cause.

If the oscilloscope check reveals off-frequency operation, adjustment of the frequency control in the receiver may be all that is required. Or it may give a lead as to the cause of trouble. When the oscilloscope check displays low or no output or an incorrect waveform, service analysis of the oscillator-afc stages must be made.

In case of low or no output from the oscillator, the output tube is likely to glow because of the heavy current drain under conditions of no drive. Continued operation in this condition may harm the tube or output transformer. To prevent this, it would be advisable to remove the output tube until the oscillator is restored to normal operation. The oscilloscope is left connected to the grid connection of the output tube socket to indicate when normal operation is restored.

The oscillator generates a voltage at the horizontal deflection frequency of 15,750 cps. As was brought out previously, stable operation at this frequency requires the use of a control stage. Here the sync pulse and a sawtooth feedback pulse from the oscillator are combined to produce a control for the oscillator stage. The feedback pulse may be taken from the oscillator stage, the output stage, or both.

In the servicing procedure to be presented, the standard technique of isolating the trouble to a stage and then checking that stage will be used. Stated briefly, the scheme is to remove the control voltage from the oscillator stage. Then, as a free-running oscillator, it should drive the deflection circuits well enough to reestablish the high voltage and a full raster. If it cannot do this, the oscillator stage is at fault; if it can, the trouble is in the control stage.

There are three basic horizontal-afc systems in general use: the synchroguide, phase detector, and synchrolock. The procedure of isolating the trouble to a stage and then checking that stage is applied to all three systems. However, differences between the systems require different stage-isolation methods. Each system will be studied separately.

THE SYNCHROGUIDE SYSTEM

The synchroguide, or pulse-width system, as it is sometimes called, uses half of a twin triode like the 6SN7 in a blocking-oscillator circuit which operates at a frequency close to 15,750 cps, the horizontal deflection rate. Usually, the other half of the same tube is connected in an afc circuit, where the sync pulse and oscillator pulse together provide a control which stabilizes the oscillator frequency at 15,750 cps, in spite of noise pulses and oscillator drift.

The Blocking Oscillator in the Synchroguide System. The blocking oscillator in the synchroguide circuit has been redrawn in Fig. 7-2*b* to resemble the familiar vertical blocking oscillator which is drawn above it for comparison purposes. The same component-numbering system has been used in both circuits to heighten the comparison. Thus *T*-1 in both circuits is the oscillator transformer. In the vertical circuit, it operates at 60 cps, whereas in the horizontal circuit, it is slug-tuned to operate at 15,750 cps. Condenser *C*-2 is the oscillator grid condenser in both circuits.

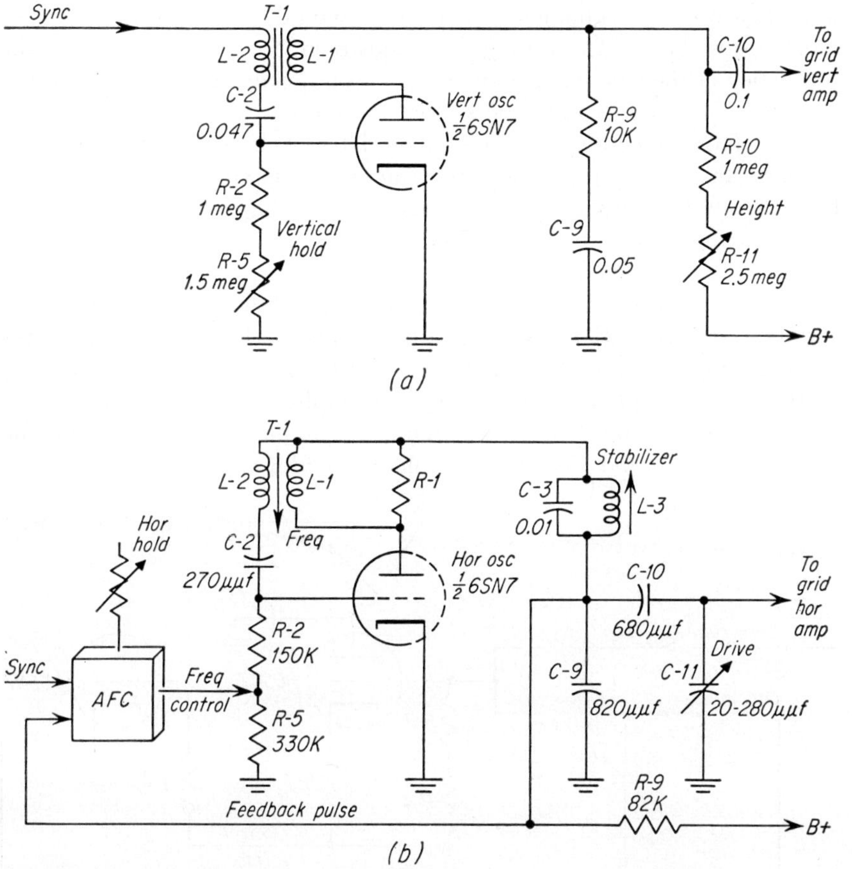

Fig. 7-2. Blocking oscillators. (*a*) Vertical. (*b*) Horizontal.

Note that grid-leak resistor *R*-5 is the hold control in the vertical circuit. The analogous grid resistor in the horizontal circuit has the control voltage impressed across it from the afc stage. In the vertical circuit, the sync pulse is applied directly to the oscillator grid, whereas the horizontal sync pulse is applied through the control stage.

Resistor *R*-1 across plate coil *L*-1 in the horizontal oscillator serves to damp shock oscillations which occur in this type of blocking oscillator. A similar resistor is sometimes found in vertical oscillators.

Resistor *R*-9 and condenser *C*-9 in the plate circuit of the vertical oscillator have a waveshaping function. Condenser *C*-9 serves a similar function in the horizontal oscillator.

Coil *L*-3 and condenser *C*-3 in the horizontal plate circuit also have a waveshaping function. A coil and condenser connected in parallel set up a sine-wave oscillation between them at their resonant frequency. This imparts a sine-wave characteristic to the sawtooth pulse generated by the blocking oscillator. The tuning-slug adjustment for coil *L*-3 controls the frequency of the sine-wave oscillation. This adjustment is best carried out with the aid of an oscilloscope so that the correct waveform may be obtained. This procedure will be given later. When the adjustment is properly made, the addition of the sine-wave characteristic to the sawtooth generated by the oscillator has a stabilizing effect, in that noise pulses are not likely to trigger the oscillator. It is because of this that coil *L*-3 is often called the "stabilizing" coil. In both circuits, the output of the oscillator is coupled to the grid of the following amplifier tube through condenser *C*-10. Note also the similarity in the position of the drive controls. In the vertical circuit, resistor *R*-11 controls the voltage to the oscillator plate, and thereby the amplitude of signal fed to the amplifier grid. In the horizontal circuit, the oscillator signal is fixed, and condensers *C*-10 and *C*-11 form a voltage divider which feeds a variable

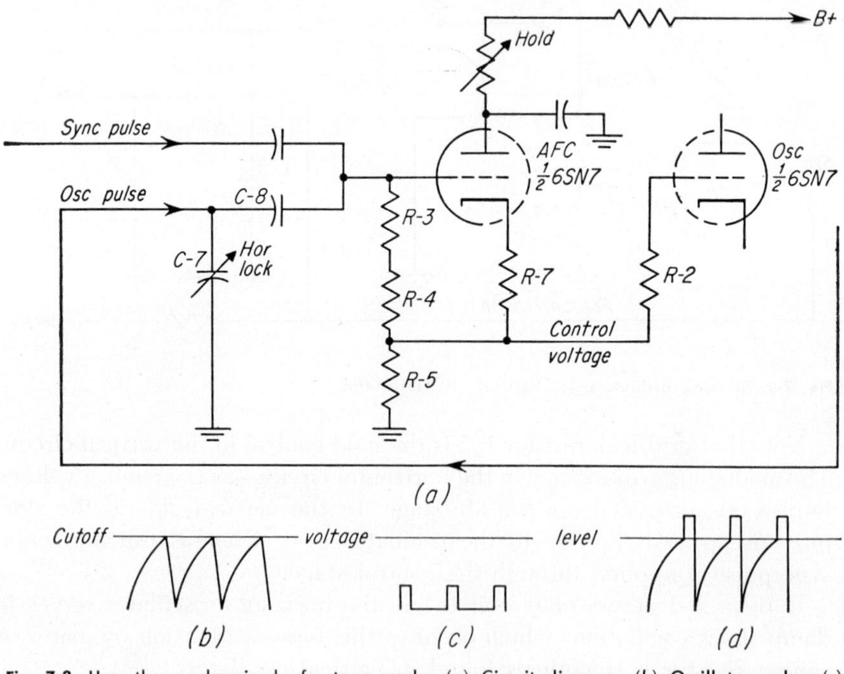

Fig. 7-3. How the synchronized afc stage works. (a) Circuit diagram. (b) Oscillator pulse. (c) Sync pulse. (d) Combined input.

amount of signal to the horizontal amplifier grid. Condenser *C*-11, the horizontal drive control, is not used to control width, as its companion control in the vertical circuit is used to control height. It must be remembered that the horizontal system has a secondary role of providing high voltage for the picture tube. While the drive control will control width, its primary function is to provide sufficient signal to drive the horizontal amplifier to produce a large high-voltage output. The actual width control in the horizontal system is an adjustment on the output transformer.

The AFC Stage in the Synchroguide. The afc stage in the synchroguide consists of a triode biased to cutoff by the sawtooth feedback pulse from the oscillator. The circuit is shown in Fig. 7-3*a*. Condenser *C*-8 is the grid condenser, and resistors *R*-3, *R*-4, and *R*-5 comprise the grid leak. Grid-leak action establishes the bias voltage on the tube. Condenser *C*-7 acts in a voltage-divider circuit to control the amplitude of the sawtooth input signal so as to establish a cutoff bias. This voltage level is indicated in Fig. 7-3*b*.

At the same time, the horizontal sync pulse from the transmitted signal is applied to the same grid. The sync pulse is indicated in Fig. 7-3*c*. The

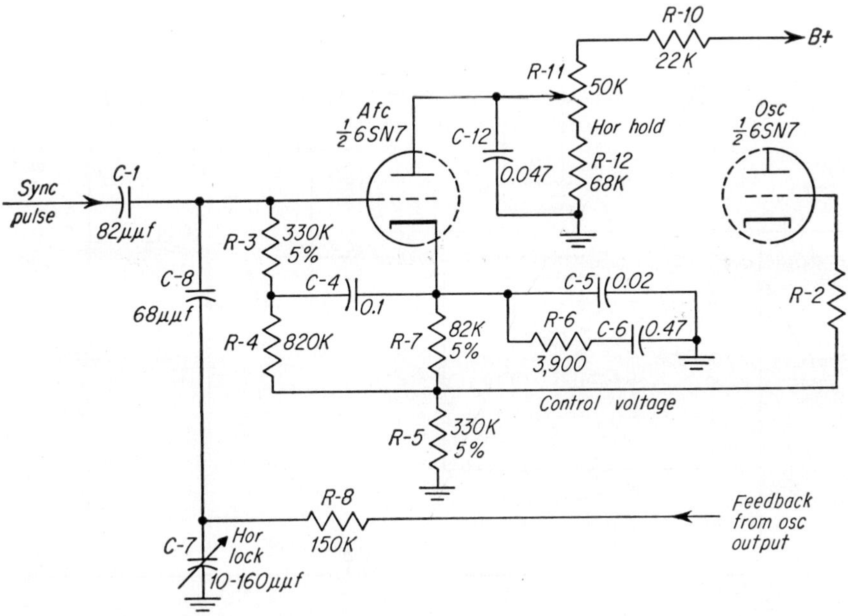

Fig. 7-4. Afc stage in the synchroguide circuit.

two voltages add, and if they are timed so that the point of the sawtooth occurs at the same time as the center of the sync pulse, the addition is above the cutoff level, as shown in Fig. 7-3*d*. Now the tube draws plate current, establishing a voltage across resistor *R*-5 in its cathode circuit, proportional to this current flow. This is the normal operating condition of the circuit.

Coming back to the blocking-oscillator circuit, its frequency may be controlled by varying the voltage applied to its grid. The voltage across resistor *R*-5 is applied to the oscillator grid by making it part of the oscillator grid leak as well. Now if the oscillator should change frequency slightly, the sawtooth feedback voltage would shift somewhat from the correct additive position with the sync pulse. This causes change in the afc-tube plate current, producing a change in the voltage across resistor *R*-5. This changes the bias on the oscillator tube, which in turn corrects the oscillator frequency drift.

The complete afc stage is shown in Fig. 7-4. Note that the sync pulse is fed to the afc grid through condenser *C*-1, and the sawtooth wave from the oscillator is fed in through condenser *C*-8. Note the use of 5 per cent resistors in the grid and cathode circuits to ensure an accurate

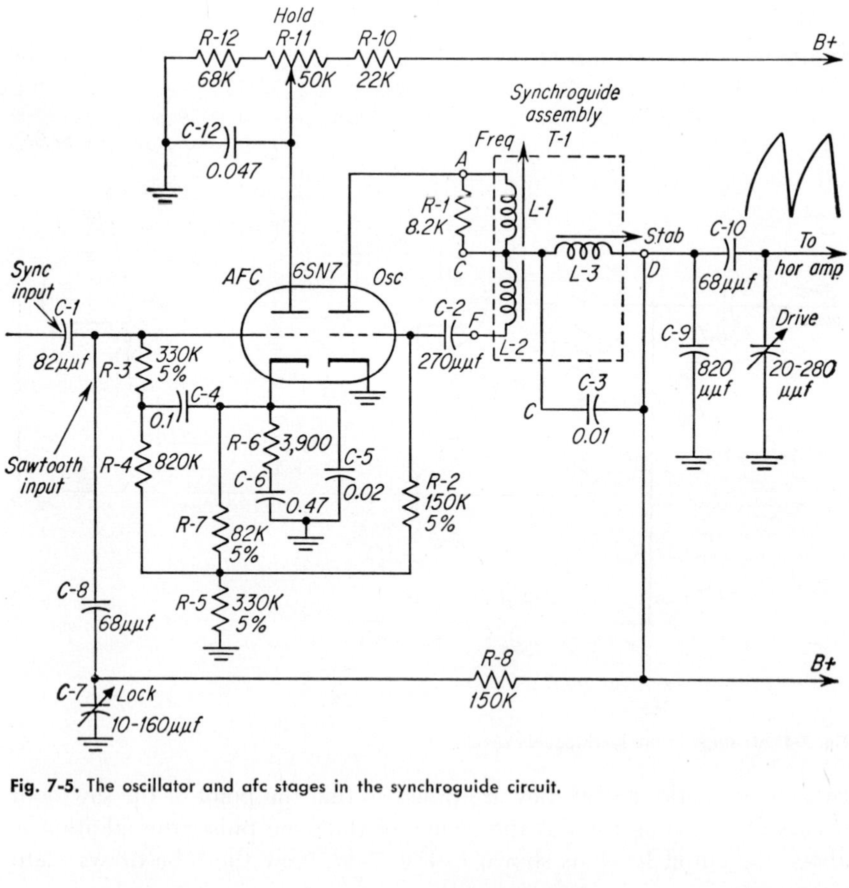

Fig. 7-5. The oscillator and afc stages in the synchroguide circuit.

voltage distribution for the control voltage to the oscillator grid. The cathode bypass network consists of condenser *C*-5, resistor *R*-6, and condenser *C*-6. This stabilizes the cathode voltage and sets limits on the control voltage swing, thereby forming an antihunt circuit. Condenser *C*-4 and resistor *R*-4 make up a decoupling network which tends to keep the sync pulses out of resistor *R*-5 and therefore out of the oscillator grid. Plate voltage for the afc tube is supplied through a voltage divider consisting of resistors *R*-10 and *R*-12, and potentiometer *R*-11. Adjusting potentiometer *R*-11 gives a variation in plate voltage from roughly 125 to 200 volts. Condenser *C*-12 serves as a bypass condenser. Adjusting the plate voltage on the tube affords a control of plate current, which in turn provides some control over the voltage across resistor *R*-5 in the cathode circuit. This gives a minor control of oscillator frequency through the control circuit, thereby making *R*-11 the manual horizontal-hold control. The complete circuit of the synchroguide system showing both oscillator and afc stages is given in Fig. 7-5.

Identification of a Synchroguide Circuit. When the servicing procedure calls for investigation of the horizontal oscillator-afc stages, it is important to know which type of circuit is being used. The synchroguide may be identified by the coil assembly, either on the schematic diagram or on the receiver chassis. On the receiver diagram, look for the characteristic three-coil arrangement at the horizontal oscillator tube which is reproduced in Fig. 7-6*a*. On the receiver, look for the aluminum can usually mounted on the rear apron near the high-voltage r-f cage. Then

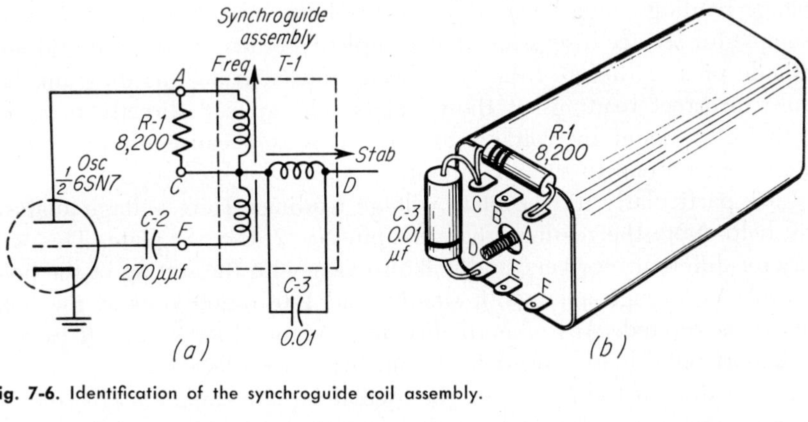

Fig. 7-6. Identification of the synchroguide coil assembly.

look for the condenser (*C*-3) and resistor (*R*-1) which are soldered directly to the terminal lugs on the can as shown in Fig. 7-6*b*. Point *C*, the junction of the coil and condenser, is an important check point.

Sometimes, the assembly can is not used, and the two unshielded coils are mounted separately on the chassis. In this case, the schematic diagram is the only sure identification.

Servicing Procedure for a Defective Synchroguide Circuit. Before isolating the trouble to a defective oscillator or afc stage, it may save time on the service bench if the tube and certain key voltage and resistance checks are made first.

Tube Replacement Notes in the Synchroguide. The tube most often used in the synchroguide circuit is the twin triode 6SN7-GT type, or its miniature equivalents, the 12AU7, 7AU7, or 6CG7. Sometimes, the 6SN7-GTB tube is used. This is similar to the 6SN7-GT, except for the heater which is designed for a controlled warm-up time, for use in series filament strings. The tubes are usually interchangeable, at least for test purposes. The first step in the servicing procedure for a defective synchroguide is to try operation with a tube known to be good. If this restores the high voltage, the repair is probably completed.

It was just mentioned that the various types of 6SN7 tubes were interchangeable for test purposes. However, if the repair is accomplished by a tube change, make sure to use the correct type for replacement. This is especially true for the 6SN7-GTB in series filament receivers because of the controlled heater warm-up time.

Voltage and Resistance Checks in the Synchroguide. If tube substitution does not restore operation, the next step is to make a voltage check of the tube pins. Use the manufacturer's service notes for normal readings, since there is considerable variation in this regard. Representative voltage readings have been added to the basic circuit in Fig. 7-7 as an example for service discussion. The complete circuits around the grid and cathode pins of the afc tube have been omitted from this diagram, because incorrect readings at these points seldom lead directly to a defective component by voltage analysis. This is because of the voltage interaction normal to the afc circuit.

Look particularly for the plate voltage readings. Here voltage analysis will help. Note the reading given for pin No. 2, the afc plate. This will vary for different receivers, and will also vary with the setting of the hold control. As a rough average, it should read 100 to 200 volts as the hold control is rotated. Any abnormality here points to resistors *R*-10 and *R*-12, horizontal hold control *R*-11, and bypass condenser *C*-12.

The reading at the No. 5 pin, the oscillator plate, calls for 175 volts. It is, of course, assumed that the voltage at B plus has been checked and found normal. Also, it must be remembered that in some receivers the B boost line is used to supply voltage to the afc plate, the oscillator plate, or both. Since the output stage is not functioning normally because of lack of drive, the boost line is reduced to approximately the B plus value.

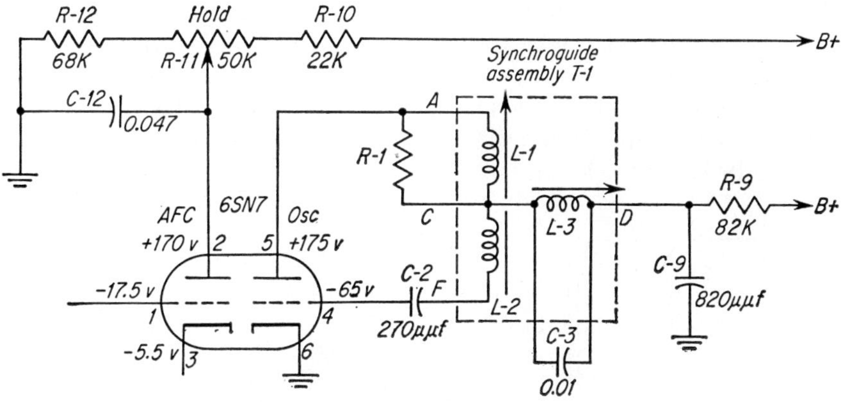

Fig. 7-7. Voltage and resistance checks in the synchroguide circuit.

Resistance data for synchroguide assembly *T*-1

A to C	Plate coil *L*-1	35 ohms
C to F	Grid coil *L*-2	90 ohms
C to D	Stabilizer coil *L*-3	35 ohms

In these sets, therefore, a low voltage reading at the plate must be taken as normal until the condition has been cleared.

Taking this exception into account, an incorrect reading at the oscillator plate calls for an examination of the line to the B plus source. This includes the coils in the synchroguide assembly, resistor *R*-9 and wave-shaping condenser *C*-9. Resistance data for the coils in the synchroguide assembly can are given in Fig. 7-7.

Voltage analysis may point to one more defective component: oscillator grid condenser *C*-2. With an inoperative oscillator, the normal voltage reading of −65 volts is not likely to be present at the No. 4 oscillator grid pin. However, a positive voltage reading here may indicate leakage in grid condenser *C*-2. To check the condenser for leakage, open the connection to the grid, and check voltage from the opened condenser lead to ground. A positive voltage indication now means leakage in the condenser, which is then replaced.

If the condenser proves to be good, with the condenser lead still open, remove the tube, and check for voltage at both grids (pins No. 1 and No. 4). Any positive voltage present now is coming from the sync line or the sawtooth feedback line. Check condensers *C*-1 and *C*-8 (see Fig. 7-5) for leakage.

Synchroguide Oscillator Stage Check. If the voltage check gives no clue as to the cause of oscillator failure, the automatic frequency control is removed from the oscillator to see if it will drive the deflection circuits in a free-running condition. This is the check for the oscillator stage.

The procedure is as follows. Remove all connections from the oscillator grid, except the grid condenser. This removes the control voltage. Then connect a 2-meg potentiometer from the grid to ground. A test potentiometer equipped with clip leads like the one shown in Fig. 7-8 is handy for this use. The circuit is now as shown in the diagram, and we have a free-running blocking oscillator.

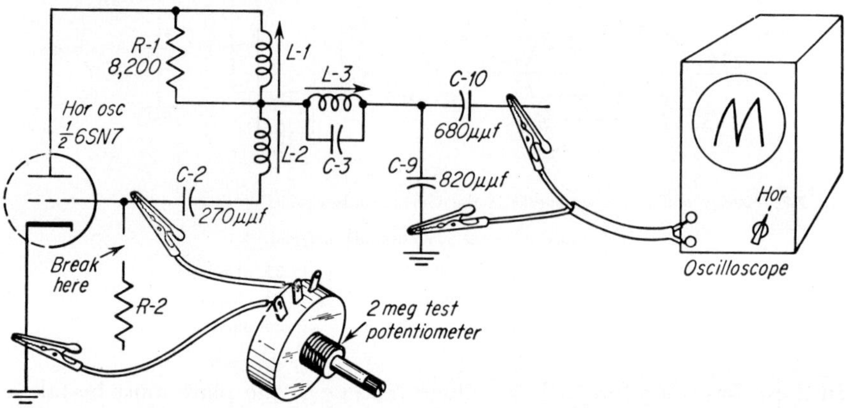

Fig. 7-8. Operating the synchroguide as a free-running blocking oscillator.

The test potentiometer is now the main frequency control, and careful adjustment should bring the oscillator close to 15,750 cps. Check to see that the oscilloscope is at its marked positions for observing the trace at the grid of the output tube, and connect it as shown in Fig. 7-8. Then slowly adjust the test potentiometer while looking for the correct trace on the oscilloscope.

If the oscillator is in working order, the correct trace will appear. Confirm the restoration of operation by replacing the output tube. Reset the ion trap, if needed. A raster should appear on the screen of the picture tube. The oscillator is now working and the trouble is in the control stage.

If the correct trace and raster do not appear, the trouble is in the oscillator stage. Check all components in the stage. Do not neglect to check resistor *R*-1 for change in value and condenser C-3 and C-10 for opens. Finally, do not overlook the possibility of shorted turns in the coil assembly which do not affect the resistance readings. Replacement of the coil assembly is the best way of checking for this. An exact replacement should be used.

When components are changed, the frequency and afc controls should be readjusted. A complete adjustment procedure for the synchroguide is given in Chap. 8.

Checking the Sync Pulse. With the oscillator operating as a free-running blocking oscillator, and a raster on the screen of the picture tube, connect an antenna to the receiver, and by careful manipulation of the test potentiometer, resolve the display on the screen into a picture.

Most likely, the picture thus obtained will not hold steady but will tear away as the oscillator drifts off frequency. It has been proved, however, that the oscillator is capable of operating at 15,750 cps, even if momentarily.

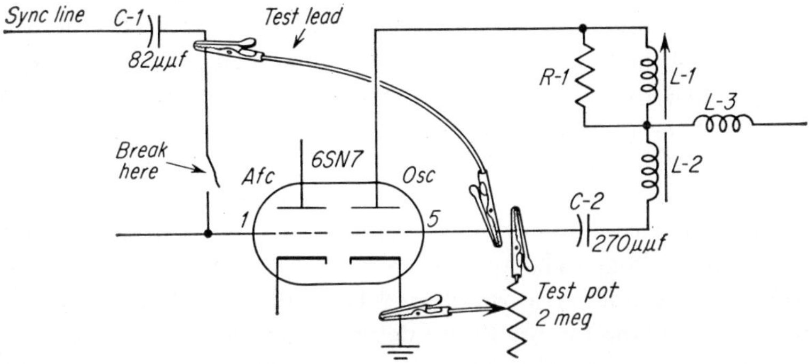

Fig. 7-9. Applying the sync pulse to the free-running horizontal oscillator.

Now remove the horizontal sync line from the No. 1 grid pin of the afc tube, and make a temporary connection to the oscillator grid at the No. 5 pin. The setup is shown in Fig. 7-9. If this causes the screen to go dark, there is trouble in the sync line, probably leakage in condenser *C*-1. If a normal horizontal sync pulse is present, the picture will hold, and the trouble is in the afc stage.

Another possibility must be mentioned at this point. The receiver may have two troubles at once, loss of sync and loss of high voltage. The former trouble is unknown to the serviceman because it cannot be seen on the dark screen. Now that the high voltage is restored, the loss of sync will be noticed because the picture cannot be made to hold. The cause for lack of high voltage is still in the afc stage, but the set must also be serviced for loss of sync. This is taken up in Chap. 8.

Servicing the AFC Stage. Having isolated the trouble to the afc stage, the problem becomes simply a check of the resistors and condensers in the stage. Refer to the manufacturer's service notes for the values of these components. Note that some of the resistors have a 5 per cent tolerance rating, and so check these carefully. As for the condensers, remember to open one lead when checking for leakage, and to substitute a test condenser of the correct value when checking for change in capacitance.

It may happen that all components seem to check normal. This is because many parts may change value under heat and operating conditions. This makes a resistance check of little value. In actual service practice it is expedient to replace all components in the stage. This procedure may be used in trouble shooting any of the afc systems when localizing tests place the trouble in that stage.

THE PHASE-DETECTOR OSCILLATOR AND AFC SYSTEM

In the phase-detector system, both halves of a twin triode like the 6SN7-GT are used as a multivibrator type of oscillator operating at 15,750 cps. The oscillator is kept at this frequency by a control voltage which is established by the action of the generated signal and the sync pulse through a twin diode tube called a phase detector. The sync input is taken from the plate and cathode of the last sync amplifier. These two pulses, equal in amplitude but opposite in phase, are then fed to the two diodes. Any phase difference between the sync pulses and the feedback signal produce a voltage in this discriminator-type detector. This voltage is the control voltage. The phase-detector system owes its name to the fact that a phase difference is detected to produce the control voltage.

The Oscillator Stage in the Phase-detector System. The horizontal oscillator in the phase-detector type of oscillator-afc system is a cathode-coupled multivibrator similar to the one used in vertical deflection circuits and described in Chap. 5. The diagram is shown in Fig. 7-10.

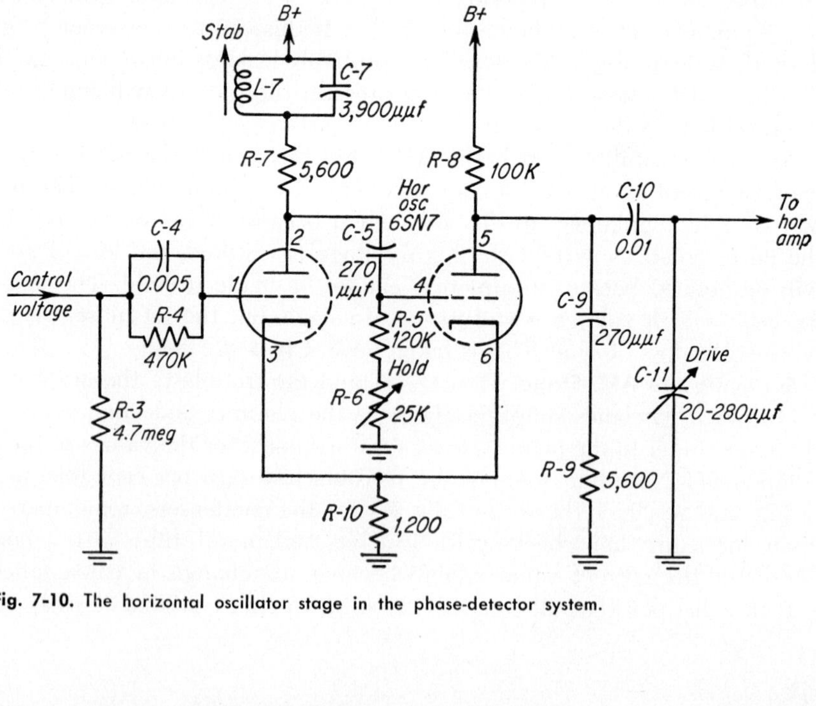

Fig. 7-10. The horizontal oscillator stage in the phase-detector system.

The tube most commonly used is a 6SN7-GT twin triode, or its miniature equivalents, the 12AU7 or 7AU7. Both halves of the tube are used in the multivibrator type of oscillator circuit. Note that both cathodes are tied together, making resistor *R*-10 common to both tube circuits. This is the feedback arrangement between the tubes, making this a cathode-coupled multivibrator.

Another familiar circuit is the coupling to the horizontal amplifier tube. Resistor *R*-8 in the plate of the second tube (pin No. 5) is the plate load. Condenser *C*-9 and resistor *R*-9 form the waveshaping circuit, and condenser *C*-10 is the coupling condenser. Condenser *C*-11 is the horizontal drive control, bypassing a portion of the oscillator signal to ground. The values shown in the diagram are representative.

The frequency controls are located in the coupling between the two tubes of the multivibrator. The value of the grid leak consisting of resistors *R*-5 and *R*-6 determines the discharge rate of coupling condenser *C*-5, and therefore the frequency of the oscillator. Resistor *R*-6 is variable and is the horizontal-hold control.

The components in the plate circuit of the first tube (pin No. 2) have a function which is partly stabilizing and partly frequency-controlling. The plate load consists of resistor *R*-7 in series with the parallel combination of coil *L*-7 and condenser *C*-7. Here again we have a parallel *L-C* circuit which, when shock-excited, will produce a sine-wave oscillation between the coil and condenser at their resonant frequency. This adds a sine-wave characteristic to the normal waveform of the multivibrator. As in the synchroguide circuit, the sine-wave characteristic is added for stabilization purposes. However, in the stabilized multivibrator, the coil is called a "ringing" coil. It has a tuning slug which makes the resonant frequency adjustable. The adjustment procedure is given in Chap. 8. The resonant frequency of coil *L*-7 and condenser *C*-7 also has an effect on the frequency of the multivibrator. In some receivers, the grid leak is kept at a fixed value, and the tuning slug in the ringing coil is used as the horizontal-hold control.

The grid of the first triode (pin No. 1) may be considered as the frequency-control grid of the multivibrator. If this grid were shorted to ground (0 volts), the multivibrator would generate a signal, the frequency of which is governed by the values in the coupling between the two triodes. A small voltage impressed on this grid has the effect of changing the oscillator frequency. A small positive voltage decreases, and a small negative voltage increases, the operating frequency of the multivibrator. The grid is connected to the afc stage which produces a correction voltage if the oscillator frequency should drift.

The AFC Stage in the Phase-detector System. The diagram of the discriminator-type phase detector commonly used to provide the control voltage for the multivibrator is shown in Fig. 7-11. Components in the oscillator and amplifier stages not needed for this discussion have been omitted to simplify the diagrams.

The sync pulses, equal in amplitude and opposite in phase, are taken from the phase splitter and fed to the twin diodes of the 6AL5 discriminator or phase detector as shown in the diagram. Diode load resistors *R*-1 and *R*-2 are made equal to balance the circuit. Because of the balanced circuit, the potential difference between points *A* and *B* is zero. With only sync pulses, therefore, the potential applied to the No. 1 control-grid pin of the multivibrator is zero.

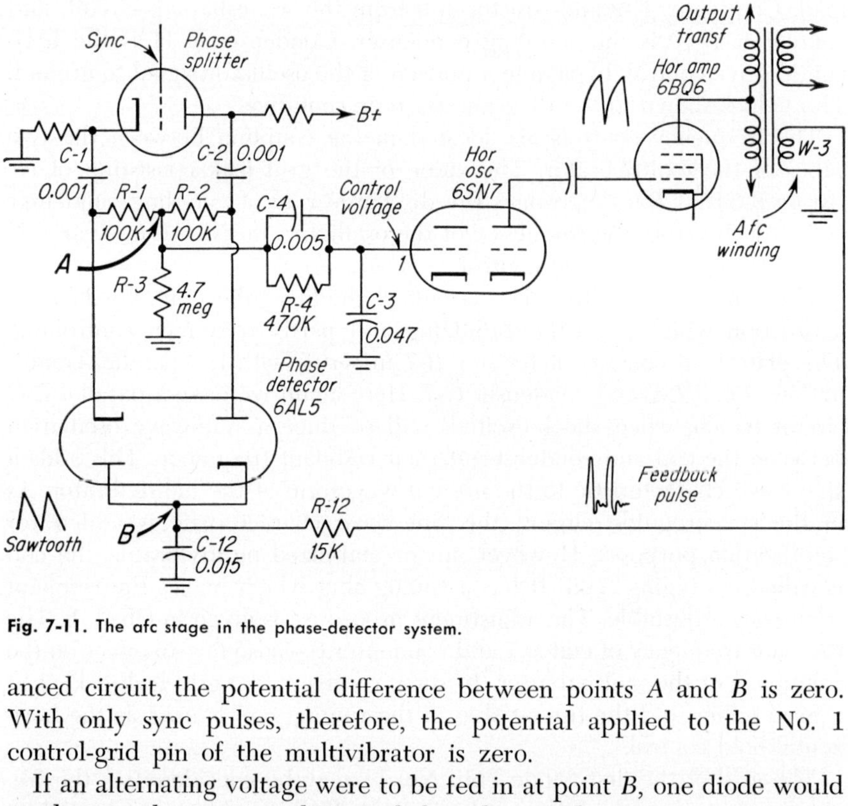

Fig. 7-11. The afc stage in the phase-detector system.

If an alternating voltage were to be fed in at point *B*, one diode would conduct on the positive phase and the other on the negative, again producing zero output at point *A* through the balanced circuit. This alternating voltage is obtained from winding *W*-3 of the horizontal output transformer, as shown in Fig. 7-11. Resistor *R*-12 and condenser *C*-12 form a waveshaping circuit which changes the output pulses to a sawtooth wave. Again, because of the balanced circuit, the sawtooth wave alone produces zero bias on the No. 1 control grid of the multivibrator.

When the multivibrator is at its correct frequency, the output pulses are occurring at the same time as the sync pulses, and the discriminator output voltages balance out for a zero control voltage. But if the multivibrator should speed up or slow down, the sawtooth wave from the output transformer will be out of step with the sync pulses. This changes the phase relationship of the combined voltages applied to the diodes. One will conduct more than the other, creating a potential at point *A* which is then applied to the oscillator control grid at pin No. 1 as a correction voltage. A slowed-down oscillator produces a negative correction voltage which will increase its frequency. If the oscillator frequency increases, the discriminator will produce a positive correction voltage which will decrease the oscillator frequency.

The correction voltage is small—approximately 1 volt positive or negative being sufficient to keep the oscillator in step with the sync pulse for normal reception conditions. Resistor *R*-4 and condensers *C*-3 and *C*-4 in

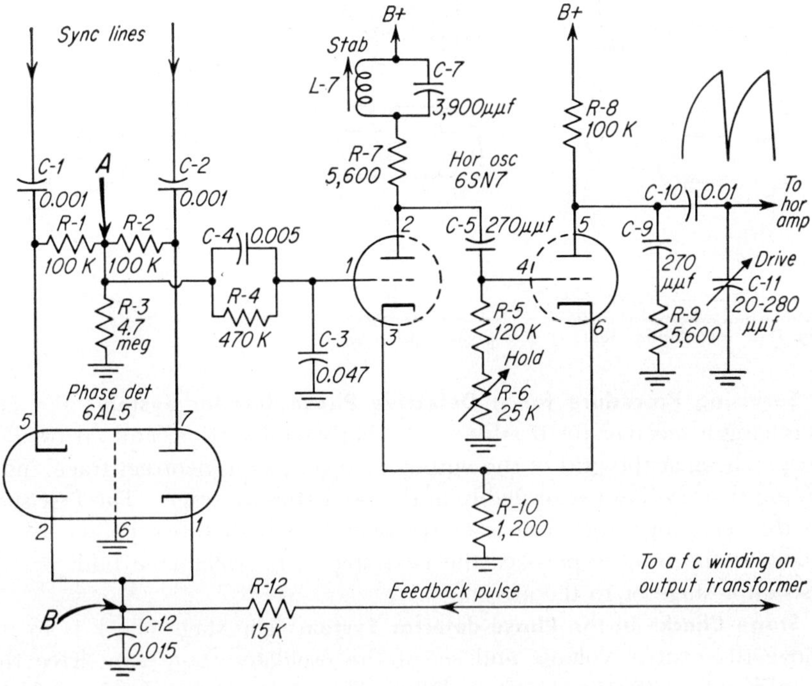

Fig. 7-12. Schematic diagram of a phase detector, stabilized multivibrator, afc oscillator circuit.

the control-grid circuit of the oscillator act as a filter on the correction voltage. The combined multivibrator-discriminator circuit of the phase-detector system is given in Fig. 7-12.

Identification of a Phase-detector System. In the schematic diagram, the phase-detector system is recognized by the use of the twin diode 6AL5 or 6H6 in the horizontal section. The multivibrator oscillator circuit is identified by the cathodes being tied together and by the ringing coil in the plate circuit.

On the chassis, the circuit is not so easily recognized. The ringing coil *L*-7 looks very much like a width or linearity coil. However, condenser *C*-7 is usually a mica condenser soldered directly across the coil terminals, and this may aid in identification. Also, the very small 6AL5 or 6H6 is easily identified in the horizontal section. On the underside of the chassis, look for the two 100,000-ohm resistors on pins No. 5 and No. 7. These are resistors *R*-1 and *R*-2. The junction of resistors *R*-1, *R*-2, and *R*-3 is an important test point. The identification features are shown in Fig. 7-13.

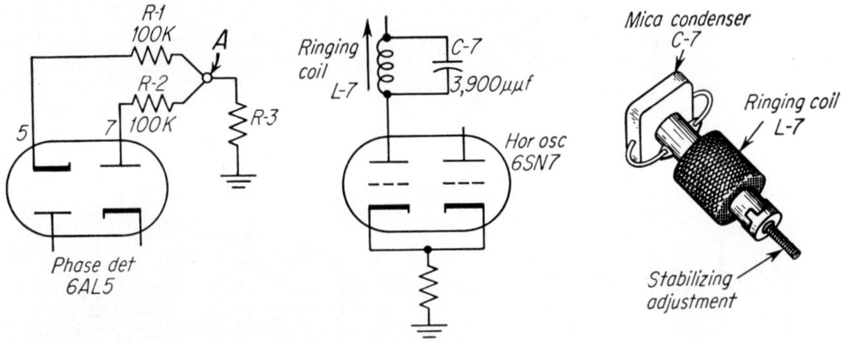

Fig. 7-13. Identification features of a phase-detector system.

Servicing Procedure for a Defective Phase-detector System. We are servicing a receiver for trouble in the horizontal section, and an oscilloscope check at the grid of the output tube shows an incorrect trace, indicating that the fault is probably in the oscillator-afc stages. The first step in the servicing procedure is to try operation with tubes known to be good. If the trouble persists, the next step is to isolate the fault to the oscillator stage or to the afc stage.

Stage Checks in the Phase-detector System. The stage check is to remove the control voltage and see if the oscillator stage can drive the amplifier in a free-running condition. The multivibrator lends itself to this procedure very easily. A clip lead from the chassis to either the No. 1 control grid or the junction of the diode-load resistors (point *A*), whichever is handier, shorts out the control voltage. Then, with zero bias on this tube, the multivibrator is running freely.

The setup is shown in Fig. 7-14. Check the oscilloscope settings to see that they are at the marked positions. The oscilloscope probe is connected to the grid of the horizontal amplifier.

If the oscillator is working, the screen of the picture tube will probably

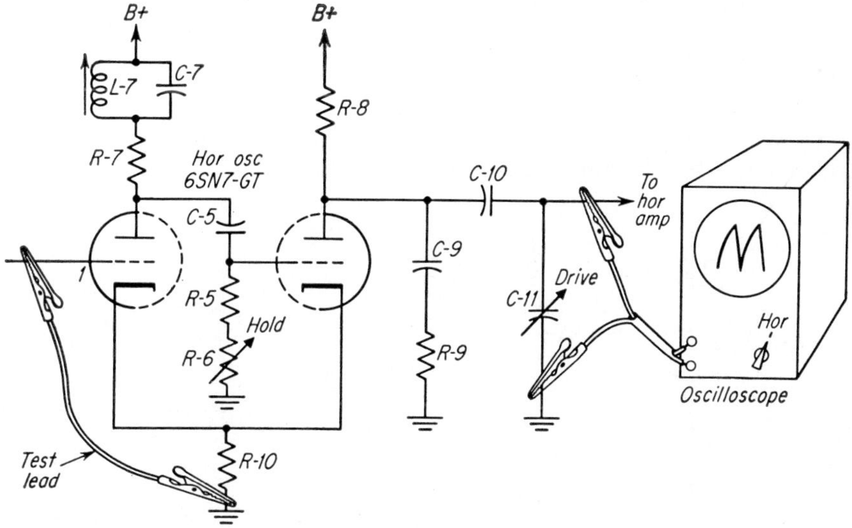

Fig. 7-14. Operating the horizontal multivibrator as a free-running oscillator.

light up as soon as the short is connected to the No. 1 grid of the 6SN7 tube. The hold control is now the main control of frequency, and adjusting it should bring the correct trace on the screen of the oscilloscope. It should also bring an almost stationary picture on the screen of the picture tube. If a raster is obtained but the display on the screen cannot be resolved into a picture by the hold control, the probability is that the ringing coil is out of adjustment. Do not attempt to adjust it at this time; short it out by connecting a jumper across its terminals.

If an almost stationary picture is now obtained, we know the oscillator stage is functioning and the fault is in the afc stage or possibly the sync line feeding it. If the grid-shorting procedure does not bring on the high voltage, the fault is in the oscillator stage.

Finding the Defective Component in the Phase-detector System. When the stage check places the trouble in the multivibrator or afc stage, voltage and resistance checks are used in determining the faulty component. The basic diagram of the phase-detector system has been repeated in Fig. 7-15, with normal voltage readings added to aid in this discussion. However, when working on an actual receiver, use the manufacturer's service notes for data on normal readings, since there is considerable variation between different receivers.

The plate supply for the multivibrator in Fig. 7-15 is taken as 300 volts for both tubes. Note the difference in the normal readings for the two plate pins. The lower reading at the No. 5 plate pin is caused by the high ohmic value of plate load *R*-8. In some receivers, the plate supply is taken from the boost line, which results in correspondingly higher readings.

An abnormal reading at either one or both of the plate pins indicates faulty components in the plate lines. These are then checked with an ohmmeter. Do not neglect to look for and check a decoupling filter (not indicated in the diagram) which may be in the B plus line. A low read-

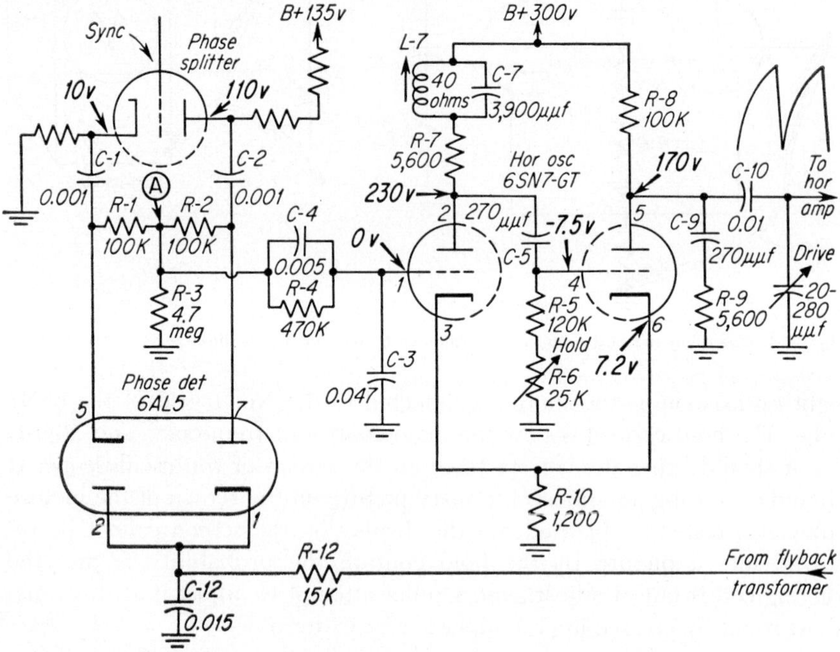

Fig. 7-15. Normal voltage readings in the phase-detector system.

ing at plate pin No. 2 may be caused by leakage in coupling condenser *C*-5. Similarly, a low reading at plate pin No. 5 may be caused by leakage in coupling condenser *C*-10.

The reading at grid pin No. 4 is given as −7.5 volts. This is developed by grid-leak action in the oscillator circuit and will be absent if the oscillator is not functioning. The reading at the No. 1 grid pin is given as zero. This is the point at which the control voltage from the afc stage is fed to the multivibrator. The reading may vary from approximately 0.5 volt positive to 0.5 volt negative in a normal receiver, depending on conditions in the control circuit.

If the trace on the oscilloscope screen shows an incorrect waveshape, condenser *C*-9 and resistor *R*-9, the waveshaping circuit should be checked. Coupling condenser *C*-10 may also cause a change in waveshape.

If the stage check places the trouble in the afc stage, it is serviced by tube replacement and ohmmeter checks. Voltage measurements are of little aid here, since the voltages present at the diode electrodes are signal voltages. Readings for these points have not been entered on the diagram. However, when checking the stage, do not neglect to check condensers *C*-1 and *C*-2 in the sync feed line for leakage. This is a common fault, which may stop the oscillator by leaking a positive voltage into the control feed line to the multivibrator. The condenser connected to the plate of the sync splitter, *C*-2 is more likely to develop the leakage. Normal voltage readings for the plate and cathode pins of the sync splitter are included in the diagram to aid in making these checks.

Component values in the phase-detector system are not so critical as in the synchroguide. In selecting replacement parts, components with a tolerance rating of 10 per cent will be satisfactory. Replacement of a tube or component usually requires readjustment of the controls. A complete adjustment procedure for the phase-detector system is given in Chap. 8.

Variations in the Phase-detector System. Variations in the phase-detector system are confined to the control stage. The oscillator stage is generally the basic multivibrator circuit of Fig. 7-10. There are two important variations in the control stage. Both involve the use of a different rectifier to produce the control voltage. In one, the 6AL5 twin diode is replaced by a twin-crystal assembly; in the other, it is replaced by a single triode.

The crystal-rectifier assembly uses the same circuit and same servicing procedure as the 6AL5 tube type. If defective crystal operation is suspected, the crystals are checked by substitution. Two single-type 1N60 crystals may be used for checking, and also for the final replacement. Polarity is important. Servicing hints and precautions to be observed when replacing crystals are given in Chap. 11.

The single-triode circuit is given in Fig. 7-16. This shows the diagram of the oscillator-afc stages in the Motorola TS 530 receiver. Note that the grid and cathode circuits of the phase-detector tube are fed oppositely phased pulses from the sync splitter. The plate is fed from a tap on the flyback transformer through a waveshaping network. Load resistors *R*-505 and *R*-506 are connected in the grid and cathode circuits of the triode, and the junction of these resistors feeds the control voltage to the conventional multivibrator.

When servicing a receiver of this type, the inoperative stage is determined in the usual way by shorting the control voltage to ground. If the control stage is at fault, it is serviced by tube substitution and resistance measurements, making due allowance for circuit differences.

Reexamine the Motorola circuit of Fig. 7-16, and note that, although

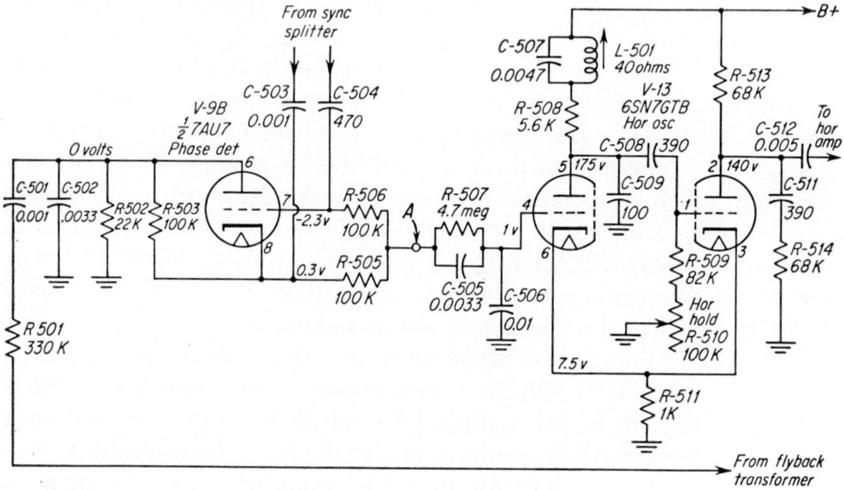

Fig. 7-16. Phase-detector circuit of the Motorola TS 530 chassis.

the oscillator is the conventional multivibrator circuit, component values differ somewhat from the basic circuit of Fig. 7-12. Always expect to find some differences in an actual receiver, since the values used in the basic circuit represent an average. Note further that the triodes in the 6SN7-GTB have been reversed and that the control voltage is applied to the grid at the No. 4 pin rather than to the grid at the No. 1 pin, as is done in the basic circuit. Because of this, it is best to refer to the diagram of the receiver in making all servicing checks.

THE SYNCHROLOCK SYSTEM

A somewhat more elaborate afc-oscillator system than the synchroguide or phase-detector circuits is found in the synchrolock method of control for the horizontal deflection oscillator. This system, originally designed for the RCA 630 receiver, was very widely used in many early receivers. Although it has been replaced in modern sets, the circuit, with minor variations, is still found in quality receivers. The synchrolock system is also used to control the 3.58-Mc subcarrier oscillator in color-television receivers.

A block diagram of the synchrolock system is shown in Fig. 7-17. The deflection generator is an oscillator stage using a 6K6 tube. The oscillator and sync signals are applied to a 6AL5 diode in a phase-discriminator circuit similar to the one used in the phase-detector system. The resultant control voltage, instead of being applied to a multivibrator, is used to operate a 6AC7 control pentode which functions as a variable reactance

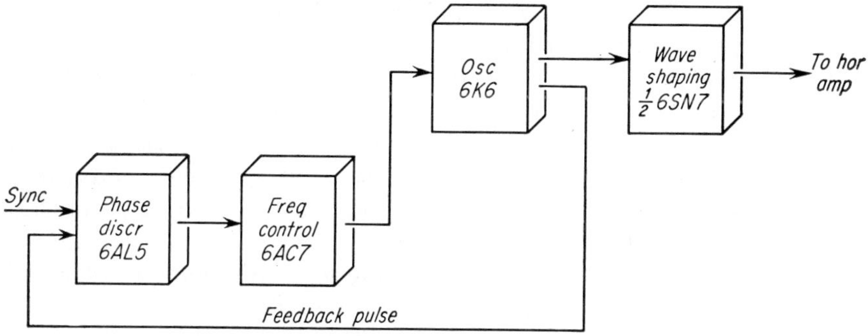

Fig. 7-17. Block diagram of the synchrolock system.

across the oscillator tank circuit. The oscillator output drives the grid of the horizontal amplifier through a triode, called a discharge tube. The function of the latter is primarily waveshaping.

The Oscillator in the Synchrolock System. The schematic diagram of the oscillator in the synchrolock system is given in Fig. 7-18. The circuit is a standard Hartley oscillator, with the cathode of the oscillator tube connected to a regeneration tap on the grid coil. The latter is a coil *AC*, tapped at *B*, in the sync-discriminator transformer assembly.

The main components in the oscillator tank which control the frequency of the oscillation are coil *AC* and condenser *C*-3. The tuning slug in this coil is the control of frequency. Condenser *C*-3*A* across main condenser *C*-3 is a temperature-compensating device which tends to keep the total tank capacitance constant over the normal range of temperatures to be expected inside a television receiver. The automatic frequency control functions through the plate and cathode of the reactance tube, which is connected across the main tank condenser *C*-3. Condenser *C*-6 insulates

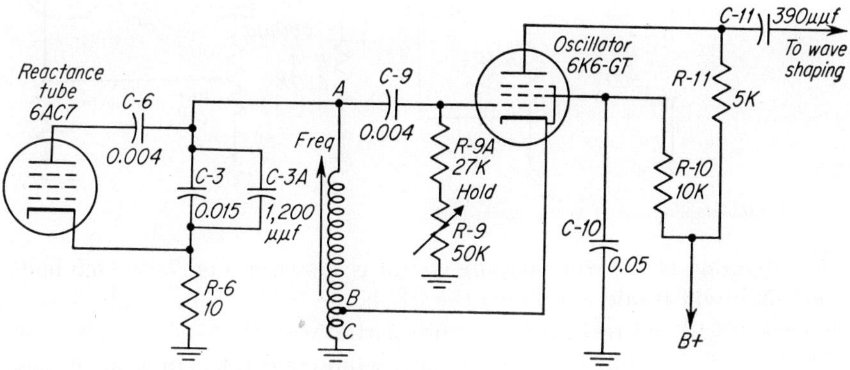

Fig. 7-18. The Hartley oscillator in the synchrolock system.

the oscillator grid from the plate voltage of the reactance tube, and resistor *R*-6 provides bias for the same tube.

Resistors *R*-9 and *R*-9*A* and condenser *C*-9 form the oscillator grid leak and condenser. The values used will not cause the oscillator to block; so that it is a sine-wave generator. Variable resistor *R*-9 has a limited effect on the frequency and is used as the horizontal-hold control. Resistor *R*-9*A* is a stop resistor, which prevents a too-small value of grid leak on manipulation of the hold control.

Resistor *R*-10 and condenser *C*-10 are the screen-dropping resistor and bypass condenser for the oscillator tube. The latter is a 6K6-GT beam power pentode in the 630 circuit. Resistor *R*-11 and condenser *C*-11 form the plate load and coupling condenser to the waveshaping circuit.

The oscillators in the synchroguide and phase-detector circuits are essentially sawtoothed generators, and a simple *R-C* circuit is sufficient for producing the linear sawtooth wave required for the amplifier grid. The synchrolock oscillator generates a sine-wave output which requires more waveshaping. The signal is fed to another triode called a discharge tube, the output of which is easily brought to the required linear sawtooth wave.

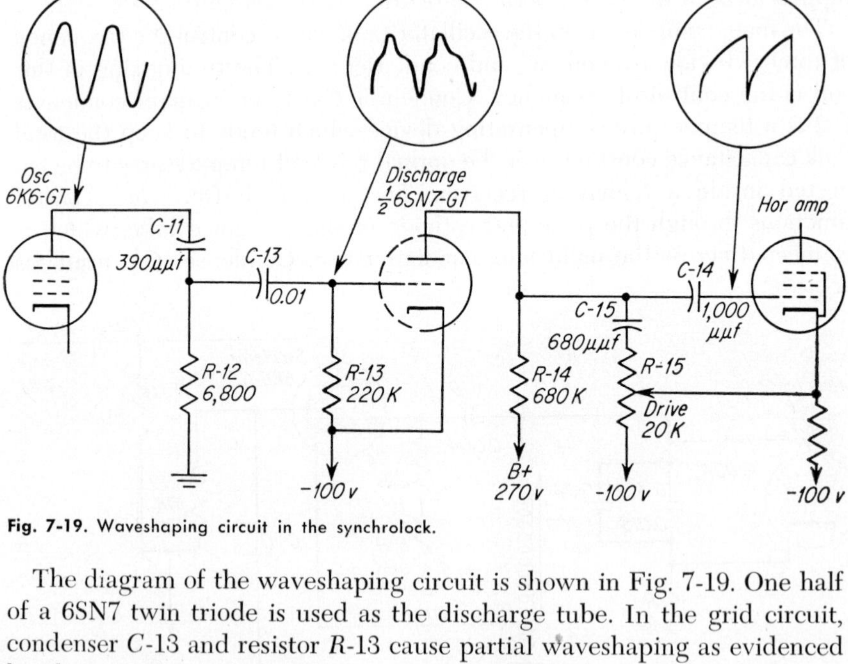

Fig. 7-19. Waveshaping circuit in the synchrolock.

The diagram of the waveshaping circuit is shown in Fig. 7-19. One half of a 6SN7 twin triode is used as the discharge tube. In the grid circuit, condenser *C*-13 and resistor *R*-13 cause partial waveshaping as evidenced by the waveform showing the normal oscilloscope trace present at the grid. The final correct sawtooth wave at the grid of the amplifier tube is produced by the series *R-C* combination of condenser *C*-15 and resistor *R*-15. The latter is made variable and acts as the horizontal drive control.

Automatic Frequency Control in the Synchrolock. The synchrolock system uses a phase discriminator to develop a control voltage in a manner similar to that in the phase-detector system. This time, signals of equal amplitude but opposite phase are derived from the oscillator. The oscillator signal is fed to the phasing circuit by transformer coupling in the synchrolock coil assembly. Winding *DEF* is tapped in the center. Acting as the transformer secondary, it feeds two signals of equal amplitude and opposite phase to the discriminator diodes, as shown in Fig. 7-20. The oscillator coil acts as the primary of this transformer.

The output circuits of the two diodes in the 6AL5 discriminator are so arranged as to produce a balance or zero output when equal sine-wave voltages are applied to the diode plates. At this time, we are considering the application of the oscillator signal only to the discriminator. The center-tapped transformer applies equal and out-of-phase voltages to the diodes. On the first half cycle, the positive phase is applied to the top diode, which therefore conducts, producing a voltage across load re-

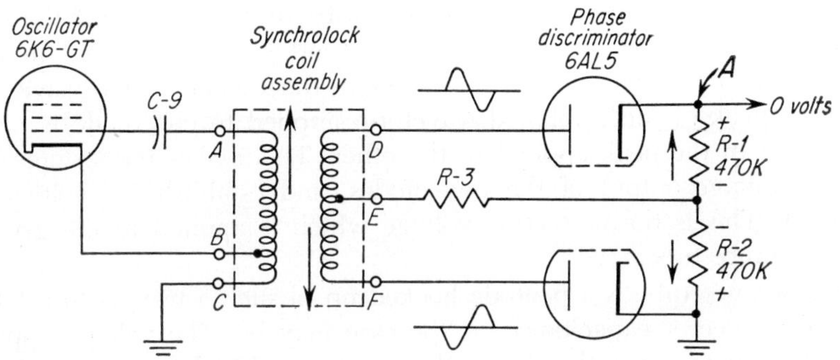

Fig. 7-20. Oscillator coil *ABC* acts as the primary of the discriminator transformer. The secondary, coil *DEF*, feeds signals of opposite phase to the discriminator diodes.

sistor *R*-1. The polarity is as shown in the diagram, and the amplitude will reflect the peak value of the applied voltage. On the next half cycle, the bottom diode conducts, driving current through its load resistor *R*-2 in the direction indicated. Again, the amplitude will depend on the peak positive value of the applied signal. However, these two voltages will be of equal value because the applied signals are of equal amplitude when coil *DEF* is tapped in the center, and load resistor *R*-1 and *R*-2 are also equal. Since the current flow is in opposite directions through the series combination of *R*-1 and *R*-2, the voltage across the pair is zero. Hooked up, as shown in Fig. 7-20, with the cathode of the bottom diode connected to ground, the voltage at point *A* is zero. In the actual circuit as used (refer to Fig. 7-21*a*), the bottom cathode is connected to a potential of −2 volts in the power supply. This places point *A* at −2 volts. This connection is made for biasing the 6AC7 reactance tube.

However, the oscillator signal is not the only signal fed to the two diodes. The afc section of the synchrolock circuit is given in Fig. 7-21*a*. Note that the sync pulse is also fed in at the center tap of coil *DEF*. Because of this connection, the sync pulse is fed to both plates in parallel through the coil windings. The sync signal therefore is added to the oscillator signal in each half winding, and the sums are applied to the diode plates. Two conditions of combined waveforms are shown in Fig. 7-21*b* and *c*.

In Fig. 7-21*b*, the two out-of-phase sine waves represent the oscillator signal in each half of winding *DEF*. In this case, the sync pulse occurs at 180° when both sine waves are passing through zero voltage. Note that in the additive waveform both positive voltage peaks are equal. As a result, equal voltages appear across each diode load, and the resultant voltage is zero (or −2 volts) at point *A*. This is the normal condition for the system, where the oscillator and sync pulse are in step.

Now suppose the oscillator frequency drifts somewhat, so that the sync pulses occur at 200° of the cycle. This condition is illustrated in Fig. 7-21*c*. When the sync pulses are added to the oscillator sine wave, the positive peak of the combined waveform applied to one diode is much greater than the peak applied to the other. This makes for a difference in the voltage output of the two diodes, and establishes a voltage at point *A*. This is the correction voltage which is applied to the grid of the reactance tube.

A reactance tube is a pentode hooked up in such a way as to act like an inductance or capacitance, as the case may be. The voltage applied to its grid determines the size of this reactance. Its plate-cathode circuit is connected across the oscillator tank; so its net effect is to add a variable reactance to the tank, the variation being determined by the grid voltage.

We have just seen that, when the sync pulse and oscillator are in step, the diode outputs cancel out, and −2 volts is applied to the grid of the reactance tube as its normal bias. Then, if the oscillator drifts out of step, the phase difference produces different diode output voltages, establishing a correction voltage at point *A*. This adds to or subtracts from the normal 2-volt bias on the reactance tube, thereby changing the amount of inductance in the tank circuit and bringing the oscillator frequency back in step with the sync pulse.

The network consisting of resistors *R*-4 and *R*-5 and condensers *C*-4 and *C*-5 acts as a filter on the correction voltage to be applied to the reactance-tube grid.

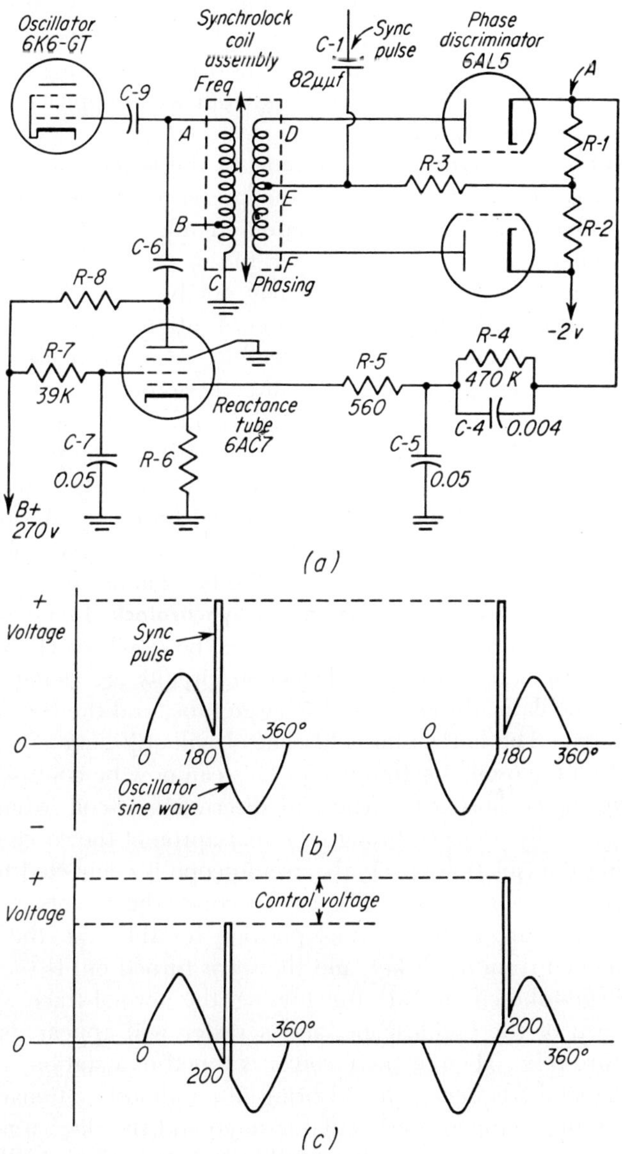

Fig. 7-21. How the afc section of the synchrolock works. (a) The afc circuit. (b) Waveforms applied to the discriminator diodes when the oscillator is on frequency. (c) Waveforms applied to the discriminator diode when the oscillator frequency drifts.

The tuning slug in the *DEF* sync grid coil in the sync-discriminator transformer has a minor effect on the oscillator frequency. Adjusting this slug throws frequency off enough to shift phase for proper operation of the discriminator circuit. It is called the horizontal phasing control. The complete diagram of the synchrolock system as it appears in the 630 receivers is given in Fig. 7-22.

Identification of a Synchrolock System. In the schematic diagram, the synchrolock is quickly identified by the presence of the reactance and discharge tube in the horizontal section in addition to the usual afc control, oscillator, output, and damper tubes. On a chassis containing 20 tubes or more, a few extra tubes may not be noticed. The sync-discriminator transformer should be an identification point, but this looks exactly like the synchroguide coil assembly. Also, both assemblies are likely to have a condenser connected directly across the transformer assembly lugs: terminals *C-D* for the synchroguide and terminals *D-F* for the synchrolock. The actual identification is made by the presence or absence of a resistor. The synchroguide assembly (see Fig. 7-6) has a resistor, usually 8,200 ohms, connected across terminals *A* and *C*; the synchrolock does not use this resistor. Confirmation is then obtained by looking for a 6AC7 reactance tube near the horizontal oscillator.

Quick Check for Stage Isolation in the Synchrolock. Removing the reactance tube is all that is needed to change the synchrolock system to a free-running oscillator. Now the deflection circuits are being driven by the Hartley oscillator through the discharge tube, and the control circuits are inoperative. The horizontal-hold control will still have comparatively little effect on the oscillator frequency. This can now be controlled by the tuning slug in the horizontal-sync and discriminator-coil assembly. The tuning slug usually projects through the rear apron of the receiver chassis.

In making the isolation check, the oscilloscope is connected to the grid of the horizontal output tube. The oscilloscope adjustments are checked to see that they are at the marked position for this test, the reactance tube is removed from its socket, and the set is turned on. If the oscillator and discharge-tube circuits are functioning, the normal trace will appear on the screen of the oscilloscope, and a raster will appear on the face of the picture tube. Then, if the receiver is tuned to a station, adjustment of the horizontal frequency should bring in an almost stationary picture.

If these things happen, the oscillator stage and the discharge tube are known to be functioning, and the trouble is before them in the control stages. If the normal trace does not appear on the screen of the oscilloscope, the oscillator or discharge stage is at fault.

Servicing the Phase-discriminator and Reactance-tube Circuits in the Synchrolock. If the removal of the reactance tube restores the high voltage, the trouble has been isolated to the discriminator or control-tube

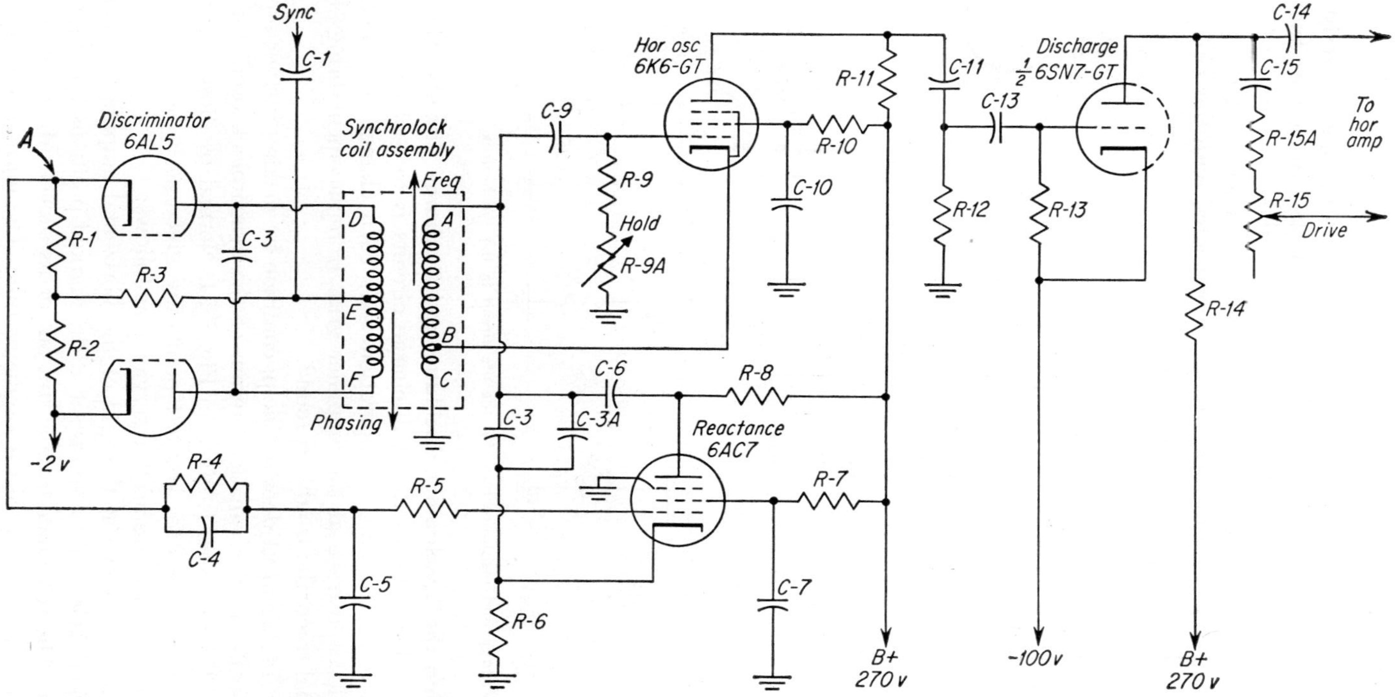

Fig. 7-22. Schematic diagram of the synchrolock system.

stages. These are checked by tube substitution and voltage and resistance checks. The reactance-tube circuits are best checked by voltage analysis. Normal readings at the pins of the reactance tube are included in Fig. 7-23. The phase-discriminator circuits are best checked by a routine ohmmeter check of all resistors and coils associated with this circuit. All condensers are checked by substitution.

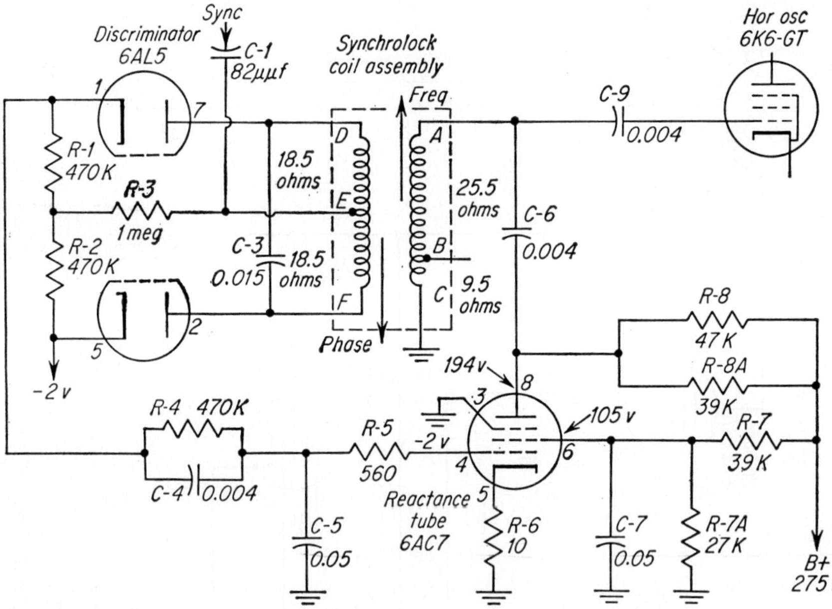

Fig. 7-23. Voltage and resistance data for the afc circuit in the synchrolock.

Servicing the Synchrolock Oscillator and Discharge Circuits. The first step in servicing the oscillator-discharge circuits is to replace the tubes one at a time, checking for operation after each change by observing the oscilloscope trace at the amplifier grid. If replacement brings back the normal trace, the trouble is found.

If tube replacement does not help, continue by making further oscilloscope checks. The schematic diagram of the synchrolock oscillator and discharge-tube circuits is given in Fig. 7-24. Normal oscilloscope traces and voltage readings for key test points are given as in manufacturers' service notes. These are used in servicing as follows:

The sawtooth wave at the grid of the horizontal amplifier is missing or incorrect. The oscilloscope is set at the marked positions for observing this trace. The reactance tube is still out of its socket. Move the oscilloscope probe to the grid of the oscillator tube. The normal trace for this point is shown in the diagram of Fig. 7-24. It may be necessary to reduce

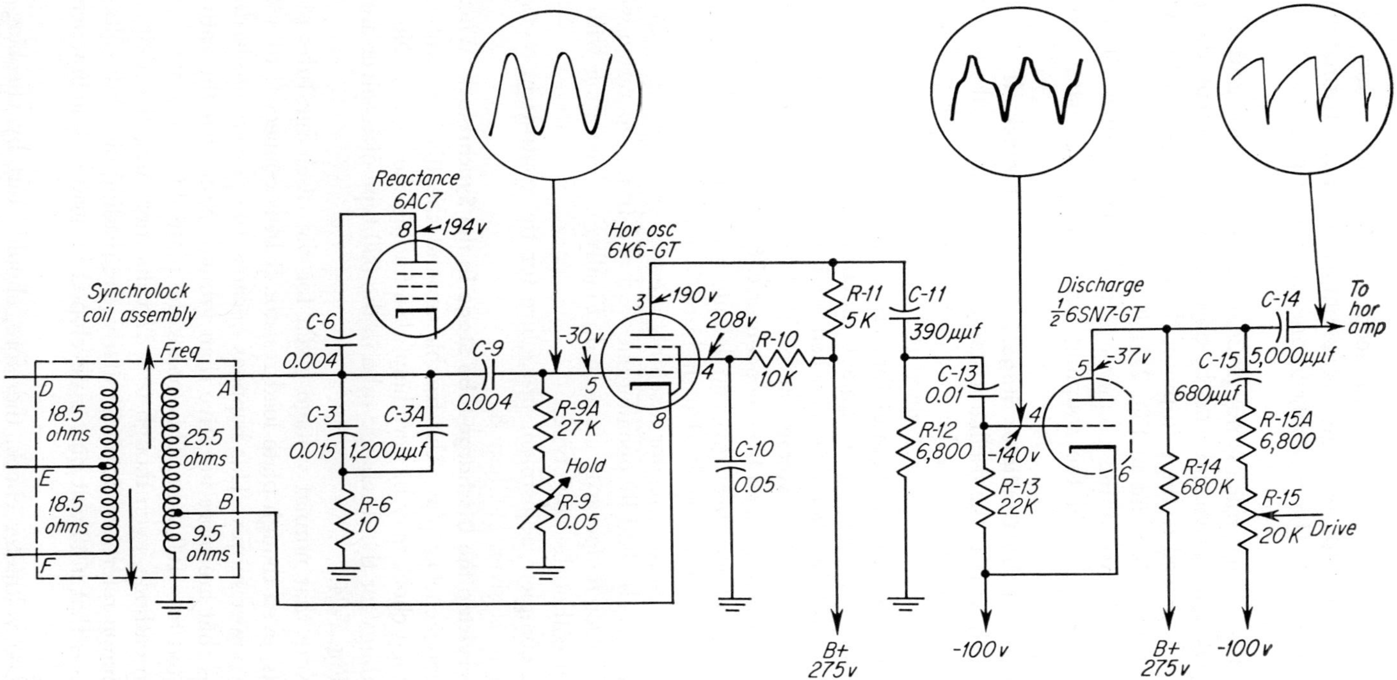

Fig. 7-24. Waveform, voltage, and resistance data for the oscillator, and waveshaping stages in the synchrolock.

the vertical-gain settings on the scope to get this trace, since the generated signal, if present, is of greater amplitude than the normal signal at the amplifier grid. The oscillator-frequency setting may also have to be adjusted to get the trace.

If the normal trace is not present at the oscillator grid, the oscillator stage is inoperative and must be checked. If it is present, the oscilloscope probe is moved to the grid pin of the discharge tube. If the normal trace shown for this point in Fig. 7-24 is not present, the trouble lies in the coupling between the oscillator and discharge tubes. If it is present, the trouble lies in the coupling between the discharge and amplifier tubes.

Servicing the Oscillator Stage in the Synchrolock. If the checks just given show that the oscillator stage is inoperative, make a voltage check of the stage. Normal voltage readings are given in Fig. 7-24. These are not likely to be normal since the oscillator is not working. However, complete lack of voltage at the screen or plate pins will give a definite lead as to the cause of trouble. Condenser *C*-6 in the plate circuit of the reactance tube should be checked for leakage with a voltmeter. If there is a high positive voltage at point *A* on the sync-discriminator transformer, open the lead from condenser *C*-6, and check again.

If the voltage check gives no definite leads for investigation, all resistors in the circuit are checked with an ohmmeter. Ohmic values for the coils in the sync-discriminator assembly are also given in Fig. 7-24. Each condenser is checked by opening one lead and trying operation with another condenser temporarily connected. Finally, try operation with a replacement coil of the same type.

A complete adjustment procedure for the controls in the synchrolock system is given in Chap. 8.

Servicing the Discharge-tube Stage in the Synchrolock. When the oscilloscope checks place the trouble around the discharge-tube circuits, a voltage check may be very helpful in finding the trouble. Normal voltage readings for the discharge tube in the 630 type of receiver are also given in Fig. 7-24.

Note that normal voltage given for the discharge tube plate is −37 volts, even though plate load resistor *R*-14 is connected to +275 volts in the power supply. This happens because the power supply delivers voltages that are more negative than ground. Note that the cathode is connected to −100 volts. This makes the plate +63 volts with respect to its own cathode, even though it is 37 volts more negative than ground, the common reference point. Similarly, the reading of −140 volts at the grid means that a 40-volt bias is developed by grid-leak action across resistor *R*-13.

The voltmeter check, therefore, should start by checking the power supply. The −100-volt reading can be taken at the cathode (pin No. 6) of the discharge tube and the +275-volt reading at the supply end of plate resistor *R*-14. If either of these voltages is incorrect, the trouble is in the low-voltage power supply.

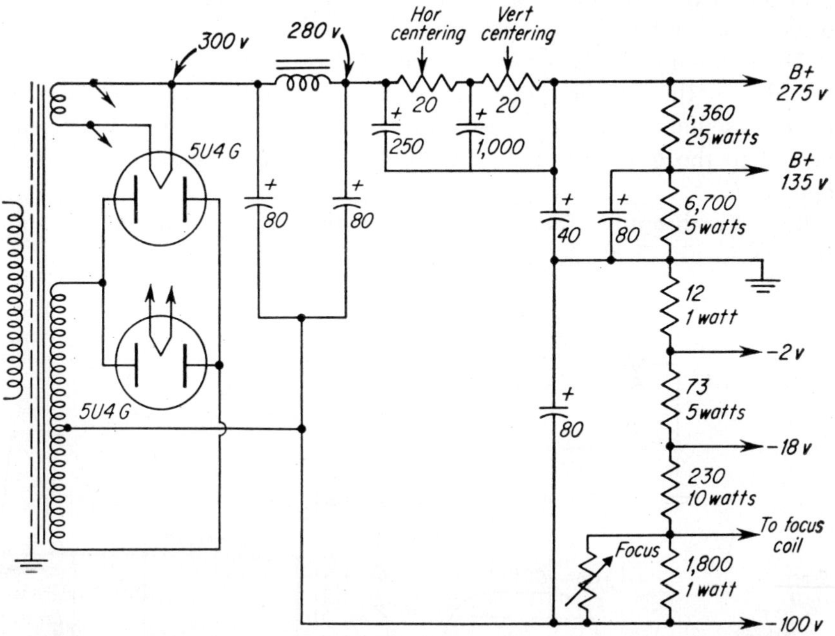

Fig. 7-25. Power supply in the 630 type receiver.

A circuit of the power supply of the 630 type of receiver is given in Fig. 7-25. Note that the center tap of the rectifier plate winding, which is the most negative point in any power supply, is not connected directly to ground. This connection is made through a string of bleeder resistors, thereby providing several steps of negative voltage. From ground, two more bleeder resistors connect to the +275-volt point, which is the equivalent of B plus for the receiver. A break in one of these resistors or a short in one of the bypass condensers will upset the entire voltage-distribution system in the receiver. Defects of this type are easily checked with an ohmmeter.

If the supply voltage readings are normal, voltage readings at the grid and plate pins of the discharge tube will indicate faults in these circuits. Finally, condensers *C*-11, *C*-13, *C*-14, and *C*-15 are checked for opens and leakage. The oscilloscope check at the grid of the discharge tube indicates whether the fault is likely to be in the grid or plate circuits as explained previously.

SUMMARY

Quick Check for Normal Operation of the Horizontal Oscillator and AFC Stages. An oscilloscope check is made of the signal present at the grid pin of the horizontal output tube. If this is normal as to amplitude, waveshape, and frequency, the oscillator and afc stages are operative.

The following charts outline the servicing procedure to be followed for troubles in the synchroguide, phase-detector, and synchrolock systems for the oscillator and afc stages. In all cases, the oscilloscope is left connected to the grid pin of the output tube as an indication of restored operation.

Basic Synchroguide Circuit. A basic circuit of the synchroguide system is given in Fig. 7-26.

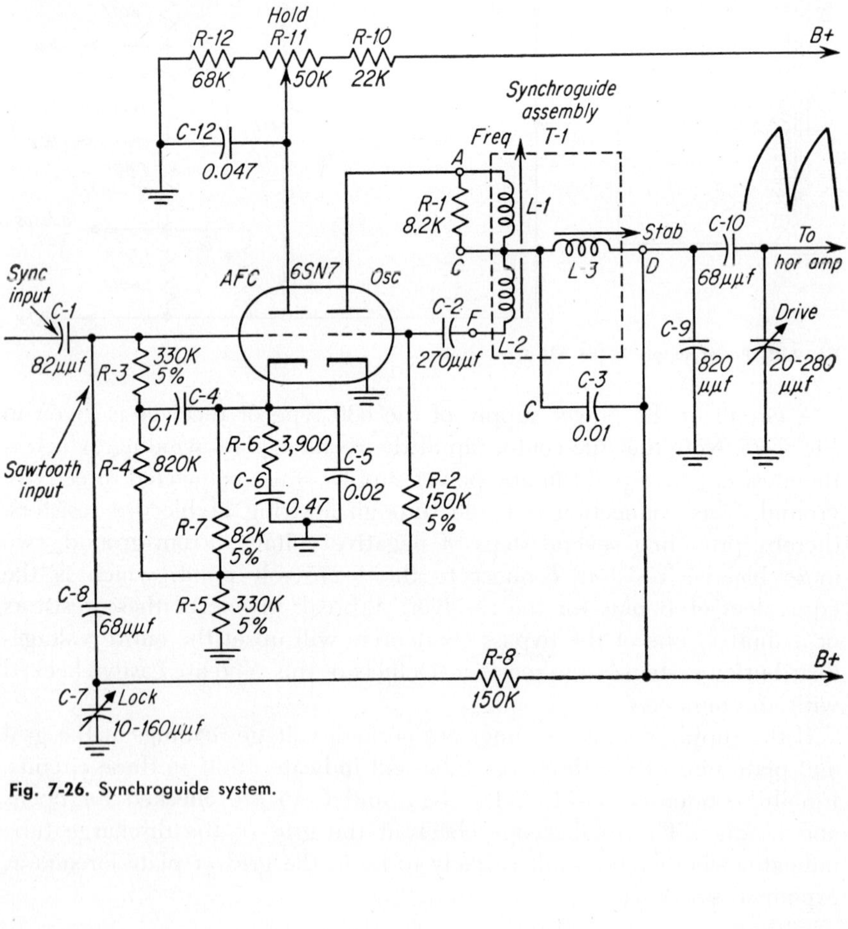

Fig. 7-26. Synchroguide system.

Service Chart for Synchroguide System

Procedure	*Result*	*Trouble and Subsequent Check*
1. Replace oscillator-afc tube	Normal trace is restored	Defective tube
	Trace remains unaltered	Proceed to step 2
2. Check plate voltages of the oscillator-afc tube	Abnormal voltage at the afc plate. (Normal varies with the hold control from 100 to 200 volts)	Check resistors *R*-10, *R*-11, and *R*-12 and condenser C-12 in hold control circuit
	Abnormal voltage at the oscillator plate (normal is 175 volts)	Check sync assembly *T*-1, resistor *R*-9, and condensers C-2, C-9, and C-10 in oscillator plate circuit
	Voltages normal	Proceed to step 3
3. Open the oscillator grid leak, and install a 2-meg test potentiometer instead (see Fig. 7-8)	Adjusting the test potentiometer produces normal trace at amplifier grid, and display on screen of picture tube	Proceed to step 4
	Oscillator will not produce trace	Check all components in oscillator circuit
4. Apply sync pulse to oscillator grid (see Fig. 7-9)	Picture holds	Check all components in afc stage
	Picture blanks out	Check coupling condenser C-1 in sync feed line

Basic Phase-detector Circuit. A basic circuit of the phase detector-stabilized multivibrator afc-oscillator system is given in Fig. 7-27.

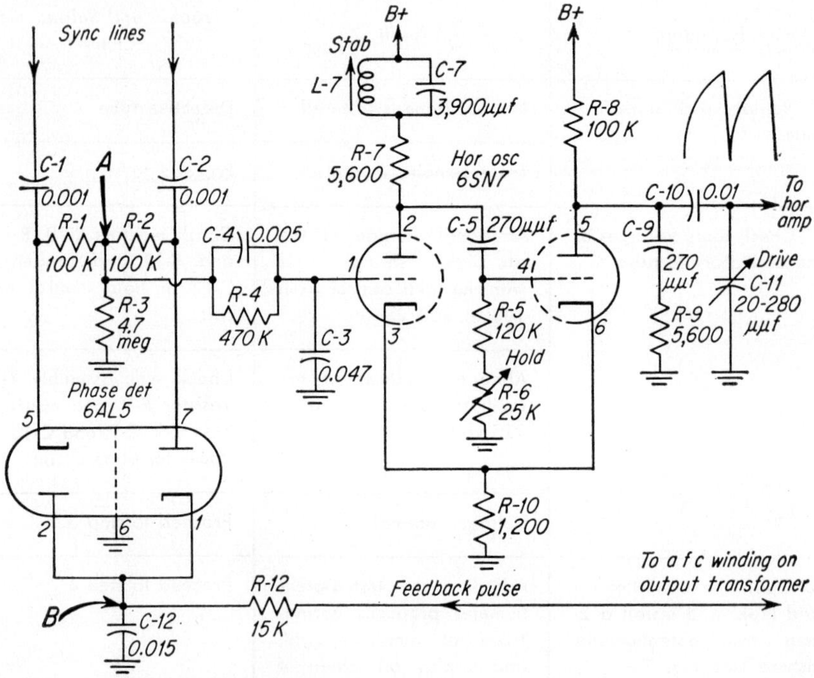

Fig. 7-27. Multivibrator afc-oscillator system.

Synchrolock Circuit Diagram. Refer to Fig. 7-24 for voltage, resistance, and waveform data for the synchrolock circuit.

Service Chart for a Phase-detector System

Procedure	*Result*	*Trouble and Subsequent Check*
1. Replace oscillator and afc tubes	Normal trace is restored at amp grid	Defective tubes
	Trace remains unaltered	Proceed to step 2
2. Operate the multivibrator as a free-running oscillator by shorting the control grid to ground (see Fig. 7-14)	Normal trace is restored Screen lights up	Check all components in afc stage including condensers C-1 and C-2 in the sync feed line
	Trace remains unaltered	Oscillator stage is inoperative
		Service stage by voltage and resistance check (see Fig. 7-15)

Servicing Chart for a Synchrolock System

Procedure	*Result*	*Trouble and Subsequent Check*
1. Remove the 6AC7 reactance tube	Normal trace is restored at amplifier grid	Trouble is in afc or reactance tube circuits. Proceed to step 2
	Trace remains unaltered	Trouble is in oscillator- or discharge-tube circuits. Proceed to step 3
2. Check discriminator and reactance tubes by substitution	Normal trace is restored	Defective tubes
	Trace remains unaltered	Make voltage and resistance checks in reactance- and discriminator-tube circuits. See Fig. 7-24
3. Replace oscillator and discharge tubes	Normal trace is restored at amplifier grid	Defective tubes
	Trace remains unaltered	Proceed to step 4
4. Make oscilloscope check at oscillator grid looking for normal trace shown in Fig. 7-24	Normal trace is present	Proceed to step 5
	Trace is not present	Check all components in oscillator stage
5. Make oscilloscope check at discharge tube grid, looking for normal trace shown in Fig. 7-24	Normal trace is present	Check components in coupling circuit between discharge tube and amplifier grid
	Trace is not present	Check components in coupling circuit between oscillator and discharge tubes

QUESTIONS

The following questions relate to the entire horizontal section as described in Chaps. 6 and 7.

1. Draw a block diagram of the stages in the horizontal section, and give a brief statement of the function of each stage.

2. Compare the horizontal deflection system with the vertical deflection system, and explain why the horizontal oscillator requires more control.

3. What is the equivalent of the damper stage in the vertical deflection system?

4. Describe the high-voltage power-supply system as obtained from the horizontal section.

5. Describe development of the boost voltage.

6. A receiver is brought in with a complaint of "no picture, no raster, sound normal." A voltage check at the second anode shows 16 kv. Where is the trouble likely to be? Outline a procedure for finding it.

7. In checking for loss of high voltage, what inference should be drawn if there is a normal oscilloscope trace at the grid of the output tube? What is the inference if the oscilloscope check shows no signal at the grid of the output tube?

8. When working on a receiver for no high voltage, describe a series of checks which establish that the output tube is working normally.

9. What is likely to be wrong if removal of the damper tube restores high voltage?

10. When working on a receiver for no high voltage, it is observed that the plate of the output tube glows red. What is likely to be wrong? How would you work on this set and prevent further damage to the output tube?

11. When working on a receiver for no high voltage, it is observed that the plate of the damper tube glows red. Voltage check shows zero at the cathode of the damper tube. List the possible causes of this trouble, and state how you would check for each.

12. When working on a receiver for no high voltage, it is observed that both the plate and cathode voltages of the damper tube are abnormally low. What is the likely cause of the trouble?

13. When working on a receiver for no high voltage, a screwdriver check draws a normal arc from the plate cap of the high-voltage rectifier. List the possible causes of this trouble, and explain how you would check for each.

14. When working on a receiver for no high voltage, high voltage from another receiver is brought over and connected to the second anode of the dead set.

a. A keystone raster appears on the screen. What is likely to be wrong?

b. A bright vertical line appears on the screen. What is likely to be wrong?

c. A narrow raster appears on the screen. What is likely to be wrong?

d. Foldover appears on the left side of the screen. What is likely to be wrong?

15. When working on a phase-detector type of receiver for no high voltage, the high voltage is restored by shorting the multivibrator control grid to ground. What steps would you take to locate the trouble?

16. When working on a synchroguide type of receiver for no high voltage, voltage check shows no voltage at the plate of the oscillator tube. Refer to Fig. 7-26 and list the defects which may cause this condition.

17. Describe the procedure by which the oscillator in a synchroguide system can be changed to a free-running oscillator.

18. How is the oscillator in a synchrolock system made free-running?

19. List the defects which can stop the oscillator in the synchrolock system of Fig. 7-28, and state how you would check for each.

20. List the defects which can stop the oscillator in the phase-detector circuit of Fig. 7-27, and state how you would check for each.

21. How would you provide free-running operation for the oscillator in Fig. 7-16?

22. What are Barkhausen oscillations? Give procedure to minimize this fault.

23. Give a procedure for reducing ringing bars on the left side of the screen.

24. List the likely causes for lack of width. How would you check for each?

25. List the likely causes for a complaint of blooming. How would you check for each?

26. Describe the adjustment of the horizontal drive, width, and linearity controls.

8

Servicing the Sync Section

When the vertical sweep section operates at an off-frequency rate, the picture rolls up or down as shown in Fig. 8-1*a*. When the horizontal sweep section operates at an off-frequency rate, the resulting horizontal roll gives the picture the "marble cake" effect shown in Fig. 8-1*b*. Manipulation of the vertical- and horizontal-hold controls may bring back a correct picture, but if it starts to break up soon, the fault is improper sync action in the receiver.

(*a*) (*b*)

Fig. 8-1. Symptoms of loss of synchronization. (*a*) Rolling. (*b*) Marble cake.

The function of the synchronization or sync section is to keep the deflection circuits operating in step with the transmitter, at their correct frequencies so as to maintain a steady picture. The sync operation is of necessity tied in with frequency-control devices in the deflection oscillators. The adjustment procedure for the horizontal frequency controls cannot be properly completed unless a normal sync pulse is present. This adjustment procedure will be presented in this chapter together with a servicing procedure for the sync section.

What Is Synchronization? A simple device which may help in the understanding of synchronization is a pantograph, illustrated in Fig. 8-2.

This is a mechanical means of reproducing a picture. It consists of a point *A*, and a pencil at point *B*, connected by a series of arms. When the point *A* is moved along the lines of a picture to be copied, the pencil at *B* reproduces the action of the pointer at *A* through the connecting arms, thereby reproducing the picture. Should the connecting arms be removed, the pencil at *B* is free to move at will, and the system breaks down. The function of the connecting arms can be likened to synchronization in a television system.

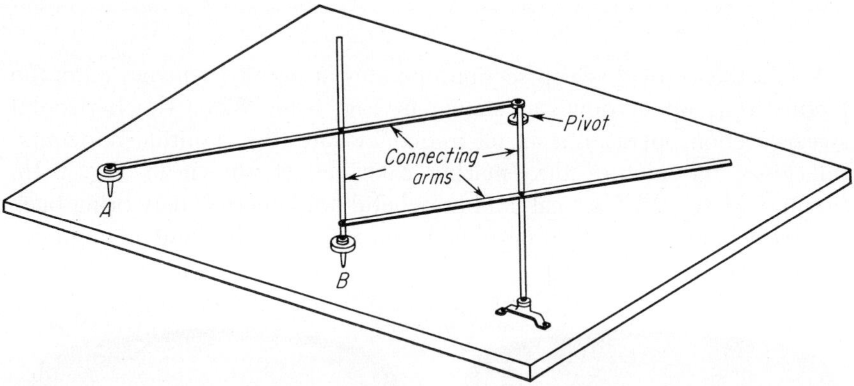

Fig. 8-2. Mechanical pantograph.

At the television studio, the scene being televised is scanned by an electronic beam in the camera. At the receiver, the screen of the picture tube is also scanned by an electronic beam. The deflection circuits which control scanning at the transmitter, and deflection circuits which control scanning at the receiver must be kept in step so that the picture at the receiver duplicates the scene at the transmitter.

The process of keeping the two scanning systems in step is called synchronization. It is accomplished by adding a series of pulses to the picture signal and transmitting them as part of the signal. These sync pulses can be likened to the arms of the pantograph. Without them, the deflection oscillators at the receiver are free to drift, and the system breaks down.

Sync Signals. The composite video signal, therefore, contains sync information in addition to the picture information. The difference between the two is a matter of amplitude. The picture information is contained in 75 per cent of the amplitude of the composite signal, the sync information in the other 25 per cent. This is shown in the diagram of Fig. 8-3. The picture information controls the shading of the beam from black to white, and varies from zero amplitude which is the white level to the 75 per cent point which is the black level. The sync information is from 75 to 100 per cent of the total amplitude and so is in the region sometimes called "blacker than black." Since the screen is dark in the black region, the sync information is not seen.

The electron beam sweeps horizontally across the picture tube 15,750 times a second. As each line is traced, the voltage of the picture information determines the shading of each point on the line. When the beam

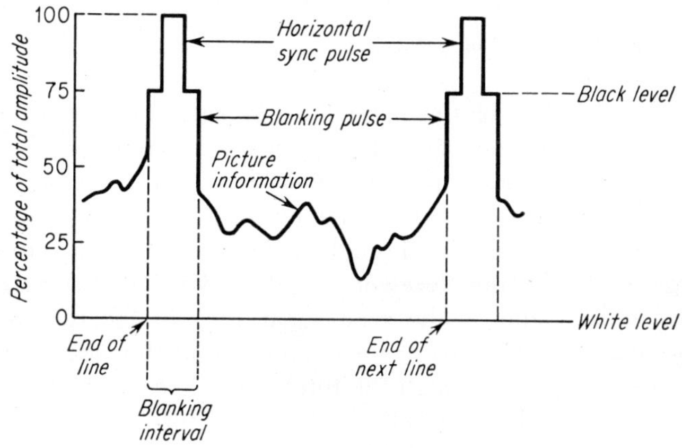

Fig. 8-3. Composite video signal.

reaches the right side of the screen, it is cut off by the transmitted blanking pulse which brings the voltage to the 75 per cent, or black, level. During this blanking period, the horizontal sawtooth oscillator snaps the beam back to the left side of the screen; the horizontal sync pulse that triggers the horizontal oscillator is transmitted during the same blanking interval.

In addition to being deflected horizontally, the beam is also deflected vertically. This deflection occurs simultaneously but at a slower rate—60 times a second. The beam, therefore, is deflected slightly downward as it traces each line. Then, when it starts the new line, the continued downward pull starts the new line at a lower position. In this way, the beam traverses the entire screen.

When one complete field composed of 262½ lines is traced out, a blanking pulse of longer duration is transmitted so that the beam may be returned to its starting point at the top of the picture. During this time, the vertical sync pulse is transmitted to trigger the vertical oscillator. The signal transmitted at the end of each field is shown in Fig. 8-4.

Since the vertical oscillator is slow as compared with the horizontal oscillator, its blanking period and sync pulse are of longer duration. To

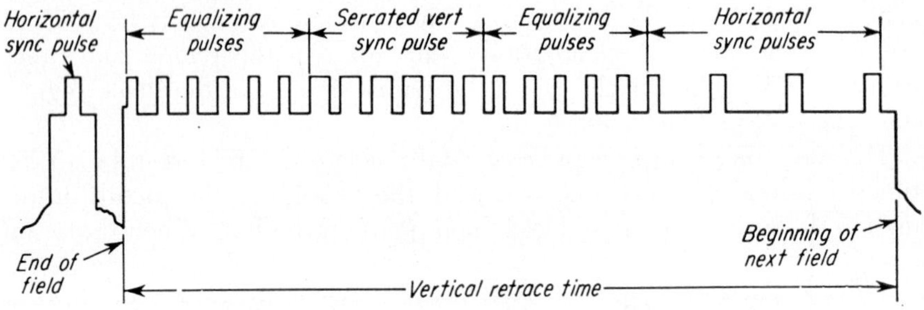

Fig. 8-4. Sync pulses at the end of each field.

prevent the possibility of the horizontal oscillator's losing sync control during the vertical retrace time, the vertical sync pulse is serrated into six individual pulses, and equalizing pulses are added before and after the serrated vertical sync pulse. The horizontal oscillator is therefore receiving a form of sync pulse throughout the vertical retrace time.

Sync-signal Path in the Receiver. See Fig. 8-5. The sync signals are transmitted pulses, and so the sync-signal path in the receiver begins at the antenna. Here it is part of the modulation of the video carrier. The composite signal passes through the tuner and i-f stages to the video detector where demodulation produces the signal form shown in Figs. 8-3 and 8-4. The signal could be fed to the sync stages any time after the video detector. However, common practice is to take advantage of the added gain and to feed the sync section from the output of the video amplifier. The sync signals are present in the picture tube, but they are

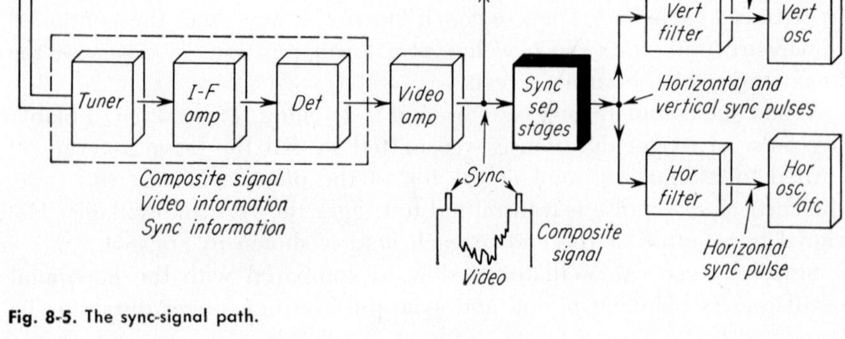

Fig. 8-5. The sync-signal path.

not seen because the beam is blanked out during the time the sync pulses are transmitted.

In the sync section, the sync pulses are removed from the composite signal and separated into vertical and horizontal sync pulses. These are then fed to their respective deflection oscillators.

In most receivers, the sync section is composed of two stages, where the sync pulses are removed from the composite signal and amplified. The separation into vertical and horizontal sync pulses is provided by filters as indicated in Fig. 8-5.

The deflection oscillators require a sync pulse of correct polarity and amplitude, and so design factors in various receivers call for additional stages for amplification or phase reversal. There may be only one stage in the sync section, and there may be three or more. As mentioned previously, the most common arrangement is two stages.

In manufacturers' diagrams and service notes, the tubes in the sync section are labeled by various terms, such as "first sync separator," "second sync clipper," and "sync noise inverter." Although these terms give an indication of the function of the tubes, there is no uniformity in the nomenclature. From the servicing point of view, it is easier to regard all the tubes simply as amplifiers, hooked up in such a way as to achieve specialized results. The end result is always sync pulses of the proper amplitude and polarity.

How the Sync Pulses Are Removed from the Composite Signal. It will be remembered that the composite video signal includes both the picture and sync information, that the picture information is contained in 75 per cent of the total amplitude and the sync signal in the remaining 25 per cent. For easy arithmetic, assume the composite signal has a total ampli-

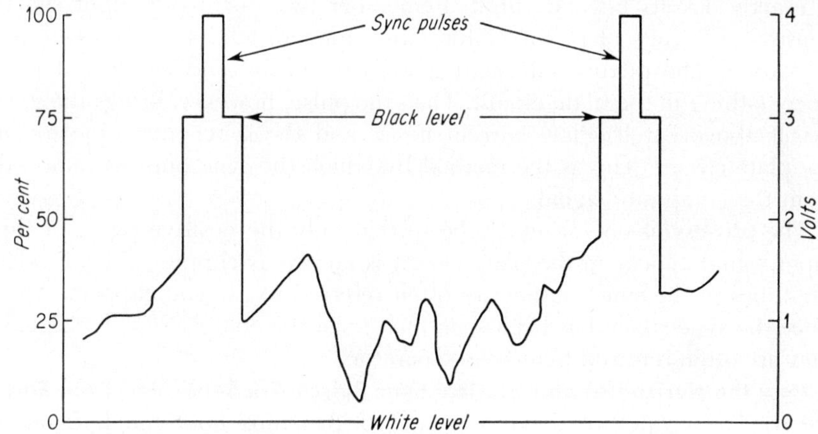

Fig. 8-6. Voltage range of video-sync signals.

tude of 4 volts as shown in Fig. 8-6. Then the picture information is a voltage varying from 0 to 3 volts. White is voltage of zero, black is 3 volts, and the sync pulses bring the voltage up to 4.

Now let us feed this signal to an amplifier biased so as to handle it without distortion. The grid voltage–plate current curve of such an amplifier is shown in Fig. 8-7*a*. The tube is operated at a bias of −4 volts so as to handle the signal, and the entire signal appears in the plate circuit. This is ordinary class A amplification, and the video-amplifier tube operates in this fashion.

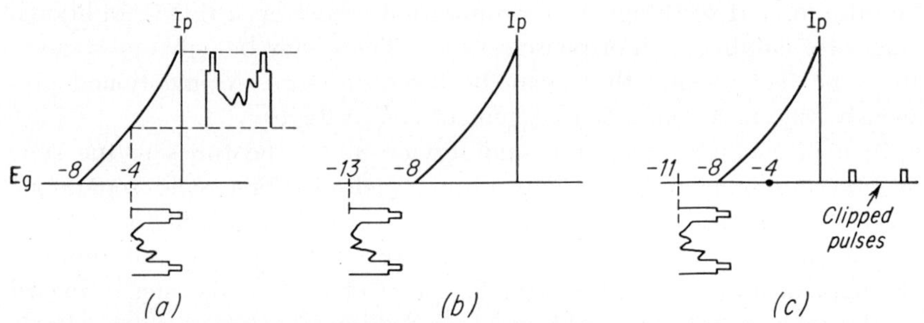

Fig. 8-7. How the sync pulse is clipped from the composite signal.

The curve shows that the cutoff point for the tube is at −8 volts. If the same tube were biased at −13 volts, and the same signal fed to its grid, the most positive point of the signal at +4 volts brings the grid to −9 volts. This is still beyond cutoff, and the plate current remains zero, with no signal in it. This condition is illustrated in Fig. 8-7*b*.

Suppose we now bias the tube to −11 volts, as shown in Fig. 8-7*c*. This is exactly 3 volts beyond cutoff. Remember that our 4-volt input signal consists of 3 volts of picture information plus an added volt of sync information. The picture information will still be beyond cutoff and produce nothing in the plate circuit. The sync pulse, however, brings the grid 1 volt above cutoff. Plate current flows, and the sync pulse appears in the plate circuit. This is the method by which the sync pulse is removed from the composite signal.

The process of operating a tube so that only the positive peaks of the input signal appear in the plate circuit is known as clipping. This is why the tubes in the sync section are often referred to as sync clippers. Also, since the sync signal has been separated from the composite video signal, they are often referred to as sync separators.

How the Horizontal and Vertical Sync Pulses Are Separated from Each Other. The output of the last sync tube contains both the horizontal and vertical sync pulses at the proper amplitude. They must now be separated from each other and fed to their respective sweep oscillators.

The horizontal sync pulses come at the rate of 15,750 per second, each one of very short duration. The vertical sync pulses come at the rate of 60 per second, and although each one is broken into six serrations, each serration is a long pulse as compared with the horizontal pulses. The length of the pulses is the basis on which the separation takes place.

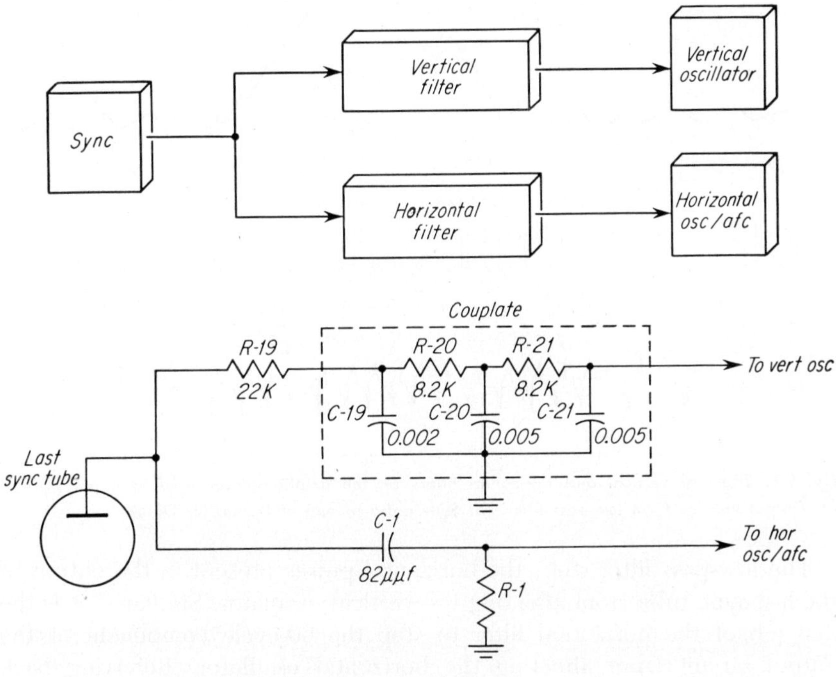

Fig. 8-8. Vertical and horizontal filters in the output of the last tube in the sync section.

The output circuit of the last sync tube is simultaneously connected to two filter circuits as shown in Fig. 8-8. The upper filter consists of three resistors and three condensers hooked up as a low-pass filter. In most receivers, all these components are contained in one printed circuit called a vertical integrator couplate. The short-duration horizontal and equalizing pulses have little effect in charging these comparatively large condensers, and the small charges leak off between pulses. This is shown in Fig. 8-9, where the middle graph represents the sync pulses transmitted for the end of a frame, and the last four horizontal lines preceding it. The top graph represents the charging effect of the sync pulses on the condensers in the low-pass filter. Note that the first of the serrated vertical sync pulses permits a large charge because of its comparatively longer duration. Also, because of the short time period between serrations, this charge does not leak off but builds up successively during the six rapid-fire serrations of the vertical pulse. The voltages thus established across the condensers at the rate of 60 per second become the vertical sync pulses which trigger the vertical oscillator.

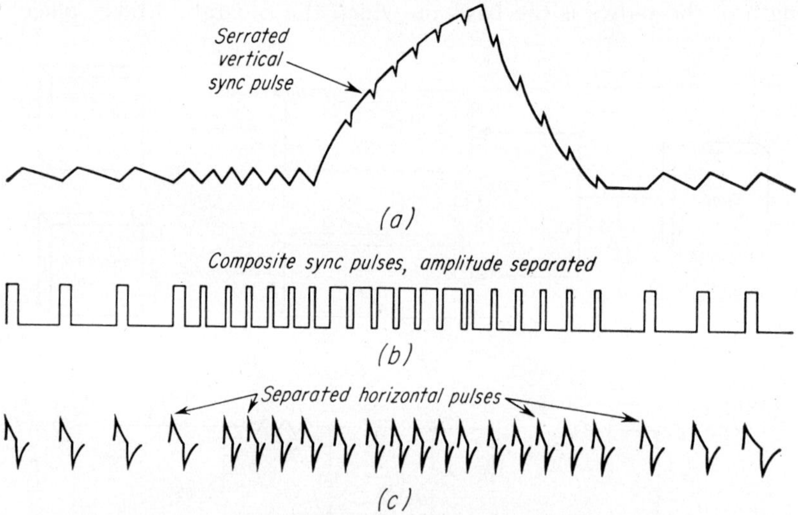

Fig. 8-9. Effect of vertical and horizontal filters on the amplitude-separated sync pulses. (*a*) Output voltage from low-pass filter. (*b*) Sync pulse at end of frame. (*c*) Output.

The low-pass filter stops the horizontal pulses present in the output of the last sync tube from affecting the vertical oscillator. Similarly, it is the first job of the horizontal filter to stop the 60-cycle component of the output circuit from affecting the horizontal oscillator. Referring back to Fig. 8-8, the horizontal filter consists only of condenser *C*-1 and resistor *R*-1. Resistor *R*-1 represents the input-grid circuit of the horizontal-afc stage. The combination is called a differentiator. It is connected as a high-pass filter, and condenser *C*-1, which has the small capacitance of 82 $\mu\mu$f, has a very high impedance to 60-cycle current, thereby blocking most of it from the horizontal oscillator. Additional filter circuits in the afc stage take care of the rest.

The sync pulses present at the plate of the output sync tube reach condenser *C*-1. At the beginning of each of these voltage pulses, condenser *C*-1 is charged. This produces a sharp pulse or spike across resistor *R*-1 which is the output of the high-pass filter. This is shown on the bottom graph of Fig. 8-9. At the end of each pulse, condenser *C*-1 discharges, producing a momentary voltage across resistor *R*-1 in the opposite direction. The negative pulse is unimportant, since it is of the wrong polarity to affect the horizontal system. In the afc stage, these positive differentiated pulses are compared with the oscillator output and a correction voltage is produced if the oscillator is not in step with the sync pulses.

Typical Sync Section Using a Dual Triode. We are now ready to study the operation of a typical sync section. The diagram of a dual-triode sync section is shown in Fig. 8-10. A 6SN7-GT or its miniature equiva-

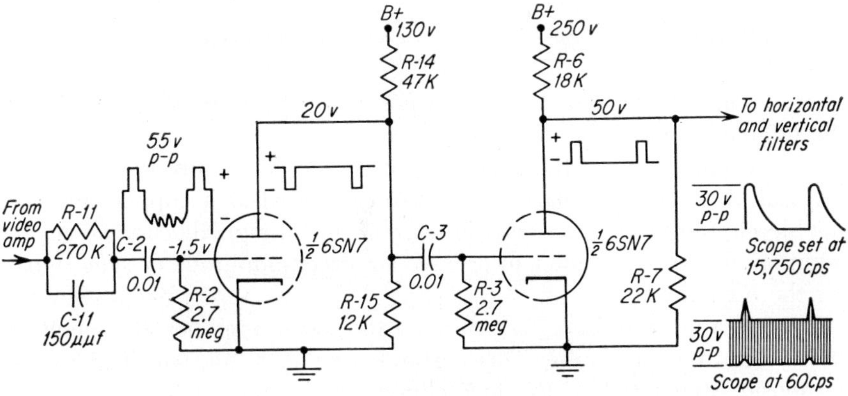

Fig. 8-10. Sync separator used with synchroguide circuit.

lent, the 12AU7, is most often used. The first triode is connected as a clipper to remove the sync signal from the composite video signal. This signal is obtained from the output of the video-amplifier tube where the signal polarity is such that the sync pulse is the most positive part of the signal. The waveform of the input signal is shown near the grid. Through grid-leak action, the input signal establishes a bias of minus 1.5 volts at the grid of the tube. The plate voltage must be very low when so small a bias is beyond cutoff. Resistors *R*-14 and *R*-15 in the plate circuit form a voltage divider which applies a potential of 20 volts to the plate of the tube.

Resistor *R*-11 and condenser *C*-11 in the grid circuit act as a noise filter in that they will prevent a high-amplitude short-duration noise pulse from affecting the 1.5-volt bias which is established by the average signal level. The 1.5-volt bias cuts off all the composite signal except for the positive peaks which correspond to the sync pulses.

The waveform at the plate of this triode shows only the sync pulses, but note that they are inverted. This is because a signal is always inverted 180° by the action of an amplifier tube. The deflection oscillators require a positive-going pulse for synchronizing, and the output of the first triode is a negative-going pulse. Now the second tube in the sync

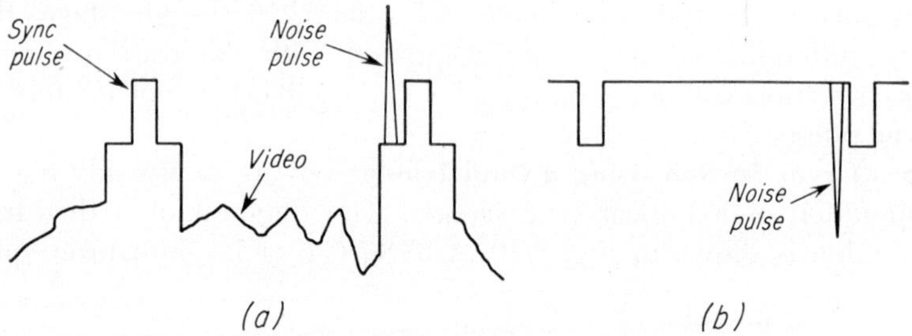

Fig. 8-11. Noise pulse on the input and output signals of the first sync clipper.

section reinverts the sync signal to a positive-going pulse so as to trigger the oscillators correctly.

The second stage in the sync section serves more purposes than the reversal of the signal polarity. It is also connected as a clipper, and the double clipping removes noise pulses that go through the first stage. In addition, it acts as a limiter to provide a correct amplitude of sync signal for the oscillators.

Assume a noise pulse comes through the video amplifier at a point in time close to a sync pulse. This is shown in Fig. 8-11*a*, which gives the composite signal fed to the first clipper grid. The action of the tube clips off the more negative portion of the signal, leaving the sync pulse and noise pulse as shown in Fig. 8-11*b*. Note that the output signal is inverted. When the output signal is fed to a second clipper tube, and it in turn removes the more negative portion of the signal, the noise pulse is reduced. This is shown in Fig. 8-12.

The second tube in the sync section is hooked up very much like the first. Referring again to Fig. 8-10, grid condenser *C*-3 and grid leak *R*-3

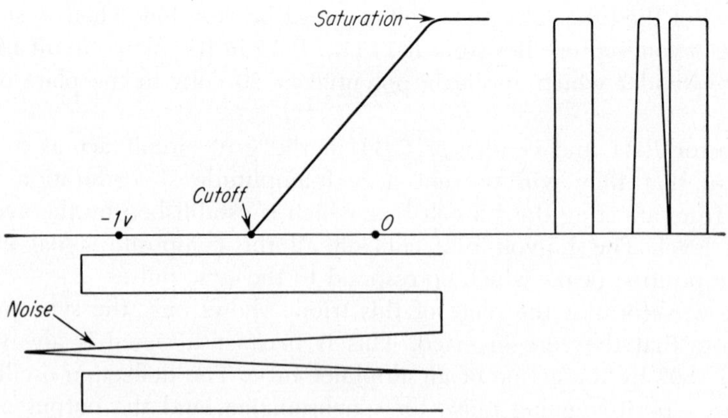

Fig. 8-12. Noise-reducing and -limiting action of the second sync tube.

provide grid-leak bias. This time the input signal amplitude establishes a bias of 1 volt. Resistors *R*-6 and *R*-7 in the plate circuit form a voltage divider providing 50 volts for the plate of the tube. When operating at low grid and plate voltages, the tube will not handle a large signal. The grid voltage–plate current curve of the tube given in Fig. 8-12 shows that a large positive signal applied to the tube brings it to the saturation point, and the output signal can be no greater than this value. When there is sufficient gain in the circuits so that this tube is always overloaded, the output signal will always be at one value. This is known as a limiting circuit. The circuits are so designed that the output of the last sync tube is limited to the correct amplitude to sync the sweep oscillators.

Sync Circuit for a Phase Detector. Proper operation for a phase-detector type of horizontal deflection system requires two horizontal sync pulses, equal in amplitude but opposite in phase. The second signal is sometimes obtained from an extra triode hooked up in a phase-inverter circuit similar to those found in push-pull audio amplifiers. The more common method, however, is to avoid the extra tube and to connect the last sync stage as a phase splitter.

A basic diagram of a phase splitter is shown in Fig. 8-13. The cathode of the triode is unbypassed so that plate-circuit variations are present

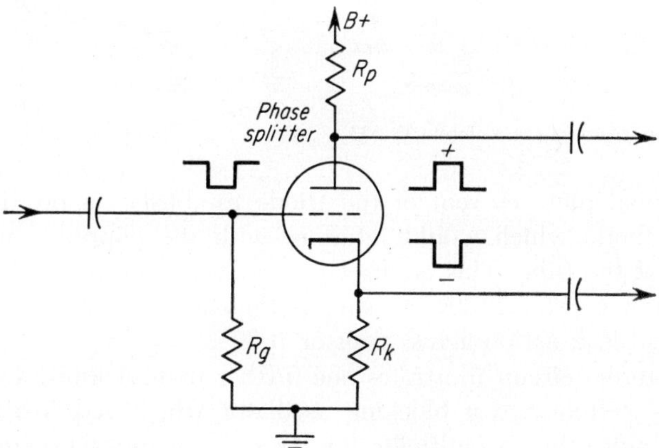

Fig. 8-13. The phase splitter.

in the cathode circuit. If the cathode resistor has the same ohmic value as the plate-load resistor, the signal voltages developed across both have the same amplitude. When a negative-going signal is applied to the control grid, as indicated in the diagram, the plate current decreases. This decreases the voltage across both the plate and cathode resistors. As a result, the cathode goes more negative with respect to ground, and the plate goes more positive with respect to ground. Thus, at the output of the plate and cathode coupling condensers, two signal voltages are developed, equal in amplitude but opposite in phase.

The sync section of the Motorola TS 533 chassis is given in Fig. 8-14 as an example of the phase-splitter type of sync output tube. A 6SN7-GT twin triode is used for both triodes in the sync section. The first triode clips the sync component from the composite video signal fed in from the video output tube. The output is fed to the second phase-splitter triode through coupling condenser *C*-303. Resistors *R*-306 in the cathode and *R*-308 in the plate circuits develop equal sync pulses of opposite polarity. These are fed to the phase detector through coupling condensers *C*-503 and *C*-504.

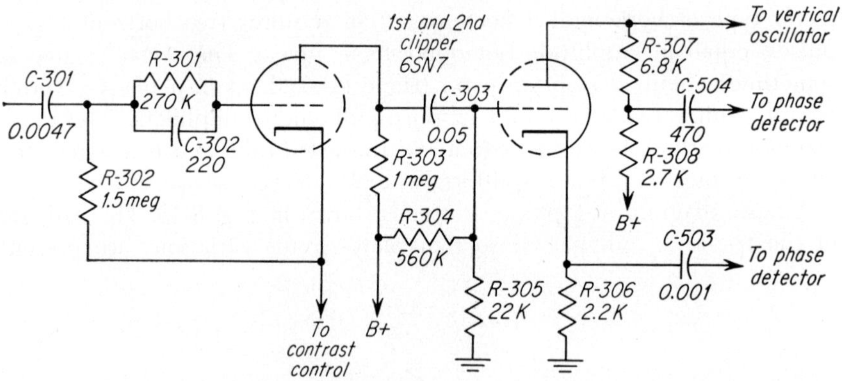

Fig. 8-14. Sync section of the Motorola TS 533.

The normal plate current of the triode establishes a positive biasing on this cathode which would interfere with the clipping and limiting functions of the tube. This is offset by feeding a positive voltage to the grid through resistor *R*-304. Biasing is then accomplished in the usual way by grid-leak action across resistor *R*-305.

The Motorola circuit illustrates one further item of interest. The vertical sweep section uses a blocking oscillator which requires a stronger triggering pulse than a multivibrator. As a result, the plate circuit of the last sync tube has been designed to give high and low output signal voltages. Note that the plate load consists of resistors *R*-307 and *R*-308 in series. The full sync output voltage is fed to the vertical section, and the tap at the junction of the two resistors provides a sync signal at lower amplitude for the horizontal section. Note that the cathode resistor matches the low-voltage tap on the plate load.

Lock-in and Pull-in Ranges. Lock-in and pull-in ranges are two commonly used terms when working on sync circuits. They can best be understood by performing the following experiment on a receiver in good working order.

1. Tune in a good picture on a strong channel.

2. Turn the vertical-hold control until the picture starts to roll. Then turn it slowly in the other direction until rolling starts again, this time in the opposite direction.

When the vertical-hold control is adjusted, it changes the frequency of the vertical oscillator. Yet there is a small degree of rotation where the hold control has no apparent effect. It is attempting to change the frequency of the vertical oscillator, but the sync pulse is holding it, and the picture remains steady. This is called the hold-in range, or lock-in range. The lock-in range is of much greater significance in the horizontal oscillator, where the afc circuit is also working to produce the same result.

3. Adjust the vertical-hold control through its lock-in range to determine the center position. Leave the control at this setting.

4. Turn the horizontal-hold control slightly, and note that the picture stays in sync. Determine the hold-in range of the horizontal system by turning the hold control back and forth in ever wider arcs until the picture begins to slip sideways at each end of the swing.

Depending on the circuit and receiver conditions, the hold-in range of the horizontal system is from one-third to two-thirds of the full rotation of the control. In some cases, it is the full rotation of the control.

5. Turn the horizontal-hold control to either end of its lock-in range. Momentarily interrupt the signal by switching off channel and back. Note that the picture is now out of sync.

When the signal is interrupted, the horizontal oscillator changes to the new frequency called for by the setting of the horizontal-hold control. When the signal comes back on, the sync pulse cannot pull the oscillator back to 15,750 cps, and the picture is out of horizontal sync.

6. Readjust the horizontal-hold control until the picture comes back on. Determine the center of the horizontal lock-in range.

7. Adjust the horizontal-hold control to a setting about halfway between the end and center of the lock-in range. Momentarily interrupt the signal by switching to the next channel and back. Note that the picture remains steady.

At this setting of the hold control, the oscillator again drifts to its new frequency when the signal is interrupted. This time, the off-frequency position is closer to 15,750 cps, and when the signal is restored, the sync pulse pulls the oscillator back to 15,750 so that the picture comes on. This is known as the pull-in range, which is, of course, smaller than the hold-in range.

Effect of Improper AGC Action on Sync Conditions. Automatic-gain-control systems provide a variable bias to control the gain of r-f and i-f tubes so that the signal level at the video amplifier is about the same for all stations that are expected to produce a good picture. Very often, the agc systems provide a variable control so that the receiver may be adjusted to meet reception conditions in widely differing areas of signal strength.

If the manual control in the agc system is improperly set or the circuit develops certain faults, there may be an adverse effect on the operation of the sync section. If the developed bias does not sufficiently reduce the signal level when a strong station is tuned in, one of the tubes in the i-f or video chain may overload. Since sync pulses represent greatest amplitude, the overloaded tube may remove all or part of the sync component by limiting action. The result is unstable operation caused by loss of sync. To help in recognition of this condition, other overloading symptoms, often occurring at the same time, are too contrasty or negative pictures. In the latter, the picture looks like a photographic negative.

The cause is confirmed when reducing the antenna strength by loose coupling cures the condition. In areas of strong signal strength, loose coupling is easily achieved by removing the lead-in from the receiver and placing it near the coupling lead between the antenna connection post and the tuner. If this clears the overload and instability, the agc system is at fault. Servicing procedures for the agc circuits are given in Chap. 12. If the receiver has a variable control of agc action, a readjustment may be all that is required. The adjustment procedure is given later in this chapter.

Noise-reducing Sync Circuits. In radio receivers, interfering signals from static discharges, ignition systems, transients on the line, etc., mar reception by producing various rattling noises. As a result, these interfering electric disturbances are known as noise pulses. Most noise pulses are an a-m type of signal and cause very few problems in the f-m sound section of television receivers. The a-m noise pulses are present in the video signal, however, and although they are usually of too short a duration to cause much interference with the transmitted picture, they may trigger the deflection oscillators at the wrong time, causing an unstable picture. Noise pulses lower in amplitude than the sync pulses do not affect sync, since they will be removed by the clipping action of the sync tubes. High-amplitude noise pulses will ride in with the sync signals and may act to trigger the deflection generators.

In fringe areas, where the strength of desired station signals is low, noise pulses may easily be stronger than the station signals with an undesirable effect on the sync stability of the receiver. As a result, many receivers incorporate special circuits designed to immunize the sync system against noise pulses. The noise-reducing circuits generally fall into two classifications. One type uses a pentagrid tube like the 6CS6 as the first tube in the sync section; the other uses a noise-canceling circuit before the sync section. Both types use a control which is adjusted for individual reception conditions. Each circuit will now be described, together with generalized control-adjustment procedures.

Pentagrid Sync Clippers. Many receivers use a pentagrid tube in a noise-gate circuit for the first tube in the sync section. The tube most often used is the 6CS6. Sometimes a 6BE6 pentgrid converter or a 6BN6 gated-beam discriminator is used. Of course, there will also be the 600-ma 3CS6 and 3BN6 series string tubes.

The five grids in the 6CS6 are labeled *G*-1 to *G*-5, as shown in Fig. 8-15*a*. Grid *G*-1, nearest the cathode, has the greatest effect on electron flow. A small negative voltage applied to this grid will cut off the tube.

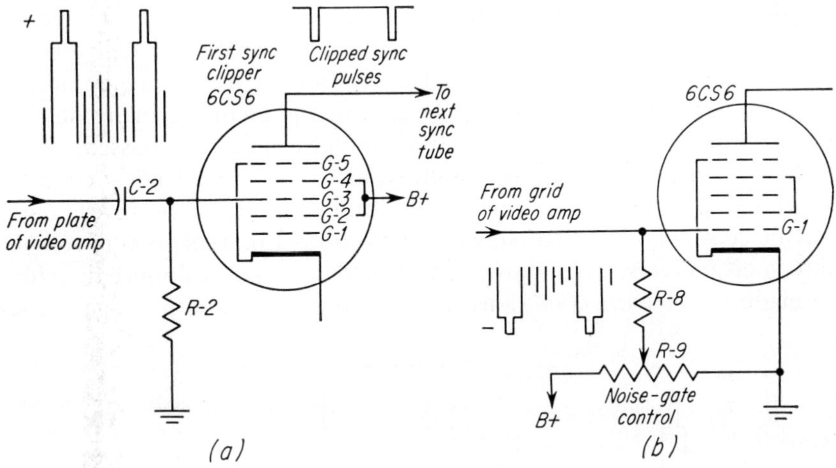

Fig. 8-15. Basic noise-gate circuit.

It is used as the noise-gate control grid, but for the moment let us disregard it except to assume a slight positive bias, so as not to hinder electron flow to the rest of the tube. Consider grid *G*-3 as the regular control grid of a first sync tube receiving the composite signal from the video output stage. Grids *G*-2 and *G*-4 serve to shield the control grids. They are connected as a screen grid going to B plus through a properly bypassed dropping resistor. Grid *G*-5 acts as a suppressor and is internally connected to the cathode.

Grid *G*-3 receives a positive-going composite signal from the output of the video-amplifier tube. The signal establishes the proper bias across grid-leak resistor *R*-2 so that cutoff occurs at about the blanking level, and only the sync pulses appear in the plate circuit. This is a standard clipping procedure.

Let us now examine the noise-gate action at grid *G*-1. It is connected to grid-leak resistor *R*-8 which in turn is connected to the arm of potentiometer *R*-9. The arm provides the small positive bias mentioned before which permits the operation of the rest of the tube. At the same time, grid *G*-1 receives a negative-going composite signal pulse of low amplitude from the grid of the video-amplifier tube. This produces a negative voltage across resistor *R*-8 which counteracts the positive voltage from control potentiometer *R*-9. The latter is correctly adjusted when it almost balances the bias produced by the normal sync signal, and the tube is just before the cutoff point (Fig. 8-15*b*).

Now consider a noise pulse greater in amplitude than the sync pulse. The high-amplitude pulse will produce a large negative voltage at *G*-1, driving it beyond cutoff. This closes the gate, stopping electron flow and therefore all action in the tube. Nothing appears in the plate circuit. For the duration of the noise pulse, no sync information is delivered to the sweep oscillators. These are not likely to drift off frequency during the short interval of the noise pulse, and so the picture remains steady. The sync pulses resume control after the noise pulse has passed.

The average level of the composite video signal is kept fairly constant in strong-signal areas by agc action. In fringe areas, however, there may be considerable variation, which will still permit reception of reasonably good pictures. The adjustment of potentiometer *R*-9 must therefore be made under local conditions. It is usually called the noise-gate con-

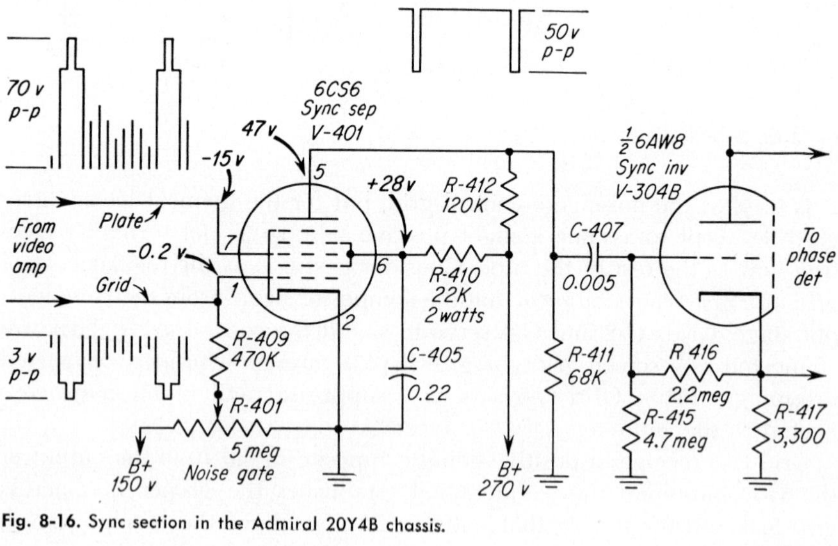

Fig. 8-16. Sync section in the Admiral 20Y4B chassis.

trol. Because of its added importance in fringe areas, it is sometimes also called a fringe control or a fringe lock.

The schematic diagram of the sync section in an Admiral 20Y4B chassis is given in Fig. 8-16 as an example of a receiver using a 6CS6 noise-gate tube. Note that the second sync tube is a phase splitter for use with a phase-detector type of horizontal oscillator.

Noise-canceling Circuit. A method of incorporating noise cancellation before the sync section is shown in Fig. 8-17. The composite signal from the detector is fed to the first video amplifier, which in turn feeds the second video amplifier and the sync tube. The detector output is a negative-going signal. By inversion in the first video amplifier, the signal input to the second video and sync tubes becomes positive-going.

The noise-canceling tube is a triode with resistors *R*-210 and *R*-202 in the cathode and plate circuits. When the grid is held constant, if the cathode is made more negative, there is increased current flow through

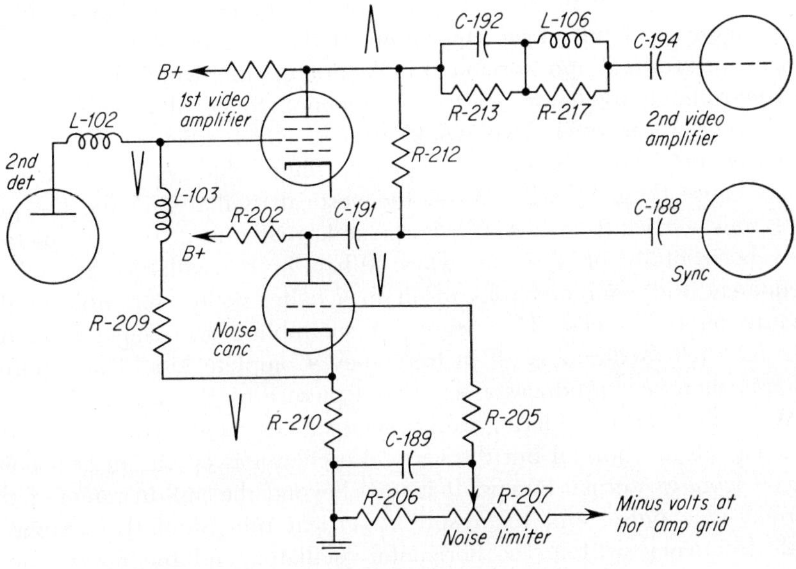

Fig. 8-17. Noise-cancellation circuit in RCA KCS 96 chassis.

the tube. This increases the voltage across plate-load resistor *R*-202, causing the voltage at the plate to decrease. Therefore, a negative pulse fed to the cathode produces a negative pulse in the plate.

The cathode is also connected to the negative-going signal from the detector plate, thereby tending to produce negative-going signals in the triode plate. The grid, however, is connected to a negative voltage obtained from the grid circuit of the horizontal output tube. The amount of grid bias is adjustable through control *R*-207, which is called the noise limiter or sync stabilizer. The correct adjustment is sufficient negative at the grid for the tube to be at the cutoff point when there is a normal signal fed to the cathode. A noise pulse of greater intensity than the sync pulse, applied to the cathode, causes the tube to conduct, producing a negative noise pulse in the plate.

The plate of the noise-canceling tube is connected to the plate of the first video amplifier through resistor *R*-212. The normal signal at this point is the positive-going composite signal. When the signal is free of noise, nothing is coming through the cutoff noise tube; but a strong noise pulse produces a strong positive pulse at the plate of the video-amplifier tube. At the same time, it drives the noise tube into conduction, producing a strong negative pulse. The noise pulse, therefore, cancels itself out at the junction of *R*-212 and *C*-191.

The operating point of the noise-canceling tube is critical. If control *R*-207 is set too negative, there will be insufficient cancellation of noise pulses. If set too far in the other direction, it will also cancel sync pulses.

Symptoms of Faulty Sync Operation. If the sync pulses were entirely absent, the vertical and horizontal deflection circuits would be operating as free-running oscillators. If these happened to be at exactly 60 and 15,750 cps, respectively, we would have a steady picture. Oscillators, however, rarely remain at one frequency. Slight changes in circuit constants cause them to drift. When the vertical oscillator drifts, the picture rolls vertically; when the horizontal oscillator drifts, the picture rolls horizontally or slips sideways. When both oscillators are off in frequency, the combined rolls result in a helter-skelter scramble of the picture on the screen. The degree of scrambling will depend on the amount each oscillator is off in frequency. Complete loss of sync information, therefore, produces a scrambled picture on the screen.

We have been considering faulty sync signals. Now, assume that the sync signals are normal but the vertical oscillator is operating at a point far off from its normal 60 cps. It is now beyond the pull-in range of the normal sync pulse, and the result is vertical roll. Similarly, assume a wide frequency drift in the horizontal oscillator, and the normal horizontal sync pulse cannot correct matters, resulting in horizontal roll. Note that, regardless of whether the condition is due to a faulty sync signal or a faulty oscillator, the symptom is the same. The sync and oscillator frequency controls are so intimately connected that any servicing procedure for off-frequency operation must take both factors into consideration.

There are other symptoms of off-frequency operation besides the rolling or scrambling already described. Sometimes, the picture starts to move vertically or horizontally but snaps back before it gets very far.

Continued repetition of this effect gives this symptom the name of jitter or bounce. Another symptom occurs when some of the horizontal lines go out of sync. Depending on the extent of out-of-sync operation, this fault may be described as pulling, bending, tearing, or flag waving. Finally, there is the type of trouble where the 60- or 120-cycle hum frequency gets into the signal chain and causes a bend in the picture.

It is intended that this chapter cover all types of off-frequency operation, as well as the bend caused by hum in the picture.

Steps in Servicing Faulty Sync Operation. The sync-signal path of a receiver has been repeated in Fig. 8-18 to show the possible locations of defects which may cause out-of-sync reception. Since the sync pulse is a transmitted signal, the chain begins at the antenna. The sync signals go through the r-f, i-f, and video sections before coming to the sync section. Finally, the sync section delivers triggering pulses to the vertical and horizontal oscillators. The cause of faulty sync operation is a break anywhere in this chain.

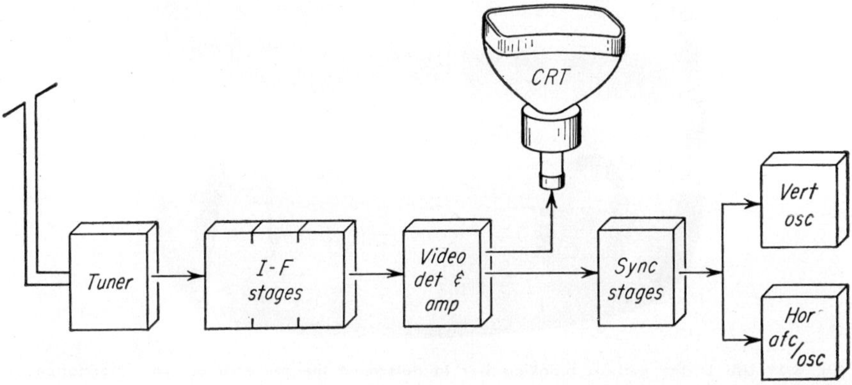

Fig. 8-18. Sync-signal path.

The chain is a long one, and for ease in servicing, it is usually subdivided into three major steps.

1. Prove that the sync pulses are present in the composite signal arriving at the sync section.

2. Prove that the deflection generators can be brought to their correct operating frequencies.

3. If steps 1 and 2 give affirmative results, service the sync section for the break in the chain.

Step 1 can be accomplished by adjusting the receiver controls to view the vertical blanking bar. Step 2 requires the adjustment of the frequency controls for the various types of horizontal deflection systems. These procedures will now be described.

The Vertical Blanking Bar as a Quick Test Indicator. An examination of the block diagram of Fig. 8-18 shows that the sync pulses are present in the signal fed to the picture tube. They are not normally seen, but they may be made to appear by the following procedure:

1. Use a receiver in good working order, and tune it for best picture on a strong channel.
2. Advance the brightness control and reduce the contrast control so that the black portions of the picture become gray.
3. Manipulate the vertical-hold control so that the picture rolls slowly. A gray bar appears between frames, as shown in Fig. 8-19. Try to hold this gray bar stationary on the screen for visual examination.

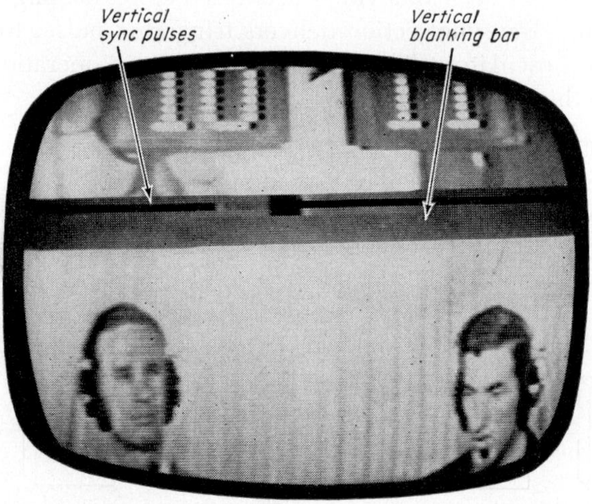

Fig. 8-19. Use of the vertical blanking bar to determine the presence of sync information in the composite signal at the picture tube.

The vertical deflection oscillator in the receiver is out of step with the vertical deflection oscillator at the transmitter. When the gray bar is in the middle of the screen, the receiver is just half a step ahead of or behind the transmitter. When the transmitter reaches the end of a frame, a long blanking pulse, extending over several lines, is sent to cover the vertical retrace time. In the receiver, the end of the transmitted frame now occurs in the middle of the screen, and the next several lines of the vertical retrace period appear as a gray bar. This is called the blanking bar. It is gray rather than black, because the receiver brightness control has been misadjusted so that all shades are lighter. Yet the gray of the blanking bar should be as dark as the darkest portion of the transmitted picture.

During the blanked retrace period, the vertical sync signal is transmitted. This consists of equalizing pulses and the serrated vertical sync pulse. These pulses are at greater amplitude than blanking and normally appear at a blacker-than-black shading level. Since our shading scale has been misadjusted so that black is gray, blacker-than-black will now appear as black. The sync signals will be seen against the gray background of the blanking bar, as shown in Fig. 8-19. The narrow black bar is formed by the serrated vertical sync pulse, and the mallet head by the equalizing pulses.

When the procedure is performed on any receiver, the appearance of the bar and mallet head in the blanking bar proves the presence of a normal sync signal at the output of the video-amplifier tube, which is the input to the sync section. If they cannot be made to appear, the sync signal has been lost in one of the stages preceding the picture tube. Service notes pertaining to suppression of sync signals will be found in the chapters dealing with the appropriate stages.

How to Hold Free-running Oscillators on Frequency. If the receiver shows a scrambled picture indicating loss of sync, our first step calls for an examination of the blanking bar to prove the presence of sync pulses at the video output. This does not seem possible since the controls have little apparent effect, yet it can be done by careful manipulation. This is an important servicing technique to acquire, and you are urged to repeat the following experiment until you gain proficiency in its use. Choose a receiver in good working order, and with separate front-panel vertical- and horizontal-hold controls.

1. Tune the receiver for normal operation on a channel that gives good reception.

2. Remove one of the sync tubes. If this stops the set because of series filament circuits, or because half of the tube is working elsewhere in the circuit, interrupt the sync chain by removing the plate lead from the last sync tube. We now have free-running oscillators whose normal drift gives the scrambled picture associated with loss of sync.

3. Adjust the horizontal-hold control very slowly through its range. A point should be reached where the scrambled display resolves itself into a picture which may be rolling in the vertical plane. If such a point cannot be found, it may be that 15,750 cps is outside the range of the hold control. In this case, set the control in the center of its range, and slowly adjust the horizontal frequency slug on the rear apron for the same result—a picture rolling vertically. This should not require more than a touch in either direction, because the oscillator must be close to correct frequency to have brought in the normal picture we started with. Having obtained the picture with the frequency slug, adjust the horizontal-hold control to keep it from sliding off in either direction.

4. This step calls for two-hand manipulation. Use one hand on the horizontal-hold control to prevent the picture from slipping sideways, and at the same time slowly adjust the vertical-hold control so as to stop the vertical roll. When you have one picture on the screen, hold it there by using either hand to correct any tendency to drift.

Bringing a picture into frame and holding it there by manipulation of the frequency controls of the sweep oscillators which are running freely without sync pulses, is known as freewheeling. In this case, we are freewheeling both sweep oscillators.

5. Advance the brightness control, and shift the vertical control slightly so as to examine the blanking bar. Note that the black bar and mallet head, indicating the sync pulses, are present.

Adjustment of Automatic Gain Control and Noise-immunity Control. When servicing for loss of sync, observation of the blanking bar by the procedures just given indicates the presence or absence of sync signals at the picture tube. If they are absent, improper agc operation is a likely cause. If the receiver uses automatic gain control or noise-immunity control, these adjustments should be checked next.

In adjusting these specialized controls, it is best to follow the manufacturer's instructions. This is especially true where the receiver uses both noise control and manual control in the agc system, since their effects are interrelated as regards sync operation. If manufacturer's service notes are not obtainable, the following generalized procedures will give satisfactory results with most receivers.

Adjustment Procedure for an Automatic Gain Control. This procedure is to be used on sets that do not have an additional noise-immunity control.

Step 1. Tune the receiver for best sound on the strongest station in your area. Freewheel the hold controls for a steady picture, and disregard tendency to slip. Set the contrast control to the maximum clockwise position.

Step 2. Rotate the manual control in the agc system through its range to determine its scope of operation.

At one end (usually counterclockwise) the greatest agc bias voltage is developed. This end is identified by seeing that the picture loses contrast and takes on a washed-out appearance as it is approached. The other end of the control is the position of least agc bias, and therefore greatest gain. When the control is turned toward this end, the picture successively gains contrast, starts to bend as sync becomes affected, becomes negative, and finally disappears. This last will happen if the signal level at some tube becomes high enough to develop a cutoff bias by grid-leak action and therefore block that tube completely. With low signal input, the bending may not occur until the control is well advanced.

Step 3. Adjust the manual control in the agc system for greatest contrast. When the picture begins to bend or become negative, back off until the bend stops and the picture becomes positive.

Step 4. Momentarily interrupt the signal by switching off channel and back. If the picture remains steady, the pull-in range of the sync circuits is satisfactory, and the adjustment is correct. If it bends, the manual control in the agc system should be backed off another fraction of a turn.

Step 5. Check operation and pull-in qualities on all channels.

Since the adjustment has been made on the strongest station, there should not be any overloading on the other channels. However, it sometimes happens that individual antenna and circuit conditions favor reception on one channel. If there is any sign of overloading on another channel, repeat the adjustment on that channel.

Adjustment Procedure for Noise-gate or Noise-limiter Controls. These controls are sometimes called noise gate, fringe lock, sync stabilizer, or noise limiter. Do not confuse them with the buzz control (sometimes also called noise limiter) which is used in connection with the 6BN6 gated-beam discriminator found in sound sections. If the set does not have a manual control in the agc system, omit step 4.

Step 1. Tune the receiver for best sound on the strongest channel. Freewheel the vertical- and horizontal-hold controls for a steady picture, and then disregard slipping. Set the contrast control to the full on position.

Step 2. Rotate the noise control through its range to determine its mode of operation.

In both the noise-gate and sync-stabilizer systems, the circuit is inoperative at one end of the control, and the set works as though it had an ordinary sync system. In the noise-gate circuit, the G-1 grid of the 6CS6 tube is made positive, allowing operation of the rest of the tube. In the sync-stabilizer system, the noise-canceling tube is cut off, allowing normal sync action through the rest of the circuit. At the opposite end, in both cases, overcontrol wipes out the sync pulse, and the set becomes unstable.

Step 3. Set the noise control to the end which gives normal sync action.

Step 4. Adjust the manual control in the agc system as follows:

a. Rotate the control back and forth through its range. Toward one end the picture loses contrast and becomes washed out. As the other end is approached, the picture successively becomes blacker, starts to bend, becomes negative, and finally disappears.

b. Set the control at the washed-out end and rotate it slowly toward the other. When the picture begins to bend or become negative, back off the control until the bend straightens and the negative becomes positive.

Step 5. In strong noise-free signal areas, the adjustment is now complete. In fringe areas, observe the picture for a time. If it begins to jitter, or roll, as noise pulses interfere with the sync action, advance the noise control slowly until the picture just becomes stable. Do not go beyond this setting, because overcontrol will wipe out the sync signals entirely.

Step 6. Momentarily interrupt the signal by switching off channel and back. If the picture remains steady, all adjustments are correct. If not, back off the noise control slightly, and try again.

Step 7. Check operation on all channels, and if necessary, repeat the adjustment.

Final adjustment of the automatic gain control and noise control are best carried out in the customer's home.

Adjustment of the Horizontal Frequency Controls. In many cases of out-of-sync operation of the receiver, after the previous procedure shows a normal mallet head and sync bar, the fault may be due to improperly adjusted horizontal frequency controls. These should be checked before the sync section is serviced. The controls should be adjusted in accordance with the manufacturer's instructions. If these are not available, generalized procedures are given for the several types of horizontal deflection systems.

In all cases, it is assumed that the receiver is in working order, except for the misadjusted controls. Sometimes the adjustment procedure is used as a servicing aid because the failure of a control to operate properly often points directly to the component at fault.

If the receiver is equipped with manual control in the agc system, noise-gate control, or sync-stabilizer controls, these should be adjusted first. If the receiver is in a badly out-of-sync condition, set the manual control in the agc system and the noise control to midway positions where they are not likely to reduce the sync signals. After the horizontal controls are adjusted, the automatic gain control and noise-elimination control are adjusted to their correct settings.

In many receivers, plate voltage for the horizontal oscillator and afc stage is taken from the B plus boost line. Boost voltage will vary with the setting of the drive control. Since a change in B voltage may cause a change in frequency, make it a general rule in your service work to complete any adjustments pertaining to width or drive before making horizontal frequency adjustments.

Adjustment Procedure for a Synchroguide System. The following procedure is for a synchroguide having four controls: horizontal hold, frequency, stabilizer, and lock range.

Preliminary Steps. Locate and identify the synchroguide adjustments. The horizontal-hold control is the potentiometer in the plate circuit of the afc tube. It is used as a fine-frequency control and is usually located on the front panel. The synchroguide coil assembly can has an adjustment at each end. The horizontal frequency or coarse-frequency control extends through the rear apron. The adjustment screw at the other end of the can tunes the stabilizer coil. The horizontal lock range trimmer is located nearby on the rear apron. If the set also has a horizontal drive trimmer, trace the wiring to make sure of the identification. The drive trimmer is connected to the output tube, and the lock trimmer is connected to the oscillator-afc tube.

Tune the receiver for best picture on a channel which normally gives good reception.

Any agc, fringe lock, or sync-stabilizer controls are set to a mid-range position. These are properly adjusted later after the horizontal frequency adjustments have been completed.

The vertical-hold control is set in the center of its lock-in range.

The horizontal-hold control is set to a mid-range position.

Step 1. Adjust the stabilizer coil as follows:

a. Set the oscilloscope to the positions marked for observing the waveform at the grid of the horizontal output tube. Connect the oscilloscope lead to the junction of the resistor and condenser on the synchroguide coil assembly can, as shown in Fig. 8-20. A 10-$\mu\mu$f isolating condenser and short leads are used to prevent undue loading of the circuit. The vertical gain setting on the oscilloscope may have to be reduced because the signal level at this point is higher than at the grid of the output tube.

b. Make sure the horizontal-hold control is at its mid-rotation setting, and adjust the coarse-frequency control for a synced picture.

c. Observe the waveform on the oscilloscope screen. Adjust the tuning slug of the stabilizer coil until the oscilloscope trace appears as shown in Fig. 8-20. The rounded and pointed peaks of the waveform must be of equal height. During the adjustment, if the picture goes out of sync, bring it back with the frequency control before continuing with the stabilizer adjustment. When the correct waveform is attained, remove the oscilloscope leads.

Step 2. Check the setting of the lock range trimmer as follows:

a. Set the horizontal-hold control to its extreme counterclockwise posi-

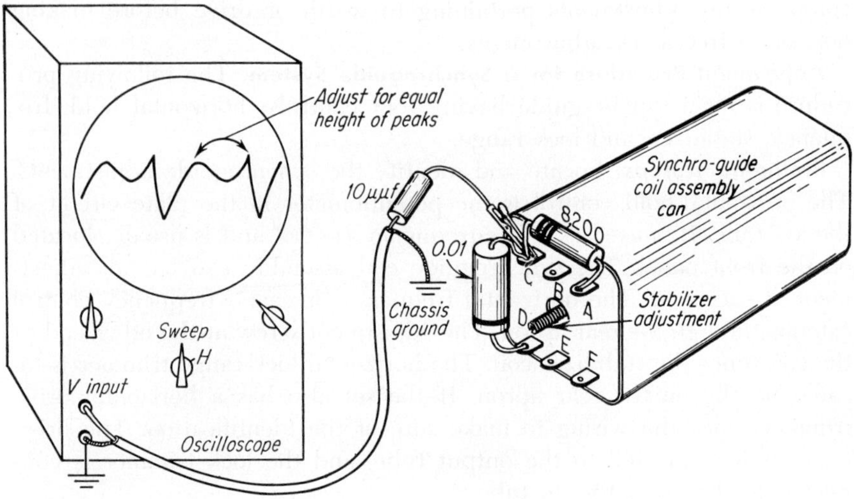

Fig. 8-20. Adjustment of the stabilizer coil in a synchroguide.

tion. If this does not cause the picture to break sync, momentarily switch off channel and back. The out-of-sync picture should show several bars sloping downward to the left.

b. Slowly turn the horizontal-hold control from its extreme counterclockwise setting and note that the number of bars decreases. Count the number of bars just before the picture falls into sync. If there are two or three bars, the lock range trimmer is at its correct setting.

c. If more than three bars are noted just before the picture falls into sync, tighten the lock range trimmer a fraction of a turn. If less, loosen it a fraction.

d. Repeat steps *a*, *b*, and *c* until two or three bars are counted before the picture falls into sync.

Step 3. Check the lock-in range as follows:

Rotate the horizontal-hold control back and forth to determine the extent of the lock-in range. On a strong station in strong-signal areas, this should cover the entire rotation of the control. In fringe areas, it should be about half the rotation of the control.

If the picture bends or breaks sync before these limits are reached, go on to step 4.

Step 4. Touch up the coarse-frequency setting as follows:

Set the horizontal-hold control at the position where the picture becomes unstable, and adjust the frequency control until the picture just falls back into sync. Recheck the lock-in range, and if it does not fulfill the conditions set down in step 3, repeat the touchup adjustment of the coarse-frequency control.

Special Fringe-area Synchroguide Adjustment. At the serviceman's discretion, it is possible to adjust a synchroguide system for improved noise stability at the expense of the pull-in range. When adjusting the stabilizer coil in step 1*c* above, adjust for a waveform where the pointed peak is approximately 10 per cent higher than the rounded peak. The rest of the adjustment procedure is the same, making allowance for a slightly decreased lock-in range. The customer should be carefully instructed in the use of the hold control, if marble-cake bars appear when changing from station to station.

Adjustment Procedure for a Synchroguide Having Two Controls. In some receivers, the horizontal-hold and lock-range controls are omitted. In these receivers, the coarse-frequency control in the synchroguide can is fitted with a knob for ease of adjustment, and also functions as a hold control.

Preliminary Steps. Locate and identify the synchroguide adjustments. In some receivers, both coils in the synchroguide circuit are contained in one can like the one shown in Fig. 8-20. In these, a plastic shaft and knurled knob extends through the rear apron as the frequency control. The stabilizer adjustment is at the other end of the can. In other receivers, the two coils are split. The oscillator tank coil is mounted on the front panel, so that the combined frequency-hold control is a front-panel adjustment. The sine-wave stabilizer coil is then mounted near the horizontal oscillator tube. Locate this coil, and also identify the lead to be used for the oscilloscope check. This is the connection which is wired to the center tap on the oscillator coil.

Tune the receiver for best picture on a channel which normally gives good reception.

Set the vertical-hold control in the center of its locked-in range.

Step 1. Check the need for adjustment. Turn the horizontal frequency control fully clockwise. The picture should be out of sync and show many bars, slanting to the left. Turn the control counterclockwise slowly. The number of horizontal bars should decrease to three or two when the picture pulls into sync. The picture should remain in sync for approximately two full turns, after which it should fall out of sync showing two to five bars.

If the picture does not hold sync for two turns beyond the pull-in point, the frequency-control system needs readjustment.

Step 2. Place a jumper across the sine-wave stabilizer coil. This is at terminals *C* and *D* if the coil is in a synchroguide can. Return the frequency control to the full clockwise position and back it off until the picture pulls into sync. Remove the jumper.

Step 3. Connect the oscilloscope through a 10-$\mu\mu$f condenser to the junction of the frequency and stabilizer coils. Turn the frequency control clockwise until the picture breaks sync, and then back it up until the picture just pulls into sync. The oscilloscope pattern should be like the waveform shown in Fig. 8-20. Adjust the tuning slug in the stabilizer coil until the two peaks are at the same height. During the adjustment, if the picture falls out of sync, pull it back in with the frequency control before continuing with the stabilizer adjustment.

Remove the oscilloscope leads.

Step 4. Check to see that the picture is held in sync for two full turns counterclockwise beyond the pull-in point. If this does not happen, repeat the adjustment.

Horizontal Adjustment Procedure for a Phase Detector

Preliminary Steps. Locate and identify the controls. The horizontal-hold control is on the front panel. The ringing coil is usually located on the rear apron. Do not confuse this coil with the width or linearity coils which may be mounted nearby. The ringing coil always has a mica condenser soldered across it, and is wired to the horizontal multivibrator.

Locate the phase-detector check point which is the junction of the two 100,000-ohm resistors in the output of the 6AL5 tube. This is point A in the basic phase-detector diagram of Fig. 7-27.

Tune the receiver for best picture on a channel which normally gives good reception.

Set the vertical-hold control in the center of its lock-in range.

Step 1. Check the need for adjustment. Rotate the horizontal-hold control to check the lock-in range. This should be about one-quarter of the full rotation. If the control is too critical, the horizontal frequency-control system needs adjustment.

Step 2. Set the tuning slug on the ringing coil to its maximum counterclockwise position, and short it out by connecting a jumper across its terminals. Short the afc check point to ground with another jumper. The multivibrator is running freely, and the stabilizing influence of the ringing coil has been removed.

Step 3. Adjust the horizontal-hold control to the point where the picture is almost stationary. Check to see that this point is near the center of rotation.

If the hold point is too far to either side, refer to the schematic diagram of the receiver, and see if it includes an auxiliary control to permit centering of the hold control. The auxiliary control is a variable stop resistor, or a variable coupling condenser between the two triodes of the multivibrator. If there is such a control, set the horizontal-hold control to its mid-range position, and adjust the auxiliary to get the picture almost stationary horizontally.

If the receiver does not have an auxiliary control, and the hold point is too far off center, the coupling condenser, the stop resistor, or the hold control are wrong in value. Check them against the values given in the schematic diagram. These components are the equivalent of *C*-5, *R*-5, and *R*-6 in the basic phase-detector diagram of Fig. 7-27. Check the resistors with an ohmmeter, and if these are normal, replace the condenser with one of correct capacitance. Adjust the hold control again to the point where the picture is almost stationary. With correct values in the circuit, this point will be close to the center of rotation.

Step 4. Remove the short from the ringing coil. Without touching the horizontal-hold control, turn in the tuning slug on the ringing coil until the picture again becomes almost stationary. Continue turning the tuning slug clockwise, noting that the picture holds for a few more turns. Back the tuning slug up to about the center of this holding range.

Step 5. Remove the shorting jumper from the afc check point, and check the lock-in range, as described in step 1.

Adjustment Procedure for a Synchrolock

Preliminary Steps. Tune the receiver for best picture on a channel which normally gives good reception. Set the vertical-hold control in the center of its lock-in range.

Locate and identify the controls. The horizontal-hold control is usually located on the front panel. Then look for the synchrolock coil assembly can. The coarse-frequency adjustment extends through the top of the synchrolock can where it can be easily reached. The phasing adjustment is on the opposite end of the can and is reached from the underside of the chassis.

Step 1. Check the need for adjustment by checking the pull-in and lock-in ranges as follows:

a. Rotate the horizontal-hold control over its full range. The picture should stay in sync.

b. Set the horizontal-hold control to its full on position. Momentarily interrupt the signal by switching off channel and back. The picture should pull in.

c. Repeat at the other end of the hold control. The picture should lock in and pull in across the full range of the hold control. If it fails to pull in at either end, the synchrolock needs adjustment.

Step 2. Set the hold control at the end where the picture failed to snap back. Break sync by interrupting the signal momentarily. Then adjust the frequency control on the synchrolock coil assembly can until the picture pulls back into sync.

Step 3. Recheck the pull-in range at both extreme positions of the hold control, and retouch the frequency control if needed. If a satisfactory adjustment cannot be made, or if the picture has a broad white arrow across it or is split by a black bar through the center, the phasing control needs resetting.

Step 4. Narrow the picture so that both edges are well within the mask opening. If the width control does not have sufficient range for this, use the drive control. Increase the brightness so that the edge of the raster is visible. Reducing the drive reduces the boost voltage and may change the horizontal frequency. If the picture goes out of sync, bring it back with the coarse-frequency control.

Step 5. Set the horizontal-hold control in the center of its rotation range, and adjust the phasing control so as to center the picture on the raster. If the picture goes out of sync during this adjustment, bring it back with the frequency control before continuing with the phasing adjustment.

Step 6. Restore the picture to normal width and brightness.

Step 7. Check to see that the picture pulls in at each end of the hold control, and retouch the frequency control if needed.

Checking Sync Pulses with an Oscilloscope. One further servicing technique must be introduced. The presence of the sync bar and mallet in the blanking bar shows that sync pulses are there, but it does not show their amplitude. A complaint like vertical roll, for example, may be due to a partially suppressed sync pulse. This is because the vertical oscillator generally requires a stronger triggering pulse than the horizontal. The amplitude of the sync pulses can be determined with an oscilloscope.

Your oscilloscope is already marked with the proper settings for observing the waveform at the grid of the horizontal output tube. These markings give a reference point to judge peak-to-peak amplitude of about 100 volts, and a 2-cycle trace at 15,750 cps. Two further reference markers should be added: the vertical gain setting for a lower amplitude, and the frequency setting which gives a 2-cycle trace when viewing a 60-cycle current.

Both jobs can be done at once by connecting the hot vertical input from the oscilloscope to a 6-volt point on the filament line in a receiver, and the ground terminal to the chassis. Use a 0.05-μf condenser for isolating purposes when working at 60 cps instead of the 200-$\mu\mu$f recommended for use at 15,750 cps. Reduce the sync control on the oscilloscope, and bring its coarse-frequency control down to a range which covers 30 cycles. Adjust the fine-frequency control to give an almost stationary 2-cycle trace. Steady the trace by advancing the sync control. Mark the fine-frequency setting, using a differently colored crayon to avoid confusion with the horizontal markers. Then adjust the vertical gain and positioning controls so that the trace fills the space between the masking tape markers on the oscilloscope screen. Mark this setting of the vertical gain control as 18 volts, which is the peak-to-peak value of a 6.3-volt rms potential.

Tune the receiver for a normal picture on a strong channel, and connect the input probe of the oscilloscope through the 0.05-μf isolating condenser to the plate lead of the video-amplifier tube. The oscilloscope is adjusted to give a 2-cycle trace at 60 cps, and the composite signal showing the two vertical sync pulses will appear on the oscilloscope screen. A representative trace is shown in Fig. 8-21*a*. Adjust the vertical gain control so

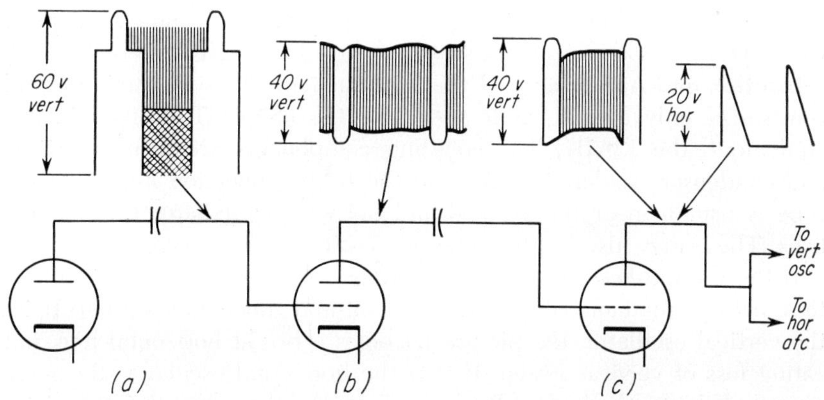

Fig. 8-21. Waveforms obtained with an oscilloscope along the sync-signal path. Scope is set at vertical rate except where noted. (*a*) Video output. (*b*) First sync clipper. (*c*) Second sync clipper.

that the trace is within the masking-tape reference levels. Note that this setting of the vertical gain control on the oscilloscope is close to the 100-volt peak-to-peak marker established at the grid of the horizontal output tube. This is because the signal amplitude at the plate of the video amplifier is 80 to 100 volts. Most receivers use cathode feed to the picture tube; so the composite signal is most likely to be positive-going with the sync pulses at the top. The smeared section between pulses is the video information which is changing with the picture scene. Look for many fine lines running up to the sync level. These are the horizontal sync pulses which are barely noticeable when the oscilloscope is adjusted to observe the vertical sync pulse.

Now compare the height of the sync pulse over the blanking level with the over-all height of the trace. It should be about one-quarter of the total height, since the sync pulse is contained in the top 25 per cent of the composite signal. If it is less than this, the sync pulse is being partially clipped in the i-f or amplifier stages, and we have the cause of the vertical roll.

The oscilloscope can also be used for tracing the sync pulses through the sync section. As shown in Fig. 8-21*b*, oscilloscope traces are often added to schematic diagrams, as an aid in servicing. The peak-to-peak voltage and frequency settings are indicated either in the manner shown or by some other method. Note the comparative height of the sync pulse at the grid of the first sync clipper tube. Note also the phase inversion through the system. The plate of the first tube shows the peaks going downward making this a negative-going pulse. It becomes positive again at the output of the second sync tube. The hash of fine lines between the vertical sync pulses is caused by the horizontal pulses. If the oscilloscope is shifted to the horizontal frequency, only the horizontal sync pulses will appear, as shown in the last waveform (Fig. 8-21*c*).

Location and Appearance of the Sync Section. The sync section usually consists of only a dual-triode tube like the 6SN7-GT or its miniature equivalent, the 12AU7. The coupling components are nearby resistors and condensers soldered directly to the tube connection lugs. The sync tube is usually near the vertical and horizontal oscillator tubes it controls. These may also be the same 6SN7-GT or 12AU7 types.

If the three tubes are close, and you are not sure as to their identification, lift one momentarily with the set on and tuned to a station. If it is the vertical oscillator, the picture becomes a bright horizontal line, indicating loss of vertical sweep. If it is the horizontal oscillator, the screen goes dark, since the high voltage stops. If it is the sync tube, the picture becomes scrambled, indicating loss of sync. If the symptom is an out-of-sync condition, removal of the sync tube may cause no change on the screen.

Servicing Procedure for the Sync Stages. When a normal mallet head can be seen in the blanking bar (see Fig. 8-19), sync pulses are present at the picture tube. If the deflection oscillators can be brought to their correct frequencies, the cause of off-frequency operation is pretty sure to be in the sync stages.

The common troubles in the sync stages are defective tubes and incorrect operating voltages. An open coupling condenser which breaks the sync signal chain is another possibility.

When trouble shooting in the sync stages, the first step is to replace the tubes, using substitutes that are known to be good. If this does not clear up the trouble, the next step is to make a careful voltage check, using the manufacturer's service notes as reference. Examine the notes to see if information is included as to the conditions under which the voltage readings are made. Set up the same conditions, since this may make considerable difference when making voltage checks in the sync section. Do not allow wide tolerances in your readings because correct clipping action is very dependent on operating potentials. Trace down the cause of any wrong voltage reading, and correct the condition.

Check coupling condensers by opening one lead and temporarily connecting in a condenser of the same size. Do not check coupling condensers by bridging with any available test condenser, because a wrong value will disturb the grid-leak bias.

The phase splitter and 6CS6 noise-gate circuits do not present any special servicing problems. Tube substitution and voltage analysis solve problems in these circuits just as in the simpler twin-triode sync sections.

The oscilloscope may be used in tracing the signal from the sync take-off point through the sync section. It provides the fastest means of finding a break such as an open coupling condenser, since there is signal on one side and none on the other. It will also trace either the horizontal pulses or the vertical pulses through the system simply by shifting the frequency setting on the oscilloscope.

General Servicing Procedure for Out-of-sync Operation. The various servicing techniques needed for working on out-of-sync troubles in a receiver have been described. They will now be applied to cure the various symptoms that have been assembled under the one heading: out-of-sync operation.

The most common complaint is usually called marble cake by the consumer, because the bars denoting loss of horizontal sync are on the screen in a pattern resembling marble cake. The picture may be out of vertical sync also, but this is not easily recognized. Lesser degrees of out-of-sync operation produce symptoms known as pulling, bending, etc.

The procedure set down is for any form of out-of-sync operation. An abbreviated procedure is given later for out-of-vertical-sync operation (rolling), followed by specialized procedures for bending and pulling, when caused by hum or other unusual factors.

Step 1. Turn the contrast control up to ensure a strong composite signal, and check the operation of the receiver on all channels. Try to sync in a picture on each station by manipulating the hold controls.

If a picture can be synced in on a station normally considered weak, look for trouble in the agc circuits. If the set has a manual control in the agc system or noise-immunity control, check their adjustment. This may be all that is wrong with the receiver. If there are no controls, or if they are correctly set, the agc circuit is checked as described in Chap. 12.

If all stations remain out of sync, we have no clue as to the cause of trouble and go on to the next step.

Step 2. Tune to a normally strong channel, and slowly adjust the horizontal frequency. In the case of a synchroguide or synchrolock, use the coarse-frequency control on the rear apron for this adjustment, after first setting the front-panel hold controls to their mid-range positions. For a phase detector, adjust the horizontal frequency by means of the front-panel hold control.

If a picture or a rolling picture comes on and holds horizontally, the misadjusted control is all that was wrong. Set the vertical-hold control. Check the lock-in and pull-in ranges of the horizontal system.

If a picture comes on but will not hold horizontally, the horizontal oscillator can operate at 15,750 cps, but the sync pulse is missing. Go on to step 3.

If the pattern on the screen cannot be resolved into a picture, the horizontal oscillator is working, but the frequency control will not bring it to 15,750 cps. It may be that the stabilizer coil prevents this. Short out the stabilizer coil in either the synchroguide or phase-detector systems, and again adjust the frequency control of the horizontal oscillator to a picture. If the picture comes in, leave the frequency control at the correct setting, remove the short, and adjust the stabilizer to bring the oscillator to frequency. Then go on to step 3. If the stabilizer control will not bring the oscillator to frequency, it is defective and must be replaced. If the pattern on the screen cannot be resolved into a picture even with the stabilizer shorted out, the horizontal oscillator frequency cannot be brought to 15,750 cps. Go on to step 4.

Step 3. Advance the brightness control, and freewheel the deflection oscillator by adjusting the vertical- and horizontal-hold controls so as to hold the blanking bar in view.

If the sync pulses do not appear in the blanking bar, the i-f, video, and agc circuits must be checked for clipping of the sync signals.

If a normal mallet head appears in the blanking bar, the sync signals are present at the picture tube. The sync section is at fault and is checked by signal tracing with the oscilloscope, tube substitution, and voltage analysis.

Step 4. In step 2, adjusting the frequency control of the horizontal system does not bring in even a momentary picture. The oscillator frequency cannot be brought to 15,750 cps. The fault is in the oscillator or afc stage. In order to further localize the trouble, the afc stage is removed from the circuit to see if the oscillator, working alone as a free-running unit, can be made to work at 15,750 cps and produce an almost stationary picture.

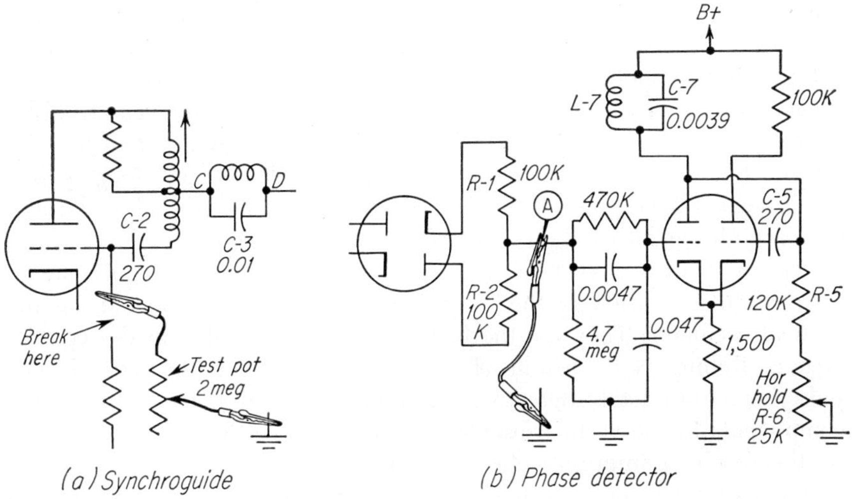

Fig. 8-22. How to disable the afc stage to see if the oscillator running freely will operate at 15,750 cps.

A different method is used to remove the afc action in each of the three common horizontal systems. In the synchroguide, it is done by disconnecting the oscillator grid leak and substituting a test potentiometer. The connections are shown in Fig. 8-22*a*. The potentiometer then becomes the main frequency-determining component. This is adjusted, and if an almost stationary picture can be brought in, the trouble is in the afc stage, which is checked in accordance with the procedure given in Chap. 7. If a picture cannot be brought in the oscillator stage is at fault. Grid condenser *C*-2 and synchroguide assembly *T*-1 are suspect.

The afc stage in the phase-detector system is disabled by shorting the diode output with a grounding jumper as shown in Fig. 8-22*b*. Also short out ringing coil *L*-7. Now, the frequency-determining components are condenser *C*-5 and resistors *R*-5 and *R*-6 between the two triodes in the multivibrator. Resistor *R*-6 is the hold control. Its adjustment should bring in an almost stationary picture. If it does not, one of these three components is at fault. If the picture comes on, remove the short from the ringing coil. Without touching the horizontal-hold control, adjust the ringing coil to hold the picture stationary. If the ringing coil throws the circuit off frequency and will not adjust to a picture, it is defective. If the almost stationary picture is brought on, the oscillator stage is functioning and the trouble is in the afc stage.

The afc stage in the synchrolock system is disabled by removing the reactance tube. This makes the frequency control in the synchrolock assembly can the main control of the free-running oscillator. If its adjustment brings in an almost stationary picture, the afc stages are at fault and must be checked. If the picture cannot be brought in, the frequency-determining components in the synchrolock must be checked.

Servicing Procedures for Vertical Sync Troubles. Vertical sync troubles include vertical rolling, jitter, or bounce. The horizontal system may be in sync but because of the extra control stage, it can get along with a much weaker sync pulse. Therefore, we must look for a weak or wrong-value condition rather than a completely defective unit.

Jitter or bounce may be caused by a weak vertical sync pulse or an intermittent condition in the frequency-determining components in the vertical oscillator. They are the first indication of trouble in the vertical system. Rolling is the result of a very weak or missing vertical sync pulse, or a completely defective condition in the frequency-determining components in the vertical oscillator. The weak vertical sync pulses can be the result of improper operation in the sync stages, or partial suppression of the sync information in the i-f and video stages. Our defective condition can be in the i-f and video stages, the sync stages, and the vertical oscillator. The following procedure is recommended for localizing the trouble.

Step 1. Set the contrast control on full, and check operation on all channels. Try the operation of the vertical-hold control on each station.

If the picture can be made to hold on a normally weak station, the trouble is a poorly operating or misadjusted agc system. Look for the manual control in the agc system and noise-immunity control and check their adjustment. If this does not help, check the agc system.

If the picture rolls on all channels, go on to the next step.

Step 2. Tune to a strong channel, and carefully adjust the vertical-hold control.

If the picture continues to roll in the same direction regardless of the position of the hold control, the vertical oscillator cannot be brought to 60 cps. If two pictures, one above the other, lock in, the oscillator is operating at 30 cps. Similarly, a double-exposure effect will appear if it is operating at 120 cps. Any of these results indicates that the vertical oscillator is operating at the wrong frequency. The defect is a wrong value in a condenser or resistor in the frequency-determining components in the vertical oscillator, or leakage in the vertical integrator couplate. A procedure for checking these is given in Chap. 5.

If adjustment of the vertical-hold control causes the direction of roll to change from up to down, or vice versa, adjust very slowly at the change point, and manually hold one picture in frame. This proves the oscillator will work at 60 cps. Go on to the next step.

Step 3. Increase the brightness, and adjust the vertical-hold control so as to observe the blanking bar. The mallet head and sync bar are probably present because we have horizontal sync, but if absent, check the video and i-f stages for clipping of the sync pulses.

If the mallet head is present, sync pulses are arriving at the output of the video amplifier. Go on to the next step.

Step 4. Set up the oscilloscope to check waveforms at the vertical frequency, and connect it to the input circuit of the first sync tube. If the tube is a 6CS6, connect the oscilloscope to the No. 7 pin. The composite video signal will appear on the oscilloscope, but it may not be normal.

First, compare the height of the sync pulse with the over-all height of the trace. If it is less than 25 per cent, check the video and i-f stages for partial clipping of the sync pulses.

If the sync pulse above the blanking level appears to be a full 25 per cent of the total height of the trace, check the peak-to-peak value of the trace against the manufacturer's service notes. If this is below normal, check the coupling to the sync take-off point. The video and i-f stages are probably working at full efficiency, because signs of a weak picture would have been noted before.

If the amplitude of the composite signal is normal, make an oscilloscope check at the vertical frequency through the sync stages, looking for incorrect clipping, as evidenced by the loss or weakening of the vertical sync pulse, or the presence of video signals.

Servicing Procedure for Gear Toothing and Christmas Treeing. Recognition of the symptom often provides a short cut to servicing. Two cases of this type are illustrated in Fig. 8-23. The symptom shown in Fig. 8-23*a* is variously called gear toothing, pie crusting, or cog wheeling. It is caused by overcontrol of the horizontal oscillator frequency by the afc circuits, a condition called hunting.

The antihunt circuits in the synchroguide and phase-detector systems are shown in Fig. 8-24. In the synchroguide circuit, the antihunt components are resistor *R*-6 and condensers *C*-5 and *C*-6 in the cathode circuit of the control tube. In the phase detector, they are resistor *R*-4 and condensers *C*-3 and *C*-4 in the control circuit of the multivibrator.

Gear toothing is usually caused by leakage or a wrong value in the

Fig. 8-23. Easily recognized symptoms of faulty sync operation. (*a*) Gear toothing. (*b*) Christmas treeing.

condenser, or a wrong value in a resistor. Servicemen usually replace all three components in each case. Choose components of the correct value, and with a 5 per cent tolerance rating. When replacing the condensers, be careful to connect the outside foil of a replacement condenser to the chassis.

The symptom shown in Fig. 8-23*b* is called Christmas treeing or gunboating. The zigzag pattern of bright horizontal lines often takes on the appearance of a Christmas tree, which accounts for one of the names applied to this symptom. The tree is usually accompanied by a staccato sound which seems to come from the high-voltage cage. This accounts for the name gunboating.

Christmas treeing occurs in the synchroguide oscillator-afc systems and is caused by a temporary failure to maintain a stable deflection fre-

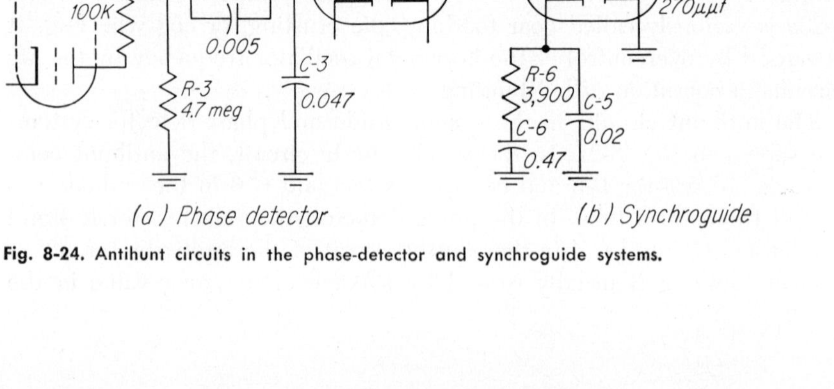

Fig. 8-24. Antihunt circuits in the phase-detector and synchroguide systems.

quency. Readjustment of the horizontal frequency controls including the stabilizer usually clears up this fault. If the symptom persists, another causative factor is a change in value of oscillator grid condenser *C*-2 in Fig. 8-24. Use a mica condenser of good quality as a replacement.

Bending, Pulling, Weaving, Flag Waving, Hook, and Whip. Depending on degree and form, horizontal disturbances in a visible picture have many names such as bending, pulling, weaving, flag waving, hook, and whip. These may be slight out-of-sync conditions, or they may be brought about by other factors, such as the introduction of hum. A knowledge of the effects of hum in the various circuits of a television receiver is important for understanding a procedure designed to locate the cause of these complaints.

Hum in the Picture. In radio receivers, when the power-line frequency gets into the signal chain, the result is a low-pitched growl called hum. The method of entry may be through cathode-heater leakage in a tube, or through the plate circuit because of inefficient filtering of the B supply. Cathode-heater leakage injects some 60-cycle current into the signal chain. If the B supply uses a full-wave rectifier, a 120-cycle ripple is added to the signals that are normally present. These are called the hum frequencies.

In a television receiver, the hum frequencies likewise enter a signal chain, either through cathode-heater leakage or by inefficient filtering of the B supply. But here there are many different signal chains, and so many different effects of the introduction of the hum frequency.

When the point of entry is the picture-sound chain, the hum frequency is added to the video signal. At the picture tube, this will cause the beam to be strengthened and weakened 60 or 120 times per second. In the case of 60-cycle hum, since the vertical sweep is also operating at 60 cycles, half the frame will be brighter than normal, and the other half will be

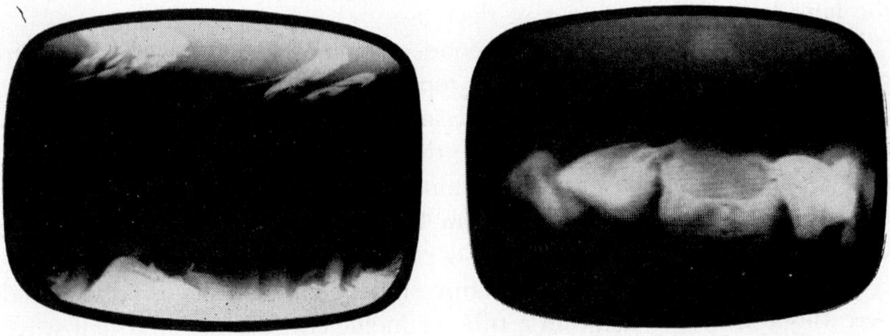

(*a*) (*b*)

Fig. 8-25. The effect of hum in the picture. (*a*) 60-cycle hum. (*b*) 120-cycle hum.

darker. This gives the effect shown in Fig. 8-25*a*. In the case of 120-cycle hum, the picture will be darkened and brightened twice in each frame, giving the effect shown in Fig. 8-25*b*. The dark sections of the screen are called hum bars.

The addition of the hum frequency to the video signal also has an effect on the horizontal sync-sweep system. Each line is displaced with respect to the last because of the hum modulation. As a result, the picture takes on the sine-wave characteristics observed on the illustration in Fig. 8-25*a*.

Servicing Procedure for Two Hum Bars. When two hum bars are seen in the picture like the illustration of Fig. 8-25*b*, the indication is that a 120-cycle hum frequency is getting into the picture-signal chain. This is caused by open filter condensers in the low-voltage power supply, or by open decoupling filters.

At the same time, there may be hum in the loudspeaker and a bend in the picture. These symptoms may be caused by the same defect which is causing the hum bars. The presence of the two hum bars is a sure indication of the inefficient filtering in the power supply, and this condition is cleared up first. It will probably clear up the other symptoms also.

The filter condensers are usually contained in electrolytic filter units, each one of which has two or more condensers in the same can. An easy check procedure is to clip the negative lead of a test condenser to the common negative connection on the condenser can, and touch the positive lead of the test condenser to each of the terminal lugs in turn, while watching for results on the picture. A test condenser of approximately 40 μf and 400 working volts will serve. Always connect to the common negative on the filter unit rather than to chassis, since the common negative is not always grounded. Repeat the check with all filter condensers found in the chassis. When the defective condenser is reached, the hum bars will disappear or decrease.

When replacing the defective condenser, it may not be necessary to replace the entire block. Choose a replacement condenser of the correct capacitance and voltage rating and solder its negative lead to the common negative on the block. Remove the leads from the defective section, and solder them to the positive terminal of the replacement condenser.

Servicing Procedure for One Hum Bar. When the picture shows one hum bar like the illustration of Fig. 8-25*a*, the inference is that a 60-cycle current is getting into the picture-signal chain. This is usually caused by cathode-heater leakage in a tube or inefficient filtering in a half-wave rectifier system. This includes sets that use the half-wave voltage-doubler circuit, and sets designed for a-c/d-c operation with a single selenium rectifier. In these receivers the ripple in the plate supply will be at 60 cps and will cause a single hum bar in the picture. If the receiver is of this type, check all filter condensers first, by the procedure just given for two hum bars.

The tubes are checked by substitution. This could be a long process in a receiver with 20 or more tubes; so a process of elimination is helpful. The tubes in the sweep circuits do not cause hum bars; so these are eliminated. The same applies to the tubes in the sound section. The hum bar is caused by leakage in any tube in the tuner, i-f, and video sections, and in the picture tube itself. To shorten this chain, remove the last i-f tube from its socket. If the set has a series filament string, short the heater terminals. If the bar disappears, the defective tube is in the tuner or i-f strip, including the last i-f tube just removed. These tubes are then replaced one at a time. If the bar remains on the screen, the defective tube is the detector, the video amplifiers, or the picture tube itself. The detector and video amplifiers are replaced first, and if these do not clear up the bar, the picture tube is at fault.

Confirmation of the defective picture tube may be obtained by connecting the same tube through an isolation type of CRT booster. This will isolate the heater in the picture tube and remove the hum bar.

Cathode-heater leakage of this type in the picture tube may be repairable. Remove the socket from the tube, and allow the heater to cool. Then charge a test filter condenser from the highest B plus point in the set, and connect the condenser leads to the cathode and filament of the picture tube directly at the No. 1 and No. 11 pins. The discharge of the condenser may burn away the leakage material in the tube.

Servicing Procedure for Bending, Etc. As mentioned before, bending, pulling, tearing, weaving, flag waving, hook, and whip may be caused by the introduction of hum, accompanied by hum bars. The procedures just given for servicing the hum bars would automatically clear any condition of bending, etc., that was brought about by the same cause. First decrease the brightness setting to make certain that the hum bar is not present.

Factors other than hum can cause horizontal distortion of the picture. Some of these are magnetic effects, video signals in the sync line, and improperly filtered vertical deflection signals in the horizontal circuit. The following procedure will trace all of these.

Step 1. Go through the horizontal adjustment procedure. The bend may be caused by the incorrect adjustment of one of the horizontal controls. Going through the complete horizontal procedure will automatically correct such a condition. If the fault is due to other causes, you may not be able to adjust some of the controls. This may give a lead to the trouble. If not, go on to step 2, and after the defect has been found and corrected, as a final step go through the complete horizontal adjustment again.

Step 2. Observe the edge of raster.

The wavy characteristic of the picture provides valuable clues as to the possible cause of the trouble. The characteristic can be most easily identified at the edge of the raster. Decrease the width or misadjust the centering device so that an edge of the picture appears on the screen. An illustration of a raster edge in a normally operating receiver is shown in Fig. 8-26*a*. Note the three labeled lines on the right side of the screen.

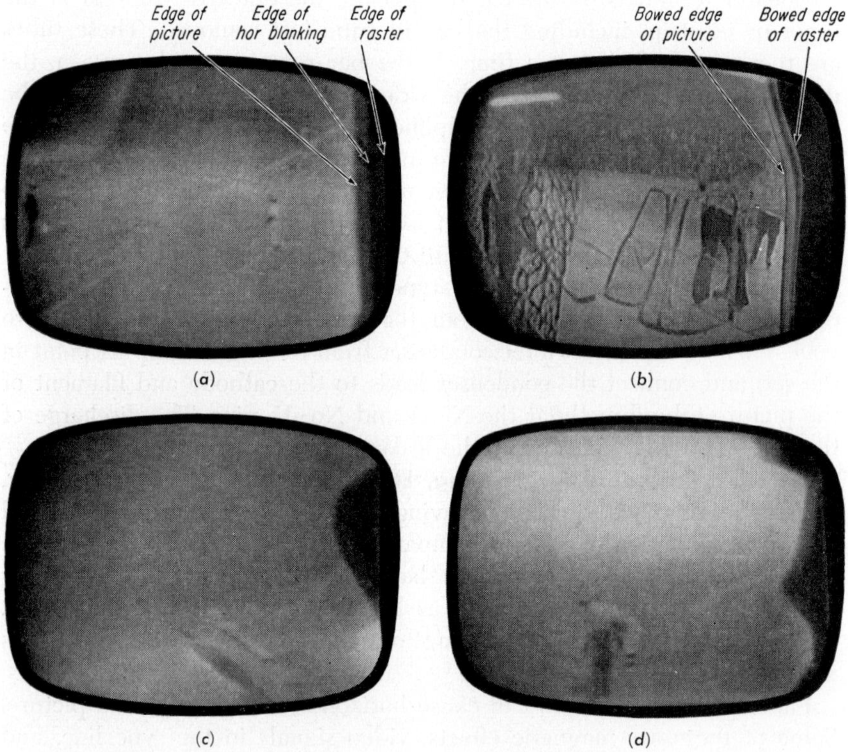

Fig. 8-26. Servicing for bend by observing edge of raster and edge of picture.

It may be necessary to increase the brightness in order to see the edge of the raster. Adjustment of the horizontal-hold control moves the edges about so that they may be more easily distinguished. The line marked "edge of horizontal blanking" may or may not be seen. This is of little importance, since the servicing discussion which follows deals only with the picture and raster edges.

The normal receiver of Fig. 8-26*a* produces straight edges for the raster and picture. In symptoms of bending or pulling, one or both edges are distorted, and the shape of the distortion is indicative of the defective condition.

If both raster and picture are bowed, as shown in Fig. 8-26*b*, the distortion is probably due to magnetic effects. The bowing may be either in or out. If the set uses a glass picture tube, look for pincushion or raster-corrector magnets extending out from both sides of the yoke housing. These are adjustable either by a sliding mount or by bending the mount. To adjust the pincushion magnets, center the picture and reduce the raster so that both sides are visible. The magnets are then moved forward or backward to straighten the sides of the raster.

In the case of a metal picture tube, the pulled raster is probably due to magnetization of the metal shell. This can be confirmed by loosening the picture tube and rotating it slightly. If the pulling on the raster moves with the rotation, the shell is magnetized and needs to be degaussed. Degaussing coils are obtainable from supply houses because of their usefulness in color-television work. Proceed as follows:

Remove the picture tube from the receiver, being careful to observe the precautions for handling tubes. Place it face down on the work bench, with a protective mat in between. The coil is held near the tube and in front of it, as shown in Fig. 8-27, and the current is turned on. The coil is

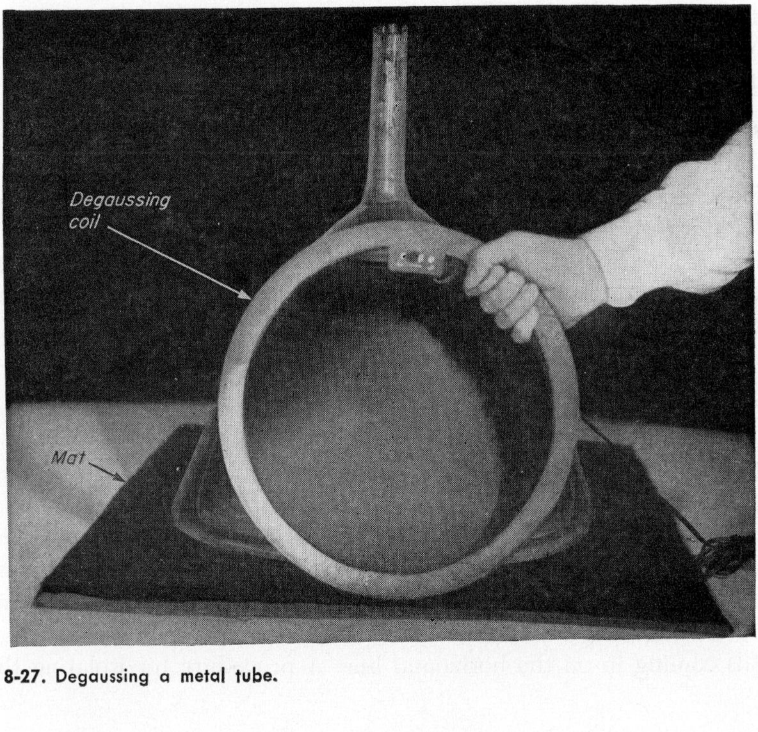

Fig. 8-27. Degaussing a metal tube.

then rotated in ever wider circles as it is moved back from the tube. At a distance of about 3 ft away, turn off the current. The tube is then turned about 45°, and the process is repeated until all areas of the metal shell have been covered.

If the raster edge shows two sine waves, the trouble is in the B supply filter circuit. Since there are no hum bars, the faulty condenser is not likely to be one of the main filter condensers, but a decoupling filter feeding the deflection or sync circuits. However, it is probably faster to locate and bridge all filter condensers than to trace the circuit; so check with a test condenser, while watching the raster edge, just as though two hum bars were visible.

If the raster edge shows a single sine wave, the indication is cathode-heater leakage in a tube. Since the hum bar is not present, the tube is not in the picture-signal chain but in the deflection circuit. Replace, one at a time, the afc, horizontal oscillator, horizontal output, and damper tubes.

If the raster edge is straight but the picture is bent, the deflection circuits are good. Since there are no obvious hum bars, the picture-signal chain is good, and the trouble is limited to the sync-afc stages. A straight raster with 60 cycles on the picture edge is shown in Fig. 8-26*c*, and one with 120 cycles on the picture edge is shown in Fig. 8-26*d*. As was described previously, the 60-cycle disturbance is likely to be cathode-heater leakage in a sync tube, and the 120-cycle indication points to filter or decoupling condensers in the low-voltage power supply. Go on to step 3 in the procedure for bending.

Step 3. Replace, one at a time, the tubes in the sync section, and the afc tubes.

The injection of the 60-cycle hum frequency by any of these tubes will add a 60-cycle hum component to the sync signal. This will pull the picture, without pulling the raster and without causing hum bars. If a sync or afc tube replacement cures the bend, the receiver is repaired. If not, the bend is still likely to be coming through the sync line but for some other cause. Go on to step 4.

Step 4. Open the horizontal sync line.

To confirm that the bend is coming through the sync line, the horizontal sync pulse is entirely removed from the horizontal sweep system. The break points for the synchroguide and phase-detector systems are shown in Fig. 8-28. The horizontal oscillator-afc systems are now running freely and will have to be held against the tendency to drift by free-wheeling the frequency controls. If the picture edge is now parallel with the raster edge, the cause of the bend is either video or vertical oscillator signals coming in on the horizontal line. A procedure for isolating these is given in step 5. If the bend persists, the trouble is an unstable condition in the oscillator-afc system.

This is likely to be in the antihunt circuit mentioned before for a symptom of gear toothing. If the symptom was not recognized previously, it now becomes evident. The components to be checked are condensers *C*-5 and *C*-6 and resistor *R*-6 in the synchroguide circuit, and condensers *C*-3 and *C*-4 and resistor *R*-4 in the phase-detector circuit.

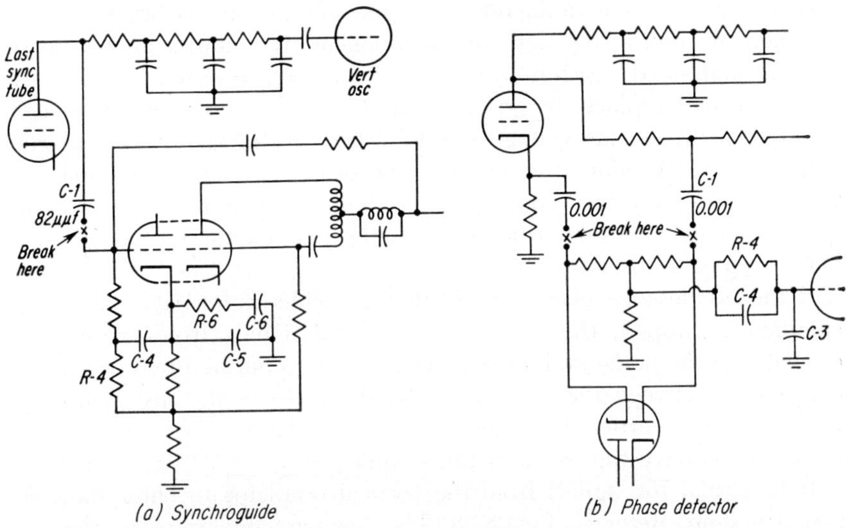

Fig. 8-28. How to remove the horizontal sync pulse.

The unstable oscillator-afc system may be caused by defects other than the antihunt circuit. If the antihunt components are good, all components in these stages must be systematically checked.

Step 5. Check for video in the horizontal sync line.

If the first sync tube is not operating at its correct bias, the clipping action may take place below the blanking level rather than slightly above it. Any video information which approaches the black level will be clipped along with the sync pulses. Because of the amplification of the following tubes, it will act like a sync pulse, tripping the horizontal oscillator out of step and causing flag waving in the picture.

The presence of video information in the sync line is suspected by experienced servicemen when the bend or pull is reshaped by a change of scene which causes a large change in the video content of the composite signal.

An oscilloscope check confirms the incorrect clipping. Normal traces at significant check points are shown in Fig. 8-21. The oscilloscope is adjusted to check the composite signal at the vertical rate and connected to the grid of the first sync tube. Here the entire composite signal, including the video information, is seen on the oscilloscope. Now shift the oscilloscope probe to the plate of the first sync tube. Only the inverted sync pulses with a straight line on top should be seen. If the straight line is fuzzy with video information, the bias is incorrect, and video is present in the sync signals.

To cure the condition, all voltages on the first sync tube are carefully checked against the manufacturer's service notes, to see if voltage analysis will help in finding the defective component. If voltages appear to be normal, plate-load resistors are checked for correct value. If these also appear normal, replace the grid condenser and grid leak, using 5 per cent components. The bias voltage is established by the grid-leak and condenser values. A value at an extreme end of the tolerance range may develop an incorrect bias voltage. Since these component values are frequently changed in later runs, refer to the manufacturer's service notes before replacing them.

If the oscilloscope check proves that the video information has been completely clipped, the bend is caused by signals from the vertical oscillator in the horizontal circuits. This can be confirmed by opening the connection between the integrator and the plate of the last sync tube. Freewheel the vertical-hold control against rolling, and if the bend is missing, you have the cause of the trouble.

It is normal for signals from the vertical oscillator to enter the horizontal system. Refer to Fig. 8-28. The low-pass integrator in the grid circuit of the vertical oscillator is a two-way road. The large signal developed by the oscillator will be present at the plate of the last sync tube, along with the vertical and horizontal sync pulses. Some of this signal will go through condenser *C*-1 to the afc circuit. In the synchroguide, condensers *C*-4 and *C*-5 and resistor *R*-4 provide a filter for the vertical component. In the phase detector, condensers *C*-3 and *C*-4 and resistor *R*-4 serve the same function.

When the signal from the vertical oscillator is removed by breaking its path at the integrator, the vertical filter is ineffective. If the bend is removed, replace the filter.

If the bend is still present, there is one other possible path by which the vertical signal may enter the horizontal oscillator-afc system. The vertical oscillator is usually connected to the B plus boost line. If the oscillator or afc stage is also connected to the boost voltage line there may be coupling between the stages if the decoupling filter condenser placed there for the purpose should open. Refer to the schematic diagram of the receiver, and check along the plate supply lines of the horizontal oscillator-afc and vertical oscillator stages, looking for a common decoupling filter condenser of about 10 μf. If there is one, check across it with a test filter condenser to see if this will remove the bend.

SUMMARY

Out-of-sync operation results in rolling, marble cake, or one of the various forms of bending of the picture. The fault may be due to a defective condition in the picture-signal chain, the sync section, or the frequency controls in the sweep sections.

Quick check for the sync section: Manipulate controls to observe the blanking bar. The presence of a normal sync bar and mallet head indicates that sync pulses are present at the input to the sync section.

Basic Sync Circuits. The schematic diagram of a basic sync section coupled to a synchroguide is given in Fig. 8-29. The phase-detector version is given in Fig. 8-30.

The following charts outline servicing procedures to be followed for the various symptoms of out-of-sync operation. Reference in the charts is to Figs. 8-29 and 8-30.

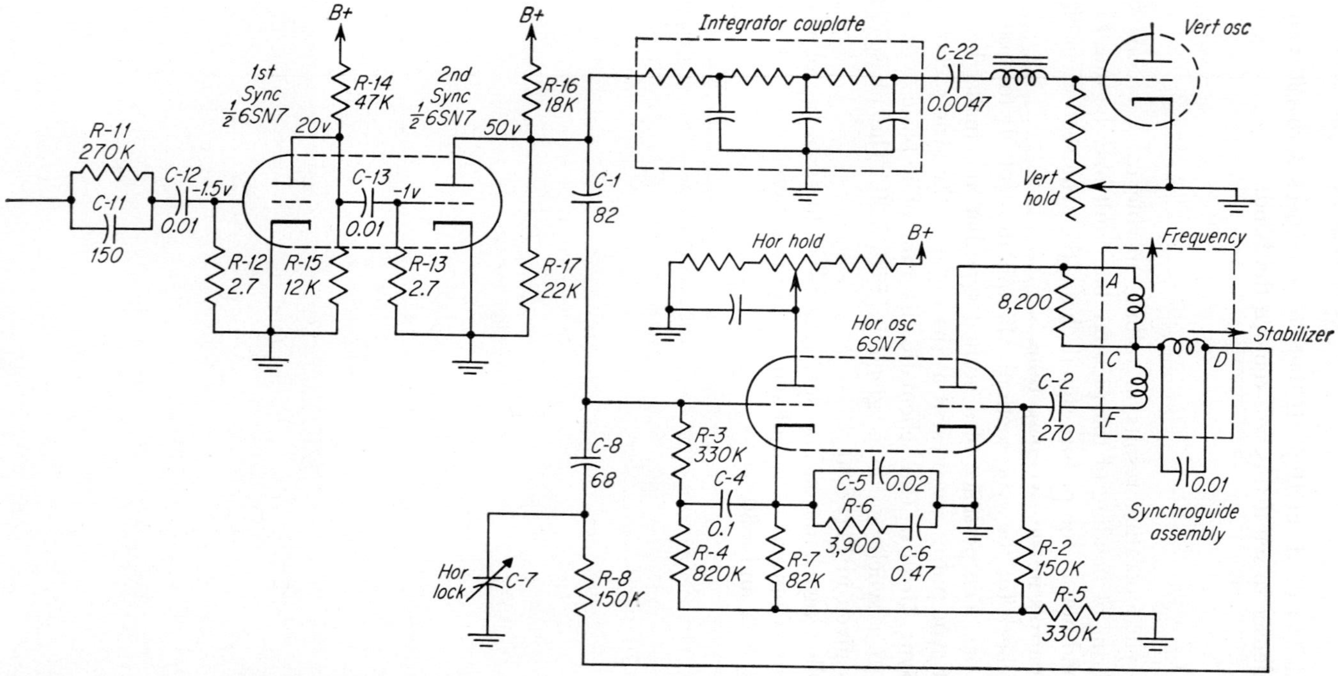

Fig. 8-29. Basic sync section coupled to a synchroguide.

Procedure for Complete Loss of Sync (Marble Cake)

Procedure	*Result*	*Trouble and Subsequent Check*
1. Check operation on all channels	Picture syncs in on weak channel	Sync clipping in picture chain Check setting of agc and noise controls Check i-f–agc system
	Set remains out of sync on all channels	Go on to step 2
2. Adjust horizontal frequency control (rear apron)	Picture syncs in	Horizontal controls out of adjustment. Make a complete horizontal readjustment
	Picture appears but will not hold	Sweep oscillator can be brought to frequency, but sync pulse is missing. Go on to step 3
	Pattern cannot be resolved into a picture	Sweep oscillators cannot be brought to frequency. Go on to step 4
3. Manipulate controls to observe vertical blanking bar	Normal sync bar is present	Sync section is at fault Replace sync tubes Make oscilloscope, voltage, and resistance checks
	Sync bar is absent	Check video and i-f stages for clipping. See Chap. 9
4. Disable the afc stage, and freewheel the oscillator control (see Fig. 8-22)	Single picture can be held in frame	Oscillator can be brought to frequency. Check components in afc stage
	Pattern cannot be resolved into picture	Check frequency-determining components in oscillator stage

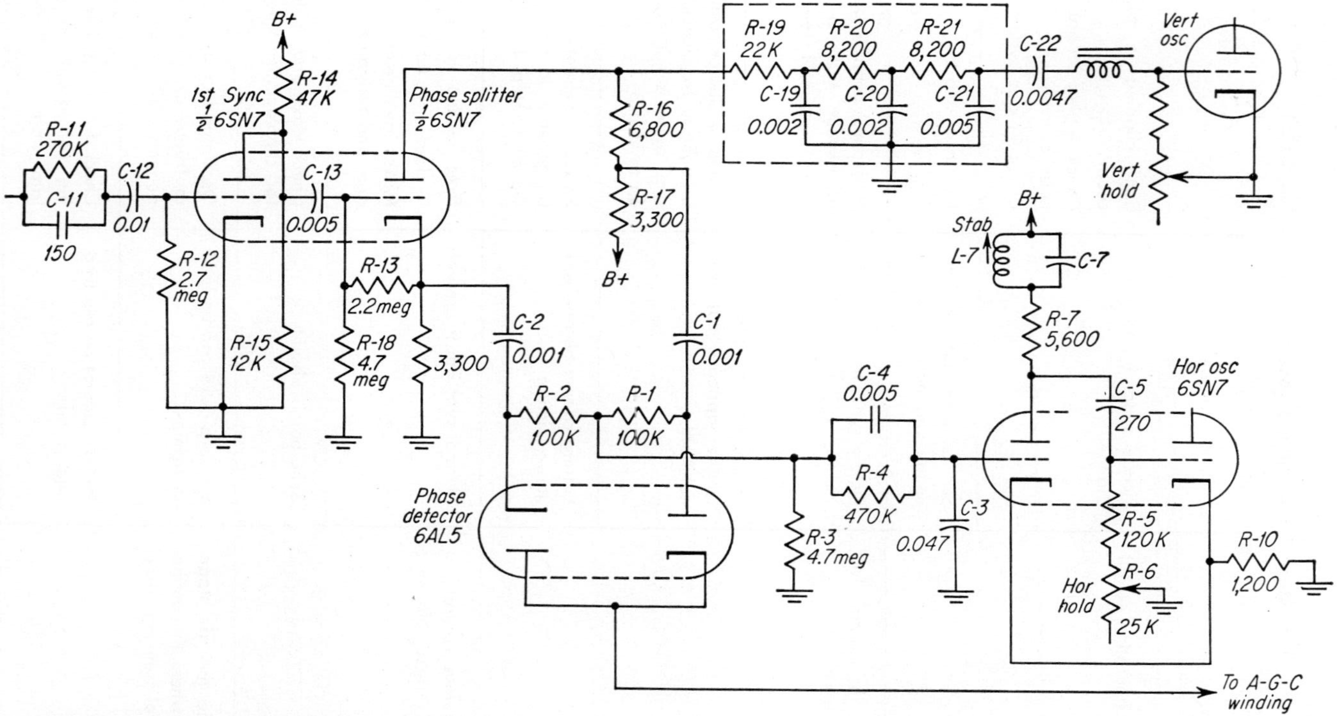

Fig. 8-30. Basic sync section coupled to a phase detector.

Servicing Procedure for Vertical Instability (Rolling)

Procedure	*Result*	*Trouble and Subsequent Check*
1. Check operation on all channels	Picture holds on weak station	Partial sync clipping in picture chain. Check i-f–agc system. See Chap. 9
	Symptom is present on all stations	Go on to step 2
2. Tune to strong station and observe operation of vertical-hold control on picture	Direction of rolling cannot be reversed	Vertical oscillator cannot reach 60 cps Check frequency-determining components in vertical oscillator and integrator
	Single picture can be manually held in frame but will not sync in	Vertical sync pulse is weak or missing. Go on to step 3
3. Make vertical oscilloscope check at input of sync section	Sync pulses are less than 25 per cent of total amplitude of composite signal trace	Check for partial sync clipping in i-f and video stages
	Normal trace	Vertical sync pulse weakened or lost in the sync section Make oscilloscope, voltage, and resistance checks of the sync section

Servicing Procedure for Partial Loss of Sync (Bending, Pulling, Tearing, Weaving, Flag Waving, Hook or Whip, and Vertical Jitter)

Make sure automatic gain control and noise control are properly set. If the symptom is accompanied by hum bars, service for the hum bars first. Decreasing the brightness control makes the hum bar more evident.

Procedure	*Result*	*Trouble and Subsequent Check*
1. Check horizontal adjustment	Symptom is cleared up	Horizontal controls out of adjustment. Go through complete readjustment procedure
	Symptom remains	Go on to step 2
2. Reduce width or misadjust centering so as to observe the edge of the raster	Raster edge is pulled or bowed	Magnetic effects Readjust pincushion magnets Degauss a metal-shell picture tube
	Raster edge shows two sine waves	Hum in the B supply of the deflection circuits Check filter condensers
	Raster edge shows one sine wave	Hum introduced by defective tube in the sweep circuits. Replace, one at a time, the horizontal-afc, horizontal output, and damper tubes
	Raster is straight but picture is bent	Defective condition in sync and afc stages. Go on to step 3

Servicing Procedures for Partial Loss of Sync (Continued)

Procedure	*Result*	*Trouble and Subsequent Check*
3. Replace one at a time the sync, afc, and horizontal output tubes.	Bend straightens	Defective tube
	Bend remains	Defective condition may be: (*a*) unstable oscillator-afc system, (*b*) video in the sync signals, (c) unfiltered vertical oscillator signal in the afc stage. Go on to step 4
4. Open the horizontal sync line, and freewheel the horizontal oscillator	Bend remains	Unstable horizontal oscillator-afc system Replace antihunt circuit: Synchroguide: *R*-6, C-5, C-6. Phase detector: *R*-4, C-3, C-4 Check all frequency-determining components in oscillator-afc stage
	Bend straightens	Video or vertical oscillator signals in the horizontal oscillator. Go on to step 5
5. Make an oscilloscope check at the plate of the first sync tube	Video signals are present	Improper biasing of first sync tube Check plate voltage, and plate load resistor. Replace grid leak and condenser C-12, *R*-12
	Trace is normal	Improper vertical filtering in afc stage Replace: Synchroguide: *R*-4, C-4, C-5. Phase detector: *R*-4, C-3, C-4

Servicing Procedure for Hum Bars

Procedure	*Result*	*Trouble and Subsequent Check*
1. Look for one or two hum bars	Two bars	Poor filtering in low-voltage power supply Check filter condensers
	One bar	Check set for single selenium rectifier. If only one, check filter condensers. If two, hum enters through defective tube. Subdivide search as suggested in step 2
2. Remove last i-f tube	Hum bar remains	Replace, one at a time, video detector, amplifier, and picture tube
	Hum bar disappears	Substitute last i-f tube, and continue, one at a time, with the rest of the i-f and tuner tubes

QUESTIONS

1. What are some of the results of out-of-sync operation of a television receiver?
2. What is meant by the composite video signal?
3. What is meant by synchronization in a television system?
4. Draw a diagram of the composite video signal indicating blanking level, sync pulses, and video information.
5. What is the proper amplitude ratio of sync pulses to the entire composite signal?
6. What signal level represents white? black?
7. What is meant by the term "blacker than black?"
8. Why is the vertical sync pulse serrated?
9. What is the function of the sync section?
10. How are the sync pulses removed from the composite signal? What is meant by clipping?
11. How are the vertical and horizontal sync pulses separated from each other? Define integration and differentiation in this regard.
12. Draw a diagram of a basic sync section using a dual triode. Indicate the waveform at the grid and plate of each tube.
13. Draw a diagram of a basic two-stage sync section, where the last stage is connected as a phase splitter for operation with a phase-detector system.
14. Describe lock-in and pull-in ranges.
15. Describe what is meant by signal-to-noise ratio.
16. What is the result of improper agc action in a strong-signal area? in a fringe area?
17. Why are noise-reducing circuits of more importance in a fringe area?
18. Out-of-sync operation may be caused by faulty sync signals or faulty sweep generators. Explain this statement.
19. What is the quick check for normal sync input to the sync section?
20. What is the indication that a sweep generator can be made to operate at its correct frequency?
21. What is meant by freewheeling a horizontal sweep generator?
22. What is the significance of the sync bar and mallet head? Why must the brightness control be advanced to observe the presence of the sync bar?
23. Describe the horizontal adjustment procedure for a synchroguide system. If such a receiver is available, run through the adjustment.
24. Describe the horizontal adjustment procedure for a phase-detector system. If such a receiver is available, run through the adjustment.
25. Describe a procedure for running down the cause of a complete loss of sync.

26. A receiver comes in for complete loss of sync. Freewheeling the oscillators produces a normal sync bar. What inferences can you draw from this? What is the next step in the procedure?

27. A receiver comes in for vertical roll. Adjusting the vertical-hold control changes the speed of the rolling, but not the direction, and a single picture cannot be obtained. What inference can be drawn from this? What is the next step in the procedure?

28. What is the significance of two hum bars in the picture? one?

29. Why is it that hum introduced into the sync or sweep circuits will not cause hum bars? What will it cause?

30. What is the advantage of removing the last i-f tube when servicing for one hum bar?

31. Why will an isolation type of CRT booster remove the hum bar when there is cathode-heater leakage in the picture tube?

32. A receiver comes in with the complaint that the picture is pulled out of shape, and none of the horizontal controls correct it. An examination of the edge of the raster shows a single bow instead of a straight line. What inference can be drawn from this?

33. Describe the degaussing procedure for a metal-shell picture tube.

34. A receiver comes in with the picture pulled out of shape horizontally. An examination of the edge of the raster shows that the raster is straight but the picture is pulled. What inference can be drawn from this? Describe a procedure for finding the trouble.

35. What is likely to happen when video information gets into the sync signals? What is this due to? Which components are likely to be at fault?

9

Servicing the Picture Chain: Video-amplifier and Picture-tube Circuits

A sectional diagram of the picture-signal path is given in Fig. 9-1. Reference to this diagram shows that a break at any point in the chain from the antenna to the picture tube will stop the picture. Of course, the picture will also be stopped by a defective condition in the low-voltage or high-voltage power supplies, but this stops the raster as well. The

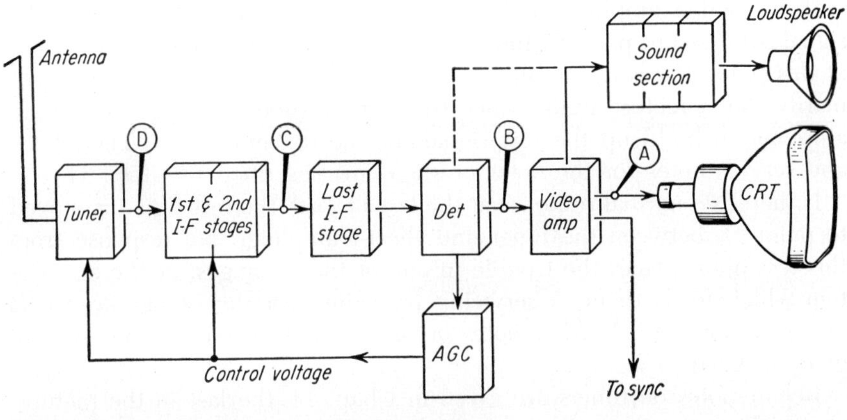

Fig. 9-1. Check points in the picture-sound signal chain.

absence of light on the screen of the picture tube takes precedence, since the picture cannot be seen on a blank screen. Conversely, the presence of a normal raster indicates normal operation of the power supply and sweep sections, and then the absence of picture puts the trouble in the picture-signal chain. Therefore, the symptom which leads to an investigation of the picture-signal chain is "no picture—raster normal."

In most cases, a break in the picture chain also stops sound. This is because the sound take-off point is well along in the chain, in the output of the video amplifier or detector. If the sound section is functioning, sound will be restored when the break in the picture chain is found and corrected. In order to present a universal servicing procedure which may be applied regardless of the location of the sound take-off point, the presence or absence of sound is neglected.

Subdivisions in the Picture-signal Chain. The picture-signal chain is primarily a superheterodyne receiver, incorporating many tubes and many functions. The servicing procedure to be presented in this and the next chapters starts at the picture tube and works backward toward the antenna. The long chain is broken up into logical servicing subdivisions. These are indicated by the circled letters in the block diagram of Fig. 9-1. Quick signal checks have been devised to place the trouble before or after each of these check points in the chain.

The quick signal check at point *B*, the input of the video amplifier, tells whether the video-amplifier and picture-tube circuits are working. If they are not, subsequent tests at point *A* locate the fault in the video amplifier or the picture tube. These checks, together with servicing hints for the video amplifier, are the business of this chapter. The next chapter deals with repair and replacement notes for the picture tube.

If the video amplifier and picture tube are working normally, the signal is moved to point *C*, the last i-f grid circuit. An abnormal response at this point places the trouble between test points *B* and *C*. This is primarily the detector circuit, since the last i-f tube is being used as an amplifier to build up the test signal for the detector stage. This check and service notes for the detector stage are described in Chap. 11.

If there is a normal response at the last i-f grid, the test signal is moved to point *D*, between the tuner and the i-f amplifier. No response from this test point places the trouble in one of the i-f stages, or the agc system which feeds them. A servicing procedure for the i-f–agc system is given in Chap. 12. Further notes on simple and keyed agc systems are given in Chap. 13.

Repair notes on tuners are given in Chap. 14, the last in the picture-chain series. For a complaint of no sound or poor sound—picture normal, a servicing procedure for the sound section is given in Chap. 15, and the picture-sound signal chain is completed.

Two very quick checks can shorten this routine procedure. One is to decouple the antenna for symptoms of no picture, reversed picture, and loss of sync stability. If this restores a degree of normalcy, we have an overloaded condition, probably caused by trouble in the agc system. Look for a misadjusted manual control in the agc system, and if there is no control or the adjustment does not improve matters, adjust the coarse horizontal frequency control through its range, even though there is no picture on the screen. This is to take care of the possibility that an out-of-sync condition is causing the overload in a keyed agc system. If this does not restore the picture, start with the procedure for checking the i-f–agc system described in Chap. 12.

The other very quick check for shortening the picture-chain servicing procedure is to look for snow on the picture or raster. Snow is a random dispersion of dots on the screen which is caused by noise pulses in the signal. Assume a picture chain with a weak or dead tube in the tuner, and all else in normal working order. The weak tube passes the signal, but at low amplitude, and it will finally appear as a weak picture. The dead tube passes nothing, and there will be no picture. However, in either case, the agc system steps up the gain of the i-f amplifier to maximum. Shot effects generated in the early i-f tubes are amplified by the rest of the chain to noise pulses, and these register as snow on the picture tube. If the weak or dead stage is farther along in the chain, shot effects generated behind it are not sufficiently amplified to produce snow. The appearance of snow, therefore, coupled with no picture or weak picture, places the fault near the antenna end of the picture-signal chain. The serviceman can neglect the routine checks and start his investigation in the tuner.

Quick Check for the Video Amplifier and Picture Tube. For a symptom of no picture without snow, feed the audio output of a signal generator to the input of the video amplifier at the diode-load test point. The appearance of dark horizontal bars across the face of the picture tube indicates that the video-amplifier and picture-tube circuits are working, and the serviceman goes on to check the detector stage.

Function of the Video Amplifier. The signal needed for proper operation of the picture tube has a peak-to-peak amplitude of approximately 80 volts. The normal detector output is a signal with a peak-to-peak amplitude of about 5 volts. The function of the video amplifier is to build up the detector output to provide sufficient signal for the operation of the picture tube.

The function is similar to that of the audio amplifier in a sound system, except that the video amplifier must operate over a much wider range of frequencies. A change in shading level as the picture is scanned from one spot in a line to the next could correspond to a frequency change of as much as 4 Mc. Also, the change from frame to frame corresponds to a frequency of 30 cps. The video signal therefore contains elements varying in frequency from 30 cps to 4 Mc, and the video stage must give reasonably flat amplification over this range.

Basic Diagram of a Video Amplifier. Most video amplifiers consist of a single stage feeding a positive-going signal to the cathode of the picture tube. This is the type chosen for study as the basic circuit. A simplified version is given in Fig. 9-2.

The stage is primarily an *R-C*–coupled amplifier with wide-band characteristics. The tube is a sharp-cutoff pentode like the 12BY7. Signal check points *A* and *B*, used in the quick checks, are indicated in the diagram. Point *A* is where the output signal is fed to the picture tube, and point *B* is where the input signal is fed in from the detector stage. Point *B*, called the diode-load check point, is of considerable importance in television service work.

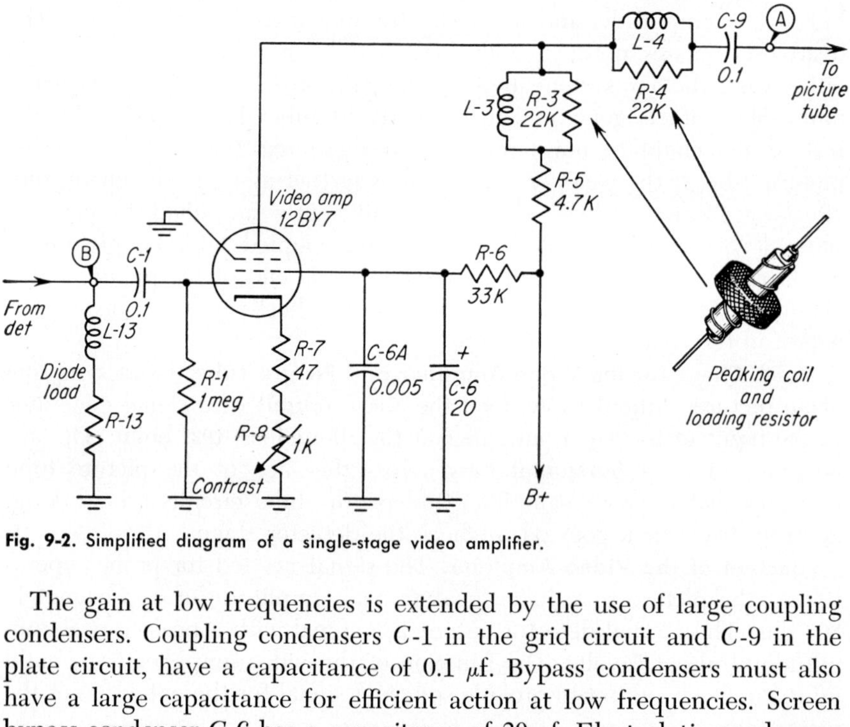

Fig. 9-2. Simplified diagram of a single-stage video amplifier.

The gain at low frequencies is extended by the use of large coupling condensers. Coupling condensers *C*-1 in the grid circuit and *C*-9 in the plate circuit, have a capacitance of 0.1 μf. Bypass condensers must also have a large capacitance for efficient action at low frequencies. Screen bypass condenser *C*-6 has a capacitance of 20 μf. Electrolytic condensers have an inductive component which makes them ineffective as bypass condensers at high frequencies. Since the amplifier is meant to work at high frequencies as well, a noninductive condenser *C*-6*A* of 0.005 μf also bypasses the screen voltage.

In an ordinary *R-C*–coupled amplifier as used for audio work, gain at high frequencies drops off because of the bypass action of the capacitance to ground of the tube electrodes and circuit wiring. In the grid circuit of the video amplifier, the gain at the high frequencies is extended by r-f coil *L*-13. Coils *L*-3 and *L*-4 counteract the output capacitance in the plate circuit of the tube. The coils are called peaking coils since they are designed to resonate (peak) with the inherent capacitance of the circuit at high frequencies. This increases the impedance at these frequencies, and therefore the gain. The peaking coils tend to produce very high gain at the resonant frequency. The sharp peak is reduced, and the response is broadened by loading the coils. Resistors *R*-3 and *R*-4 in Fig. 9-2 serve this function. A peaking coil and loading resistor combination is usually constructed as one component, with the coil mounted on the resistor, as shown by the drawing in Fig. 9-2.

Another factor which helps maintain equal response over a wide band of frequencies is the use of a low value for the plate-load resistor. This reduces the over-all gain of the amplifier, but the reduced level is extended to higher frequencies before the shunting action of the output capacitance cuts the gain. The plate load varies from 4,000 to 6,000 ohms in different receivers. The average value of 4,700 ohms has been chosen for plate-load resistor *R*-5 in the standard circuit.

Contrast Control R-8. Resistor *R*-7 and variable resistor *R*-8 in the cathode circuit act as a gain control for the video amplifier in a dual manner. Since the cathode is unbypassed, the circuit is degenerative, thereby reducing gain. Decreasing the cathode resistance decreases the degenerative effect, and so increases the gain. At the same time, it decreases the bias, which also tends to produce more output. Resistor *R*-7 is a stop resistor.

The gain control will increase and decrease the amplitude of the video signal fed to the picture tube. A picture element meant to be black should be at a signal level of about 60 volts. If the video signal is at a low level, the black elements will be at a decreased voltage, let us say, 30 volts. This will register as gray. A picture where maximum black registers as gray will have a washed-out appearance and lack contrast. Video-amplifier gain control *R*-8, in providing a variation of output signal level, acts as the contrast control for the receiver.

Actually, there is only one correct setting of the contrast control—the setting which places the black level at the cutoff bias of the picture tube. The contrast control is a front-panel unit since it will compensate for changes in over-all gain due to aging, and in addition will give the consumer the opportunity to adjust the average contrast to his preference.

Sound and Sync Take-off Points. A standardized diagram of a video-amplifier and picture-tube circuit is given in Fig. 9-3 to add the features omitted from the simplified version of Fig. 9-2. The sync take-off point is shown at the junction of plate-load resistor *R*-5 and peaking coil *L*-3. Thus the sync feed is at the output of the video amplifier, a connection almost universally employed.

The sound take-off point is at the plate of the video-amplifier tube through condenser *C*-7. Coil *L*-5 and condenser *C*-5 form a sound trap to keep the 4.5-Mc sound signal out of the picture-tube circuit. The adjustment procedure for this trap is given in Chap. 15. In some receivers, the sound take-off point is in the output of the detector circuit.

Picture-tube Connections. Condenser *C*-9 couples the amplified video signal to the cathode of the picture tube. In a cathode-fed picture tube, the control grid is usually at ground potential. Also connected at the cathode are resistors *R*-9 and variable resistor *R*-10, going to a B plus point. These establish an average bias on the picture tube, thereby

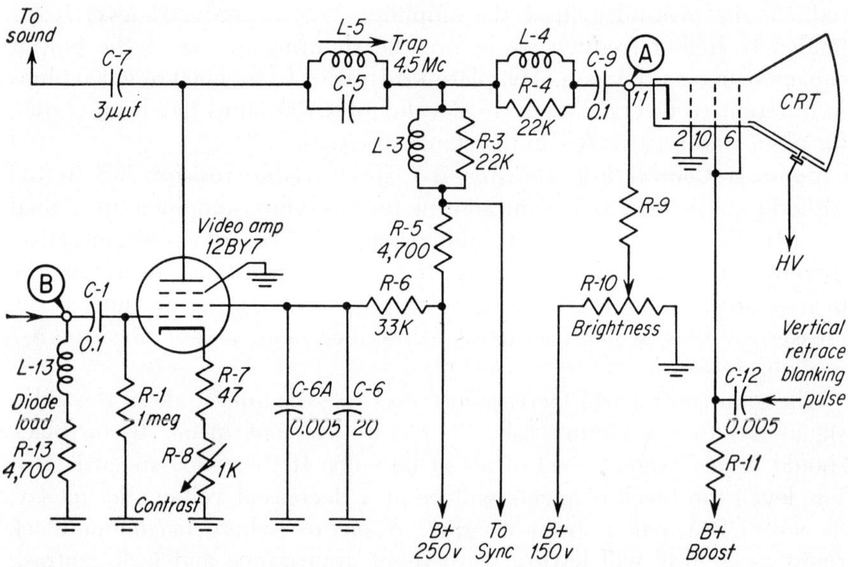

Fig. 9-3. Schematic diagram of basic video-amplifier and picture-tube circuits.

controlling the average beam current. The video signal fed in through condenser *C*-9 causes the bias to change, increasing and decreasing the beam current. These changes in turn are translated into light and dark picture elements on the screen of the tube. Resistor *R*-10 is made variable so as to give a control of the average beam intensity, and so becomes the brightness control for the receiver.

The first anode at the No. 10 pin is connected to a positive voltage from the boost line. A negative-going pulse from the vertical output transformer is also coupled to the first anode through condenser *C*-12 for vertical retrace blanking. This negative pulse cuts off the tube during retrace time by canceling the positive voltage from the boost line. Older receivers do not include provision for vertical retrace blanking. In these, at advanced positions of the brightness control, the beam can be seen on its retrace trip to the top of the tube as a series of bright slanting lines.

Pin No. 6 in the picture tube is the focusing anode. In the basic circuit of Fig. 9-3, it is shown connected to the first anode, a very common arrangement. Other connections for this electrode are described in Chap. 10.

Location and Appearance of the Video Amplifier. The characteristic which identifies the video amplifier is the peaking coils. A drawing of a peaking coil is included in Fig. 9-2, and two or three of these near a socket identify the video amplifier. For confirmation, see that this tube is located near the neck of the picture tube.

NORMAL TEST DATA FOR THE VIDEO AMPLIFIER

Quick Check for the Video Amplifier and Picture Tube. The symptom of normal raster and no picture may be caused by a break at any point in the long picture-signal chain from the antenna to the picture tube. The first step in servicing for this symptom is to determine whether the video amplifier and picture tube are working.

Locate point B (see Fig. 9-4), the top end of the diode load. Some manufacturers bring this point out to a special terminal for convenience in servicing.

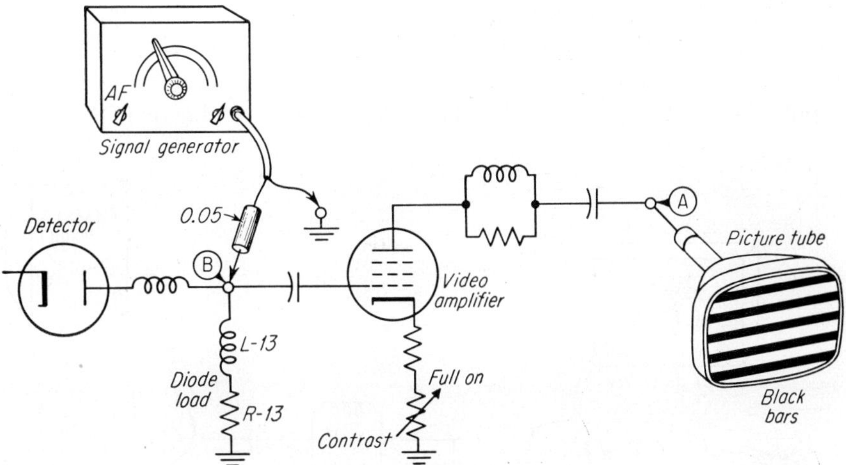

Fig. 9-4. Quick check for the video-amplifier and picture-tube circuit.

Tune the receiver to an unused channel, and adjust the controls for a normal raster. Feed the audio output of a signal generator to the diode-load test point through a 0.05-μf isolating condenser. If the generator has a control of the audio output, set it to maximum.

The audio output of the average signal generator is at a frequency of approximately 400 cps and at an amplitude usually greater than the

5-volt peak-to-peak signal normally found at the diode load. If the video amplifier is working, the amplified signal will appear in the plate circuit and cause the electron beam in the picture tube to be strengthened and weakened 400 times a second. Since the vertical deflection system is working at 60 cps, the changing beam current at 400 cps will produce six or seven (400/60) alternate dark and light horizontal bars across the face of the picture tube. It may be necessary to adjust the vertical-hold control to keep the bars steady. The bars should be a rich black because the audio output of the signal generator is usually greater than the average video signal at the diode load.

The appearance of the bars shows that the video amplifier and picture tube are working, and the serviceman goes on to check the next section in the picture chain, the detector. If the bars do not appear, either the video amplifier or the picture tube is not functioning, and the serviceman goes on to further localizing checks.

Quick Check to Determine Whether the Lack of Picture Is Caused by the Video Amplifier or Picture Tube. In the previous check, the output of the signal generator was fed to the diode load of the receiver, which is adjusted for normal raster on an unused channel. Leave these connections undisturbed, and trace the output lead of the video amplifier from the plate pin to the picture tube (point *A*). Adjust the oscilloscope to observe traces at the vertical frequency (30 cps), set the vertical gain control of the oscilloscope to zero, and connect the probe to the signal-input lead on the picture tube. The connections are shown in Fig. 9-5.

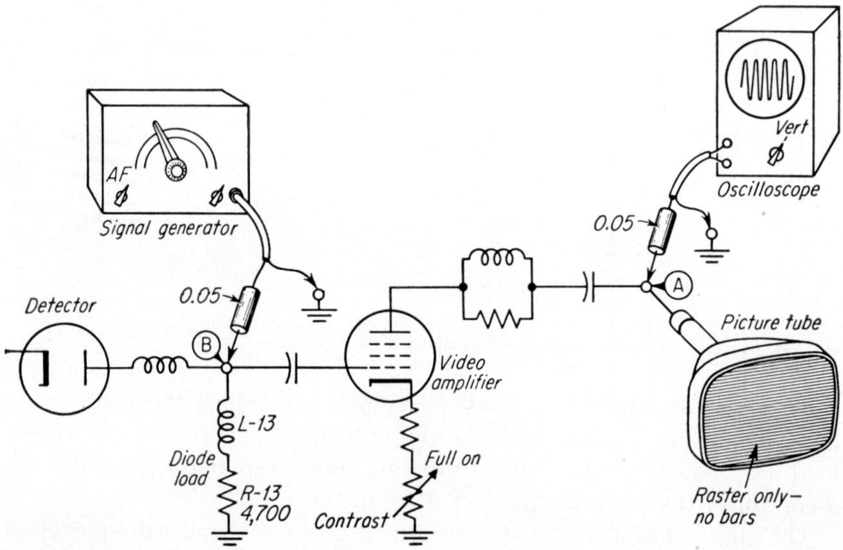

Fig. 9-5. Quick check to determine if trouble is located in video-amplifier stage or in the picture tube.

Then increase the vertical gain adjustment on the oscilloscope while looking for a trace on its screen.

The audio output of most signal generators is a sine wave, or near sine wave, at about 400 cps. If the video amplifier and picture tube were working normally, the picture tube would register the audio input as black horizontal bars, and the oscilloscope would show a trace of the amplified signal. Since the oscilloscope is adjusted to sweep at 30 cps, the audio input would cause a trace of 12 or more sine waves on its screen. These would probably be moving across the screen and could be steadied by a slight readjustment of the sweep frequency control on the oscilloscope. For the present check, we are interested in the appearance of the trace, moving or steady. If it is seen, it is proof that the video amplifier is working. The absence of bars proves that the picture tube is defective.

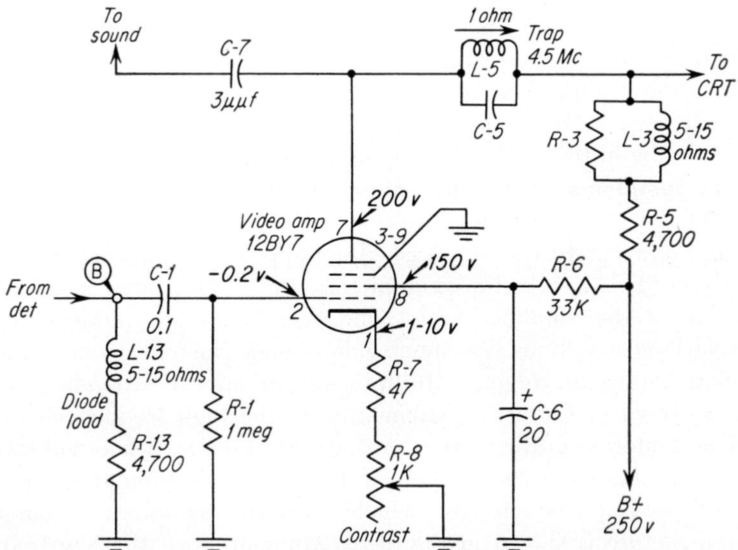

Fig. 9-6. Normal voltage data for the 12BY7 used as a video amplifier.

If the audio sine-wave trace does not appear on the oscilloscope, either the video amplifier is not working, the test signal is being shorted out, or the picture tube is shorting the output signal. To resolve these questions, first check for the short at the input. An ohmmeter check from the diode-load check point to ground should read 3,000 to 5,000 ohms. If this reading is normal, remove the socket from the picture tube. In case the receiver stops because of series filament operation, close the filament circuit by plugging in a CRT booster, or your test 8AXP4. If the sine-wave trace appears now, the picture tube has been shorting the signal.

Before replacing the picture tube see repair notes in Chap. 10. If the sine-wave trace is still absent, there is a fault in the video-amplifier stage.

The video amplifier is first checked by trying a new tube. If this does not restore the picture, the fault is located by voltage and resistance analysis in the stage.

Normal-voltage Data in the Video Amplifier. Normal-voltage data for a 12BY7 tube in a video amplifier are given in Fig. 9-6. The readings are approximations and will vary with the setting of the contrast control. The variation is indicative of tube operation, and the control should be adjusted through its range, as each measurement is taken. The plate voltage will normally vary from about 180 to 250 volts. If it remains at about 250 volts, regardless of the setting of the contrast control, the tube is not drawing plate current. The screen voltage will likewise vary from about 150 to 200 volts. If the screen is connected to a B plus 150-volt point in the receiver without a dropping resistor, the screen voltage will not change. The cathode measurement will vary from about 1 to 10 volts.

When the screen is fed from a B plus 150-volt point in the low-voltage power supply, an incorrect reading leads to an examination of the voltage source. It must be remembered that this may be a weak or dead audio-amplifier tube in the shelf type of power supply.

Other Symptoms Caused by Failure in the Video Section. Complete failure of the video amplifier results in no picture. Insufficient amplification produces a picture with a washed-out appearance, lacking rich blacks and whites. This is called a weak picture and is illustrated in Fig. 9-7*a*.

Failure of the amplifier to respond equally to all frequencies produces varying symptoms. Too much gain at high video frequencies gives a symptom known as ringing. Multiple images appear, extending to the right as shown in Fig. 9-7*b*. Attenuation of the high frequencies causes a lack of definition in the picture. A sharp black vertical line will trail off to the right in decreasing shades of gray. It is as though the picture were smeared—a characteristic which gives this symptom its name. A smeared picture is shown in Fig. 9-7*c*. Attenuation of the low frequencies causes a white section to trail behind black lines in the picture, giving it an engraved effect. This symptom is known as trailing whites and is illustrated in Fig. 9-7*d*.

Another symptom is sync instability. This is caused by overloading in the amplifier which clips off amplitude peaks by limiting action. Since the sync information is at peak amplitude, its suppression due to this cause results in the unstable picture.

A final symptom which may be caused by defects in the video amplifier or picture tube is a reversed picture which looks like a photographic negative. This is also caused by an overload condition.

It is important to recognize that all these symptoms may be caused by

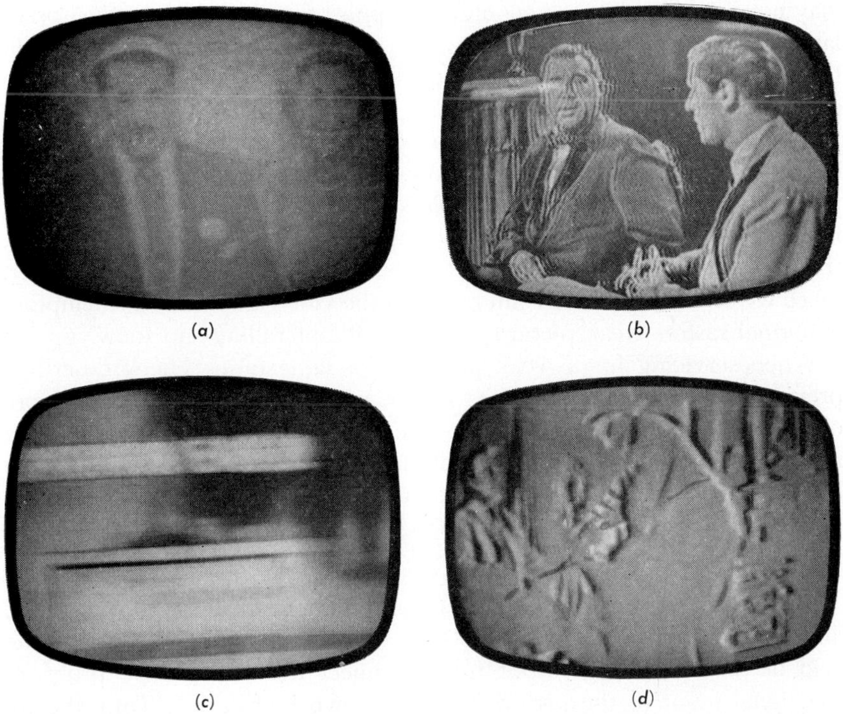

Fig. 9-7. Other symptoms caused by defects in the picture chain. (*a*) Weak picture. (*b*) Ringing. (*c*) Smeared picture. (*d*) Trailing whites.

failure in other parts of the signal chain. In the servicing procedures which follow, it will first be definitely established that the defective condition is located in the video section or elsewhere in the chain. If it is localized to the video section, tube substitution and voltage and resistance checks within the stage will find the defective components.

Servicing Procedure for a Weak Picture. A weak picture has previously been described as one with a washed-out appearance, lacking in rich blacks and whites. The symptom, sometimes also called poor contrast, must not be confused with a lack of brightness in the picture, which is described in Chap. 6. In the true weak picture, the raster is normal, and there is sufficient brightness on the picture tube. One of the stages in the picture chain is operating at low efficiency, so that the signal amplitude, on reaching the picture tube, is too low to cause proper beam modulation. The maximum signal amplitude produces gray instead of black, resulting in a picture which has a washed-out appearance and lacks contrast.

The weak picture symptom is often accompanied by rolling. This is because the sync signal is weakened along with the picture signal and may not produce a strong enough sync pulse to trigger the vertical oscillator properly. When this is the case, we service for the weakness in the picture chain, because the sync condition will be rectified when the cause of the weak signal is found and corrected.

Another symptom often accompanying a weak picture is a dispersion of white dots commonly called "snow." Its presence or absence is of importance because it gives a valuable servicing clue. Snow in the picture is the result of noise pulses in the composite signal. As explained before, the presence of snow places the trouble in the front end of the receiver. Since we are now primarily interested in the video amplifier, the symptom is normal raster—weak picture, with or without rolling—no snow.

When servicing for a weak picture, the same quick check described previously for the video amplifier can be used to place the trouble before or after the diode load. This time, the video amplifier and picture tube are working, though perhaps not at full efficiency, and we shall get the bars in either event. We need some quantitative data to determine if the bars are black enough. The following check is tried first on a receiver in good working order.

Step 1. Adjust the oscilloscope to observe traces at the vertical rate, and set the gain control to the 80- to 100-volt mark established for the grid of the horizontal output tube. Connect the oscilloscope probe to the signal input of the picture tube as shown in Fig. 9-8. Tune the re-

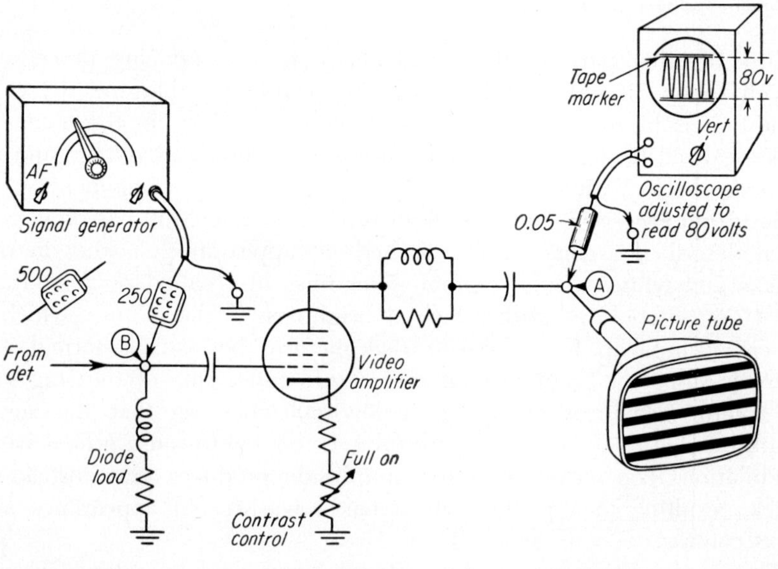

Fig. 9-8. Establishing a normal input signal needed to produce full output from the video amplifier for working on weak receivers.

ceiver to an unused channel, set the contrast control to the full on position, and adjust the brightness control for a normal raster. Feed the audio output of the signal generator to the diode load through a 0.05-μf isolating condenser. If the generator has a control of the audio output, this is adjusted so that the height of the oscilloscope trace is within the screen markers to show an 80-volt peak-to-peak series of sine waves. If the signal generator is not equipped with an audio attenuator, the trace will probably be too large for the oscilloscope screen. It can be brought down to the markers by using a smaller isolation condenser as an attenuator. Try 200 or 500 $\mu\mu$f. Since the normal signal at the input of the picture tube is approximately 80 volts, the audio input at the diode load is now at about the same amplitude as the normal video input signal. The bars on the screen of the picture tube should be a rich black, since this is a good receiver.

The contrast control on the receiver is at the full on position. Reducing it causes the scope trace to diminish and the bars to lose blackness, becoming lighter and lighter shades of gray. Since the contrast control does not go down to zero gain at the minimum position, the oscilloscope trace will show a small amplitude, and the bars will be discernible.

This check should be performed on a number of receivers in good working order. This will give you an average value of attenuator setting or isolation capacitance to produce normal output from the video amplifier, together with some comparison values of output signal and blackness of the bars.

When working on any receiver for weak picture, feed in the correct amplitude of audio signal, and observe the trace and bars. If the trace is at full amplitude and the bars are black, the video amplifier and picture tube are working, and you proceed to check the detector stage. If the trace is at full amplitude and the bars are gray, the video amplifier is working, and the picture tube is defective. If the trace is at low amplitude, the bars will be gray, and either the video stage is weak or the picture tube is loading it down.

Step 2. Without changing any of the equipment, remove the socket from the picture tube. If the scope trace comes up to its normal 80-volt amplitude, the video amplifier is good and the picture tube is bad. If the scope trace remains low, the video amplifier does not have sufficient gain.

Step 3. Replace the video-amplifier tube, and if this does not clear up the trouble, check the stage by voltage and resistance analysis. Do not forget to check the coupling condensers for leakage, and the grid- and plate-load resistors for change in value.

Servicing Procedure for Ringing. In the symptom called ringing, sharp vertical lines in the picture are repeated at regular intervals, forming multiple images, as shown in Fig. 9-9. Since this may be confused with ghost effects caused by reflections of the transmitted signal from surrounding objects, the first step in the procedure is to resolve this question.

In addition, the multiple images which are the result of oscillation or ringing at the high video frequencies may be caused by too much high-frequency gain in either the video or i-f sections. The second step in the procedure localizes the trouble to the video amplifier.

Fig. 9-9. Ringing.

Step 1. Tune in a picture on all channels normally received. If equally spaced multiple images are observed on all channels, the symptom is ringing. Ghosts will not be the same on all channels, nor will they be equally spaced.

Step 2. Tune in a channel that clearly shows the multiple images, and observe the effect of rotating the fine tuner. If the images move about while the fine tuner is adjusted, the trouble is a poor frequency-response characteristic in the i-f amplifier. This is cleared up by realignment of the i-f amplifier as described in Chap. 12. If the images remain fixed in position but decrease in number as the picture fades, the trouble is too much high-frequency gain in the video amplifier.

Step 3. Ringing in the video amplifier is generally caused by increase in value or opens in the loading resistors across the peaking coils. The resistors are not easily checked because the bridging coil gives a resistance reading of 20 ohms or less, regardless of the condition of the resistor. In the usual coil-resistor component, the coil lead cannot be opened to check the resistor separately. The best check procedure is to bridge each coil with a resistor of the correct value while watching for the effect on the picture. If the coil and loading resistor are good, the effect of the added test resistor will be to cut highs slightly. This may effect a partial cure of the symptom, since cutting the high-frequency response in one circuit will compensate for overemphasis of highs in the defective circuit. For this reason, it is necessary to disregard a partial cure and to keep on bridging coils with the test resistor until the ringing stops. The resistor is then soldered permanently across this coil.

Servicing Procedure for Loss of Definition, Smearing, and Trailing Whites. The symptoms of loss of definition, smearing, and trailing whites are classified together because of a similar causative factor—insufficient amplification of the low or high video frequencies. It should be remembered that the video section is designed as a wide-band amplifier with equal gain characteristics from about 30 cycles to 4 Mc. Failure to operate over this extended range produces the symptoms mentioned, depending on the frequencies attenuated and the degree of attenuation.

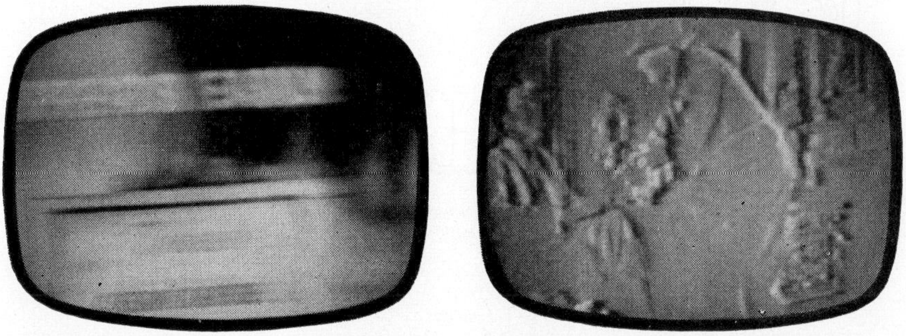

(*b*) (*a*)

Fig. 9-10. Symptoms caused by unequal amplification of video frequencies. (*a*) Smear. (*b*) Trailing whites.

Attenuation of some of the video frequencies is not confined to the video amplifier. Insufficient bandpass or an improperly shaped response curve in the i-f amplifier can cause the same losses, and therefore the same symptoms. Furthermore, in the symptoms of lack of definition, or poor picture detail and smearing, the picture looks as if it were out of focus. The servicing procedure to be set down eliminates these possibilities before the video amplifier is checked.

Step 1. Check the focus adjustment. (This step need not be followed for trailing whites.) The best check for focus is to tune the receiver to an unused channel and observe the raster lines over the entire face of the picture tube. These should be sharp and clear across the center area of the tube since this is the usual center of interest. A certain amount of fuzziness around the edges is permissible. If the receiver is out of focus, readjust the control in accordance with the instructions given in Chap. 10.

Step 2. Determine the bandpass characteristic of the video amplifier. Tune the receiver to an unused channel, set the contrast control to the full on position, and adjust the brightness control for a normal raster. Adjust the signal generator for an unmodulated r-f signal at full output, at the lowest frequency on the lowest band. Feed it to the diode-load check point through a 200-$\mu\mu$f isolation condenser. The setup is shown in Fig. 9-11. Most signal generators go down to about 100 kc, which is considered a comparatively low video signal. Start at the lowest frequency and rotate the generator control toward the higher frequencies.

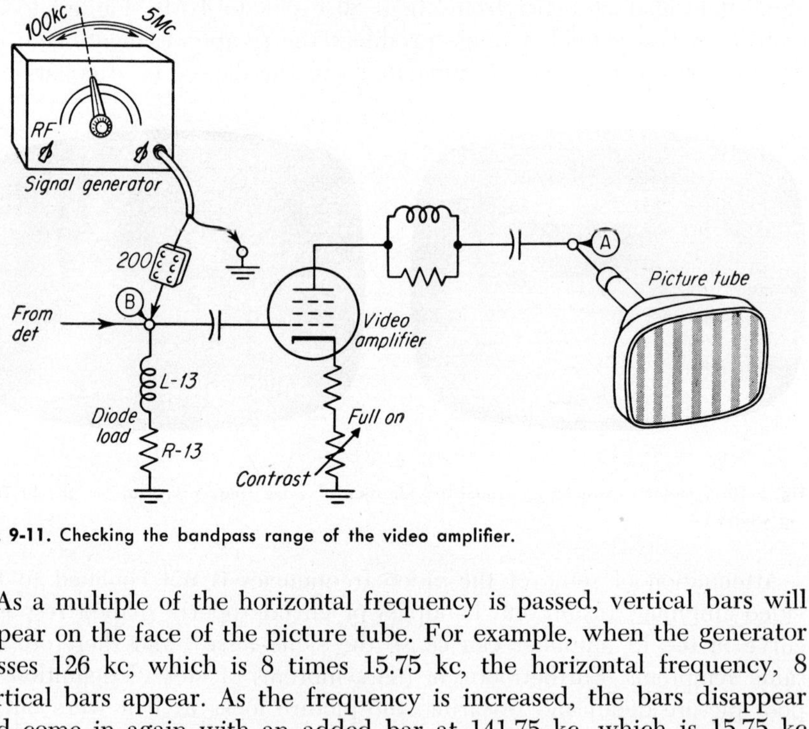

Fig. 9-11. Checking the bandpass range of the video amplifier.

As a multiple of the horizontal frequency is passed, vertical bars will appear on the face of the picture tube. For example, when the generator passes 126 kc, which is 8 times 15.75 kc, the horizontal frequency, 8 vertical bars appear. As the frequency is increased, the bars disappear and come in again with an added bar at 141.75 kc, which is 15.75 kc higher. This is repeated every 15.75 kc. Each time a bar is added, all the bars get narrower, so that they are practically lines at the higher frequencies. If it is desired to hold the lines at any one frequency setting, the horizontal-hold control may be used as a vernier adjustment.

The full r-f output of the usual signal generator is a few tenths of a volt, which is considerably less than the normal signal level at the diode load. As a result, the bars in this check will be gray rather than black, but discernible. The same applies to the lines at the higher frequencies. The bars become lighter and disappear as the frequency of the generator is increased toward 4.5 Mc. This is because the gain characteristic of a normal video amplifier falls off at about 4 Mc.

When checking a receiver by this method, if the bars or lines can no longer be seen above 1.5 Mc, this is the upper frequency limit of that video amplifier and the cause of the smearing. If a check on gain is desired for the very low frequencies, the audio output of the signal generator may be used as in the quick check for the stage.

Step 3. The video amplifier is checked for the trouble by voltage and resistance analysis. The likely faults are the components meant to extend the frequency range. In particular, check the peaking coils for opens and shorts, bypass condensers *C*-6 and *C*-6A and coupling condensers *C*-1 and *C*-9 for reduced capacitance or leakage, and resistors *R*-13 and *R*-5 for incorrect value. These components are labeled in Fig. 9-12. When

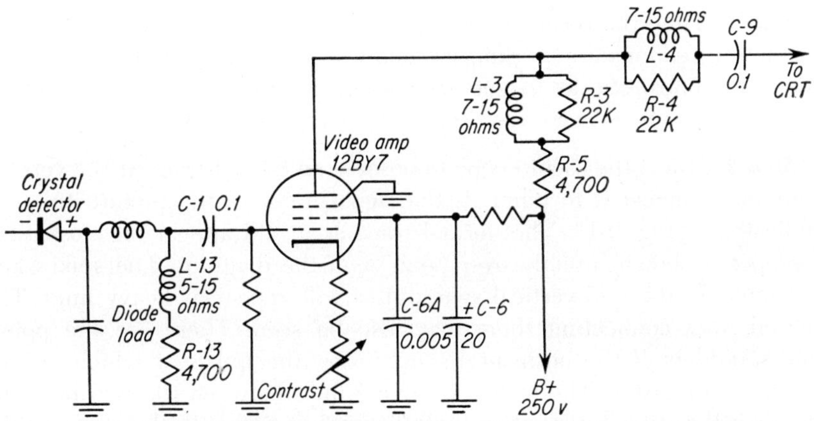

Fig. 9-12. Components in the video amplifier which are likely to cause a smeared picture.

checking diode-load resistor *R*-13, if the receiver uses a crystal detector, an erroneous reading may be obtained because of conductivity of the crystal. Always reverse the ohmmeter leads and take the higher reading as the true value.

Another possibility is added capacitance in the grid circuit caused by a defective crystal detector. This is checked by replacing the crystal, being careful to observe proper polarity. The crystal may be mounted in the last i-f coil assembly can. See notes on replacing crystals in Chap. 11.

Added capacitance to ground in the grid or plate circuits of the video-amplifier tube may be caused by physical misplacement of coupling condensers *C*-1 and *C*-9. These may have been moved while checking, or as a result of a replacement. See that the condensers are mounted at least 1 in. from the chassis pan. The above point illustrated the general rule which should always be followed in television-service work. Do not disturb the physical placement of components or lead dress when checking or replacing.

Servicing Procedure for Loss of Sync. Loss of gain or frequency discrimination in the video stages can cause loss of sync as well as a weak or smeared picture. When either of these symptoms is coupled with loss of sync, the latter is disregarded as being normal under these conditions, and the receiver is serviced for the weak or smeared picture. When the causative factor is found and corrected, the sync condition is automatically cleared up.

Sync instability can also be caused by suppression of the sync information in the picture-signal chain. The following procedure is designed to limit the fault to the video amplifier.

Step 1. Tune the receiver for best picture on the strongest channel. If the sweep sections are working normally, you will be able to manipulate the hold controls so as to get one picture in frame. Leave the controls at these settings, even though the picture slips out of frame.

Step 2. Adjust the oscilloscope to observe an 80-volt trace at the vertical rate, and connect it to point *A*, the signal input of the picture tube, as indicated in Fig. 9-13. The normal trace as it will appear on most oscilloscopes is shown directly over point *A* in the diagram. The spikes and shoulders of the two vertical sync pulses will appear as heavy lines. The dotted lines connecting them may also be seen. These are the points and shoulders of the horizontal sync pulses, the spikes of which will be too fine to register. The maze of moving lines between the two blanking pulses is the video information, constantly changing with the scene on the picture tube. Note that the two vertical spikes are a full 25 per cent of the total amplitude.

If the normal trace, as described above, is seen on the oscilloscope, the picture-signal chain is working correctly, and the trouble is behind the sync take-off point in the sync section. This is serviced in accordance with the procedure described in Chap. 8. If the trace is similar to the one drawn on the oscilloscope screen in Fig. 9-13, where the sync pulses are chopped off close to the shoulder, or if they are missing entirely, the trouble is in the picture-signal chain.

Step 3. Without changing any adjustments on the receiver or oscilloscope, remove the socket from the picture tube. If this stops the set be-

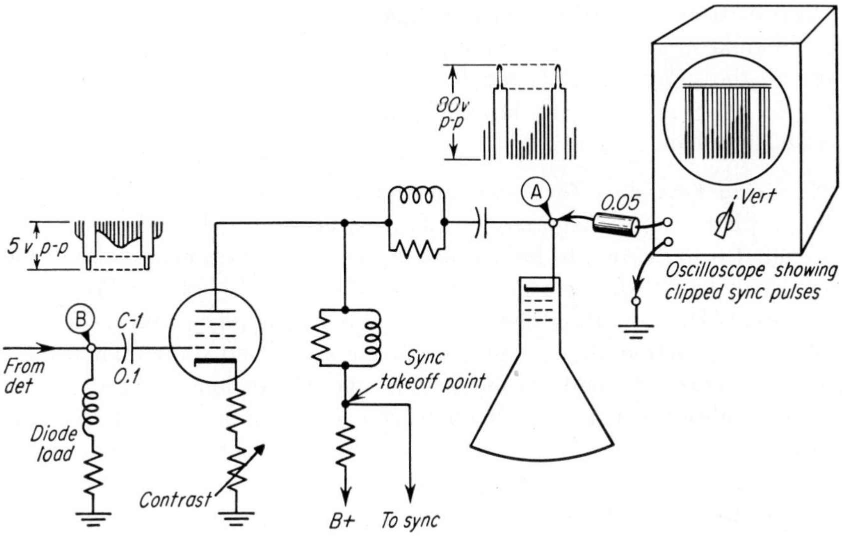

Fig. 9-13. Using an oscilloscope to check the video amplifier for suppression of sync information.

cause of series filament operation, complete the filament circuit with a CRT booster or your test 8AXP4 tube. Now observe the trace, and if the sync pulses are restored, the original picture tube is loading the video amplifier in such a way as to remove sync information. Replace the tube. If the trace remains abnormal, with the sync pulses cut off, the trouble is still in the picture-signal chain.

Step 4. Remove the oscilloscope probe, and connect it to the diode load at test point *B* in Fig. 9-13. This shows the normal trace directly above point *B*. Note that the trace is inverted and at a lower amplitude but that the sync spikes are a full 25 per cent of the total amplitude. Readjust the vertical gain on the oscilloscope to enlarge the trace, and note if the sync spikes are present at their correct proportionate height.

If the sync information is cut off at the diode load, it is being suppressed in either the i-f or detector stages or the grid circuit of the video-amplifier tube. Remove the video-amplifier tube, and if this restores the sync pulse, the trouble is in its grid circuit. If the pulse remains cut off, the i-f and detector stages are checked. If the sync pulse is normal at the diode load, the video stage is at fault.

Step 5. Change the video-amplifier tube, and if this does not correct the sync stability, make a careful voltage and resistance check of the stage. Pay particular attention to input coupling condenser *C*-1. If this component is leaky, it will distort the bias, causing nonlinear operation of the tube which will act to cut off the sync signals.

Step 6. If all components in the stage appear to be normal, decouple the antenna so as to reduce the input signal. If the picture now appears normal, the input signal is too large, and the video amplifier is chopping off the sync pulses by limiting action. Check the agc circuit as described in Chap. 12.

Servicing Procedure for Reversed Picture. A reversed picture, resembling a photographic negative, results from an overload condition where the signal is too strong to be handled by a normally operating circuit, or the circuit has deteriorated to the point where it cannot handle a normal signal. The condition of too much signal generally results when insufficient agc action allows high gain in the controlled stages. The signal then builds up sufficiently to overload a tube in the chain.

The condition can be simulated in receivers that incorporate a manual control in the agc system. The receiver is tuned to a strong station, and the control is rotated through its range. As the control voltage is reduced, there is first loss of sync, then a reversed picture, and finally no picture at all, when the signal is so large that a tube in the picture chain blocks for video information. The correct adjustment of the manual control in the agc system is, of course, to attain maximum gain before loss of sync occurs. A detailed description of this adjustment is given in Chap. 8. Similarly, in any receiver, faults in the agc system which reduce the control voltage cause loss of sync, reversed picture, or no picture, depending on the degree of the fault.

The overloaded tube may be the video amplifier. It would give normal results with a normal signal but cannot handle the large signal fed to it. Conversely, the input signal may be normal, but the video amplifier is at fault, in that it cannot handle the normal signal and causes the same symptoms. The following servicing procedure locates the trouble before or after the detector stage and traces it to the defective component if the fault is in the video amplifier or picture tube.

Step 1. Obtain the reversed picture, and rotate the fine tuner through its range. This shifts the oscillator frequency, so that more or less of the picture carrier is fed to the i-f amplifier. If reducing the input signal in this way causes the picture to change back to normal black and white, even though a good picture with good sound is unobtainable, the trouble is in the agc, tuner, or i-f circuits. The reduced signal relieves the overload, restoring a normal black-and-white picture. The continued presence of the overloading factor prevents reaching the point of best picture and sound. The i-f–agc circuits are now checked as described in Chap. 12.

If the picture on the screen remains negative throughout the adjustment of the fine tuner until it disappears entirely, the trouble is in the picture tube, video amplifier, or detector. Go on to step 2.

Step 2. Adjust the oscilloscope to observe the composite signal at the vertical rate of 30 cps and connect its probe to the signal input of the picture tube, shown as point *A*, in Fig. 9-14. The trace on the oscilloscope will appear as either one of the two traces shown as *X* and *Y* in Fig. 9-14, depending on the circuit of the receiver.

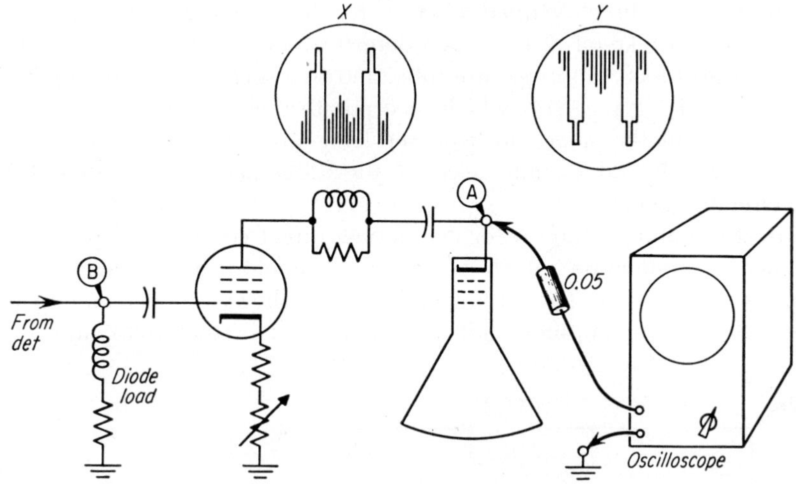

Fig. 9-14. Using the oscilloscope to determine the defective stage that causes reversed picture.

Without changing any adjustments on the receiver or oscilloscope, remove the socket from the picture tube while watching the trace. If the trace changes from *X* to *Y*, or vice versa, the picture tube is causing the trouble. Replace it in accordance with the instructions given in Chap. 10. If the trace remains unchanged, the trouble is in the video amplifier or detector. These are now checked for the abnormal condition. First look for a defective tube, or a decreased value of a plate-load resistor, since these are most commonly at fault. If the video-amplifier stage appears to be normal, check to see whether the receiver uses a crystal as the detector. Try a new crystal, being careful to observe polarity. Replacement notes on crystals are given in Chap. 11.

At this point we can confirm that the cause of the negative picture is not in the video amplifier but before it in the i-f–agc system. Connect the oscilloscope to the diode-load check point. Either of the two traces shown as *X* and *Y* in Fig. 9-14 will appear but at reduced amplitude. Increase the amplitude of the trace with the vertical gain control on the oscilloscope so that it is about 1 in. high. Then adjust the fine tuner through its range. The amplitude of the trace will change, but do not watch for this. Instead, watch for the trace to change from *X* to *Y*, or vice versa.

If it does, the fault is in the agc–i-f system. If the trace does not change, the trouble is still in the video amplifier.

VARIATIONS IN THE VIDEO SECTION

There are a number of variations of the basic amplifier of Fig. 9-3. These include direct-coupled circuits for improved low-frequency response, sound take-off at the video detector, potentiometer-type contrast controls, grid feed to the picture tube, and two stages of video amplification. In servicing receivers which incorporate these variations, the quick checks remain the same, and the servicing procedures can be applied equally well by taking into account the differences brought about by the circuit changes.

Direct Coupling. Many receivers employ direct coupling to avoid low-frequency attenuation which occurs even though large coupling condensers are utilized. The circuit of Fig. 9-15 shows direct coupling between the video amplifier and picture tube. Resistors *R*-15 and *R*-14

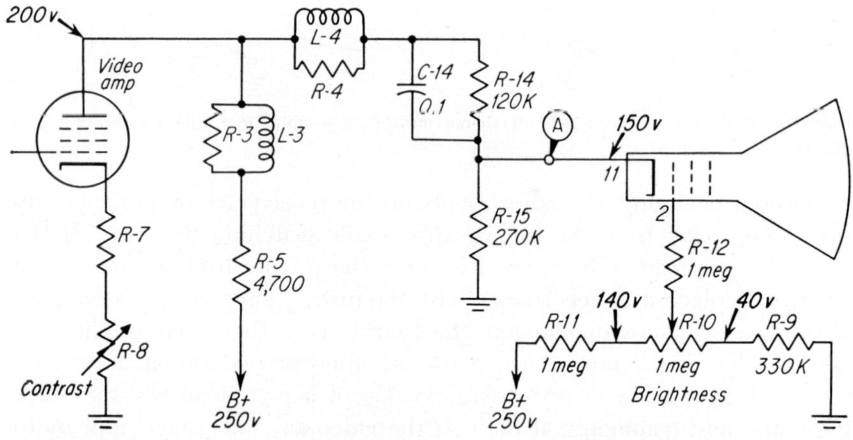

Fig. 9-15. Direct coupling between video amplifier and picture tube.

in the cathode of the picture tube form a voltage divider across the plate voltage of the video-amplifier tube, placing an average potential of 150 volts on the cathode. As the plate voltage varies with signal input, the cathode varies around its average potential of 150 volts. The control grid of the picture tube is kept at a potential somewhat lower than 150 volts to establish the average bias and beam current. Then the cathode variations produce the change in beam current forming darker and lighter elements in the picture. The control-grid potential is made variable through resistor *R*-10. This establishes the average bias and therefore acts as the brightness control. Resistors *R*-9 and *R*-11 are connected in series with brightness control *R*-10 to lower the voltage range.

When servicing this type of circuit, all stage-localizing checks operate in the same manner. Voltage analysis, however, must take into account the closer relationship of the picture tube and video amplifier. For example, when servicing for no raster, it may be found that an abnormally high cathode-grid voltage is cutting off the beam in the picture tube. The grid voltage is found to be normal, and the cathode voltage high. The serviceman must recognize that, in addition to more obvious factors, the condition may be caused by a weak video-amplifier tube, or an open contrast control. Similarly, when servicing for weak picture or loss of sync, remember that heavy loading by the picture tube can change the operating voltages on the video-amplifier tube, thereby causing the trouble.

Direct coupling may also be used between the detector tube and video amplifier. Such a circuit is shown in Fig. 9-16. Note the continuous line

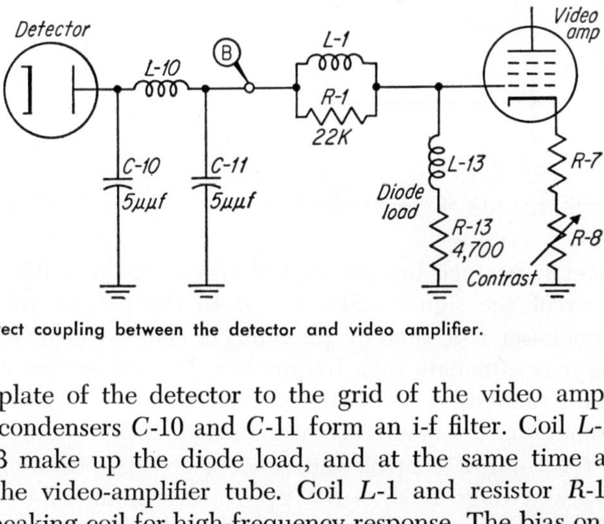

Fig. 9-16. Direct coupling between the detector and video amplifier.

from the plate of the detector to the grid of the video amplifier. Coil *L*-10 and condensers *C*-10 and *C*-11 form an i-f filter. Coil *L*-13 and resistor *R*-13 make up the diode load, and at the same time act as grid leak for the video-amplifier tube. Coil *L*-1 and resistor *R*-1 comprise a loaded peaking coil for high-frequency response. The bias on the video-amplifier tube is partly obtained from the negative polarity of the demodulated signal, and partly from the setting of the contrast control.

When servicing this circuit, test point *B* can be used to measure the output signal of the detector and to inject a signal into the video amplifier. It offers no servicing problems other than different readings when making point-to-point resistance measurements brought about by the different circuit.

Potentiometer Type of Contrast Control. In some receivers, the video amplifier is operated at full gain, and the output signal is connected across a potentiometer. Then, as in an audio volume control, the arm of the potentiometer supplies a variable portion of the signal to the load. The circuit is shown in Fig. 9-17. Resistor *R*-7 in the cathode circuit of the video amplifier provides bias for the tube. Since variable degeneration is not used to control the gain, the resistor is bypassed by condenser *C*-7, and the tube is allowed to operate at full gain. Resistor *R*-5, together with peaking coil *L*-3 and its loading resistor *R*-3, make up the plate load for the tube. The output signal appears across the plate load and across

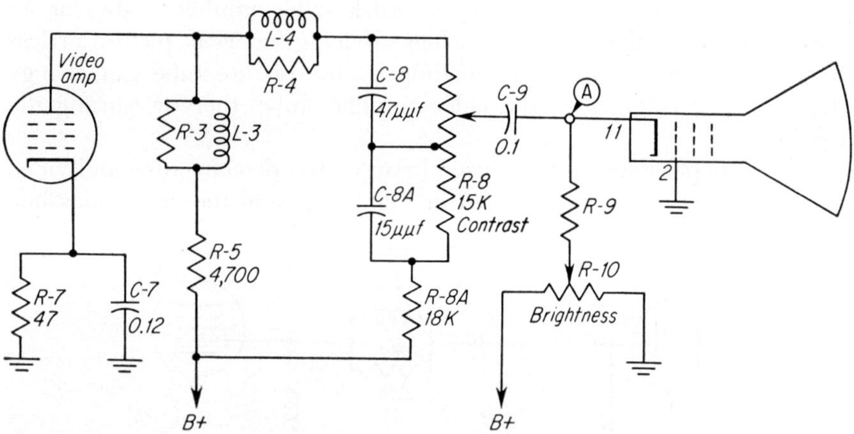

Fig. 9-17. Potentiometer type of contrast control.

potentiometer *R*-8 which bridges it. The arm of the potentiometer takes off a portion of the signal and feeds it to the picture tube through coupling condenser *C*-9, thereby providing a contrast control. A control of this type may attenuate high frequencies. For this reason, the control is tapped and condensers *C*-8 and *C*-8*A* are connected as shown in the diagram. Condenser *C*-8 maintains the high-frequency response at medium settings of the contrast control, and condenser *C*-8*A* performs the same function at low settings. Resistor *R*-8*A* provides a minimum-contrast position and also prevents the bypassing of high video frequencies to B plus by condensers *C*-8 and *C*-8*A*.

When servicing this type of receiver, it is important to keep in mind that the contrast control will not affect voltage readings. Also, an open contrast control will not stop reception since there is a signal path through condensers *C*-8 and *C*-8*A* across the control. The low capacitance of these condensers will cause a loss of low video frequencies with trailing whites as the result. See Fig. 9-10*b* for identification of this symptom.

The schematic diagram of the video section in the DuMont RA160 chassis is given in Fig. 9-18 as an example of a receiver using all the

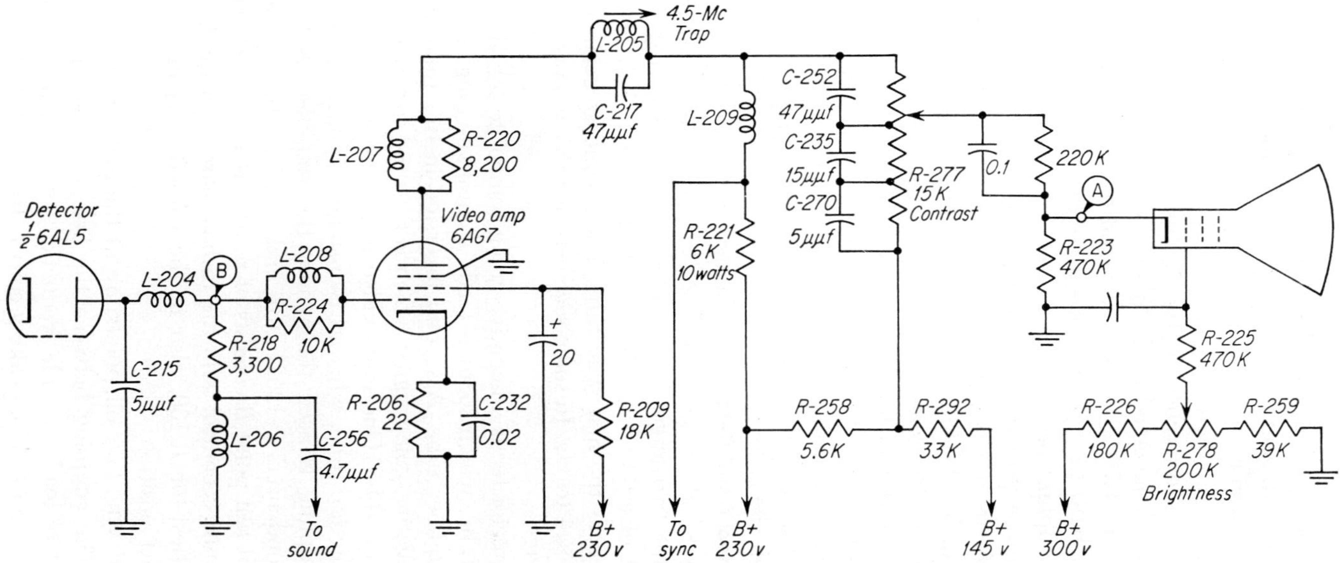

Fig. 9-18. Schematic diagram of the video-amplifier and picture-tube circuits in the Dumont RA160 receiver.

variations mentioned. It also includes one not yet mentioned. Note that the sound take-off point is at the junction of *R*-218 and *L*-206 in the diode load. The video amplifier is therefore not being used to amplify the sound signal. However, there is some sound signal present which may cause herringbone interference in the picture. The trap consisting of *L*-205 and *C*-207 is connected in the plate circuit of the video-amplifier tube and is tuned to 4.5 Mc to take the sound beat out of the picture.

Note also that both the grid and plate circuits of the video amplifier employ direct coupling for good low-frequency response. Good high-frequency response is provided by peaking coils *L*-208 in the grid circuit and *L*-207 and *L*-209 in the plate circuit. The contrast control is *R*-277 in the plate circuit of the video amplifier. It is provided with two taps, and compensating condensers *C*-252, *C*-235, and *C*-270 are connected to the taps to maintain proper high-frequency response at all settings of the control. The lower end of the contrast control is connected to the tap of a voltage divider consisting of resistors *R*-258 and *R*-292. This serves to stabilize the cathode of the picture tube against too wide a fluctuation of voltage at high signal levels.

Two-stage Video Amplifiers. The schematic diagram of the video section of the RCA KCS87 is given in Fig. 9-19 as a typical example of a receiver employing two stages in the video amplifier. The first stage uses the pentode half of a 6AN8 pentode-triode. The triode half of the same tube is connected as the first sync tube. The second stage in the video amplifier uses a 6C4 triode. In some receivers, a 6SN7 twin triode is used for both video stages.

The crystal detector feeds a negative-going composite signal to the first video-amplifier tube, which in turn feeds a positive-going pulse to the second video-amplifier tube. A second inversion in this tube makes the final output a negative-going signal which is then fed to test point *A*, the grid of the picture tube. One-stage video amplifiers invariably feed the cathode of the picture tube, whereas two-stage amplifiers feed the grid.

Capacitive coupling is used throughout the amplifier. Note that all three coupling condensers *C*-129, *C*-131, and *C*-135 are 0.1-μf condensers. The diode load at test point *B* consists of resistor *R*-129 and choke *L*-104. Note that the contrast control is in the cathode circuit of the first video-amplifier tube. Resistor *R*-130, also in the cathode circuit, provides a minimum contrast position.

The sound and sync take-off points are at the plate of the first video-amplifier tube. The coupling between the first and second video tubes has a complex plate load for wide frequency response. It consists of resistor *R*-132, condenser *C*-130, peaking coils *L*-105 and *L*-106, and resistor *R*-136. The junction of the two peaking coils feeds the coupling condenser. Condenser *C*-159 and coil *L*-109 in the same circuit form a wave

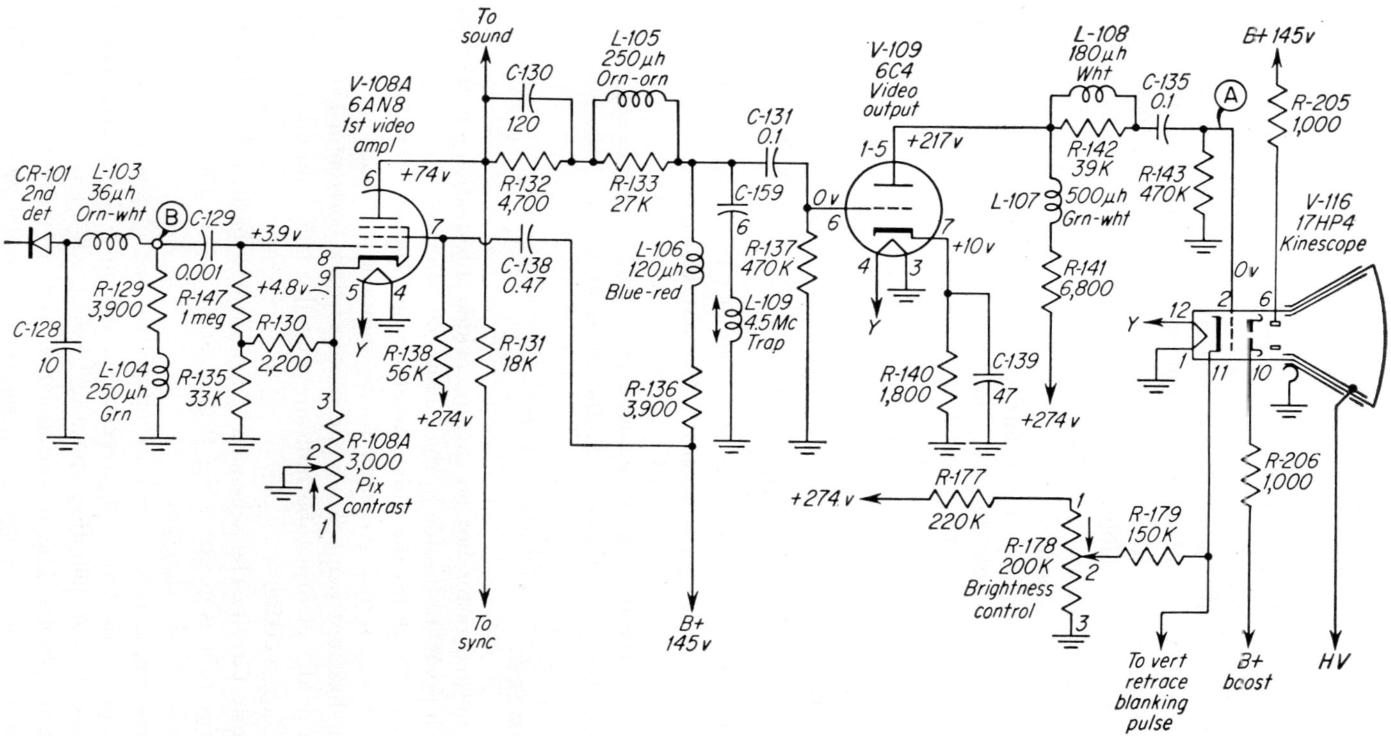

Fig. 9-19. Schematic diagram of the video section of the RCA KCS87 chassis.

trap to keep the sound beat out of the video signal fed to the second tube.

The circuit of the second video amplifier is almost an exact replica of the single-stage amplifier of the basic circuit. The cathode resistor is fixed instead of variable since the contrast control is in the first stage. Also, because of the negative-going output signal, the feed is to the grid of the picture tube. Note that the grid is still kept at zero potential, and the cathode gets a variable positive voltage from the brightness control. A vertical retrace blanking pulse is also fed to the cathode from the vertical output transformer.

When a video amplifier of this type is serviced, the signal check is the same as for the single-stage amplifier. A 400-cycle signal fed in at the diode load should produce black bars on the picture tube and an 80-volt trace at point *A*. Since we are checking a sine-wave signal, the inversion of the output signal will not be noticed. If the amplifier is inoperative, there will be no signal at point *A*. This time, since there are two stages to check, shift the oscilloscope probe to the plate of the first video tube. No trace at this point places the trouble in the first video stage, and a normal trace places it in the second stage.

When making a voltage check on the first video stage, with the normal voltage readings as given on the diagram, first connect the voltmeter to the cathode pin, and adjust the contrast control until it reads 4.8 volts. Then take the readings at the other pins, and investigate any reading that varies 20 per cent or more from the given norm.

SUMMARY

A normal raster but no picture (video information) is the symptom which points to trouble in the picture-signal chain.

Quick Check for the Video Amplifier and Picture Tube. Feed the audio output of a signal generator to the input of the video amplifier at the diode-load test point. The appearance of dark horizontal bars across the face of the picture tube indicates that the video amplifier and picture tube are working.

Basic Circuit. The schematic diagram of a basic video amplifier and picture tube is given in Fig. 9-20. Voltage and oscilloscope test data are included in the diagram.

Servicing Charts. Symptoms other than loss of picture may be caused by faults in the video amplifier and picture tube. These are weak picture, ringing, loss of definition, smearing, trailing whites, negative picture, and sync instability. The following charts outline servicing procedures for these symptoms. In each case, the trouble is first localized to the video amplifier or picture tube. Reference in the charts is to Fig. 9-20.

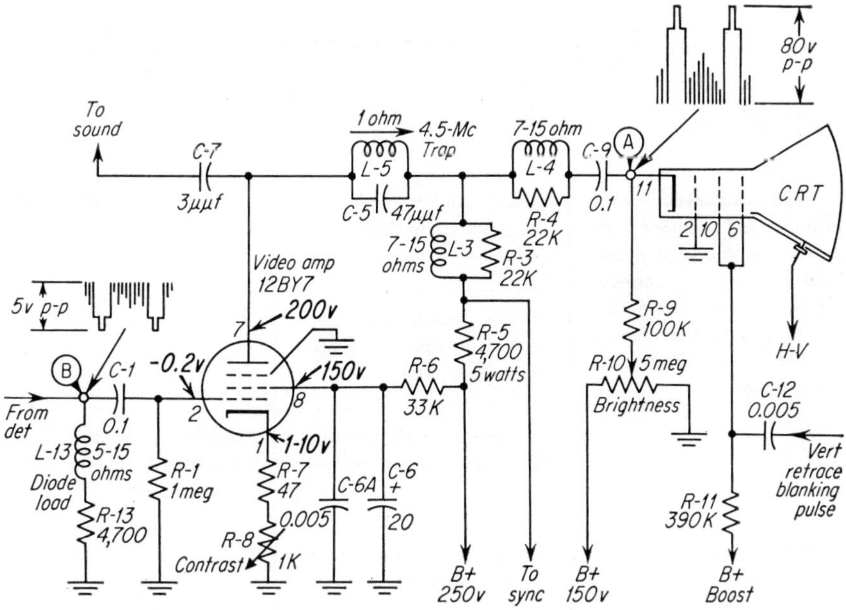

Fig. 9-20. Schematic diagram of basic video-amplifier and picture-tube circuits.

Servicing Procedure for Normal Raster—No Picture

Procedure	*Result*	*Trouble and Subsequent Check*
1. Feed calibrated 400-cycle a-f signal to diode-load check point (point *B*). Set contrast control to full on position	Black bars on picture tube	Trouble is before the video stage. Check the detector (Chap. 11)
	No response	Trouble is in the video amplifier or picture tube. Go on to step 2
2. Connect oscilloscope to signal input of picture tube (point *A*).	400 sine-wave trace on oscilloscope 80 volts peak to peak	Picture tube is bad. Replace it. See Chap. 10
	No response	Video amplifier is bad, or picture tube is shorting the output. Go on to step 3
3. Remove socket from picture tube	Sine-wave trace appears on oscilloscope	Picture tube is bad. Replace it. See Chap. 10
	No response	Video amplifier is bad. Check stage by tube substitution, voltage, and resistance analysis

Servicing Procedure for Normal Raster—Weak Picture—No Snow

Procedure	*Result*	*Trouble and Subsequent Check*
1. Feed 400-cycle calibrated a-f signal at proper amplitude to diode load (point *B*). Connect oscilloscope to signal input of picture tube (point *A*). Set contrast control to full on position	80-volt peak-to-peak trace on oscilloscope. Black bars on picture tube	Trouble is before the video stage. Check the detector (Chap. 11)
	80-volt peak-to-peak trace on oscilloscope. Gray bars on picture tube	Picture tube is bad
	Low trace on oscilloscope. Gray bars on picture tube	Trouble is in the video amplifier or picture tube. Go on to step 2
2. Remove socket from picture tube	80-volt peak-to-peak trace on oscilloscope	Picture tube is bad
	Low trace on oscilloscope	Video amplifier is bad. Check stage by tube substitution, voltage, and resistance analysis

Servicing Procedure for Ringing

Procedure	*Result*	*Trouble and Subsequent Check*
1. Check operation on all channels	Symptom is not present on all channels	Symptom is probably ghosts. See Chap. 16
	Symptom is present on all channels	Symptom is ringing. Go on to step 2
2. Rotate fine tuner adjustment and observe effect on symptom	Multiple images move about	Incorrect response in i-f amplifier. See Chap. 12 for realignment
	Multiple images remain fixed	Trouble is in the video amplifier. Check the video stage, giving particular attention to peaking coils and loading resistors *L*-3, *L*-4, *R*-3, and *R*-4

Servicing Procedure for Loss of Definition, Smearing, and Trailing Whites

Procedure	*Result*	*Trouble and Subsequent Check*
1. Check focus off channel	Focus poor	Readjust focus control as described in Chap. 10
	Focus good	Poor frequency response in i-f or video stages. Go on to step 2
2. Feed unmodulated r-f to diode load (check point *B*) and shift frequency from 100 kc to 5 Mc	Vertical bars and lines appear on the picture tube up to 4.5 Mc	Poor frequency response in i-f stages. Realign as described in Chap. 12
	Vertical bars fade out at low or high frequency	Poor frequency response in the video amplifier. Check the video amplifier, giving particular attention to: (*a*) peaking coils *L*-13, *L*-3, *L*-4; (*b*) bypass condensers C-6, C-6A; (c) coupling condensers C-1, C-9; (*d*) load resistors, *R*-13, *R*-5; (e) crystal detector

Servicing Procedure for Loss of Sync

Procedure	*Result*	*Trouble and Subsequent Check*
1. Check operation on all channels	Picture syncs in on weak channels	Sync clipping caused by improper agc. Check setting of automatic gain control or noise control. Check i-f–agc circuit; see Chap. 12
	Sync instability on all channels	Go on to step 2
2. Adjust the receiver for best picture, disregarding sync instability. Connect oscilloscope probe to signal input of the picture tube (point A) and observe trace at the vertical rate	Normal 80-volt peak-to-peak trace with sync pulses a full 25 per cent of total amplitude	Trouble is in the sync section. See Chap. 8
	Sync pulses are clipped or distorted	Sync pulses are being clipped in the picture chain. Go on to step 3

Chart continued on next page.

Servicing Procedure for Loss of Sync (Continued)

Procedure	*Result*	*Trouble and Subsequent Check*
3. Remove socket from picture tube	Normal composite trace appears	Picture tube is bad. Replace it
	Sync pulses remain clipped or distorted	Trouble is in the picture chain. Go on to step 4
4. Connect the oscilloscope probe to the diode load (point *B*)	Sync pulses are clipped or distorted	Check the i-f stages for sync suppression. See Chap. 12
	Normal sync pulses are present	Sync pulses are being clipped in video amplifier. Check the stage by tube substitution, voltage, and resistance analysis. Pay particular attention to condenser C-1 for leakage

Servicing Procedure for Reversed Picture

Procedure	*Result*	*Trouble and Subsequent Check*
1. Adjust fine tuning on receiver	Normal picture can be obtained, but not best picture with best sound	Trouble is in agc, i-f, or tuner. See Chap. 12
	Picture remains reversed	Trouble in picture tube, video amplifier, or detector. Go on to step 2
2. Connect oscilloscope probe to signal input of the picture tube (point *A*), and observe trace at vertical rate. Remove socket from picture tube	Trace on scope reverses polarity	Trouble in picture tube. Replace it. See Chap. 10
	Trace on scope remains unchanged	Trouble is in video amplifier or detector. Proceed as follows: (*a*) Change video-amplifier tube. (*b*) Check video-amplifier plate-load resistor for correct value. (*c*) Replace crystal detector if used

QUESTIONS

1. Name the stages in the picture-signal chain.
2. What is the result of an inoperative stage in the picture-signal chain?
3. If the symptom is no picture—no raster, which stages are checked first? Why?
4. What is the position of the video-amplifier section in the picture-signal chain? What is its function?
5. What is the frequency range of the video amplifier?
6. What is the likely result of a loss of high-frequency signals in the video amplifier? low-frequency signals?
7. What are some of the arrangements for good gain at low frequencies in the video amplifier?
8. What circuit designs are incorporated for good high-frequency response?
9. Explain the action of the contrast control in a receiver.
10. Explain the action of the brightness control in a receiver.
11. In a normal receiver, if the brightness is set too high, black elements in the picture will register gray. Explain how this happens.
12. Draw a diagram of a one-stage video amplifier. Give approximate values for all components, and give the function of each.
13. Draw a diagram of a two-stage video amplifier coupled to a picture tube. Give approximate values for all components and give the function of each.
14. What is meant by the term diode-load check point? If a receiver is available, locate the diode check point.
15. What is the purpose of loading resistors across peaking coils?
16. What is the purpose of direct coupling in video amplifiers?
17. Draw a diagram of a video amplifier and picture tube which are directly coupled.
18. Describe a quick check for a video amplifier and picture tube.
19. Describe a localizing check which places the trouble in the video amplifier or picture tube.
20. A weak stage in the picture chain produces a picture which is lacking in contrast. Describe a localizing check which proves that the weak stage is the video amplifier.
21. Make up a list of the various conditions which can cause low gain in the video amplifier, and describe how you would check for each.
22. What causes snow in a picture?
23. What is the significance of snow when coupled with a weak picture?
24. Describe a servicing procedure which places the cause of a smeared picture in the video amplifier. Which components would you check first? Why?
25. What is meant by ringing as applied to the video amplifier? How could

you establish that the ringing fault is in the video amplifier? Which components are likely to be defective?

26. A receiver comes in with the complaint of poor vertical stability (tendency to roll). An oscilloscope check at the signal input to the picture tube shows clipped sync pulses. What should your next check be?

27. If the oscilloscope check in question 26 showed a normal composite signal, what should the next check be?

10

Servicing the Picture Chain: Replacement Notes on Picture Tubes

The image-reproducing device in a television receiver is a cathode-ray tube. This is usually abbreviated to CRT. The tube is also called a kinescope. The common terms used by servicemen are picture tube and CRT.

Parts of the Picture Tube. The photograph in Fig. 10-1 shows the physical appearance of the CRT and points out the important parts. The tube is shaped like a funnel with the tube base at the narrow end. The neck

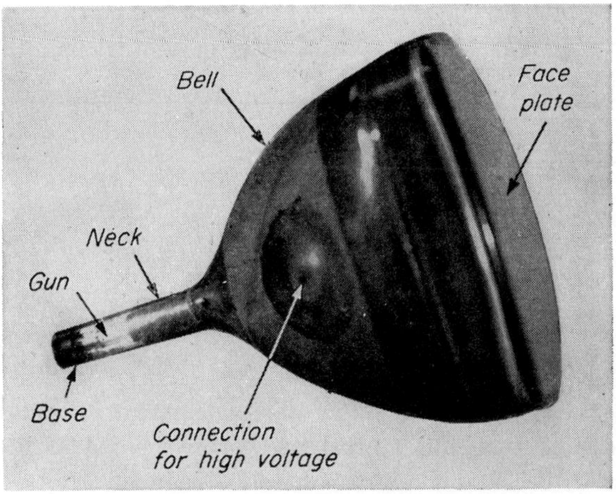

Fig. 10-1. Parts of the picture tube.

is made of glass while the bell may be either glass or metal. The wide end of the funnel is covered with a glass faceplate which is coated on the inside with fluorescent phosphors.

The tube electrodes extend from the base into the neck. This assembly is called an electron gun since its function is to send a stream of electrons in the form of a beam toward the screen. When the phosphor-coated screen is struck by the electron beam the phosphors glow, producing light.

Operation of the Electron Gun. A typical electron gun is illustrated in Fig. 10-2. The cathode is heated in the conventional manner by a 6.3-volt, 600-ma heater. The cathode is followed by a series of cylinders. The

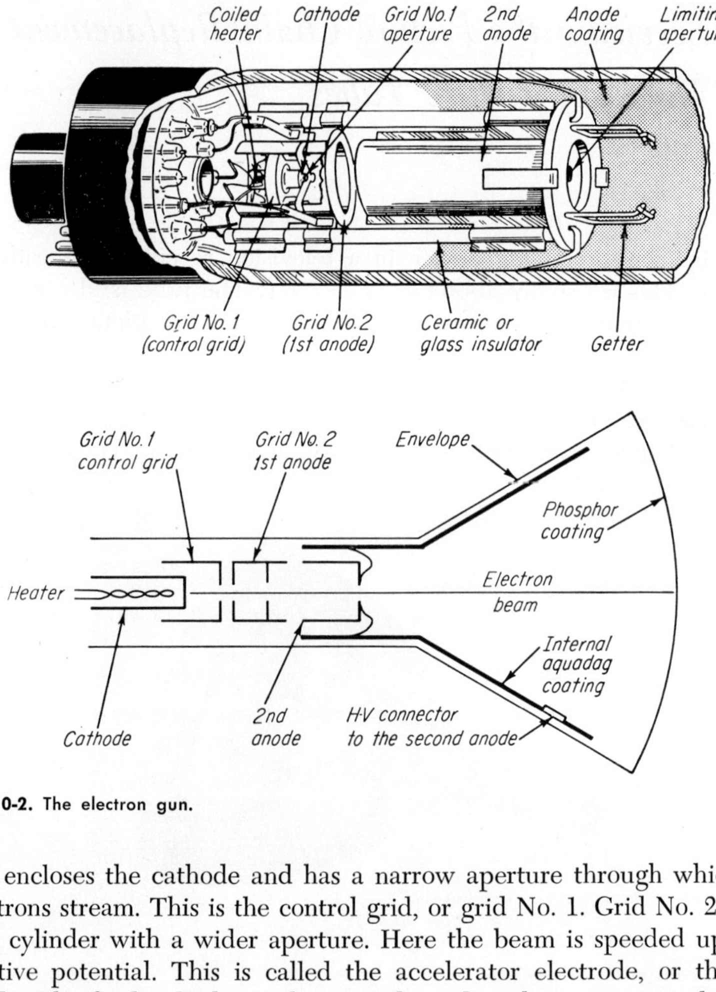

Fig. 10-2. The electron gun.

first encloses the cathode and has a narrow aperture through which the electrons stream. This is the control grid, or grid No. 1. Grid No. 2 is the next cylinder with a wider aperture. Here the beam is speeded up by a positive potential. This is called the accelerator electrode, or the first anode. The final cylinder is the second anode, where a potential of approximately 15 kv imparts a high velocity to the beam.

The inside of the bell is covered with a conductive graphite coating called aquadag. The high-voltage connection is made to this coating through a recessed contact in the side of the bell. The inside coating is continued into the neck of the tube, where a metal strip touching the aquadag serves as a contact to the second anode. In a metal tube, the aquadag coating is painted only on that portion of the tube which is made of glass. This is a relatively small area where the glass neck joins the metal bell. The high-voltage connection is made to the metal shell, usually at the rim. The dag coating then completes the circuit to the second anode.

Ion Trap. The stream of electrons which are emitted from the cathode contain heavy negative charges called ions. These heavy charges would tend to concentrate in the center of the screen, causing a brown stain called "ion burn." This is illustrated in Fig. 10-3*a*. The electron gun is

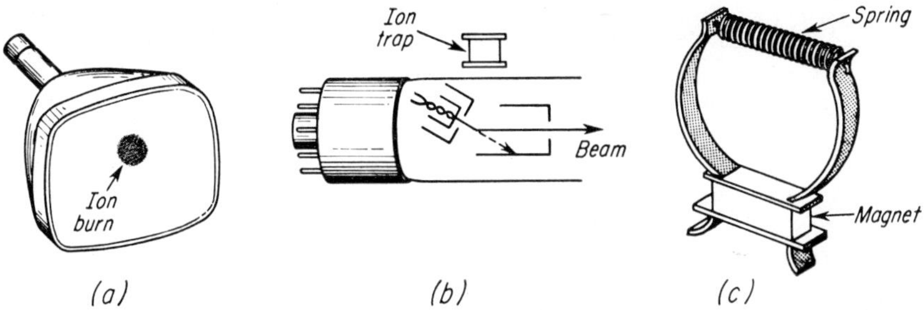

Fig. 10-3. An ion burn, and the beam bender (ion trap) which prevents it.

bent to eliminate the burn. An exaggerated view of the bent-gun construction is shown in Fig. 10-3*b*. The cathode and first grid are bent at an angle which sends the stream of electrons and ions toward the neck of the tube. The stream of electrons is then bent back to the apertures in the remaining straight electrodes of the gun by a magnetic field. This field is provided by a device known as a beam bender or ion trap. It is clamped to the neck of the tube directly over the space between the first and second grids. Since the ions are heavier than the electrons, they are not bent so much and are absorbed by the first anode. A typical ion trap is shown in Fig. 10-3*c*.

The positioning of the ion trap is adjustable to compensate for differences in gun structure. A slight misadjustment of the trap would cause some or all of the electron beam to miss the aperture, causing lack of brightness, or no light on the screen at all. At the same time electronic bombardment may harm the first anode and result in a shortened life for the picture tube. Adjustment notes for the ion trap are given in the CRT setup procedure at the end of this chapter.

Focus. Upon leaving the last aperture of the electron gun the beam tends to spread. In order to obtain picture clarity the electron stream must be formed into a fine beam which produces a minimum-sized spot on the face of the CRT.

A focus device which acts as an electronic lens and concentrates the beam to a converged spot is brought into play. The device may accomplish its purpose magnetically or electrostatically (see Fig. 10-4).

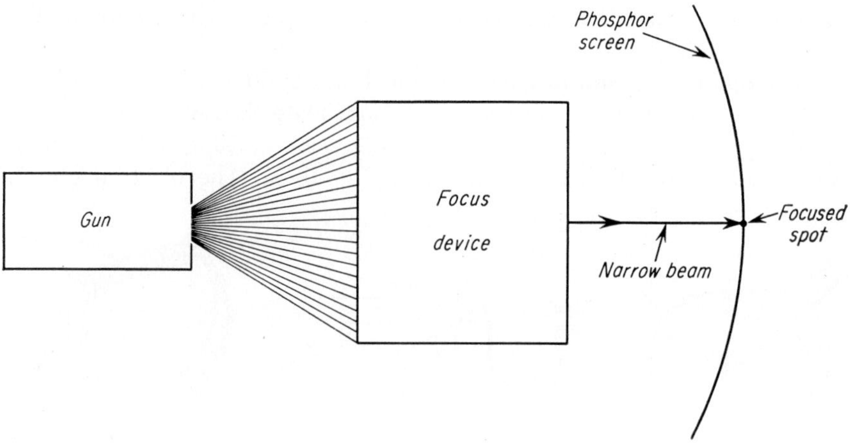

Fig. 10-4. Focusing the electron beam.

Deflection. We have seen how the electron gun shoots out a stream of electrons which are focused into a fine beam. This strikes the center of the screen, producing a spot of light. If we now place a large horseshoe magnet over the neck of the tube as shown in Fig. 10-5, its magnetic field will bend the beam, moving it horizontally off center, as indicated in the diagram. If the poles of the horseshoe were reversed, the beam would move in the opposite direction. Note that the poles of the magnet are on the top and bottom of the neck of the tube and that this results in a horizontal bending of the beam.

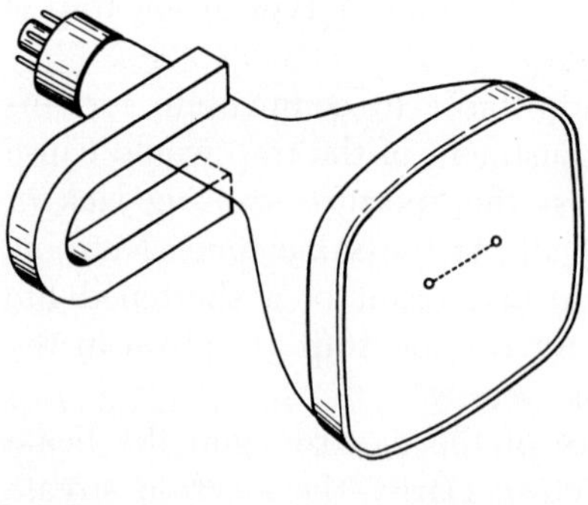

Fig. 10-5. How a magnetic field deflects the electron beam.

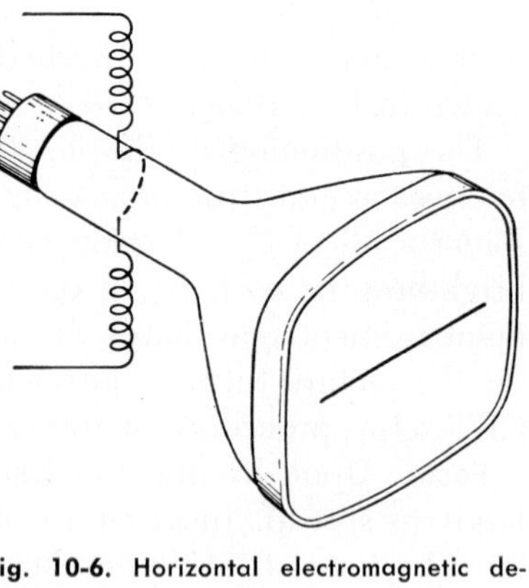

Fig. 10-6. Horizontal electromagnetic deflection.

If the horseshoe were to be replaced by a pair of electromagnets as shown in Fig. 10-6, bending of the beam would result. Now if these electromagnets are fed by an alternating current, the magnetic poles would reverse polarity with each alternation of the supply voltage. This will cause the beam to move back and forth in a horizontal line. The moving beam makes a continuous line because the phosphors remain in a glowing condition for a time after the beam has moved on. This characteristic of the phosphors is known as persistence.

The electromagnet coils are the horizontal deflection coils, and the energizing current is obtained from the horizontal deflection stages. A similar set of coils placed on the sides of the tube neck deflects the beam vertically. These are the vertical deflection coils, powered by the vertical section. Both sets of coils are wound on one form, known as the deflection yoke. A representation of the two windings and a typical deflection yoke are shown in Fig. 10-7. The yoke is slipped over the neck of the

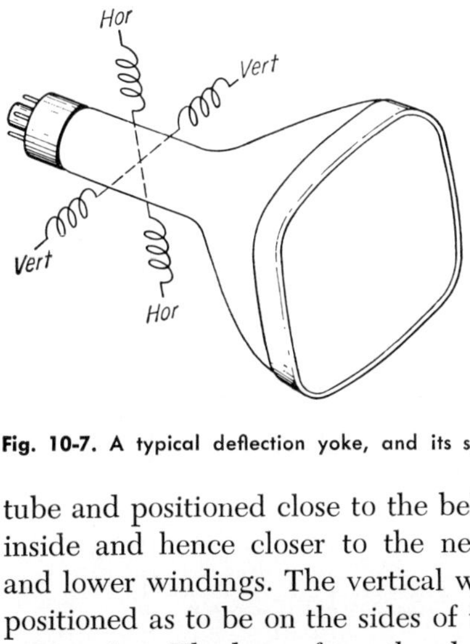

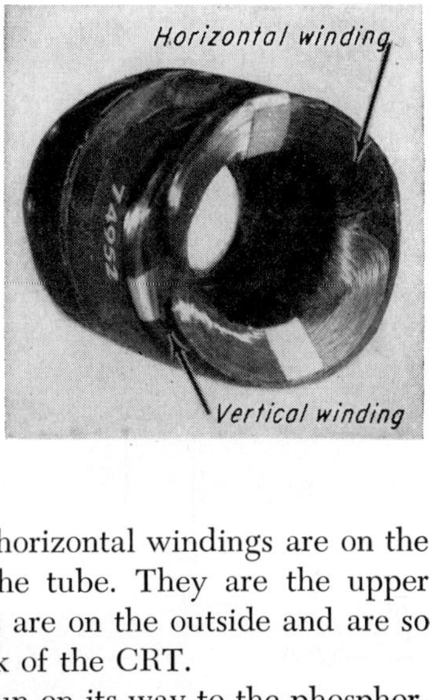

Fig. 10-7. A typical deflection yoke, and its symbol.

tube and positioned close to the bell. The horizontal windings are on the inside and hence closer to the neck of the tube. They are the upper and lower windings. The vertical windings are on the outside and are so positioned as to be on the sides of the neck of the CRT.

Scanning. The beam from the electron gun on its way to the phosphor-coated screen is simultaneously deflected from side to side by the current in the horizontal winding, and up and down by the current in the vertical winding. The combined forces make the beam trace a zigzag pattern on the screen, the pattern being determined by the amount of current and its frequency in each of the deflection windings. The deflection circuits are adjusted and regulated so that the beam sweeps or scans the screen in the prescribed trace for the American television system.

The regulation of the Federal Communications Commission is to transmit 30 pictures per second, each one consisting of 525 lines. This places the horizontal scanning rate at 15,750 cycles per second ($30 \times 525 = 15{,}750$), and the horizontal oscillator is therefore made to operate at 15,750 cps. In order to avoid flicker, a system known as interlaced scanning is used. In this system, alternate lines are scanned, the first, third, fifth, etc. The beam therefore covers half of the 525 lines, or $262\frac{1}{2}$, lines in one vertical sweep, called a field. Then, in the next field, the second, fourth, sixth, etc., lines are scanned to complete the frame. Thus it takes two vertical sweeps for each picture frame, and the vertical rate must be 60 per second in order to produce 30 complete pictures in a second.

When the receiver is operating without picture information, as is the case when it is tuned to an unused channel, the scanning lines may be observed on the screen. Although each line is traced separately, all the lines can be seen at once because of the persistence of vision. The display of lines thus produced is called a raster.

Brightness. The brightness of an individual raster line is determined by the strength of the electron beam. More electrons striking the phosphor will cause a brighter glow. The number of electrons in the beam is easily controlled by the cathode-grid voltage. When the control grid is made more negative with respect to the cathode, the electron stream is reduced just as it is in any tube, and the phosphors glow with less brilliance.

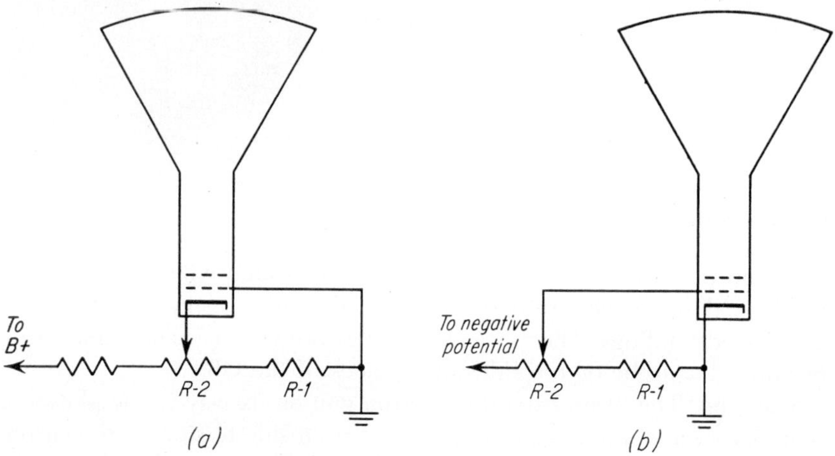

Fig. 10-8. Two methods of controlling brightness.

The cathode-grid voltage is made variable and is usually a front-panel control labeled brightness so that the user can adjust the brightness to his preference. Two brightness circuits are shown in Fig. 10-8. The circuit of Fig. 10-8*a* is more commonly found. Here the first grid is at ground potential, and potentiometer *R*-2 applies a variable positive voltage to the cathode. In Fig. 10-8*b*, the cathode is at ground potential and a variable negative voltage is fed to the grid. In either circuit, adjustment of potentiometer *R*-2 makes the grid more negative than the cathode by approximately 20 to 100 volts. The 20-volt end is the position of maximum brightness. Usually the tube is cut off, extinguishing the beam before the position of minimum brightness at −100 volts is reached.

How the Picture Is Formed. The scanning of the picture-tube screen is done in synchronization with a similar system at the transmitter where the camera is scanning the scene being televised. As the camera traces over each line, a variable voltage is obtained which corresponds to light and dark picture elements in that line. This is the video signal which is used to modulate the picture carrier.

In the receiver the video information is extracted from the modulated r-f signal at the detector and, after passing through the video amplifier, is fed to the cathode-grid circuit of the picture tube. Here the signal adds to or subtracts from the average brightness level set by the brightness control, strengthening or weakening the beam in accordance with the video signal level. Thus as each line is traced on the CRT screen, portions of the line are made brighter or darker, re-creating the similar line being scanned at the studio.

Summary of Picture-tube Operation. Refer to Fig. 10-9. The operation of the picture tube may be summarized as follows: The electron gun in the neck of the CRT shoots out a stream of electrons, the intensity of which is controlled by the bias set up by the brightness circuit. The stream of electrons is focused into a beam which then strikes a point on the face of the picture tube where it causes the phosphors in a fluorescent screen to glow.

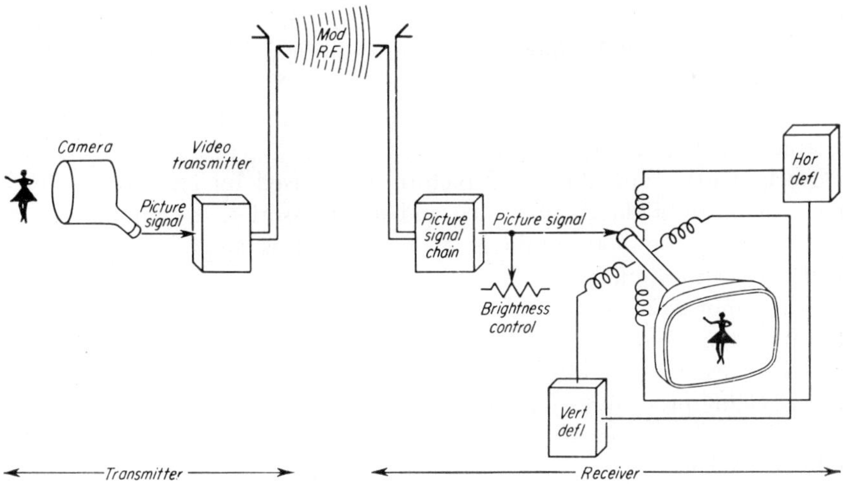

Fig. 10-9. How the picture is formed on the screen.

Vertical and horizontal deflection circuits cause the beam to scan the face of the picture tube in synchronism with a similar scanning process at the transmitting camera. The camera transforms the light and dark elements of a line in the scene into a signal voltage. This signal voltage, which is picture information, is then used to modulate the transmitter.

The receiving antenna picks up the modulated r-f carrier and amplifies it through the r-f and i-f stages. At the detector, demodulation takes place, and the signal is again in the form of picture information, as picked up by the camera. After further amplification in the video section, the picture signal is fed to the electron gun in the CRT where it strengthens or weakens the emitted beam. Hence the line being scanned on the picture tube re-creates the light and dark elements of the line being scanned in the original scene.

Symptoms Caused by Defects in the Picture Tube or Its Accessories. Picture-tube accessories include items like the ion trap and focus device which are clamped to or around the picture tube. The brightness control is also considered an accessory because of its close association with the picture-tube circuit. The treatment of the deflection yoke will be limited to mechanical adjustments in this chapter, since other servicing notes on this device are given in Chaps. 5 and 6 on the vertical and horizontal deflection circuits.

As brought out in previous chapters, a defective picture tube may cause no picture, weak picture, loss of sync, or negative picture. The servicing procedure set down for these symptoms quickly localizes the trouble to the picture tube, if it is at fault.

There are other symptoms of poor picture reproduction which point directly to the picture tube or an accessory as the faulty component, without requiring checks in the rest of the picture chain. These are improper focus, no control of brightness, ion burn, internal arcing, neck shadow, tilted picture, or gassy (ionized) picture tube. Most of these terms are familiar or self-explanatory. Neck shadow and the tilted picture are illustrated in Fig. 10-10.

Neck shadow and the tilted picture are caused by improper adjustment or setup of the picture tube and its accessories. A complete setup procedure for the picture tube is given later in this chapter. Proper setup will correct these defects.

The symptom of no control of brightness may be caused by defects in the brightness circuit or in the picture tube. Similarly, improper focus may be caused by a faulty focusing device or a defective picture tube. In the servicing procedures for these symptoms, the brightness and focusing controls or circuits are checked first, leaving the picture tube for last. In all cases when the investigation is narrowed down to the picture tube, the fault may be cleared up by a repair procedure.

The repair procedures will now be described. This will be followed by servicing procedures for the symptoms of no control of brightness and improper focus. Finally there will be replacement notes on picture tubes, and a generalized setup procedure for installing a new picture tube or making readjustments where needed.

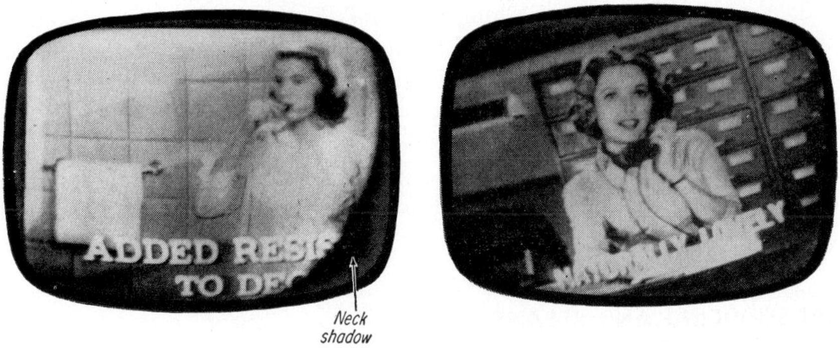

(a) (b)

Fig. 10-10. Picture defects. (*a*) Neck shadow. (*b*) Tilted picture.

Safety Precautions to Be Observed When Handling Picture Tubes. Before attempting any service work on the picture tube, read the following instructions very carefully. Mishandling the CRT can be the cause of a serious accident from the implosion which results when a vacuum of the size used in picture tubes is broken. Therefore, when service work calls for handling the picture tube, the following precautions should be observed:

1. Disconnect the line cord.
2. Disconnect the high-voltage lead and discharge the picture tube by touching the second-anode-connection receptacle with a grounding lead. The high-voltage charge retained by the tube itself is not dangerous, but an unexpected jolt from this source may cause the serviceman to drop the picture tube.
3. Wear gloves and safety goggles when handling the tube.
4. To remove the picture tube, first remove all accessories from the base and neck. This includes the CRT socket and ion trap but does not include the yoke. There may be other accessories like a clamp-type centering device, a shim-type width control, or a clamp holding the yoke in position. Any of these when present must be removed. Next remove the holding device on the rim of the tube. Then grasp the tube by the rim, and pull it toward you gently with a rotary motion. Do not use force. If the tube does not move easily, check to make sure that you have not overlooked a clamp on the neck of the tube. If the neck is clear, the

horizontal winding of the deflection yoke has probably cemented itself to the glass. The application of heat from an infrared lamp is helpful in loosening a yoke which is binding. When the tube moves freely, withdraw it from the yoke, and set it face down on a protective mat on a flat surface. Do not carry the tube by its neck, since this is too weak to withstand the pressures caused by the weight of the tube.

5. Avoid striking or scratching the CRT. Extra care must be taken to avoid scratching the faceplate of frosted-glass tubes such as the 21AP4, since this type of glass is more susceptible to damage. Fingerprints on the faceplate of a tube may be removed with carbon tetrachloride.

6. Keep bystanders at a safe distance when handling a picture tube.

7. A defective picture tube should not be left with a consumer or trash collector, since they may not know how to dispose of it safely. The dud has trade-in value.

CRT SERVICING AND REPAIRS

When the servicing procedure points out a defective picture tube, it may be that the fault can be repaired, and replacement may not be necessary. This applies to open heater connections, loose bases, low emission, internal arcing, and loading of circuits caused by interelectrode shorts in the picture tube.

Open Heater. When the heater of the CRT does not light, the fault is often due to a poor connection at a tube pin. Before removing the picture tube, sweat the No. 1 and No. 12 pins with a hot soldering iron and try operation. If the heater now lights, a repair has been effected.

Rebasing a CRT. When inspection shows that the base of a picture tube is loose, there may be a break or short in the wires passing through the base. If wiggling the base restores operation, this is sure to be the case, and a rebasing procedure is indicated.

1. Mark the position of the key on the neck of the CRT with a sliver of tape, as shown in Fig. 10-11*a*.

2. Apply a soldering iron to each tube pin in turn, while maintaining a steady pull on the bakelite portion of the base. This will free the base.

3. Straighten the wires coming up from the glass base, sand them clean, and tin them. Sand and tin six pieces of No. 28 copper wire about 4 in. long. The strands in a piece of 300-ohm lead-in wire will serve for this purpose. Cut the tube leads so that they are about 1 in. long. Lengthen each lead by wrapping an end of the No. 28 wire around it, leaving the free end extending away from the base. This step is shown in Fig. 10-11*b*. Solder each splice.

4. Heat each pin on the tube base, and shake out the excess solder.

5. The wire on each side of the tape marker on the neck should be the

No. 1 and No. 12 heater leads. This can be confirmed with an ohmmeter, since the heater leads are the only two to show continuity. Line up the key with the tape marker, and thread the No. 1 and No. 12 leads into their respective pins. This step is shown in Fig. 10-11*c*. The remaining wires will then be properly oriented to slip into their correct pins.

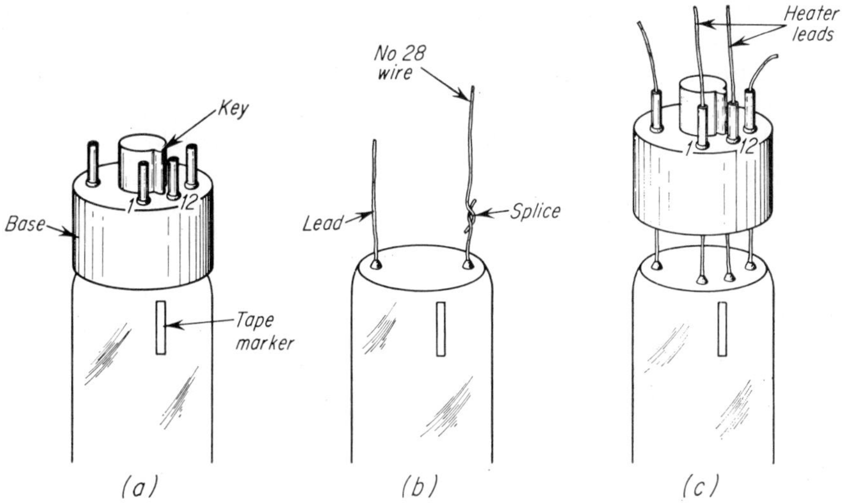

Fig. 10-11. Rebasing a picture tube.

6. Paint a layer of radio cement on the glass neck and on the inside of the base where they will come in contact. Slip the base up on the wires a little at a time, pulling on each wire end in turn, to make sure there is no slack inside the base. Finally press the base firmly onto the glass neck. Clip each lead at the point where it comes out of the tube pins, and solder, making certain that some solder is drawn into the pin.

7. Allow the cement to set, and check the tube.

Low Emission and Gassy Picture Tubes. When the symptom is insufficient brightness or a dim picture of full size, and advancing the brightness control produces an engraved effect or a silvery sheen rather than increased brightness, the fault is probably in the picture tube. It is usually due to low emission, high gas content, or both. Make sure that the voltages applied to the CRT electrodes are normal, and then try operation with a CRT booster. Note, if the set has a series filament string be sure to use a series-type booster. If this does not restore normal operation, the CRT must be replaced.

Internal Arcing and Interelectrode Shorts. Sometimes a blue spark is observed jumping between the elements in the gun of the picture tube. This is internal arcing. It is usually caused when a piece of foreign material becomes lodged in such a position as to narrow the spacing between elements. Arcing then occurs at this point.

The foreign material is usually a sliver of the inner aquadag coating which breaks away and accidentally lands between electrodes, causing arcing, leakage, or a short, depending on contact conditions. The leakage shorts out the signal or disturbs an interelectrode voltage relationship, causing symptoms of no picture, unstable sync, etc. When servicing for these symptoms by the procedure set down in the last chapter, we are brought to the picture tube when removal of the CRT socket restores a normal oscilloscope trace at the output of the video amplifier.

Whether the condition is internal arcing or interelectrode shorts, it is usually possible to burn away the sliver of foreign material, thereby effecting a repair. The leakage or arcing may be between any two of the electrodes in the gun. It is not necessary to check which two electrodes are at fault. The burn-out procedure which follows is quickly applied to all electrodes.

Step 1. Remove the CRT socket.

Step 2. Locate the high B plus point in the low-voltage power supply. Place the chassis so that the picture-tube pins and the B plus point are both accessible. Secure a test filter condenser of about 100 μf, 400 volts, equipped with a clip and test prod. Observing polarity, charge it from the high B plus point.

Step 3. Clip the negative lead of the charged condenser to the CRT cathode at the No. 11 pin. Touch the prod momentarily to a heater connection at the No. 1 pin, and then to the grid connection at the No. 2 pin. If a spark is observed when the prod is touched to either tube pin, a short is present, and it may or may not have been burned away. Recharge the condenser at the B plus point and repeat, this time tapping the neck of the tube gently as the prod is touched to the pins. The pins should be touched again while tapping even if no spark is observed. Two applications should burn out the short.

Step 4. Recharge the condenser and connect the negative clip to the CRT grid at the No. 2 pin. Repeat the procedure by momentarily touching the cathode and first anode connections at the No. 11 and No. 10 pins, respectively. Repeat with tapping.

Step 5. If the CRT has a focusing electrode, the base will be equipped with a No. 6 pin. Remove any short that may be present at this electrode by applying the charged test condenser between the No. 6 pin and the No. 10 pin.

Step 6. Replace the CRT socket, and try operation of the receiver to see if the short has been removed and normal operation restored. If not, the picture tube must be replaced. Remember that the effects of cathode-heater leakage can be avoided by the use of an isolating type of CRT booster.

Ion Burns. Under normal raster conditions an abnormal brown stain about the size of a half dollar will be observed on the screen. This is an ion burn and it was caused by poor placement of the ion trap. Tube replacement must be made for this condition. Upon insertion of a new tube, carefully follow the setup procedure given in this chapter, paying particular attention to the placement of the ion trap.

On rectangular tubes the ion burn takes on the appearance of a skull and crossbones. When a tube of this type is replaced, the position of the ion trap as well as the operating voltages on all elements should be carefully checked.

Servicing for No Control of Brightness. In a normal operating receiver, when tuned to an unused channel, varying the brightness control should vary the illumination on the screen from complete extinction to a very bright raster. When the general illumination on the screen remains unchanged regardless of the position of the brightness control, we have the symptom aptly named "no control of brightness." Intermediate conditions where there is some control, but not enough, can be described at the same time because it is due to the same defect, but in a different degree. It can be serviced by the same procedure.

A commonly used brightness circuit is shown in Fig. 10-12, where the brightness control provides a variable bias controlling the general illumi-

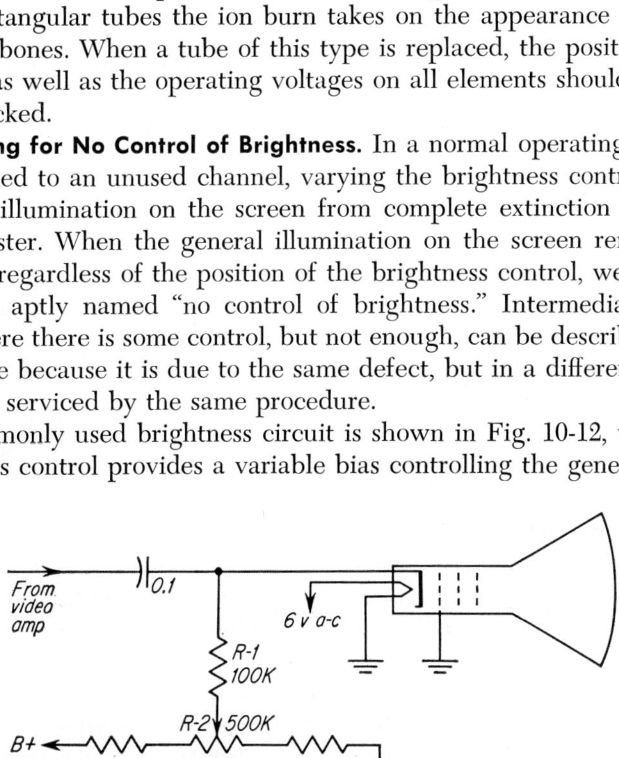

Fig. 10-12. Servicing the brightness circuit.

nation of the screen. The video signal adds to and subtracts from the general illumination to produce the light and dark elements of the original transmitted picture. Improper operation of the brightness control can be caused by defects in the *R*-1, *R*-2 brightness circuit, or by cathode-to-ground leakage in the picture tube. Obviously, the first step in a servicing procedure for this trouble is to determine which of these two factors is at fault.

Step 1. Remove the picture-tube socket, connect the positive prod of a voltmeter to the cathode lead at socket pin No. 11, the negative to the grid at the No. 2 pin, and turn the set on. Read the cathode-grid voltage while rotating the brightness control. If the reading varies from approximately 20 to 150 volts as the control is varied, the brightness circuit is functioning normally, and the picture tube is probably at fault. If the normal voltage variations are not observed, the brightness circuit is bad. Refer to the schematic diagram of the receiver, and check all components in the brightness circuit. Further servicing notes and hints on the brightness circuit are given in Chap. 6.

Step 2. If the previous step indicates that the picture tube is at fault, it is advisable to confirm this, and to see if a repair is possible before replacing the tube. The confirming check is performed with a test 8XP4 tube. Turn the set off. Ground the high-voltage second-anode lead to discharge it, and connect the CRT socket and high-voltage lead to the 8XP4. Set the brightness control to the counterclockwise end of its rotation, and turn the set on. Allow about a half minute for the high voltage to come on, and advance the brightness control. Since the deflection yoke is not being used in the test setup, a white spot will appear in the center of the screen of the 8XP4. Increasing the brightness-control setting increases the intensity of the spot. This check confirms that the receiver will vary the brightness of a normal CRT, and so the original tube is at fault.

Step 3. The cathode-to-ground short in the picture tube is probably a sliver of cathode coating or internal dag coating between the cathode and grid or heater, both of which are at ground potential, as shown in Fig. 10-12. The short is not likely to appear in an ohmmeter check because of the cold tube and the low battery voltage of the ohmmeter, but its presence has been verified by the confirming check of step 2. It may be possible to burn it away by the procedure previously given for removing interelectrode shorts. If the burning procedure proves ineffective, try an isolation transformer on the CRT heater to remove the effects of cathode-heater leakage. If all efforts fail, the picture tube is then replaced.

Servicing for Improper Focus. It will be remembered that focusing in the picture tube is an electrostatic or magnetic means of converging the electron beam in the tube so as to make a spot of minimum size. A small spot makes each picture element distinct, giving the effect of a sharply focused picture. Failure of the focusing system makes for a larger spot and a defocused picture.

When servicing for improper focus, the serviceman must keep in mind two other similar symptoms which may be confused with it. These are blooming and lack of detail in the picture. Since these are due to other causes and call for different servicing techniques, it is important to distinguish between them.

Blooming is checked for by observing the effect of varying the brightness control through its range. Blooming is usually caused by a decrease in the output of the high-voltage supply. In this condition adjusting the brightness control causes the spot size to increase, giving the blooming and defocusing effect. The picture, however, although not bright enough, will be in focus at low positions of the brightness control. A servicing procedure for blooming is given in Chap. 6.

The indistinct picture may also be caused by unequal amplification of high and low video frequencies in the i-f amplifier or video section. This possibility is eliminated by tuning to an unused channel and focusing the raster lines. If adjusting the focus control produces sharp raster lines, the i-f or video section is at fault. The receiver is serviced for loss of detail as described in Chap. 9. If the raster lines cannot be focused, the focusing system is at fault. When checking sharpness of raster lines it must be remembered that a certain amount of defocusing is normal, especially in the case of a large picture tube. The raster lines should be sharp and clear across the center portion of the tube, and a small amount of fuzziness at the edges is permissible.

There are three basic types of focusing systems in general use. One of these is the electromagnetic type which uses a coil and front-panel focus control. Another is magnetic, using a permanent magnet on the neck of the tube with a mechanical focusing adjustment reached from the rear of the cabinet. The third type is electrostatic, with an extra focusing grid in the gun brought out to the No. 6 pin on the CRT base. The three types require different servicing procedures, and each will be taken up separately.

Electromagnetic Focusing. In electromagnetic focusing, an electromagnet called a focus coil is positioned over the neck of the tube so that the electron stream is within the field of the coil. A representative focus coil and its position on the tube are shown in Fig. 10-13. The focus coil is attached to the yoke housing.

A direct current of approximately 100 ma is required to energize the focus coil. This is obtained from the low-voltage power supply, the coil being used as a choke or voltage-dropping resistor. The current through the coil is increased or decreased within limits by a variable resistor in a shunt network. The variable resistor is usually a front-panel adjustment functioning as the focus control. Two representative focus-control circuits are shown in Fig. 10-14.

When the focusing trouble is traced to the focusing circuit it can be caused by a number of faults. The coil may be open or have shorted turns, the focusing control or associated resistors may have changed in

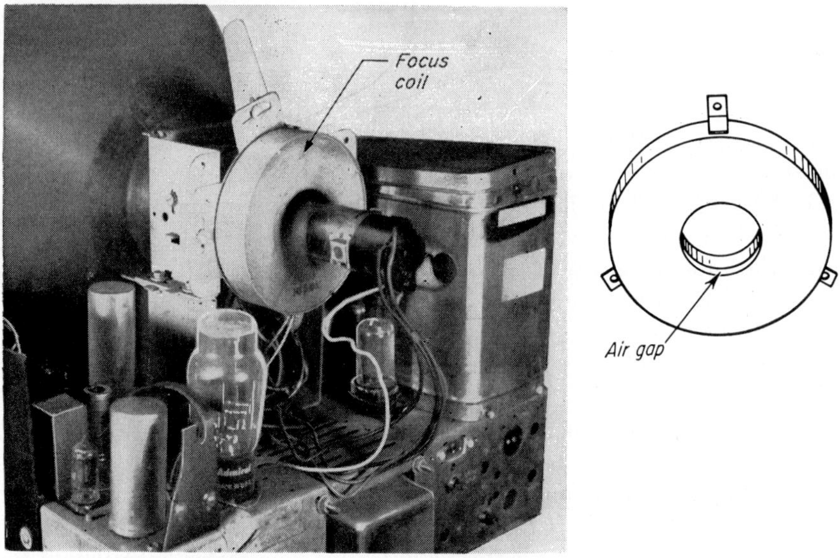

Fig. 10-13. A focus coil and its position on the neck of the picture tube.

value, or the current through the system may have changed. All these factors must be checked.

When checking resistance readings in the circuit, refer to the manufacturer's notes for correct values, and remember to open the circuit to avoid incorrect readings because of shunting action. The rectifier is checked for full output, even though a reduced raster size is usually the first sign of aging of this component. Another possibility to be checked is decreased current drain taken by an aging sound output tube.

If all the components appear to be normal, try shifting the position of the focus coil on the neck of the tube, and see if this brings about a better focusing ability. In the correct position, the yoke is snugly placed against the bell of the CRT, and the focus coil is approximately 3/8 in. behind the yoke. If a clear focus is still unobtainable, replace the picture tube.

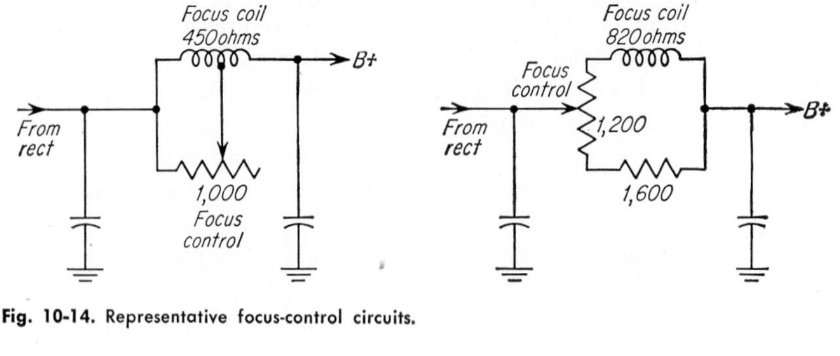

Fig. 10-14. Representative focus-control circuits.

If the focus coil is defective and must be replaced, try to obtain an exact replacement in regard to electrical and mechanical mounting specifications. When installing the new focusing coil, see that the air gap which is labeled in Fig. 10-13 is mounted toward the yoke. If the focus potentiometer is defective, be sure that the replacement has the same wattage rating as the original. A 5-watt potentiometer is the size commonly used.

Servicing Permanent-magnet Focusing Devices. In some receivers, the focus coil is replaced by a permanent magnet called a focalizer. This can be likened to the use of a p-m dynamic speaker in a sound system in place of a speaker with a field coil. Two commonly used permanent-magnet types of focusing units are shown in Fig. 10-15. As shown in the diagram, the magnet assembly is mounted in approximately the same position as the focus coil. It usually consists of four alnico slugs mounted around a rim.

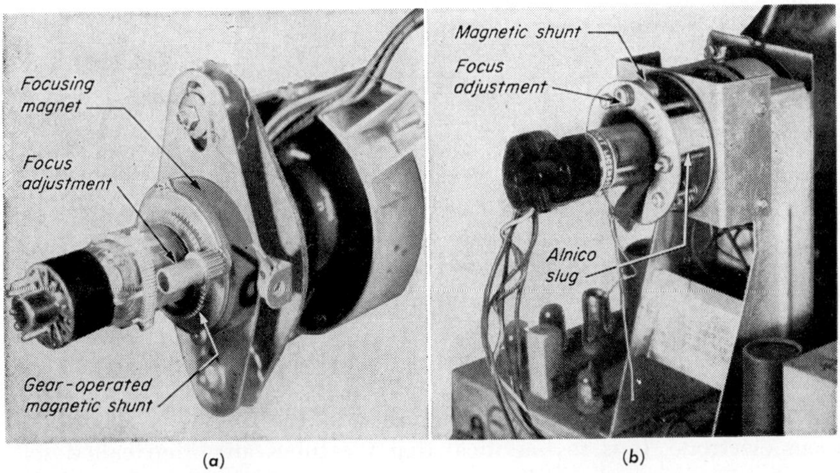

Fig. 10-15. Permanent-magnet focusing devices.

To obtain control of the focusing action, the strength of the magnetic field must be varied. This is accomplished by shorting out some of the magnetic field. In the system illustrated in Fig. 10-15*a*, a magnetic shunt is in the form of a shim. It is threaded and pushed in and out of the focalizer by turning the adjustment screw. In the type shown in Fig. 10-15*b* the magnetic shunt looks like a large screw with a slotted head. Turning the screw gets more or less of this shunt in the magnetic field and therefore acts as a focus control. This adjustment should be made with a nonmagnetic screwdriver since the correct adjustment will be lost when the steel of an ordinary tool is removed. A special focalizer-adjustment screwdriver made of beryllium copper is put out by the Xcelite Company (catalogue No. BR 1410 FC).

When the symptom of improper focus is traced to the focusing device or picture tube, there is very little to check in a focalizer system. The holding bolts can be loosened to see if a better focus is obtainable with a slight shift in position on the neck of the tube. If this fails, focalizer mounting arrangements are such that it is easier to replace the CRT than the focalizer; so operation with a new CRT is tried first. If sharply defined raster lines are still unobtainable, the focalizer assembly is replaced. Make sure to choose a replacement with correct mounting facilities. Positioning of the ion trap should be checked after any focus adjustment has been made.

Servicing Electrostatic Focusing Systems. Another focusing system makes use of an electrostatic field within the tube. The electron gun in these tubes is equipped with an extra grid called the focus electrode. This is G-4 in Fig. 10-16. The tube is easily identified by the presence of a pin in the No. 6 position on the base.

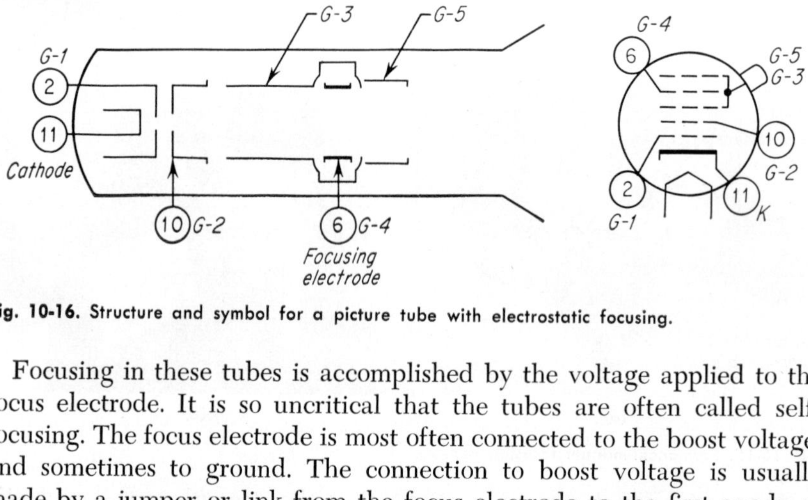

Fig. 10-16. Structure and symbol for a picture tube with electrostatic focusing.

Focusing in these tubes is accomplished by the voltage applied to the focus electrode. It is so uncritical that the tubes are often called self-focusing. The focus electrode is most often connected to the boost voltage, and sometimes to ground. The connection to boost voltage is usually made by a jumper or link from the focus electrode to the first anode at the CRT base. The focus link is shown making this connection in Fig. 10-17*a*.

In cathode-fed picture tubes, the G-1 grid is often connected to ground. This provides a ground connection for the focus electrode by the use of the same link turned upside down and connected to the grid at the No. 2 pin as shown in Fig. 10-17*b*. Instructions on some receivers suggest reversing the focus link and using the connection which provides better focusing.

In some receivers a separate wire is connected to the No. 6 pin on the CRT socket. This wire is brought to some point on the B plus or boost line. In some few cases, a two- or three-resistor voltage divider on the boost voltage is provided, and the focus electrode is connected to one of the taps. Data on these receivers include instructions to try one of the other tap positions in case of difficulty with focusing.

When servicing the electrostatic system for lack of focus, about the only possibility is an open circuit in the supply to the focus electrode. If a

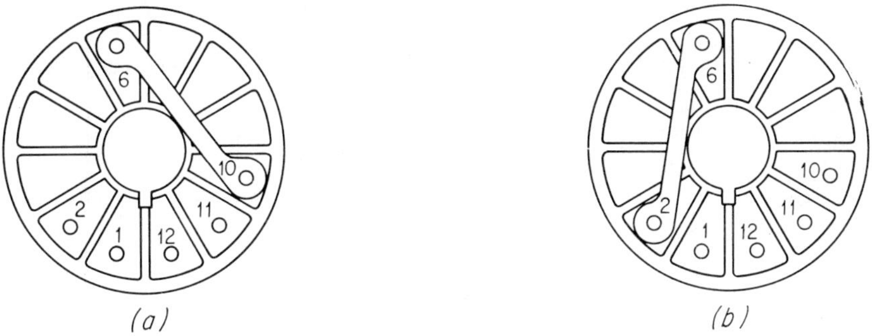

Fig. 10-17. Alternate positions of the focus strap on the base of the picture tube for selecting best focus.

focus link is used, check it for good contact to the tube pins. If a separate lead is used to a voltage-divider system, check the voltage divider for an open resistor.

How to Identify an Unmarked Picture Tube. When the numbers or letters which give the picture-tube type have become eradicated, the tube can be identified by its physical dimensions and characteristics. These in turn are compared with the listings in the RCA Picture Tube Characteristics Chart. A page made up from the chart in the booklet, "RCA Picture Tubes," is illustrated in Fig. 10-18. Arrows have been added at the heads of the columns which are useful in tube identification. It is unsafe to use the receiver schematic diagrams for identification purposes since these often list several picture tubes for a given chassis. Although all the listed tubes will work, not all of them will fit a particular cabinet and mask opening.

The first important dimension is under the last arrow in the column labeled "Envelope diagonal or diameter." This is measured as shown in Fig. 10-19. A carpenter's folding rule is good for the purpose. It is sufficiently flexible to follow the curvature of the tube. Do not use a tape measure, which might stretch. The diameter of a round tube is measured as shown in Fig. 10-19*a* and the diagonal for a rectangular tube as in Fig. 10-19*b*.

This measurement gives us the inch rating of the tube and the first number in the tube-designation system. For example, a reading of 21⅜ in. gives us a 21-in. tube, which now may be any one of the many tubes

RCA PICTURE TUBE

Type	Envelope	Aluminized Screen (Asterisk (*) denotes "Silverama" type)	Faceplate φ	External Conductive Coating Max. μμf	External Conductive Coating Min. μμf	Focusing Method	Deflection Method	Approx. Deflection Angle Degrees Diag.	Horiz.	Vert.	Maximum Dimensions Inches Overall Length	Envelope Dia. or Diagonal	Width	Height
Black-and-White Types (Cont'd)														
20CP4	G	No	FG	None	None	M	M	70	66	50	21 13/16	20 9/32	18 7/8	15 1/8
20DP4-A/ 20CP4-A	G	No	FG	750	500	M	M	70	66	50	21 7/8	20 7/32	18 13/16	15 1/16
20DP4-C/ 20CP4-D	G	* Yes	FG	750	500	M	M	70	66	50	21 7/8	20 7/32	18 13/16	15 1/16
20HP4-A/ 20MP4	G	No	FG	1500	750	E	M	70	66	50	22 1/8	20 7/32	18 13/16	15 1/16
20HP4-D	G	* Yes	FG	1500	750	E	M	70	66	50	22 1/8	20 7/32	18 13/16	15 1/16
20MP4	G	No	FG	750	500	E	M	70	66	50	22 1/8	20 9/32	18 7/8	15 1/8
21ACP4-A	G	* Yes	FG	750	500	M	M	90	85	68	20 3/8	21 1/2	20 3/8	16 1/2
21ALP4-A	G	* Yes	FG	750	500	E	M	90	85	68	20 3/8	21 1/2	20 3/8	16 1/2
21ALP4-B	G	* Yes	FG	750	500	E	M	90	85	68	20 3/8	21 1/2	20 3/8	16 1/2
21AMP4-A	G	* Yes	FG	750	500	M	M	90	85	68	20 3/8	21 1/2	20 3/8	16 1/2
21AP4	M	No	FFG	None	None	M	M	70	66	50	22 5/8	21	19 27/32	15 7/16
21ATP4	G	* Yes	FG	1500	1200	E	M	90	85	68	20 3/8	21 1/2	20 3/8	16 1/2
21AVP4/ 21AUP4	G	No	FG	1500	1200	E	M	72	67	53	23 13/32	21 1/2	20 3/8	16 1/2
21AVP4-A/ 21AUP4-A	G	* Yes	Same as 21AVP4/21AUP4, except has aluminized screen.											
21AWP4	G	* Yes	FG	1500	1200	M	M	72	67	53	23 13/32	21 1/2	20 3/8	16 1/2
21EP4	G	No	Same as 21EP4-A, except has no external conductive coating.											
21EP4-A	G	No	FG▲▲	750	500	M	M	70	65	50	23 3/8	21 11/32	20 3/8	15 11/16
21EP4-B	G	* Yes	Same as 21EP4-A, except has aluminized screen.											
21FP4-A	G	No	FG▲▲	750	500	E	M	70	65	50	23 3/8	21 11/32	20 3/8	15 11/16
21FP4-C	G	* Yes	Same as 21FP4-A, except has aluminized screen.											
21MP4	M	No	FFG	None	None	E	M	70	66	50	22 5/8	21	19 27/32	15 7/16
21YP4	G	No	FG	750	500	E	M	70	65	50	23 13/32	21 11/32	20 3/8	15 11/16
21YP4-A	G	* Yes	Same as 21YP4, except has aluminized screen.											
21ZP4-A	G	No	FG	750	500	M	M	70	65	50	23 13/32	21 11/32	20 3/8	15 11/16
21ZP4-B	G	* Yes	Same as 21ZP4-A, except has aluminized screen.											

Light face = Discontinued type.
[G] = Glass rectangular.
[M] = Metal rectangular.
CL = Clear glass.
FFG = Frosted Filterglass.

(G) = Glass round.
(M) = Metal round.
FG = Filterglass.
M = Magnetic.
E = Electrostatic.

Note: All picture tubes shown have 6.3-volt/0.6-ampere heaters except types 9AP4 and 12AP4 which have 2.5-volt/2.1-ampere heaters.
φ Spherical, unless otherwise specified.
▲▲ Cylindrical faceplate.

Fig. 10-18. Excerpts from the RCA Picture Tube Characteristics Chart.

listed from the 21ACP4-A to the 21ZP4-B. Do not use the fine distinctions such as the $21\frac{1}{2}$ and $21\frac{11}{32}$ listings in the column for identification purposes. These are maximum dimensions with a tolerance value of $\frac{1}{4}$ in. Besides, the round edge of the glass makes it difficult to obtain an accurate reading. At this point, it is sufficient to know that we have a 21-in. tube, and not a 19, 20, or 24.

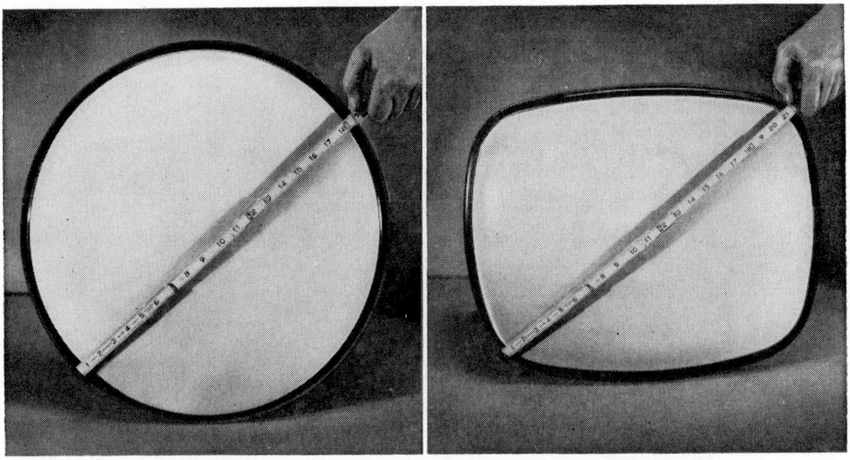

(a) (b)

Fig. 10-19. How to measure the diameter or diagonal of the envelope.

Next check the first column, which is labeled "Envelope." This describes the tube as metal or glass and whether it has a round or rectangular face. This provides further differentiation. In the case of our 21-in. example, note that all the tubes are rectangular, and there are only two metal tubes.

The next important check is to determine the type of faceplate. A filter glass FG faceplate or a frosted filter glass FFG faceplate is of little value for identification purposes, but spherical and cylindrical faceplates are of great importance, since they will require differently shaped masks in the cabinet. The type of faceplate is indicated in the third column of the chart of Fig. 10-18. Note the two black triangles in the column for the 21EP4-A. This is the symbol for the cylindrical faceplate. All the others without the triangle are spherical.

The faceplate shape is determined by holding a short straightedge vertically against the tube, as shown in Fig. 10-20. The spherical faceplate curves back away from the straightedge, whereas the cylindrical type is parallel to it.

Another identification point is the type of focusing. Some tubes use magnetic focusing by means of a focus coil or focalizer on the neck of the tube. Others use electrostatic focusing where a voltage is applied to a focusing electrode in the tube. These tubes have a No. 6 base pin for this connection. Tubes with magnetic focusing do not have a pin in the No. 6 position. So check for the presence or absence of a pin in the No. 6 position of the CRT base. If the pin is present, the tube will have an E for electrostatic in the column labeled "Focusing Method" in the

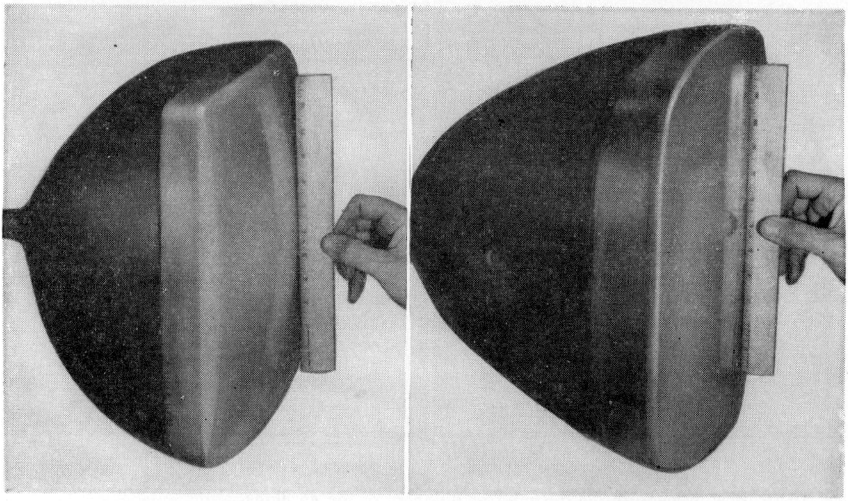

(*a*) (*b*)

Fig. 10-20. How to distinguish between spherical and cylindrical faceplates. (*a*) Spherical. (*b*) Cylindrical.

chart of Fig. 10-18. If the base does not have a pin in the No. 6 pin position, the tube will have an M for magnetic in that column. In our 21-in. example the metal and cylindrical tubes are now positively identified. There is still a choice if it is a spherical glass tube.

The last important means of identification is to determine the angle of deflection. Early picture tubes employed a deflection angle of 70° or less. As the tubes became larger, for increased screen size, the excessive tube length created a problem. The length of the tube generally determines the depth of the cabinet, and when these become more than 2 ft deep, cabinet size becomes an important consideration because of its effect on cost and appearance. One method of reducing cabinet size was to provide a hole in the rear cabinet cover so that the neck could protrude. It was then protected by a cover. Another method was to shorten the tube by employing a wider deflection angle. Tubes with 90° and 110° deflection angles were developed.

With the exception of the 110° tube it is not easy to determine the deflection angle by looking at the unmarked tube, but this can be determined by measuring the over-all length. Refer to Fig. 10-18, run down the deflection angle and over-all length columns, and note that a 70° 21-in. tube is 3 in. longer than a 90° tube. Tube length may be measured by setting it face down on a mat on a bench top. If it is a 21-in. tube, the carpenter's folding rule is particularly useful. It folds at a natural right angle, and one of these fold points is at $23\frac{5}{8}$ in. As shown in Fig. 10-21, this will just clear a 70° 21-in. picture tube with an over-all length of $23\frac{13}{32}$ in. and will leave a 3-in. gap over a 90° 21-in. tube whose over-all length is $20\frac{3}{8}$ in.

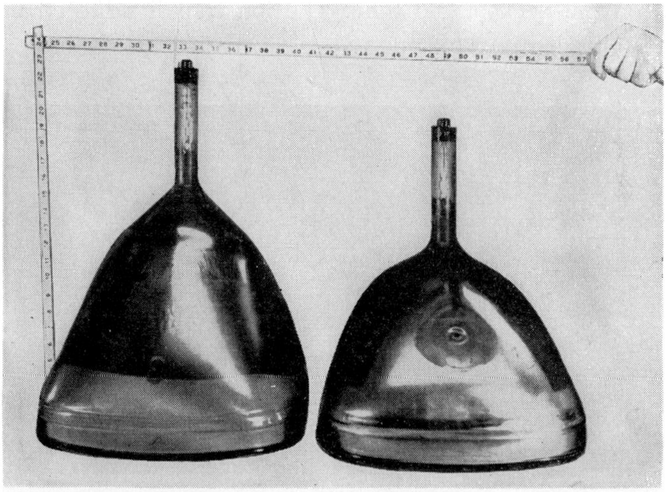

Fig. 10-21. How to identify picture tubes by length and deflection angle. A 21-in. tube with a 90° deflection angle is 2 in. shorter than a 70° tube with the same diagonal.

An aluminized screen is easily recognized by the spray of metallic aluminum on the inside of the tube, and a frosted faceplate by its ground-glass appearance. It is recommended that the same type be used for replacement purposes, since the consumer is accustomed to the picture tones provided by these features.

The 110° tube is easily recognized by its extremely short neck. The diameter of the neck is also less than the conventional 70 and 90° tubes.

SETUP PROCEDURE FOR THE CRT

Setup procedure for the picture tube is usually included in the service notes as a guide for the original installation of the receiver since some of the adjustments may have been jarred in shipping. The same instructions should be followed when the servicing procedure calls for a readjustment or when the picture tube is replaced. The setup procedure deals with the adjustment of the ion trap, picture-centering device, and focus. The following generalized setup procedure should be of advantage where the manufacturer's notes are not available or are not in sufficient detail.

When replacing the picture tube, remember to orient it so that the high-voltage connection is on the correct side. The deflection yoke should fit snugly against the bell of the picture tube. In some cases the yoke is held in position by retaining springs. If it is difficult to get the last of these springs back into position, an automotive tool called brake spring pliers, illustrated in Fig. 10-22, may prove helpful. All accessories are replaced on the neck of the tube. The ion trap is adjusted first, then centering, and finally focus.

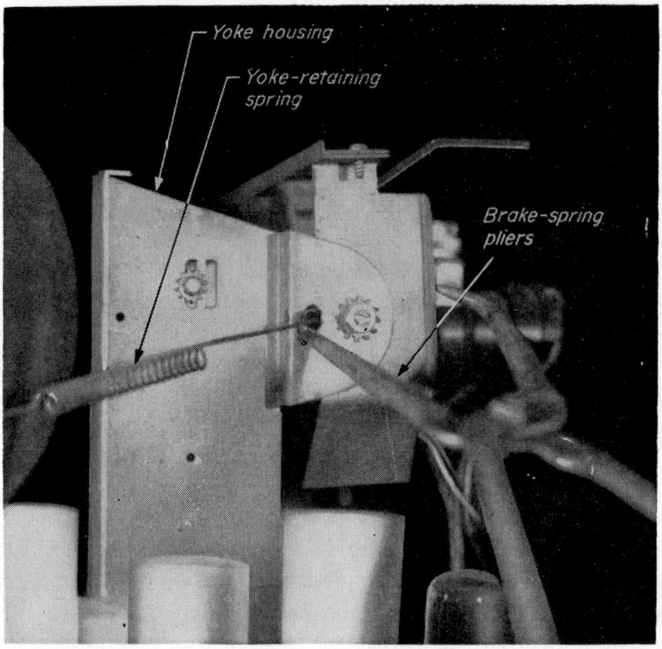

Fig. 10-22. Brake-spring pliers being used to pull a yoke-retaining spring into position.

Adjusting the Ion Trap. The ion trap is placed approximately in its correct position before the set is turned on. This is done to reduce the amount of time the set is operating with an incorrect ion-trap adjustment. Electron and ion bombardment may damage the gun.

For the preliminary adjustment, the ion-trap magnet is aligned in the space between G-1 and G-2 and parallel with the glass gun supports, as shown in Fig. 10-23. Then set the brightness control to its mid-rotation position, and turn the set on. If the screen does not light, the magnetic polarity of the ion trap is reversed. Turn the set off and rotate the ion trap 180°, but still in the position indicated in Fig. 10-23. Then turn the set on, and after the screen lights up, move the ion trap slightly forward

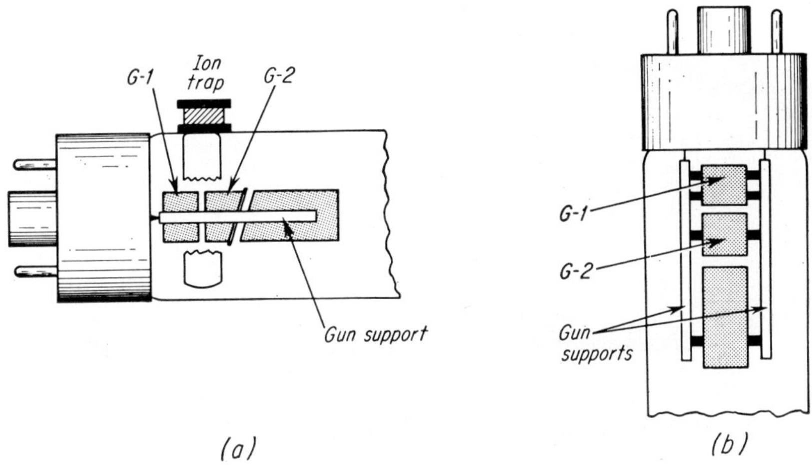

Fig. 10-23. Preliminary adjustment of the ion trap.

and backward from the initial position, rotating it slightly at the same time. Adjust for maximum brightness on the picture tube. Reduce the brightness-control setting so that adjustment of the ion trap shows a definite point of maximum brightness.

Adjustment of the ion trap will have some effect on neck shadow, centering, and focus. Do not adjust for compromise positions that seem to improve neck shadow, etc. Simply adjust for maximum brightness. After centering and focusing are adjusted you should make a final check of the ion-trap setting.

Most ion traps use a spring for self-clamping, and the spring allows sufficient slippage for adjusting. In some cases the clamping arrangement is a long screw with a wing or knurled nut. When these are adjusted, the nut must be loosened first, and care taken so that the final tightening does not move the ion-trap adjustment. In either type, if the spring or nut is lost the replacement must be of nonmagnetic material like brass or aluminum.

Leveling the Picture. If the deflection yoke is not correctly positioned, the picture will be tilted. The picture in Fig. 10-24*a* has been reduced in height so that the tilt is more obvious against the mask opening. Under conditions of full height the tilt is not so obvious, and the customer's complaint is likely to be that the lettering across the picture tube as seen in a cast of characters seems to slant uphill or down as the case may be.

When servicing for this complaint, replacing a yoke, or going through a complete setup procedure, the yoke must be leveled. The yoke is usually locked into place by a thumbscrew as shown in Fig. 10-24*b*. This screw is loosened, thereby allowing the yoke to be rotated by moving the screw

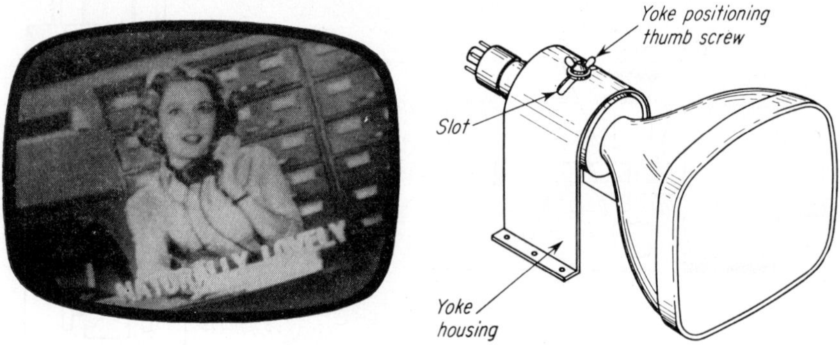

(*a*) (*b*)

Fig. 10-24. A tilted picture and yoke-leveling adjustments.

from side to side in the slotted hole in the yoke mount. If the height is reduced somewhat, the level position is easily found by making the bottom of the raster parallel to the mask opening. Before locking the thumbscrew be sure that the deflection yoke is snugly seated against the bell of the picture tube. The yoke can be pulled forward by applying pressure to the same thumbscrew since the nut is also movable in its slot in the yoke.

In some receivers, the yoke is held in place by spring tension on a clamp collar. In these receivers, the springs or clamps are removed, the yoke is rotated for a level picture, and then the fastening is restored to clamp the yoke in position.

Centering Devices. The undeflected beam from the electron gun should produce a bright spot of light in the exact center of the screen. Then, in ideal conditions, when deflection of perfect height, width, and linearity is added the picture will exactly fit the mask opening. If the beam is slightly off center, the mask opening is not completely filled, let us say on top and left side, while the bottom and some of the right side are lost behind the mask. In practical conditions with an off-center beam, the mask is usually filled by an overdriven height and width, but with a loss of some picture and linearity. The symptom which leads to a check of centering has to do with loss at the extremes of the picture, poor linearity, or neck shadow. Also any procedure for proper centering adjustments must include height, width, and linearity adjustments. The centering adjustments are a necessary part of the setup procedure when a new tube is installed.

Centering the beam is usually accomplished by adjustments on the focusing coil or magnet. When electrostatic focusing is used, special adjustable magnets are provided for centering purposes. The various centering devices will be described first and then a generalized centering procedure will be presented.

Typical mountings for magnetic focusing devices are shown in Fig. 10-25. In all cases, the hole for the neck of the tube is made oversize so that the focusing magnetic field may be shifted with respect to the beam in the tube. This deflects the beam, thereby shifting the position of the

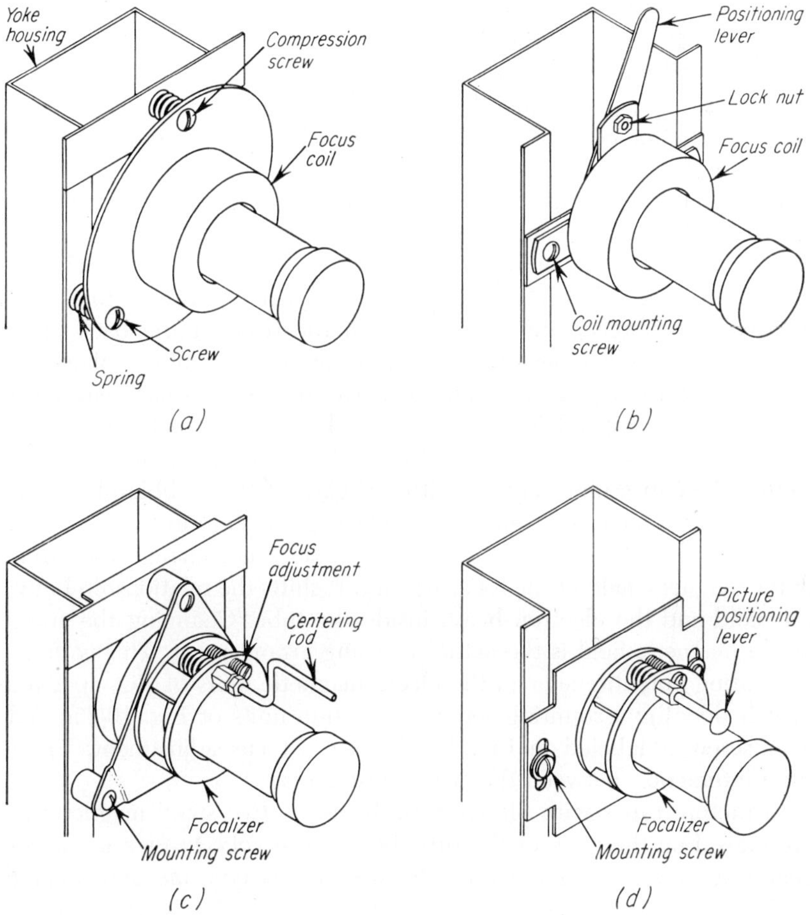

Fig. 10-25. Centering devices on tubes with magnetic focusing.

picture on the screen and giving a centering adjustment. In Fig. 10-25*a* the focus coil is attached to a plate which in turn is attached to the yoke housing by three spring compression screws. Loosening or tightening these screws against the spring tension shifts the entire focus coil, thereby positioning the beam and thus the entire raster.

The centering arrangement illustrated in Fig. 10-25*b* shows the focusing coil attached to the yoke housing by two screws. A plate with a hole for the neck of the tube is mounted on the focus coil. This is called a wobble plate and acts as a movable pole piece for the focus coil in that it changes the relationship between the beam in the neck of the tube and the magnetic field of the focus coil. The wobble plate extends upward in a handle called a positioning lever. To adjust centering on this type, the lock nut which holds the wobble plate to the focus coil is loosened, and the positioning lever is moved about, thus changing the position of the pole piece. If true centering is beyond the range of the positioning lever, the focus-coil mounting screws are loosened. The holes in the mounting legs are slotted, permitting the entire focus-coil assembly to be moved about for better centering. When this is achieved, all screws are tightened.

The centering arrangements for two permanent-magnet focusing units are shown in Fig. 10-25*c* and *d*. In both cases focusing devices consists of four alnico slugs mounted between two circular plates. The back plates are attached to a mounting plate which is screwed to the yoke housing. A wobble plate is mounted on the front of the back circular plate, and the adjustment lever for positioning the wobble plate comes through the front circular plate. The wobble-plate adjustment lever is called a centering rod in Fig. 10-25*c* and a picture-positioning lever in Fig. 10-25*d*. The mounting holes for the focalizer are slotted or oversize so that the entire unit may be shifted slightly if the wobble-plate adjustment does not have sufficient range to give good centering.

In all cases, the centering arrangement does not change the strength of the magnetic field of the focusing unit. It shifts the relationship between this field and the electron beam inside the tube. Changing the strength of the magnetic field is the actual focusing arrangement. This is done by a shunting potentiometer in the electromagnetic units of Fig. 10-25*a* and *b*; it is done by a shunting screw in the p-m units of Fig. 10-25*c* and *d*. This screw is labeled in Fig. 10-25*c* as the focus adjustment. There is little interaction between the two adjustments.

In picture tubes with electrostatic focusing, two small magnetic rings are placed on the neck of the tube behind the yoke to give a control of centering. Two commonly used devices of this type are shown in Fig. 10-26. The unit illustrated in Fig. 10-26*a* shows the centering ring attached to the yoke cover. Each ring has a small tab protruding from it for adjustment purposes. The tabs can be rotated together or singly. The centering unit shown in Fig. 10-26*b* operates in a similar fashion but is held in place by a clamp. When adjusting this type, the clamp is loosened just enough to permit movement in a forward or backward direction in

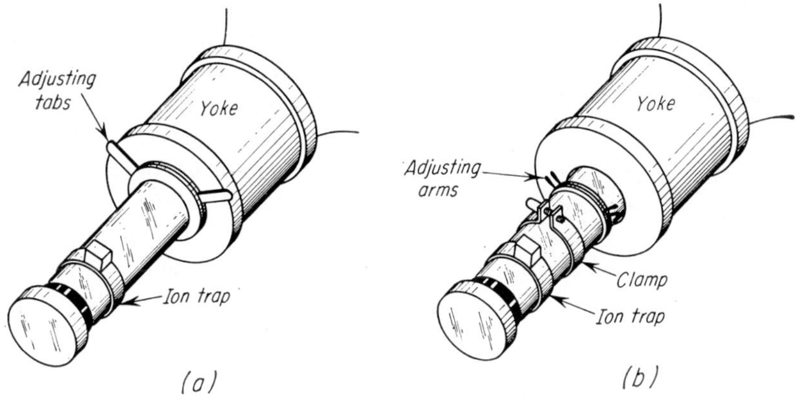

Fig. 10-26. Centering devices on tubes with electrostatic focusing.

case the movement of the adjustment arms alone is insufficient for good centering. An approximately correct position from which to start the centering procedure places the rings about ¼ in. behind the yoke, and the adjusting arms about 120° apart.

Procedure for Centering. Test patterns are of considerable help in the setup procedure, but since they are infrequently on the air, the following procedure has been devised to give satisfactory results. The set has already been adjusted for proper positioning of the ion trap, tilt, and focus.

Step 1. If the set is equipped with a control of horizontal linearity, this is adjusted now. The high-voltage fuse is removed and replaced by a milliammeter on a range of approximately 200 ma. The linearity coil is then adjusted for a minimum reading on the milliammeter.

Step 2. If the set is equipped with a width control, this is adjusted so that both edges of the raster can be seen within the mask opening. The horizontal drive control may be used for this purpose if there is no width control. If neither control is present the width may be reduced by operating the receiver on a low line voltage as obtained from an isolation transformer like the RCA WP-25A with low-voltage taps. If there is no other method of reducing the width, a high-wattage unit like a heating coil may be wired in series with the line cord.

Step 3. Stop vertical deflection by removing the vertical oscillator tube or shorting its grid to ground. We now have a horizontal line which does not quite stretch across the screen. Mark off the center of the screen or mask opening by drawing diagonals with a china marking pencil as shown in Fig. 10-27. The horizontal line will probably be off center in some manner similar to the one shown in Fig. 10-27.

Step 4. The centering adjustment will be one of the six types illustrated in Figs. 10-25 and 10-26. Adjust the centering controls first to bring the line up to the center mark, and then for equal dark spaces from each end of the line to the edge of the picture tube or mask opening. It will be noted that adjusting for equal spacing causes the line to leave the center mark, and readjustment is needed. If proper centering cannot be obtained by the positioning lever of Fig. 10-25*b*, *c*, *d*, the focus-unit mounting screws are loosened, and the entire unit is shifted for better centering. The screws are then tightened, and the centering is completed with the positioning lever. The adjustment tabs of Fig. 10-26 will be found to give better vertical movement when turned together, and better horizontal movement when turned separately.

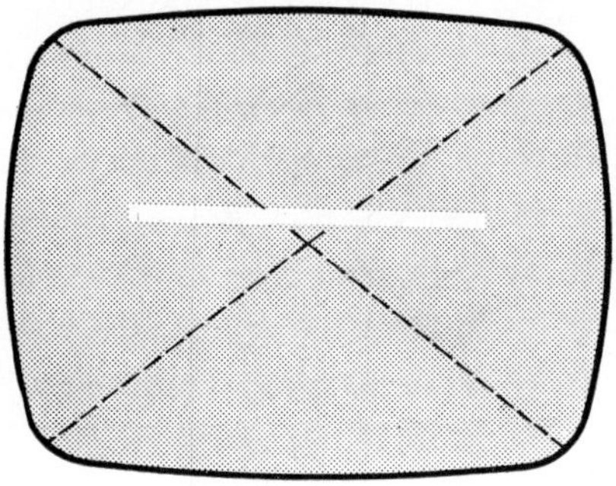

Fig. 10-27. Centering may be adjusted by using a horizontal line that is reduced in width.

Do not depend on a mirror for final adjustments because of errors due to parallax. When it appears that you have the proper adjustment, stand directly in front of the picture tube to see the relationship between the horizontal line, ends of picture tube, and center mark. Then correct for any difference noted by the direct viewing angle. When correct centering is obtained, tighten all holding clamps.

Step 5. Restore correct width and vertical deflection. If the set has a width control, extend the width to just beyond the mask opening.

Step 6. Readjust the vertical height and vertical linearity controls. This procedure is given in Chap 5.

Step 7. Readjust the ion trap for maximum brightness. A shadow may appear in one corner of the picture tube. A very slight retouch of the ion trap will cause the shadow to disappear. However, the ion-trap retouch should not affect the brightness. If it does, the ion-trap magnet may be weak. Try operation with another ion trap. If the symptom remains repeat the setup procedure.

Adjustment of pincushioning magnets. Some receivers particularly those using cylindrical tubes such as the 21EP4, are equipped with raster-corrector or pincushioning magnets. An improperly adjusted magnet will cause a bowing of the raster edge as shown in Fig. 10-28. The bowing may be either in or out, on either edge, or top or bottom. The corrector magnets are mounted on the sides of the picture tube as shown in the diagram. Sometimes they are mounted above and below the picture tube, and sometimes four magnets are employed.

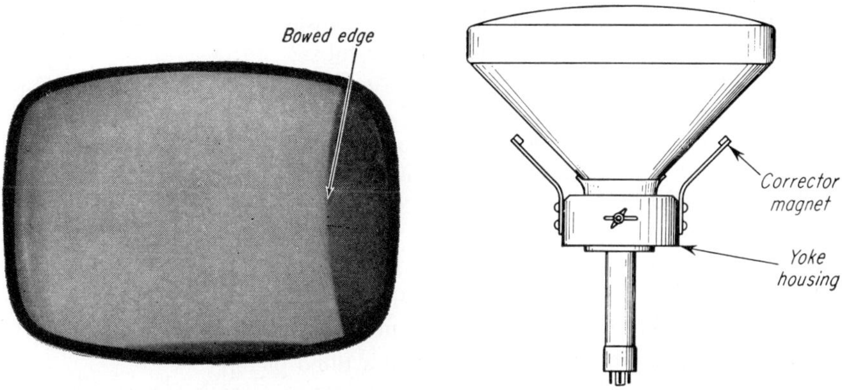

Fig. 10-28. Pincushion effect, and the location of corrector magnets.

To adjust the corrector magnets, reduce the size of the raster so that the edges are just visible. Bend the magnet arms forward or backward until the raster sides are straightened.

QUESTIONS

1. Draw a diagram of the electron gun in a CRT and give the names and functions of each of the grids.
2. Describe the bent-gun type of CRT and the operation of the ion trap.
3. Describe how the raster is produced by the action of the deflection yoke on the electron beam.
4. Draw a brightness-control circuit where a variable positive voltage is applied to the cathode of the CRT, and explain how this controls the average brightness on the screen.
5. Describe a servicing procedure which traces the cause of a no-picture symptom to a defective picture tube.
6. List symptoms which are caused by improper setup of the picture tube.
7. What is the probable cause of internal arcing in the CRT? How can this defect be repaired?
8. List the safety precautions to be observed when handling a picture tube.
9. Describe a repair procedure for a picture tube whose filament does not light.

10. What is an ion burn? How is it caused?

11. Name the three types of focusing systems.

12. How would you determine that an apparently unfocused picture is due to blooming, a defective focus device, or insufficient bandpass in the amplifiers in the picture-signal chain?

13. Describe how the focus control operates in an electromagnetic system. In a p-m type.

14. Explain why tubes with electrostatic focusing are called self-focusing. What provisions are made for focusing these tubes?

15. Use the chart in Fig. 10-18 to identify a tube with the following characteristics and measurements:

The tube is glass with a rectangular shape and an aluminized screen.
The faceplate is clear glass and spherical.
The CRT socket has wiring on the No. 6 pin.
The diagonal measures 21⅜ in.
The over-all height (length) measures 20½ in.
Centering tabs are located on the rear cover of the yoke.

16. What are the steps in setting up a new picture tube?

17. What is the procedure for correcting a tilted picture?

18. Describe the centering controls used with magnetic and electrostatic focusing devices.

11

Servicing the Picture Chain: Detector Stage

In our servicing procedure for picture faults, we start at the picture tube and work our way back through the picture chain toward the antenna. A partial block diagram of the picture-signal chain is given in Fig. 11-1.

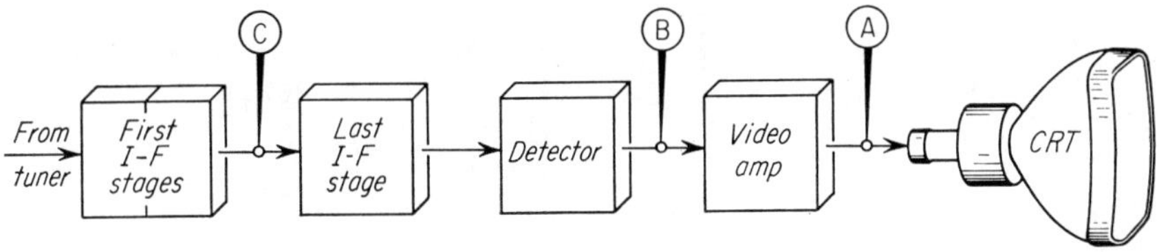

Fig. 11-1. Position of the detector stage in the servicing procedure for the picture-signal chain.

As described in previous chapters, checks made at the diode load (test point *B*) prove that the video amplifier and CRT are functioning normally. The next test point in the chain is the grid of the last i-f amplifier tube, test point *C*. This takes in the detector stage, sometimes called the video detector to distinguish it from the demodulator in the sound section. It also takes in the last i-f tube working as an amplifier. The servicing procedure presented in this chapter is mainly concerned with the detector stage but will include checks on the last i-f tube because of the location of the check point.

Quick Check for the Video Detector Stage. An a-m signal generator is set to give maximum modulated r-f output at 43 Mc and fed to the grid pin of the last i-f tube. The frequency control on the generator is rocked from 40 to 46 Mc while looking for the appearance of bars on the picture tube. If none appears, the generator is shifted to 23 Mc and rocked from 20 to 26 Mc. The appearance of the modulation bars on the face of the picture tube proves that the last i-f and detector stages are working, and the serviceman goes on to check the remaining i-f stages.

Function of the Video Detector. The video detector functions in much the same manner as the second detector in a radio receiver. It demodulates the i-f signal by rectification and removes the high-frequency i-f components by filtering. One of its by-products is the 4.5-Mc beat signal which becomes the sound i-f signal.

The detector and its output circuit must be able to pass all the video and sync signals as well as the 4.5-Mc sound beat. This is a band of frequencies from 30 cycles to 4.5 Mc.

Basic Circuit. The schematic diagram of a basic video detector is given in Fig. 11-2. It is the second detector of the picture chain which is basically a superheterodyne receiver. The modulated i-f signal is coupled

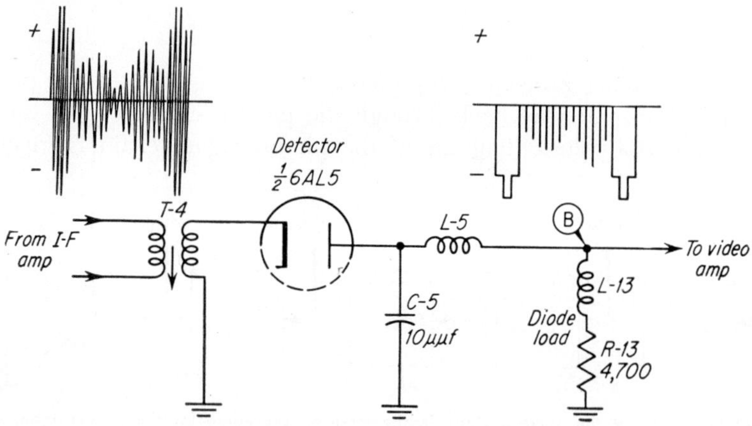

Fig. 11-2. Basic video-detector circuit.

into the detector circuit through transformer *T*-4, which is the last i-f transformer. A waveform diagram of the input signal is directly above transformer *T*-4 in Fig. 11-2. The amplitude variations represent the sync pulses and the video information.

The detector itself is a diode, so connected as to rectify the incoming signal. Coil *L*-5 and condenser *C*-5 in the detector output circuit act as an i-f filter, removing the r-f variations and leaving the composite signal across coil *L*-13 and resistor *R*-13. These last two components make up the diode load.

Note the diode-load connections, with the a-c input connected to the cathode. Since a diode conducts only from cathode to plate, a negative voltage will be established across the diode load on the negative phase of the input signal. The filtered negative-going waveform of the composite signal is shown above the diode load in Fig. 11-2.

In our basic picture chain, there is a single video-amplifier stage feeding a positive-going signal to the cathode of the picture tube. Since the video amplifier, like any amplifier, reverses the signal phase, it must be fed a negative-going signal to produce a properly polarized signal for the cathode of the CRT. As shown previously, the detector feeds a negative-going signal to the video amplifier.

If the signal is fed to the grid of the picture tube, it must be negative-going. Then, if there is only one video-amplifier stage, the detector output signal must be positive-going. This is easily arranged by reversing the cathode and plate connections of the diode detector. Two video-amplifier stages give another phase reversal. Therefore, in various receivers the detector connections are determined by the number of video stages and whether the signal is fed to the grid or the cathode of the CRT.

Some receivers use a germanium crystal instead of a diode as the detector. The crystal operates as a rectifier in much the same manner as the diode.

Symptoms Caused by the Failure of the Detector Stage. Complete failure of the detector is a break in the picture chain and results in complete loss of picture. A weak detector causes attenuation of the video signal, which results in a picture which has a washed-out appearance, lacking rich blacks. There may also be unstable sync, since the sync pulses may not be strong enough to activate the sync section. Servicing procedures for no picture and weak picture are given in this chapter.

Other picture faults such as ringing, smearing, trailing whites, loss of definition and reversed picture are rarely caused by faults in the detector stage, other than an incorrect diode load or defective crystal detector, where used. The servicing procedure for these symptoms presented in Chap. 9 includes a check of these components. These servicing procedures will not be repeated in this chapter.

NORMAL TEST DATA FOR THE VIDEO DETECTOR

Quick Check. As shown in Fig. 11-2, the input to the detector stage is the primary of i-f transformer *T*-4. When the stage is checked, however, the signal generator is connected to the grid of the last i-f tube, as shown in Fig. 11-3. This is done in order to obtain the gain of the last i-f tube as an amplifier, since the r-f output of the signal generator might not be sufficient to give an indication when connected to the input of the detector stage. Another reason for this connection is to avoid the detuning effect of the generator on the transformer.

The signal generator is adjusted for maximum output and a modulated signal at about 43 Mc. The test lead is connected to the last i-f grid which is the No. 1 pin for most i-f tubes. Use an isolation condenser of about 250 $\mu\mu$f. The receiver is set to an unused channel, adjusted for normal raster, and the contrast control is turned to the maximum clockwise position. The frequency control on the generator is then rocked back and forth through the 40-Mc i-f band (40 to 46 Mc) while looking for the modulation bars on the face of the picture tube. If the bars do not appear, the signal generator is adjusted to wobble through the 20-Mc i-f band (20 to 26 Mc).

Most receivers use a staggered tuning system in the i-f amplifiers, and there is little standardization for the frequency allotted to the last i-f transformer. As a result, the peak indication for the detector check may

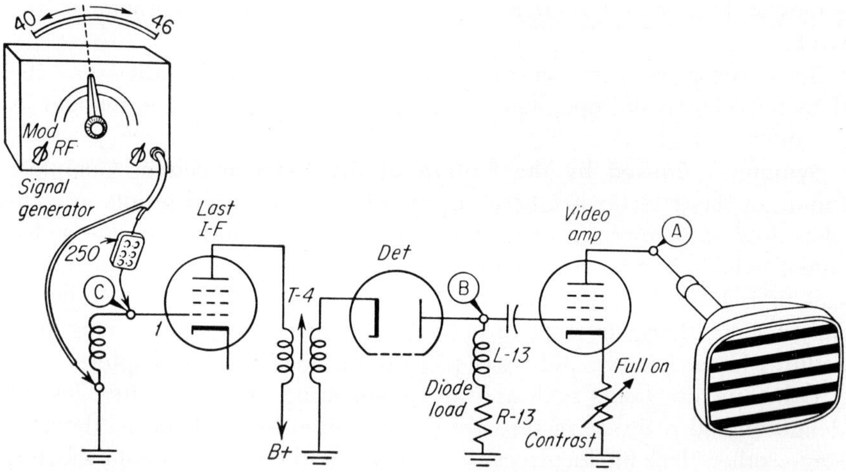

Fig. 11-3. Quick check for operation of the detector stage. A normal response from test point *B* and no response from test point C places the trouble in the last i-f or detector-tube circuits.

be at any frequency in the i-f band. Therefore, if the modulation bars appear on the picture tube at any frequency in the proper i-f band, the receiver is operating from the last i-f grid on, and the trouble must be before this point in the picture-signal chain. Conversely, if the bars do not appear, the trouble must be behind the last i-f grid. Previous checks made at the diode load (point *B*) prove that the video amplifier and CRT are working. This places the trouble between test points *C* and *B*, in the last i-f tube and detector stages.

Normal Test Data for the Detector and I-F Stages. A basic circuit for a detector and last i-f stage is given in Fig. 11-4. The test data and servicing procedures are based on this circuit but include additional information to take care of common variations.

Servicing Procedure for a Complaint of No Picture—Raster Normal. The last i-f and detector stages are proved inoperative by the quick check when a modulated signal is fed to the last i-f grid and wobbled through the ranges of the intermediate frequencies without producing bars on the picture tube.

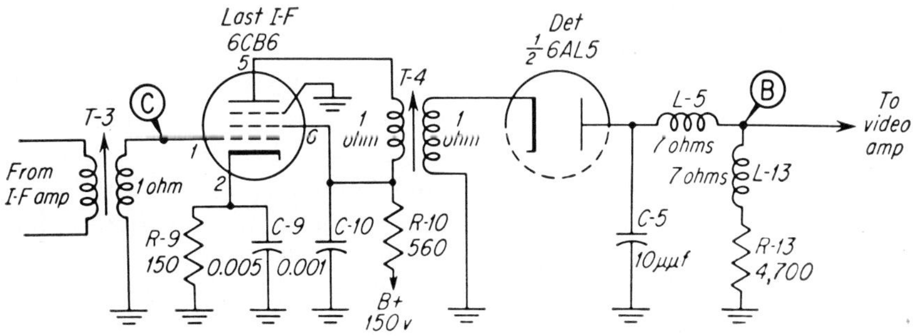

Fig. 11-4. Basic last i-f and video-detector stages.

Normal-voltage Data for the Last I-F Tube

	Pin No. 6CB6	*Pin No. 6AM8*	*Normal Voltage*
Grid	1	2	0
Cathode	2	1	1–2
Plate	5	6	130–150
Screen	6	3	130–150

Resistance Data for the Last I-F and Detector Stages

		Ohms
Transformer windings	*T*-3, *T*-4	0.1–2
Coil windings	*L*-5, *L*-13	3–12
Cathode resistor	*R*-9	100–180
Plate-decoupling resistor	*R*-10	300–1,000
Diode-load resistor	*R*-13	3,300–4,700

Step 1. Check the last i-f tube by replacement with one known to be good. If this restores operation, the original tube is bad, and the set is fixed. If there is no change go on to the next step.

Step 2. Make a voltage check of the last i-f stage. Refer to manufacturer's service notes for normal voltage readings, and investigate the cause of any abnormal reading. If the readings are normal, check condensers *C*-9 and *C*-10 for opens by trying operation with a test condenser across each. If this causes no change, go on to step 3.

Step 3. Make a resistance check of the last i-f and detector stages. There is always the possibility that the test signal may be shorted out at its input point. Since the normal reading at this point may be as low as a fraction of an ohm, it is necessary to open the secondary connection of transformer *T*-3 to check for the short. Next make an ohmmeter check of transformer *T*-4, i-f filter coil *L*-5, and the diode load consisting of coil *L*-13 and resistor *R*-13. There is little likelihood of a short in i-f filter condenser *C*-5 or the diode load, since any short in the output circuit of the detector would have interfered with the operation of the video amplifier, and this has previously been checked as normal. If the ohmmeter checks show normal results, go on to step 4.

Step 4. Replace the detector. If the detector is a diode tube, simply try operation with a new one. If the detector is a crystal, polarity must be observed when making the replacement, because reversed polarity will cause no picture or a reversed picture. Locate the crystal by tracing the wiring. In appearance it looks like a small resistor or condenser. It may be under the cover of the last i-f can as shown in Fig. 11-7, or it may be mounted on a tie-point strip between the last i-f and video tubes. Polarity on the crystal is indicated by a black bar or minus sign near one end, as shown in Fig. 11-5*a* and *b*. This represents the crystal terminal which corresponds to the cathode of the diode. The symbol for a germanium crystal is given in Fig. 11-5*c*. The black bar in the symbol also represents the crystal terminal which is the equivalent of the diode cathode. It is suggested that a crystal be set up with clips as shown in Fig. 11-5*d* for test purposes. One lead of the original crystal is opened, and the test crystal is clipped in, observing the same polarity.

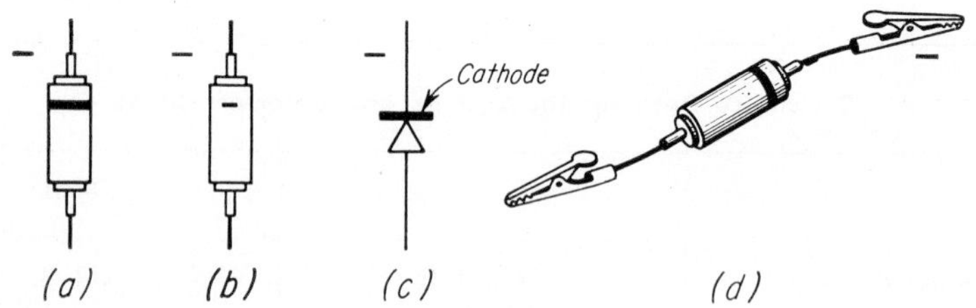

Fig. 11-5. Germanium crystals used as detectors.

Replacement Notes on Germanium-Crystal Detectors. Regardless of the original crystal used in the receiver, a GE or CBS type 1N64 or a Sylvania type 1N60 may be used as a replacement. If the crystal in the set is unmarked, or if there is doubt as to the polarization, refer to the diagram of the receiver to determine the correct polarity. Check to see if the video signal is fed to the cathode or the grid of the CRT. Next check the number of video stages. Then refer to Fig. 11-6 which shows correct polarization of the crystal for various combinations. In each case, the bar of the crystal symbol is marked with a minus sign just as it is on the replacement crystal.

Germanium crystals require considerable care in handling. The leads may be easily pulled out of the body of the crystal, and the crystal will deteriorate if subjected to heat. Failure due to these causes can best be avoided by adhering to the following wiring precautions. When replacing a crystal, leave a small amount of slack in the connecting leads. This relieves the tension that might otherwise cause the lead to pull out of the crystal. When soldering the connection, use low heat, and grasp the lead with long nose pliers between the crystal and solder point as shown in Fig. 11-7. The mass of the tool absorbs the heat of the soldering, and the lead going into the body of the crystal remains comparatively cool.

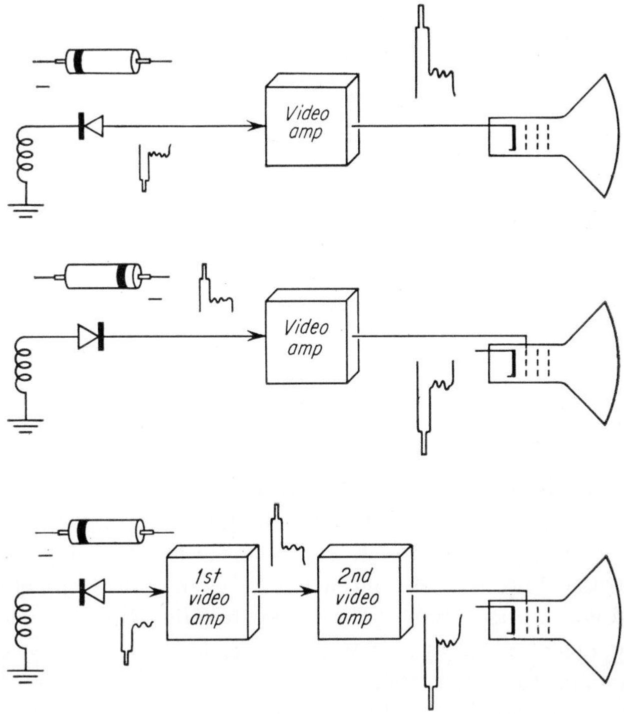

Fig. 11-6. How to determine crystal polarity.

General Replacement Notes on the Last I-F and Detector Stages. When the servicing procedure reveals the defective component, note its position, wire dress, lead length, etc., before removing and discarding it. This is because these items are very important for the correct operation of high-frequency circuits. Make sure the replacement part is the same type and size as the original and is placed in the circuit in exactly the same way. When the defective component is an i-f transformer, and the manufacturer's original part is unobtainable, a satisfactory replacement

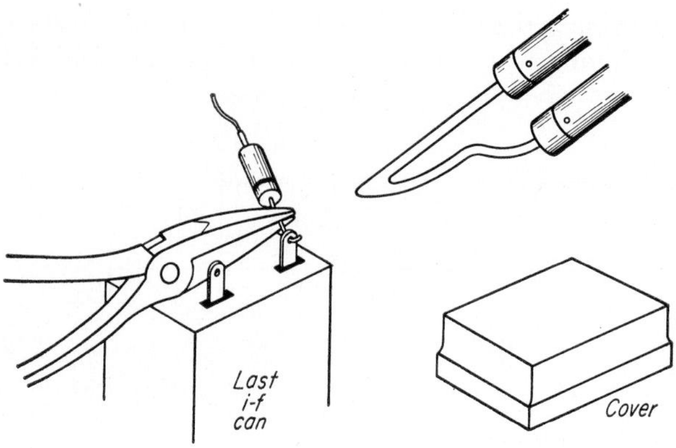

Fig. 11-7. Using long-nose pliers as a heat shunt when soldering or unsoldering a crystal. The crystal is often found under the cover of the last i-f can.

may be made by choosing a standard replacement coil of the correct frequency range from a parts catalogue. The i-f amplifier must be realigned. Alignment notes are given in Chap. 12.

Servicing Procedure for a Weak Picture. A weak picture lacking in contrast is caused by a loss of efficiency in one of the stages of the picture chain. When servicing for this defect, quantitative data are necessary to discover which stage is operating at low efficiency. Previous checks have established normal operation of the video amplifier. Low efficiency in the last i-f and detector stages is determined as follows.

When making the quick check for the last i-f and detector stages, the modulation bars will appear, but they will be gray rather than black if one of these stages is weak. The grayness or blackness of the bars, however, is not sufficiently defined to judge whether the stages are operating at full efficiency. A better quantitative check is obtained by measuring the signal developed at the diode load. Do this first on a number of sets known to be in good operating condition.

Connect the signal generator to the last i-f grid, in the same manner as when making the quick check, that is, modulated r-f output at the receiver i-f band, attenuator full on, a 250-$\mu\mu$f isolation condenser in the grid lead, and the set adjusted for normal raster on an unused channel. Contrast control should be set to the maximum clockwise position. Connect a voltmeter on its lowest voltage range to the diode-load test point, observing the polarity required by the diode connections. The setup is shown in Fig. 11-8.

Rock the frequency control on the signal generator through the i-f band, leaving it at the point which gives peak reading on the voltmeter. Note the voltmeter reading. Also notice that the modulation bars are blackest at the peak voltmeter reading.

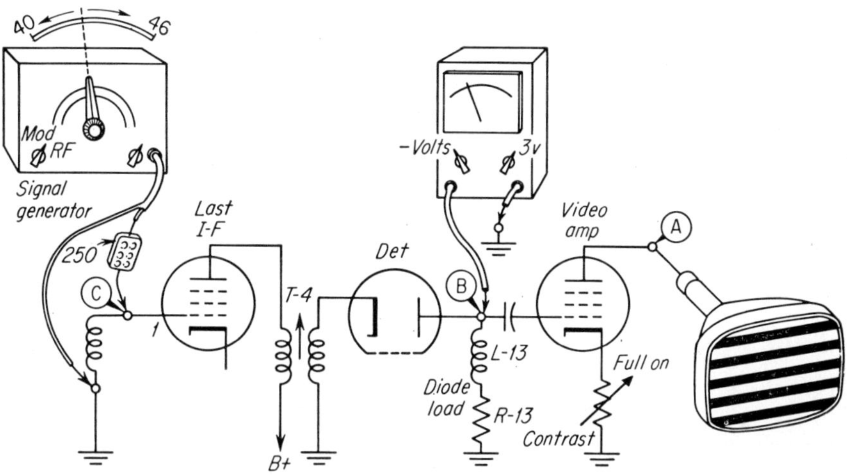

Fig. 11-8. Checking the last i-f and detector stages for normal output.

The voltmeter reading will be between 1 and 2 volts, depending on the output of your generator, and the gain characteristics of the last i-f and detector section. Perform the same check on a number of receivers in good operating condition, establishing an average voltage reading at the diode load when the last i-f stage of any receiver is fed from your signal generator at full output. Having established this figure, you are now in a position to determine that the last i-f and detector stages in any receiver are working at full efficiency, or that they are below normal, and the cause of insufficient contrast.

If the fault is in the last i-f and detector stages, they are serviced in the same manner as for no picture. The fault will turn out to be a weak i-f tube, an incorrect operating voltage in the last i-f stage, an open i-f bypass condenser, a weak diode or crystal, wrong-value diode-load resistor, or leakage in the i-f filter. Some receivers connect a resistor across i-f filter coil *L*-5. In this event, check coil *L*-5 for an open.

SUMMARY

Quick Check for the Last I-F and Detector Stages. A modulated r-f signal at full generator output is fed to the last i-f grid and wobbled through the i-f range of the receiver. The appearance of modulation bars on the screen of the picture tube proves that the last i-f and detector stages are working.

Basic Circuit. The schematic diagram of a basic circuit for the last i-f and detector stages is given in Fig. 11-9.

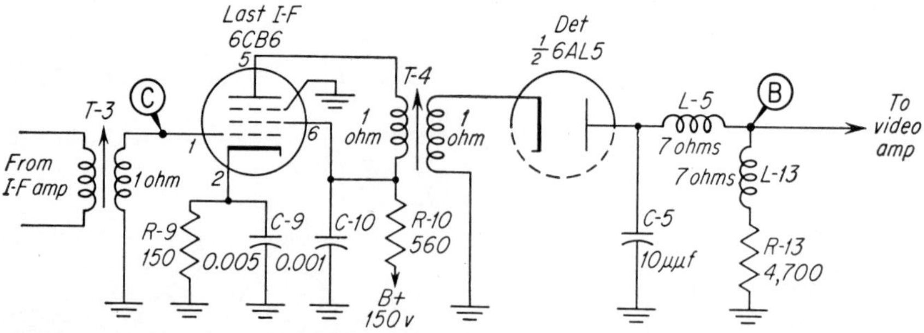

Fig. 11-9. Schematic diagram of a basic circuit for the last i-f and detector stages.

Normal-voltage Data for the Last I-F Tube

	Pin No. 6CB6	Pin No. 6AM8	Normal Voltage
Grid	1	2	0
Cathode	2	1	1–2
Plate	5	6	130–150
Screen	6	3	130–150

Resistance Data for the Last I-F and Detector Stages

		Ohms
Transformer windings	T-3, T-4	0.1–2
Coil windings	L-5, L-13	3–12
Cathode resistor	R-9	100–180
Plate-decoupling resistor	R-10	300–1,000
Diode-load resistor	R-13	3,300–4,700

Servicing Charts. Servicing charts are presented for the symptoms of no picture and weak picture. The latter may be accompanied by a loss of sync. Reference in the chart is to Fig. 11-9. The picture chain is being serviced for these symptoms. In each case, it is assumed that the video amplifier and picture tube have been checked as described in Chap. 9 and found to be normal.

Servicing Procedure for No Picture, Raster Normal, No Snow

Procedure	*Result*	*Trouble and Subsequent Check*
1. Make the quick check for the last i-f and detector stages	Modulation bars appear on the CRT	Last i-f and detector stages are working. Check the i-f amplifier. See Chap. 12
	No response	Last i-f or detector stage is at fault. Go on to step 2
2. Replace the last i-f tube	Bars appear	Tube is bad
	No change	Go on to step 3
3. Make a voltage and resistance check of the last i-f stage	Abnormal reading	Locate the defective component and replace it
	Normal readings	Go on to step 4
4. Make an ohmmeter check of components in the detector stage	Open coil or resistor	Replace the defective component
	Components read normal	Replace detector diode or crystal. Observe polarity when changing crystal detector

Servicing Procedure for Weak Picture, No Snow

Procedure	*Result*	*Trouble and Subsequent Check*
1. Make quick check and measure output at diode load. See Fig. 11-8	Normal 1- or 2-volt reading	Check remaining i-f amplifiers. See Chap. 12
	Low reading	Go on to step 2
2. Check the last i-f tube by replacement	Normal modulation bars return	Original last i-f tube is defective
	No change	Go on to step 3
3. Make voltage and resistance check of the last i-f stage	Abnormal reading	Locate and replace the defective component
	Normal readings	Check the detector stage as follows: diode-load resistor *R*-13 for lowered value; i-f filter condenser C-5 for leakage; replace detector tube or crystal

QUESTIONS

1. Describe the normal input signal to the detector stage; describe the normal output signal from the detector stage.

2. A receiver with one video-amplifier stage is so connected as to feed its signal to the cathode of the CRT. Draw a block diagram of the detector, video amplifier, and CRT, showing signal waveforms at the input and output of the video stage.

3. Draw a diagram of a detector stage using a germanium crystal so connected as to produce a negative-going output signal.

4. A receiver with the symptom of no picture is dead from the last i-f grid on. Checks show that the video amplifier and CRT are working normally. Using the basic circuit of Fig. 11-9, make a list of the possible faults that could cause this trouble symptom and how you would check each.

5. The symptom on the screen is a lack of contrast or weak picture. Make a list of the defects in the detector stage of Fig. 11-9 that could cause this symptom, and tell how you would check for each.

6. Describe the precautions to be observed when replacing a germanium-crystal detector.

12

Servicing the Picture Chain: I-F—AGC Systems

In servicing the picture-signal chain for loss of picture, previous checks at test point *C* in Fig. 12-1 proved that the detector, video amplifier, and picture tube are operating normally. The i-f–agc system is to be checked next for the location of the fault.

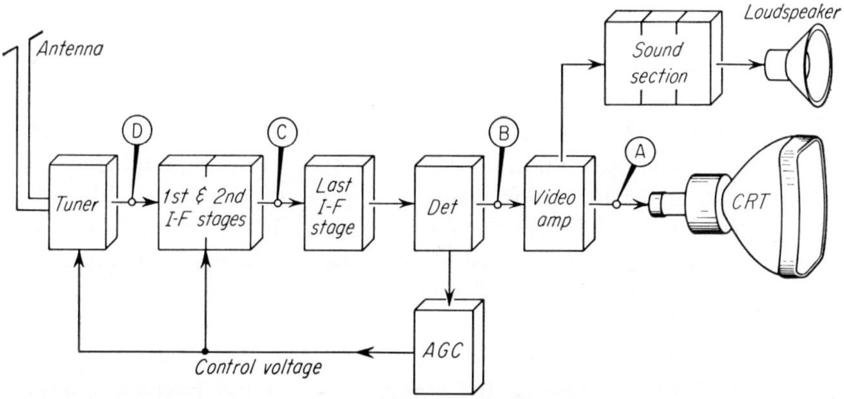

Fig. 12-1. The position of the i-f–agc system in the picture-signal chain.

Note that we are near the end of the chain in our servicing checks from the picture tube backward toward the antenna. Only the tuner is left, and whereas defects in the tuner may cause complete loss of picture or weak picture with snow, they will rarely cause any of the other symptoms of a faulty picture. These are: lack of definition, smearing, ringing, trailing whites, unstable synchronization, and negative picture. Therefore, when the checks show that the causes of these symptoms are not located in the video-amplifier or detector sections, they must be found and cleared in the i-f section.

In servicing the picture chain, the agc system and the i-f stages it controls are considered as one unit. This can be seen from the following point of view. The control voltage supplied by the agc system is simply a supply voltage, like the B plus supply, and incorrect agc voltage will cause malfunction of the i-f amplifier just like incorrect B voltage. However, the B supply is checked as a separate unit, since it derives its power from the 120-volt line. This cannot be done with the agc supply, because its power is obtained from the signal itself. Thus, when the servicing procedure brings us to an inoperative i-f amplifier, we may have a good amplifier not working because of trouble in the agc system, or we may have a good agc system not developing proper voltage because of a trouble in the i-f amplifier. The servicing procedure presented later in this chapter approaches the i-f–agc system as one unit and then locates the defective component in either half.

Quick Check for the I-F Section. The signal generator is adjusted to give a modulated r-f signal on the 40-Mc band and is fed to the input of the i-f section of the receiver. The frequency control on the signal generator is wobbled back and forth from 40 to 46 Mc while looking for modulation bars on the screen of the picture tube. If none appear, the frequency control is wobbled from 20 to 26 Mc. The appearance of dark horizontal bars on the face of the picture tube proves that the i-f amplifier is working.

The appearance of the bars places the trouble in the tuner. There is one exception to this. In a keyed agc system, the lack of picture may be caused by an overload condition due to an incorrect horizontal frequency setting. The generator quick check will produce the bars, although a normal picture signal blocks the picture chain. Before starting checks in the tuner, investigate this possibility if the receiver has a keyed agc system. Tune the receiver to a channel which normally gives good reception, and even though there is no picture on the screen, run the coarse horizontal frequency control through its range. If this restores a picture, go through the complete horizontal frequency adjustment.

Function of the I-F Amplifier. The input signal to the i-f section includes the picture i-f signal and the sound i-f signal which are produced in the converter by the beating action of the local oscillator on the transmitted picture and sound carriers of the desired station. In addition, there may be unwanted beats produced by the local oscillator and any other signals which may get through the very wide-band r-f stages. This interference will be mainly from the adjacent channels above and below the desired station. The function of the i-f amplifier is to amplify the desired picture and sound signals to produce good pictures and sound and to tune out the beats from adjacent channels. It is to be remembered that good picture reproduction requires a bandwidth of 3.5 to 4 Mc, so that the i-f tuning circuits must be of the broad-band variety.

In broad-band amplifiers, gain per stage is low. As a result, some receivers use as many as four stages to attain the required gain and bandwidth. Most receivers use three stages, and some only two. The latter are usually portable receivers where a smaller bandwidth is permissible since the reduced definition is not noticed on the smaller picture tube.

Intercarrier systems impose another problem on the common i-f section. Not only must the sound be passed, but the relative strengths of the sound and picture carriers must be kept at a favorable ratio for the production of a good intercarrier beat in the detector. When two signals are mixed, the resulting beat contains the modulation of both, but there will be more modulation from the weaker signal. For this reason, the picture carrier is made about ten times stronger than the sound. In this way, the 4.5-Mc intercarrier beat contains more of the desired sound modulation, and less of the picture modulation.

Frequencies in the I-F Amplifier. Most receivers use an intermediate frequency of 45.75 Mc for the picture carrier and 41.25 Mc for the sound carrier. Since the i-f amplifier must be wide enough to pass both these frequencies, it is usually called an i-f amplifier in the 40-Mc band. Older receivers employ an intermediate frequency in the 20-Mc band, where the picture carrier is at 25.75 Mc and the sound at 21.25 Mc. The other important frequencies to be considered are the possible interference signals from adjacent channels.

To see the frequency relationships in an i-f amplifier, examine the chart given in Fig. 12-2. This gives the frequency limits and carriers in

	Channel 2		*Channel 3 (desired station)*		*Channel 4*	
Frequency limits	54		60		66	72
Picture carrier	55.25		61.25		67.25	
Sound carrier		59.75		65.75		71.75
20-Mc i-f beat		27.25*	25.75	21.25	19.75*	
40-Mc i-f beat		47.25*	45.75	41.25	39.75*	

* Unwanted beats, possible interference.

Fig. 12-2. Important frequencies to be considered in an i-f amplifier when the receiver is tuned to Channel 3.

Channels 2, 3, and 4. Assume that the receiver is tuned to Channel 3. This places the local oscillator at 87 Mc for receivers using the 20-Mc i-f band, and 107 Mc for those using the 40-Mc i-f band. Note that, when the sound carrier from the channel below beats with the same oscillator, the interfering beat occurs at 27.25 Mc or 47.25 Mc, depending on the system in use. Similarly, the picture carrier from Channel 4 can cause a beat at 19.75 Mc or 39.75 Mc. The picture carrier of Channel 2 and the sound carrier of Channel 4 are too far off in frequency to require consideration. The figures are given for Channel 3 with its two neighboring channels. The same sequence of figures is true for the i-f beats developed, regardless of which channel is being tuned in.

The FCC does not allot adjacent channels to one community; therefore, in large metropolitan areas, adjacent-channel interference is no problem. But in fringe areas, reception conditions are different. Here, a signal from a different area too weak to give good reception can spoil reception from a stronger station on an adjacent channel. Since receivers are made for country-wide distribution, most of them are designed with some provision for adjacent-channel interference.

Desirable characteristics of an i-f amplifier are shown in Fig. 12-3. The response to the interfering beat from the adjacent picture carrier

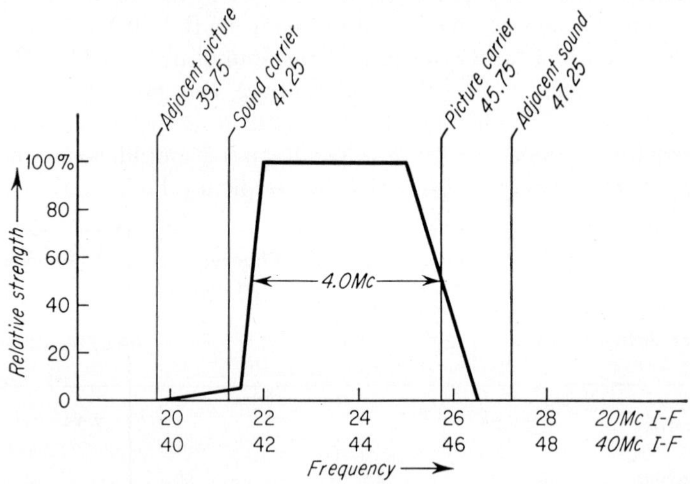

Fig. 12-3. Idealized response curve for an i-f amplifier.

at 39.75 or 19.75 Mc is zero. The response to the desired sound carrier at 41.25 Mc is about a tenth of the response of the desired picture carrier at 45.75 Mc. The response to the interfering beat from the adjacent sound carrier is zero.

Note that the response to the desired picture carrier is about half of the possible gain of the amplifier. This situation is caused by the nature of the transmitted picture signal. The television picture transmitted uses a type of modulation known as vestigial-side-band transmission. This means that the transmitted signal does not contain both side bands for all modulating frequencies. At low modulating frequencies, the carrier and both side bands are transmitted. The high-frequency components of the signal which are necessary for good picture detail are transmitted on one side band only. Now note that, in our idealized response curve, the gain 2 or 3 Mc away from the picture carrier is about twice as high as at the carrier. The gradual slope of the curve near the carrier is compensated for by the vestigial-side-band transmission. The slope of the curve near the sound carrier has to be very steep so as to maintain good signal strength 4 Mc from the picture carrier and yet drop to a very small fraction of this strength at the sound carrier 4.5 Mc from the picture carrier.

Stagger Tuning for Wide-band I-F Amplifiers. The usual i-f amplifier employs three stages requiring four coupling devices: one at the input, two interstage, and one at the output. All the couplings are tuned. The present practice is to use transformer or impedance coupling in a stagger-tuned system. This means the four circuits are tuned to different resonance points in the required 4-Mc bandpass of the i-f amplifier in order to attain a response close to the ideal of Fig. 12-3.

Let us see how this works. Consider the tuned impedance-coupled circuit of Fig. 12-4. The coil and condenser in the plate circuit are

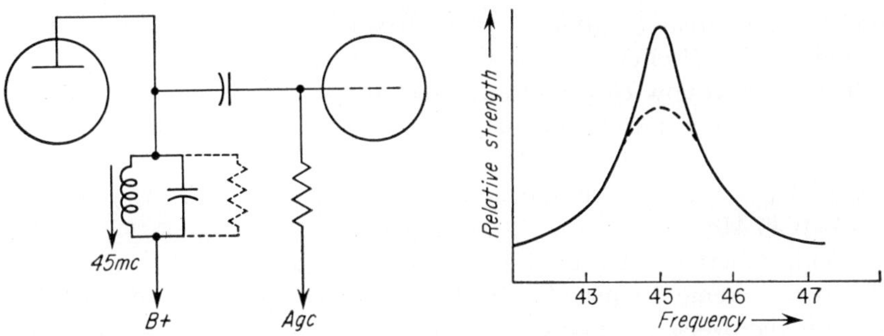

Fig. 12-4. A single-tuned coupling circuit between two i-f tubes and its response curve.

tuned to resonance at 45 Mc. Alongside the circuit is its frequency-response curve. At 45 Mc the output is greatest, decreasing sharply at frequencies close to the resonant peak, and then gradually as the frequency from resonance increases. Note that it does not go down to zero. If the resonant circuit is loaded down by connecting a resistor across it, the output at the resonant peak is reduced. The effect of the loading resistor is shown in the response curve of Fig. 12-4 by the dotted line. The loading resistor increases the bandwidth of the circuit, but at the expense of gain. Loading resistors are often used in i-f amplifiers to broaden response and flatten unwanted peaks. A transformer with a single tuning slug will have about the same type of tuning curve as the one shown in Fig. 12-4.

A second tuned coupling in the i-f amplifier would of itself have the same tuning curve as the one previously described, but if this circuit were tuned to 43 Mc, both the 43- and 45-Mc peaks would appear in the output. Each circuit would contribute its share to the over-all output. The two individual curves are shown by the dotted lines in Fig. 12-5. Where

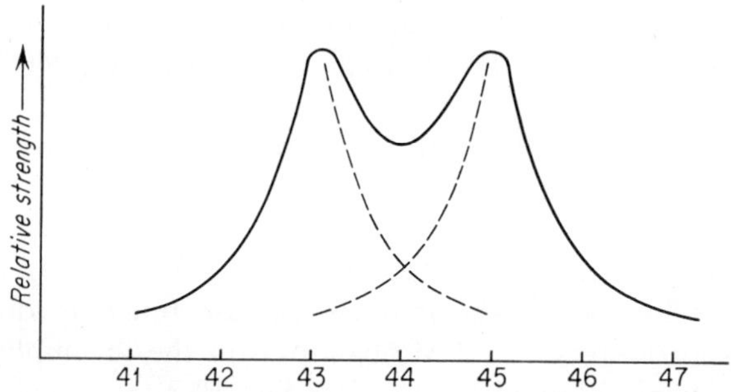

Fig. 12-5. Response curve showing the combined effect of stagger-tuning two circuits, one at 43 Mc, and the other at 45 Mc.

the curves intersect, since each circuit is contributing to the total output, the result is greater output at these frequencies. The combined effect of the two tuned circuits is shown by the heavy line.

The curve is now close to the correct shape. Some manufacturers use only two frequencies in their i-f amplifier. Two circuits, usually the first and third, are tuned to some frequency close to 43 Mc (23 Mc for a 20-Mc i-f band) and the second and fourth are tuned to some frequency close to 45 Mc.

Another method often used to give a wider band spread is to separate the two extreme frequencies further, say to 45.3 and 42.8 Mc. This spreads the over-all response, and the resulting greater dip in the center is compensated for by tuning one or both of the other circuits to an in-between frequency.

Use of Wave Traps. A wave trap is also a coil and condenser type of tuning circuit, hooked up in such a way as to remove an unwanted frequency. In the circuit of Fig. 12-6*a*, transformer *T*-1 couples energy from one tube to the next. Coil *L* is inductively coupled to the transformer windings and absorbs energy at its resonant frequency with condenser *C*. This is known as an absorption type of wave trap.

In the circuit of Fig. 12-6*b*, coil *L* and condenser *C* are connected in series. At the resonant frequency, this type of circuit draws heavy current, thereby removing currents at and near this frequency from the grid circuit. This type is known as a series wave trap.

Either type of trap working by itself produces a response like the one shown in Fig. 12-6*c*. Note that it is the same shape as the response of

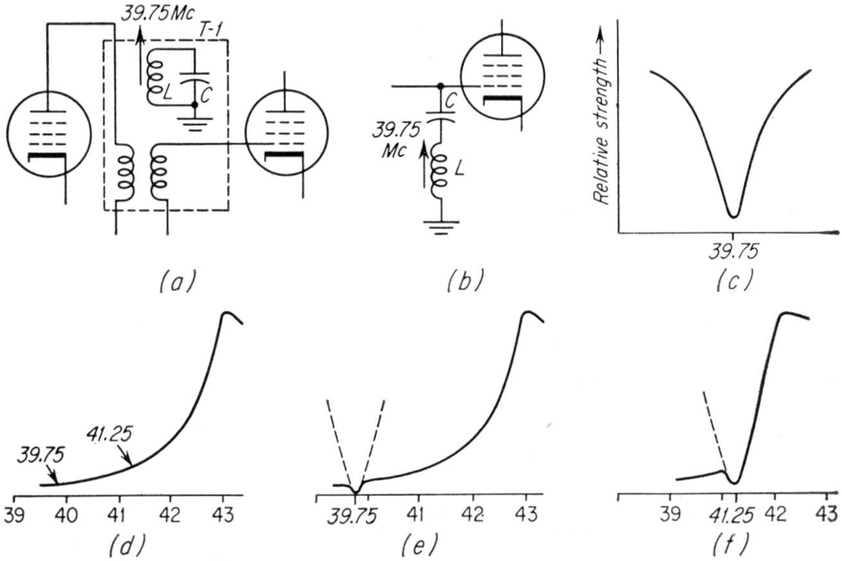

Fig. 12-6. Wave traps and their effect on the response curve of an i-f amplifier.

the single tuning circuit shown in Fig. 12-4, but this time the curve is upside down, since it is being used to remove an unwanted frequency.

Note that the null of the curve does not go down to zero. However, if the trap is used at a frequency where the signal is already low, the further reduction caused by the trap brings the low signal to a negligible level. This is the case at 39.75 Mc, the frequency of the adjacent picture carrier. The curve of Fig. 12-6*d* shows the normal response slope of a circuit tuned to approximately 43 Mc. Note that there is some response at 39.75 Mc. This may bring in interference in fringe areas where transmission on an adjacent channel has signal strength comparable with transmission on the desired channel. The inclusion of a trap tuned to 39.75 Mc further reduces the low-level signal at this frequency, as shown in Fig. 12-6*e*.

A similar condition exists at 47.25 Mc which is the frequency of the sound carrier of the adjacent channel on the other side. Here again, the circuits tuned to approximately 45 Mc may leave too high a skirt, which will be reduced by a trap tuned to 47.25 Mc.

Reexamine the curve of Fig. 12-6*d*, which shows the normal response slope of a circuit tuned to approximately 43 Mc. Note that the output signal at 41.25, the sound carrier of the desired station, is quite high. It may be too high for correct intercarrier operation, which demands that the sound carrier be about a tenth of the strength of the picture carrier. This can be corrected by including a trap tuned to 41.25 Mc. The result of such a trap is shown in Fig. 12-6*f*. Note that there is still some response at 41.25 Mc, sufficient for the production of a good intercarrier sound beat, and that the curve now slopes up more sharply toward 42 Mc, a desirable condition, if you go back and study the ideal response curve of Fig. 12-3.

Some very few receivers do not include any traps in the i-f section. Most receivers employ at least one trap, this one being usually for accompanying sound at 41.25 Mc. Many receivers employ two traps, one each for accompanying sound and adjacent sound at 47.25 Mc. Some receivers, particularly color sets, employ all three types of traps. Any one receiver may use any combination of the three types.

Interstage Coupling in the I-F Section. The most commonly used coupling between stages in the i-f section is a transformer tuned to its

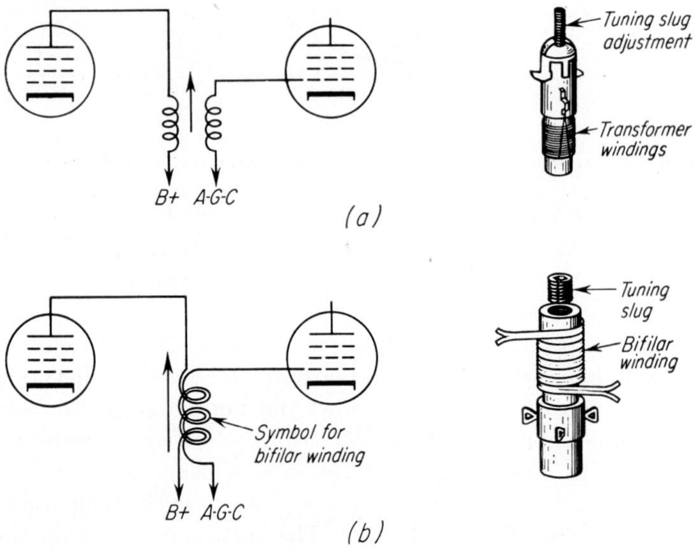

Fig. 12-7. Common interstage i-f transformers.

resonant peak by a single core slug. The circuit and a typical transformer for the purpose are shown in Fig. 12-7*a*. The two coils are wound on the same form, and the position of the core slug is set by the adjustment screw for tuning purposes. Close coupling between the primary and secondary coils is desirable for broad-band effects. A common method of obtaining close coupling is by the use of a bifilar winding, and its circuit representation is shown in Fig. 12-7*b*.

A tuning circuit consists of a coil and condenser, and the tuned circuits of Fig. 12-7 do not show the condenser. Condensers are very often omitted and the coils are peaked with the distributed capacitance of the circuit. Sometimes this is insufficient, and a small condenser is added as illustrated by condenser *C*-2 in Fig. 12-8*a*. These usually have a capacitance of 10 or 15 μμf. Loading resistors were mentioned previously as a method of reducing peaks and broadening the frequency response. Resistor *R*-8 in Fig. 12-8*a* shows how a loading resistor is connected into the circuit.

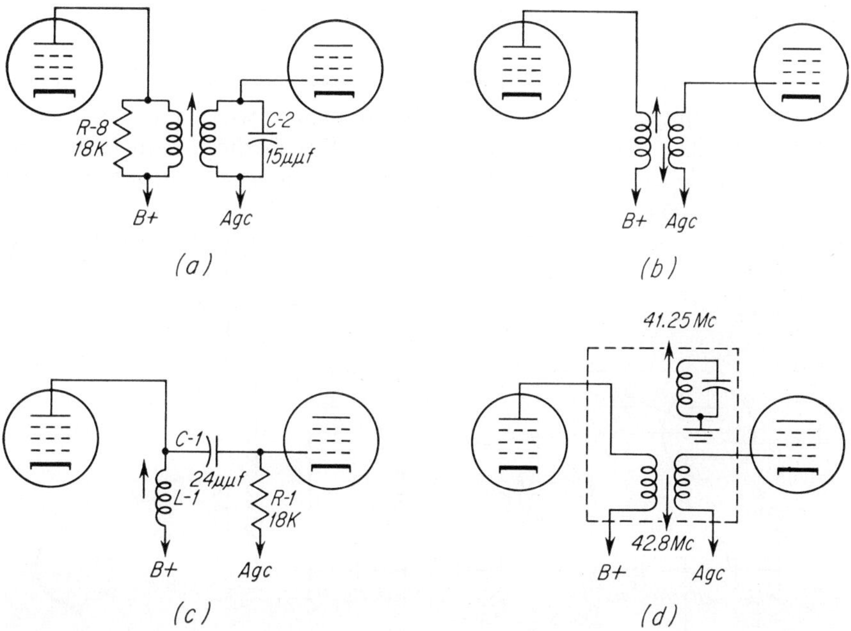

Fig. 12-8. Types of couplings between i-f stages.

Another type of transformer has the primary winding on one end of the coil form, and the secondary on the other. Separate core slugs are inserted in each end of the coil form so that the two windings are tunable individually. These are known as double-tuned i-f transformers. The schematic representation is shown in Fig. 12-8*b*. Note that one of the arrows representing the tuning slugs is pointed up and the other down. This indicates that the primary adjustment is the top tuning slug, and the secondary adjustment is the bottom. A set using double-tuned transformers requires very careful alignment and close adherence to the manufacturer's alignment instructions.

Impedance coupling, illustrated in Fig. 12-8*c*, is also found between stages in i-f amplifiers. Plate impedance *L*-1 is peaked with its distributed capacitance to the operating frequency in the 20- or 40-Mc i-f band. Condenser *C*-1 and resistor *R*-1 are the coupling condenser and grid load, respectively. The values shown in the diagram are an average, since there is considerable variance in this regard in different receivers.

A coupling circuit which includes a wave trap is illustrated in Fig. 12-8*d*. Transformer assembly can *T*-2 contains a single-tuned coupling transformer and an absorption trap. Note that the diagram indicates that the bottom slug adjustment is for the transformer, and the top one is for the trap. Trap frequencies are almost always indicated on schematic diagrams. The 41.25-Mc adjustment indicated in Fig. 12-8*d* reduces the amplitude of the associated sound.

Provisions for Decoupling in the I-F Stages. Three or more amplifier stages operating at close to the same frequency will have a tendency to oscillate. The precautions taken against this in the design of i-f amplifiers include careful layout, short leads, the use of miniature tubes like the 6CB6, degeneration, and decoupling filters. The circuit of two stages of a standardized i-f amplifier is given in Fig. 12-9 to illustrate some of these points.

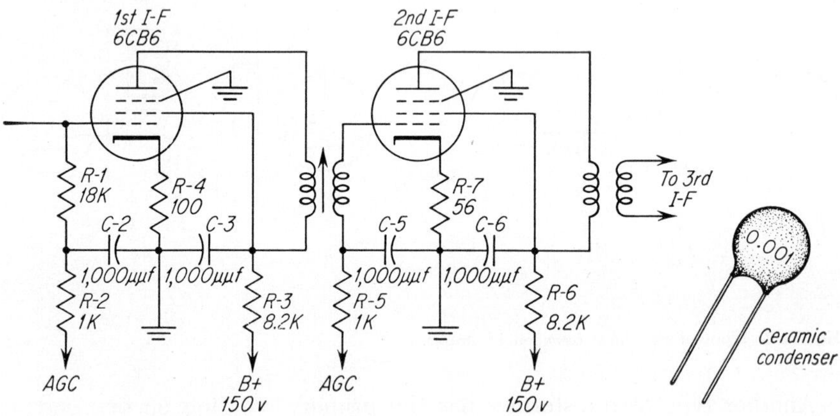

Fig. 12-9. Decoupling filters in the i-f amplifier.

The 6CB6 tube and its 600-ma equivalent, the 3CB6, have been designed as high-frequency amplifiers and are the most commonly used tubes in television i-f sections. Separate base pins are provided for the suppressor grid and cathode so that the cathode resistor may be unbypassed in a degenerative circuit. Resistors *R*-4 and *R*-7 serve this function in the circuit, and the suppressor grids are separately connected to ground.

Both stages in the diagram are controlled from the agc bus. To prevent interstage coupling from this common source, resistor *R*-2 and condenser *C*-2 act as a decoupling filter for the first i-f stage, and resistor *R*-5 and condenser *C*-5 perform the same function for the second. There are similar decoupling filters for the common plate supply. Resistor *R*-3 and condenser *C*-3 form the plate and screen decoupling filter for the first i-f stage, and resistor *R*-6 and condenser *C*-6 are in the B plus feed line to the second.

The decoupling condensers are the noninductive ceramic type usually called buttons. A drawing of one of these has been added to the diagram in Fig. 12-9 for identification purposes. The leads on these condensers are rarely more than a quarter of an inch long for close bypass action.

Standard Circuit of an I-F Section. A circuit incorporating the most widely used practices in television i-f sections is drawn in Fig. 12-10,

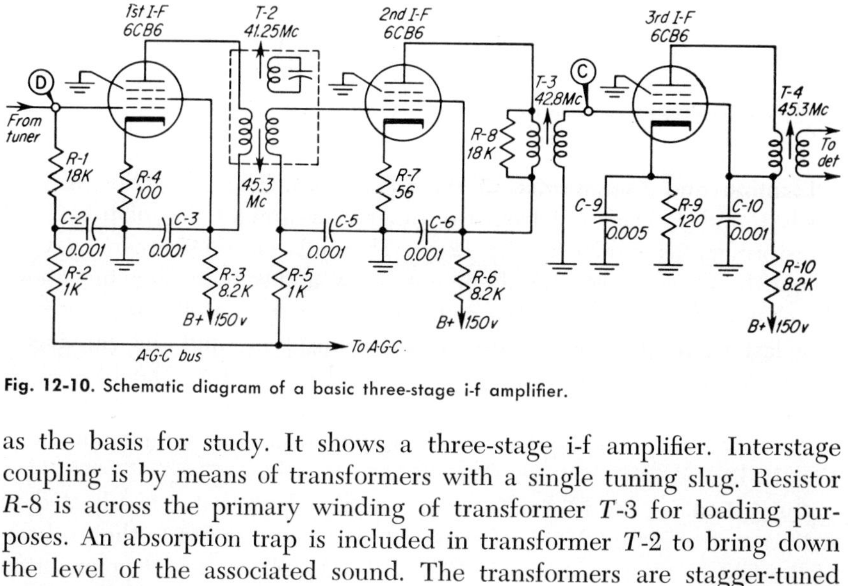

Fig. 12-10. Schematic diagram of a basic three-stage i-f amplifier.

as the basis for study. It shows a three-stage i-f amplifier. Interstage coupling is by means of transformers with a single tuning slug. Resistor *R*-8 is across the primary winding of transformer *T*-3 for loading purposes. An absorption trap is included in transformer *T*-2 to bring down the level of the associated sound. The transformers are stagger-tuned in the 40-Mc i-f band to achieve a 4-Mc bandpass.

The decoupling filters have been described in the previous section. Note the addition of resistor *R*-10 and condenser *C*-10 as the plate and screen decoupling filter for the third i-f stage. Also note that the last i-f stage is not controlled from the agc bus. This is almost universal practice. The last i-f stage is permitted to operate at full gain, even including bypass condenser *C*-9 for the cathode circuit, to prevent degeneration.

Interstage coupling from the tuner is of the impedance type. This was purposely omitted from the i-f section diagram of Fig. 12-10, because the tuning circuit adjustment is contained in the tuner. The circuit showing the connections is given in Fig. 12-11. Plate load *L*-1 and coupling condenser *C*-1 are in the tuner. Grid load *R*-1 is in the i-f section.

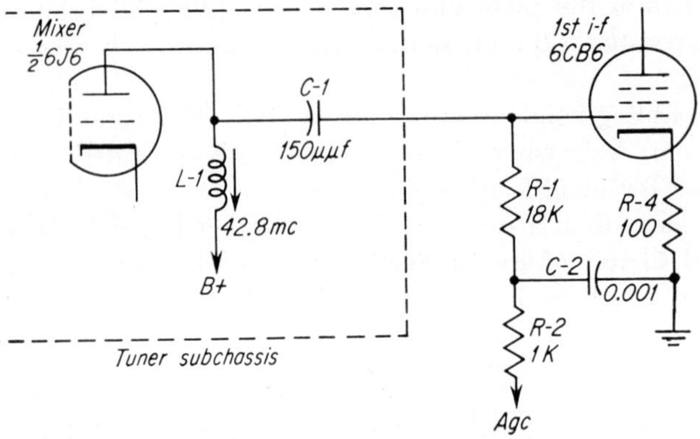

Fig. 12-11. Coupling between the tuner and the i-f amplifier.

Location and Appearance of the I-F Section. The i-f section is very easily found. It leads out from the tuner in a straight line of tubes and transformers, as shown in Fig. 12-12. The coils or transformers may be below the chassis with the alignment screw adjustments showing above the chassis. In some cases, the coils may be in the familiar square cans. The last i-f transformer is very often of this type, and the can has a removable cover for access to a crystal video detector. The i-f strip is very often a printed-circuit board, but this will retain approximately the same appearance and layout of the individual chassis mounts shown in Fig. 12-12.

Quick Test for the I-F Section. An a-m signal generator is adjusted to deliver a strong modulated r-f signal at 40 Mc and is fed to the grid pin of the first i-f tube as shown in Fig. 12-13. An alternate test point, often easily identified, is the i-f output lead from the tuner. The receiver is tuned to an unused channel, the brightness control is adjusted for a normal raster, and the contrast is turned to the full on position. The frequency control on the signal generator is then wobbled back and forth from 40 to 46 Mc while looking for bars on the face of the picture tube. If none appears, the generator is shifted to the 20-Mc i-f band and is wobbled from 20 to 26 Mc.

If the i-f section is working, the audio modulation bars will appear

Fig. 12-12. Location and appearance of the i-f strip.

and remain steady on the screen at a rich black shade as the frequency control on the generator is varied over 2 to 3 Mc of the i-f band. The check must produce rich black bars over at least a 2-Mc swing of the generator. Gray bars or gray changing to black as the generator is wobbled indicate trouble in the i-f amplifier.

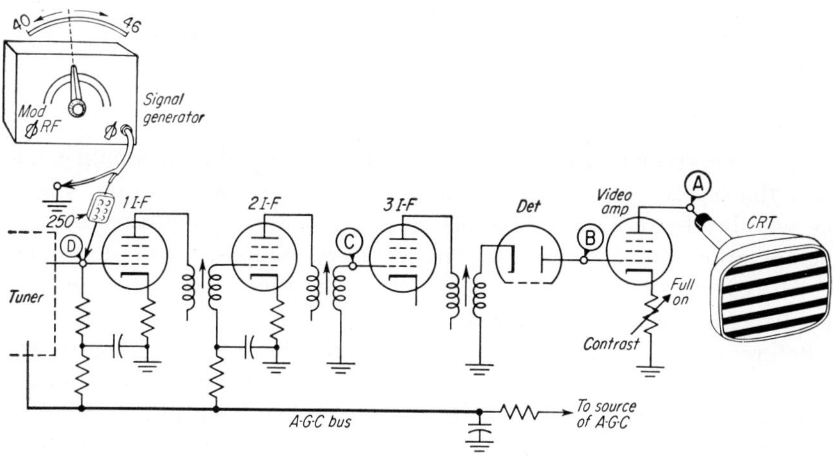

Fig. 12-13. Quick check for the i-f—agc system. A normal response from test point C and no response from test point *D* places the trouble in the i-f—agc system.

SERVICING THE I-F–AGC SECTION FOR NO PICTURE

When the quick check places the cause of no picture in the i-f–agc section, the defective component may be in any of the stages from the mixer plate to the last i-f grid. From the last i-f grid on, the receiver is operating, as proved by the last quick check of the detector. The trouble may also be in a voltage-supply source to one of the i-f stages. This includes the agc source and its decoupling filters.

In servicing for the trouble, there is little to be gained by making a stage-by-stage signal check because of the interaction between the i-f and agc systems. Operation of the agc circuit is dependent on the signal, and the signal is dependent on the agc voltage, so that a defect in either system may make both inoperative. The symptom of no picture is caused by a break in the picture-signal chain. However, the break in turn may occur because a defective agc system allows the buildup of so large a signal that it blocks an otherwise normal stage. Service work can therefore be quite difficult because of misleading interpretations of test results.

The following procedures present a series of checks which have been proved by experience to be most effective in tracing i-f–agc troubles to their source. The agc checks take in the decoupling filters to the agc bus and may be applied to either simple or keyed agc systems. The checks will also indicate whether the trouble is beyond the bus, in the agc tube or tubes. Further service notes on simple and keyed agc circuits are given in Chap. 13.

Servicing the I-F–AGC System. It is assumed that the serviceman has already checked for a misadjusted agc control and that the control, if present, is in a mid-rotation position.

Step 1. Check the i-f tubes.

Tune the receiver to a channel where a strong station would come on if the receiver were working, and adjust for a normal raster. Check the i-f tubes by replacing them one at a time and watching the effect on the operation of the receiver. It is unnecessary to replace the last i-f tube, which is known to be good because of the positive result when making the detector check at the last i-f grid.

Before replacing the tubes, refer to the schematic diagram of the receiver to make sure that the correct tube is being used in each socket. In previous service work, a 6AU6 tube may have been used to replace a 6CB6, or vice versa. At the time, no difference in operation may have been noticed, but after aging, the difference in characteristics may be enough to upset the i-f–agc system.

If a normal picture comes on, the original tube is defective, and the trouble is found. If a picture comes on momentarily after a tube has been replaced, only to disappear as the tube warms up, we have an indication that the i-f chain is operative, and a source voltage, most likely the agc, is incorrect. However, it is not a positive indication; so go on to the next step, just as though there were no change after replacing the tubes.

Step 2. Check Voltage on the AGC Bus.

Tune the receiver to an unused channel, and check the voltage on the agc bus as shown in Fig. 12-14. The agc check point is found by tracing the grid return of the first i-f tube. Either a vacuum-tube or a 20,000-ohms-per-volt voltmeter may be used for this check.

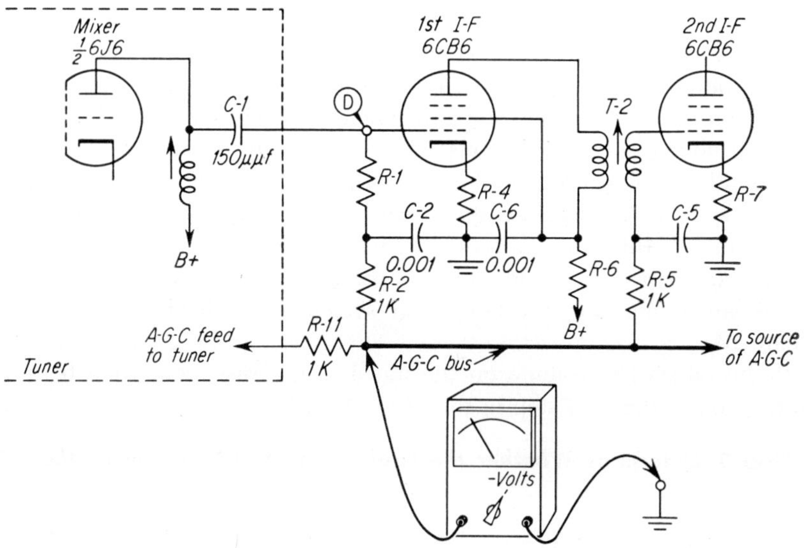

Fig. 12-14. Checking voltage on the agc bus.

Since there is no signal input, there should be no agc voltage, and zero, or a slight negative reading up to −1 volt is normal for this point. If there is even a slight positive voltage indication, there is leakage from the B plus line into the agc circuit. Refer to Fig. 12-14, and note that the possible causes for such a condition are leakage or a short in coupling condenser *C*-1, or leakage or a short between the primary and secondary windings in interstage transformer *T*-2. These are then checked separately for the fault.

If the fault turns out to be a short in coupling condenser *C*-1, and this component is in an inaccessible corner of the tuner, a repair may be effected as follows: Leave the shorted condenser intact as a conductor, and insert a replacement condenser of the correct capacitance in the lead between the tuner and the first i-f grid.

If the r-f stage in the tuner uses a triode tube, as is the case in a cascode tuner, there is another possible source of B plus voltage on the agc bus. This is a shorted neutralizing condenser from plate to grid return in the first r-f tube. To determine if this is the fault, unsolder the agc feed line to the tuner, and measure the voltage on the tuner lead, and also the agc bus. If the agc voltage now drops to its normal 0 to −1 volt, and the tuner connection now shows the positive voltage, the neutralizing condenser is shorted. Replacement notes on the neutralizing condenser are found in Chap. 14.

If the reading on the agc bus shows a high negative voltage, the condition is brought on by oscillation in an i-f stage, or by improper operation of a keyed agc system. Leave the voltmeter connected to the agc bus for monitoring purposes, and connect a 0.001-μf noninductive condenser from ground to each of the i-f control grids in turn. If a condition of oscillation exists, it will be stopped when the oscillating grid is reached, and the voltage on the agc bus will return to its normal zero reading. The oscillation is caused by an open bypass condenser or misalignment. Use the same 0.001-μf condenser to bridge the condensers in the suspected stage, and if this does not help, go over the alignment by the procedure described later in this chapter. If the negative voltage remains on the agc bus, the keyed agc system is serviced as described in Chap. 13.

If the off-station voltage on the agc bus measures a normal 0 to −1 volt, go on to the next step in the procedure.

Step 3. Determine Whether the Fault Is in the I-F System or the AGC System.

The quick check for the detector stage proved that the receiver was working from the last i-f grid on by producing bars on the picture tube when a modulated test signal at the intermediate frequency was fed in at that point. The full output of a signal generator when tuned to a fundamental and connected to the last i-f grid (test point *C*) should develop a control voltage of 0.5 volt or more if the agc system is operating normally. This applies to both simple and keyed agc systems and can be used to determine that the system is working.

The procedure is illustrated in Fig. 12-15. The set is tuned to an unused channel, and the voltmeter is connected to the agc bus to read a negative voltage. The signal generator is adjusted for a maximum modulated output on a fundamental frequency range that covers the i-f band of the receiver. The generator test lead is connected to the last i-f grid at point *C*, and the frequency control is wobbled through the intermediate frequency of the receiver until the blackest bars show on the picture tube. The generator test lead is then removed temporarily.

With no signal input, the bars disappear, and the voltage on the agc bus is its normal 0 to −1 volt. Note the exact reading. Then reconnect

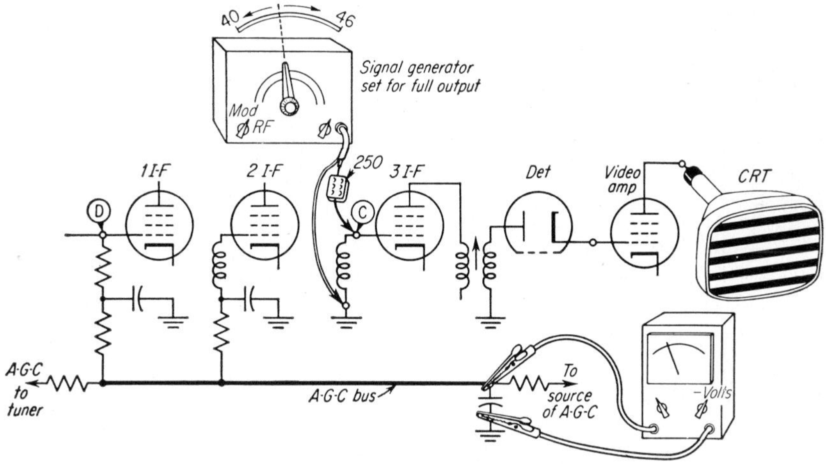

Fig. 12-15. Determining that the agc system is operative.

the generator lead to the last i-f grid, while watching for a new reading on the voltmeter, as well as the reappearance of the bars.

If the voltmeter reading becomes more negative by 0.5 volt or more, when the signal comes on, the agc system is capable of developing a control voltage, and is working. The absence of picture must therefore be caused by an inoperative first or second i-f stage. Go to step 5.

If the voltmeter reading remains the same, and the bars reappear, the agc system is inoperative. Either the agc system is not producing the control voltage, or it is being loaded down by leakage on the agc bus.

When the above check places the trouble in the agc system, it now becomes necessary to determine whether the agc tube or tubes are inoperative, or whether the line is shorted. This is done in step 4.

Step 4. Make a Resistance Check on the AGC Line.

Refer to the schematic diagram of the receiver, and trace the circuit from the agc bus to ground, making note of resistance values. The basic i-f amplifier is connected to a simple agc system in Fig. 12-16*a* and to a keyed agc system in Fig. 12-16*b* to illustrate how this is done.

In the simple agc system, the agc bus returns to ground through filter resistor *R*-12 and diode-load resistor *R*-13. The normal resistance reading from agc bus to ground is therefore 1 meg (filter resistor *R*-12 and diode load *R*-13 in series). When measuring the agc bus to ground, if a reading of open circuit is obtained, filter resistor *R*-12 is open. The diode load cannot be open because of the production of bars on the picture tube, and the diode detector and i-f filter are operative for the same reason. If the reading is well below the normal 1 meg, there is leakage on the agc bus.

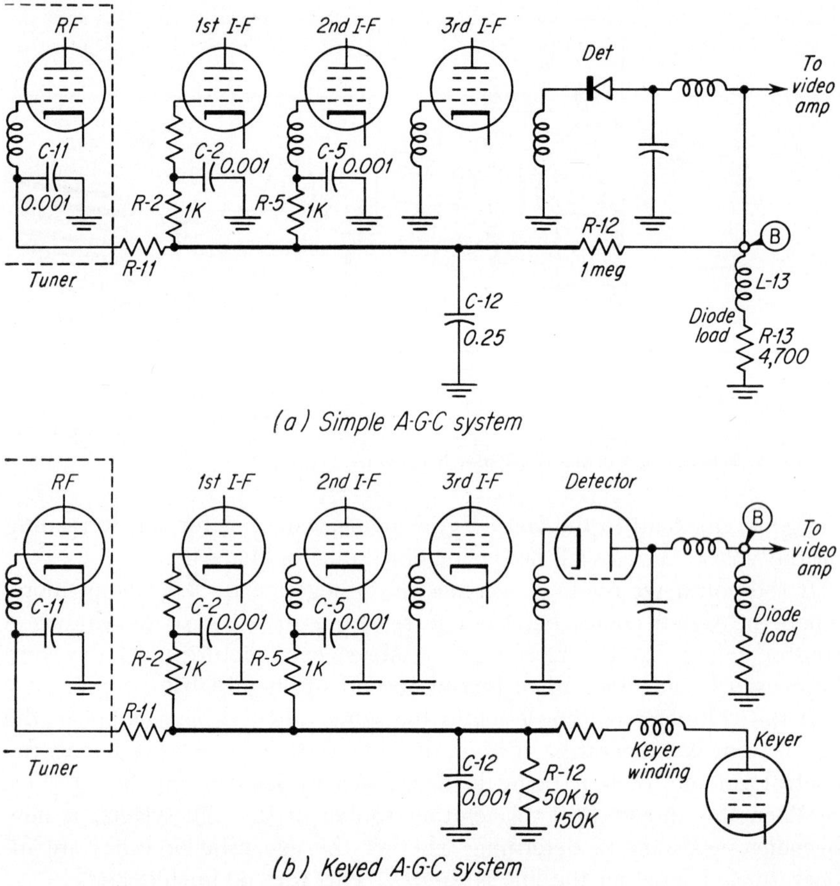

Fig. 12-16. The agc line is checked by resistance measurements.

The leakage is probably in one of the agc bypass condensers *C*-11, *C*-2, *C*-5, and *C*-12. Since these are fed in parallel, it is necessary to open each circuit in checking for the short. Because of short leads on the condensers, it is not practical to unsolder a condenser lead for check purposes. Instead, it is suggested that each decoupling filter circuit be broken by removing each filter resistor from the agc bus in turn, until the short is found. The defective condenser is then replaced.

In the keyed agc system of Fig. 12-16*b*, the agc bus is returned to ground through a resistor represented by *R*-12 in the plate of the keyer tube. This varies from 50,000 to 150,000 ohms in different circuits and gives the normal resistance value to be expected from the agc bus to ground. If this reading is normal, the keyed agc circuit is inoperative. This is then serviced in accordance with the procedure given in Chap. 13. If the reading is below normal, there is leakage in one of the agc by-pass condensers, which are then opened and checked as described before for the simple agc system.

Step 5. If the check in step 3 shows a normal agc voltage, the agc system is operative and the cause of the no-picture symptom is failure of the first or second i-f stages. These are then checked by voltage and resistance analysis. It is not necessary to check the tubes or the third i-f stage, since these checks were made earlier in the procedure.

Normal Voltage Readings in the I-F Section. The basic diagram of an i-f section has been repeated in Fig. 12-17 with normal voltage readings and tube pin numbers added, as an aid in making voltage checks.

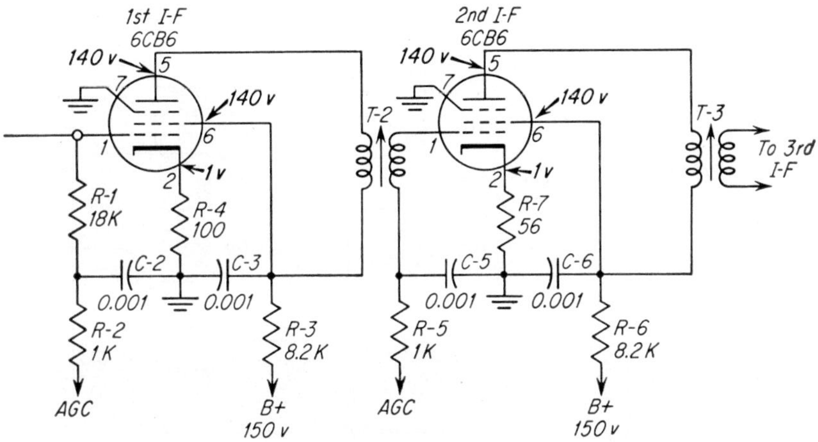

Fig. 12-17. Normal voltage readings in the i-f stages.

In the case of incorrect plate and screen voltages which call for a check of the B plus supply line, refer to the schematic diagram of the receiver to see if the source point is the cathode of the audio output tube in a shelf type of power supply. A defective audio tube or circuit may be the trouble. Further notes on this type of power supply are given in Chap. 4.

Another type of voltage-distribution system may be encountered where the plate circuits of the first and second i-f tubes are connected in series. This circuit will be described now.

Voltage Distribution in Receivers Where the Plate Circuits of Two I-F Stages Are in Series. In some receivers, voltage checking in the i-f stages is complicated because the plate circuits of two tubes are connected in series. The circuit is shown in skeleton form in Fig. 12-18*a*. Each tube is acting as the voltage-dropping resistor for the other tube, on the 240-volt B plus line.

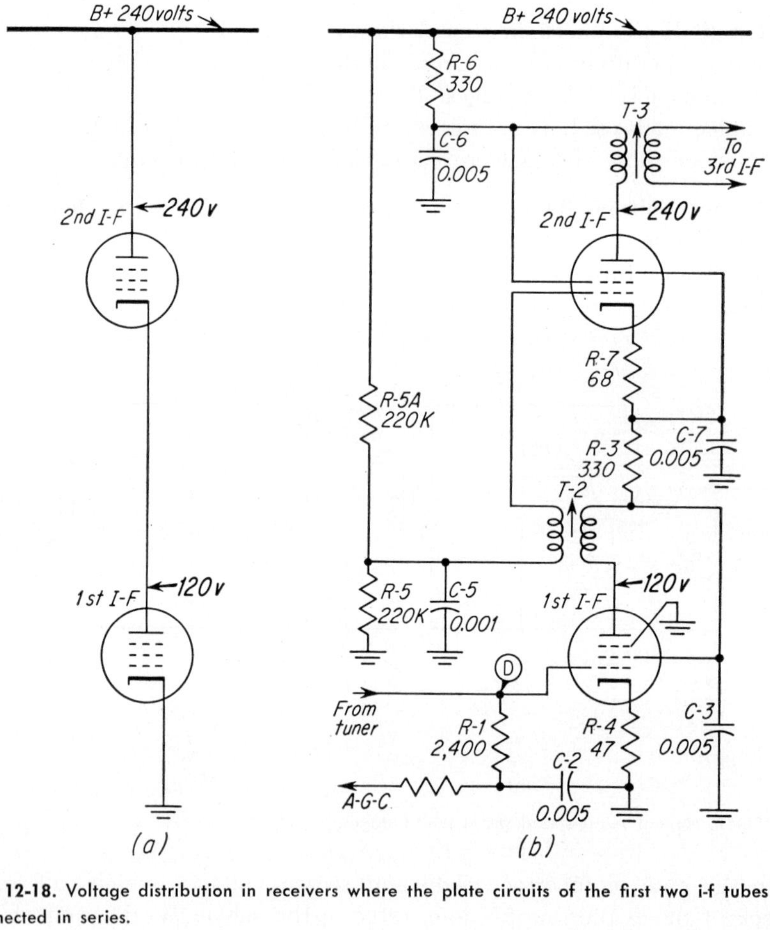

Fig. 12-18. Voltage distribution in receivers where the plate circuits of the first two i-f tubes are connected in series.

A complete circuit of this type has been redrawn from the schematic diagram of the Admiral 17XP3 chassis in Fig. 12-18*b*. Note that the circuit of the first i-f stage is identical with the basic circuit, except for the values of the components. Beginning at the plate of the first i-f tube, trace the wiring through the primary of interstage coupling transformer *T*-2 to resistor *R*-3. This component, together with condenser *C*-3, forms the B plus decoupling filter for the first i-f tube. In the basic circuit of Fig. 12-17, resistor *R*-3 goes on to the B plus 150-volt source; in the series circuit, it goes on to the cathode of the second i-f tube through degeneration resistor *R*-7. Note that the suppressor is connected to this same point which is the equivalent of B minus or common negative for the second i-f tube. Condenser *C*-7 keeps the point at chassis potential for signal voltages. The plate of the second i-f tube goes to the high B plus of the receiver through decoupling filter *R*-6 and *C*-6.

The secondary winding of interstage coupling transformer *T*-2 feeds the signal from the plate of the first i-f tube to the grid of the second. The grid return is connected to a point which is at approximately +120 volts because of the voltage-divider action of resistors *R*-5 and *R*-5A. Condenser *C*-5 bypasses this point to ground. The series connection of the two tubes places the cathode of the second i-f tube at a potential of approximately +120 volts. It is actually a little higher because of the voltage drop across resistor *R*-7. Since the grid return is at +120 volts, the actual bias on the tube is the voltage drop across resistor *R*-7. Agc bias is not applied to the second i-f tube.

When working on this type of receiver, it is important for the serviceman to recognize that any defect which disturbs the voltage distribution in either stage will affect the operation of both. For example, a short in condenser *C*-6 will probably burn out resistor *R*-6, and both stages will be inoperative because of the open circuit in the plate line. The signal checks will quickly bring us to the inoperative i-f stages, and then voltage checks would show a lack of potential on both tubes except for the +120-volt reading at the control grid of the second i-f tube. The positive reading at this point should be the signpost to the type of circuit involved, and then reference to the receiver schematic diagram, and careful resistance analysis should find the actual fault.

Replacement Notes on Components in the I-F Section. When making replacements in the i-f section, the serviceman must remember that this is an amplifier which operates at a comparatively high frequency and that certain precautions are necessary. Before replacing any component, study the wire dress so that the new component is mounted with the same length and positioning of leads.

If a decoupling filter condenser becomes defective, it must be replaced with a similar ceramic type. A paper tubular condenser of the same capacitance will not make a satisfactory replacement, since it has a slight inductive component which makes it useless as a bypass condenser at 45 Mc. Soldering in a ceramic condenser must be carefully done, especially if the lead is short. The heat of soldering will travel through the short lead into the body of the condenser with possible harm. The best way to solder is to hold the lead with the long-nose pliers close to the condenser during the soldering operation. The mass of the tool absorbs the heat so that the lead going into the condenser is comparatively cool.

When replacing an i-f coil, if an exact replacement is not available, satisfactory results may be achieved by using a general replacement coil of the correct frequency range. These are listed in the catalogues of coil manufacturers and are obtainable with and without traps. Of course, the entire amplifier should be realigned after a coil is replaced.

When replacing a defective decoupling filter resistor, if the original circuit used two resistors mounted alongside each other and connected in parallel, use the same system of two resistors for your replacement. This maintains the original characteristics of the circuit in regard to capacitance and inductance.

The i-f amplifiers are often mounted on printed-circuit strips. Pointers for working on this type of receiver construction are given in Chap. 4.

Servicing Procedure for Weak Picture—No Snow. The symptom of weak picture or insufficient contrast is caused by insufficient amplification in one of the stages in the picture-signal chain. The calibrated check of the video detector described in Chap. 11 proved the efficiency of the last i-f stage and detector. The absence of snow clears the tuner of suspicion, and so the cause of the weak picture symptom must be found in the first or second i-f stages. When the weak picture is accompanied by unstable sync action, disregard the latter symptom because it is most likely to be cleared up when the picture signal is restored to full amplitude.

Stage-by-stage signal checking to find a weak stage may not be effective, since bandwidth is increased more than gain as each successive stage is added to the chain in a stagger-tuned i-f system. Furthermore, capacity and loading effects of the generator lead may mask the gain as it is connected to successive grids. It is therefore better to handle the i-f stages as a unit when servicing for weak picture.

The low gain is usually caused by a weak tube, an improper operating voltage, or misalignment. The causes suggest the servicing procedure. Tune the receiver for best picture on a known strong station, and check the i-f tubes by replacing them one at a time and watching for the contrast to come up.

If tubes do not correct the condition, make a voltage check of the i-f stages, looking for a below-normal indication. Voltage checking in the i-f amplifier was described earlier in this chapter.

If voltage checks give normal readings, the next step is a check of the alignment. Misalignment can cause the decrease in the over-all gain of the amplifier. In addition, during the realignment procedure, if an adjustment fails to come to a peak, the coil or decoupling filter condenser associated with that adjustment is open. Either of these conditions will also cause the weak-picture symptom. Alignment instructions are given later in this chapter.

Servicing Procedure for Loss of Sync and Negative Picture. When sync instability is coupled with a weak picture, we service for the weak picture, because the loss of sync information is caused by the same factor and is cleared up at the same time. This is also true when the loss of sync is coupled with the symptom of trailing whites; we service for the trailing whites. When the picture is normal except for the unstable sync condition, the fault is isolated to the i-f–agc system by an oscilloscope check at the diode-load check point, which shows suppressed sync pulses in the composite video signal. The condition is the result of a slight overload in one of the stages. The signal is just large enough so that the sync pulses which represent peak signal are clipped by limiting action in one of the amplifier tubes. The common fault is slightly reduced agc action which permits too much gain in the controlled stages.

A negative picture, where blacks register as whites and vice versa, as in a photograph negative, is closely associated with sync instability when due to an overload condition. The negative picture is caused by a slightly greater degree of overload. As a matter of fact, in some receivers we can change from rolling picture to negative picture by a slight adjustment of the fine tuner. Both symptoms are serviced by the same procedure when the trouble is isolated to the i-f–agc section.

An early step in servicing for loss of sync is to look for and check adjustment of agc and noise controls. In the following procedure it is assumed that the fault is not due to misadjustment of these controls and that they are in a mid-rotation position.

Step 1. Check the I-F Tubes.

Tune the receiver for best picture and sound on a known strong station. Disregard the sync instability. A gassy tube in an i-f stage can cause the trouble. Check for this by changing the i-f tubes one at a time, and note the stability after each replacement.

Step 2. Check Voltage on the AGC Bus.

Turn the station selector to an unused channel, adjust brightness for a normal raster, and check the voltage on the agc bus. The normal reading is 0 to −1 volt. A positive reading of even 1 volt or less is enough to unbalance the i-f–agc system to give the overload condition. The positive voltage is caused by leakage in coupling condenser *C*-1 or interstage transformer *T*-2 in Fig. 12-14. The text accompanying this diagram gives this check in more detail, if needed. If the normal reading of 0 to −1 volt is obtained, go on to the next step.

Step 3. Check the AGC Line for Leakage.

Slight leakage in the decoupling filter condensers on the agc line can also cause a reduction in the control voltage, and thus the overload condition. Therefore, the next check is for this leakage. Refer to Fig. 12-16 which shows normal resistance conditions on the agc line in both the simple and keyed systems. When checking the same circuit for no picture, we were looking for low-resistance leakage which loaded down most of the control voltage. Now, in checking for unstable sync or negative picture, we are looking for a higher-resistance leakage which is only reducing the developed control voltage. The resistance checks must therefore be made more carefully.

If there is no sign of leakage on the agc bus, the trouble is in the agc circuit itself. In the case of a simple agc system, the trouble may be an open 0.25-μf filter condenser *C*-12 in Fig. 12-16*a*. Try operation with another condenser for check purposes. In a keyed agc system, insufficient control voltage is being developed. This circuit is then checked as described in Chap. 13.

Other Symptoms: Lack of Definition, Smearing, Ringing, and Trailing Whites. The video signal in the i-f amplifier is an r-f signal at 25.75 or 45.75 Mc. This is amplitude-modulated by the video information which contains signal elements with a frequency range extending from approximately 30 cps to 4 Mc. The video modulation is extracted in the detector, amplified in the video amplifier, and finally fed to the CRT to produce the picture.

As we have seen in Chap. 9, unequal amplification over the frequency range in the video amplifier causes picture defects like lack of definition, smearing, ringing, and trailing whites, depending on whether low or high video frequencies are amplified too much or too little. An improper response curve in the i-f amplifier can likewise give unequal amplification of the high or low modulating frequencies and can therefore cause the same symptoms.

The servicing procedure for these picture defects is to check the frequency response of the video amplifier as described in Chap 9. If this proves satisfactory, look for the trouble in the i-f amplifier. In the video amplifier, the individual symptoms are caused by different factors such as open peaking coils and incorrect loading resistors. In the i-f amplifier, although different symptoms result from whether low or high frequencies are not sufficiently amplified, the cure is always the same—realignment.

Sync information is part of the composite video signal. The signals are at comparatively low video frequencies, which may be inadequately amplified in the i-f amplifier. However, the resultant unstable sync will be accompanied by trailing whites which are also caused by poor low-frequency response. Therefore, the combination of the two symptoms calls for a check of the low-frequency response, disregarding all the other factors which may cause unstable sync.

The alignment of an i-f amplifier is checked by going through the realignment procedure, checking to see that each coil peaks at its proper frequency.

ALIGNMENT OF THE I-F AMPLIFIER

The schematic diagram or service notes of a receiver will indicate the number of tuned circuits in the i-f section and the presence of traps. It will also indicate the resonant frequencies of the coils and traps, and may include a diagram of their location. If the frequencies of all the transformers are given, they should be used. If this information is incomplete, disregard it, except to ascertain whether the receiver uses a 20- or 40-Mc i-f band, and use the alignment frequencies recommended in the generalized procedure which follows.

Equipment Needed for Alignment. The equipment needed for alignment is an a-m signal generator, a multimeter with a-c voltage ranges, and the necessary assortment of alignment tools. The latter should include a recessed-nib alignment screwdriver, and several widths of hex-stud alignment tools, such as the Walsco No. 2544. The hex stud on one end of this tool has a thin nylon shank which permits the stud to be slipped through a hollow-core top tuning slug to engage a bottom core. This permits both tuning slugs to be adjusted from the top. The hex stud on the other end of the tool has a shank wider than the stud. This is useful in aligning the top slug because the wide shank acts as a shoulder to keep the tool from slipping through the hollow core. The nylon shank is flexible, allowing adjustments to be made in tight quarters.

The a-m generator should have a large r-f output. The frequency coverage should include the 40-Mc i-f band on a fundamental. The Precision E-200-C signal-marking generator, illustrated in Fig. 12-19, reaches 60 Mc on a fundamental and has been found satisfactory for this work.

Calibration of the Signal Generator. The accuracy of the average signal generator is usually satisfactory for i-f alignment work, but it should be checked. This is especially important for the trap frequencies at 21.25 and 41.25 Mc.

The equipment needed is a communication receiver capable of picking up signals from WWV, the station maintained by the Bureau of Standards at 20, 25, 30, and 35 Mc. The transmission is a steady audio note or the familiar time signal. This is tuned in at 20 Mc on the communications receiver. The signal generator is loosely coupled to the receiver, set near 20 Mc, and then wobbled for zero beat, which will occur when the generator is at exactly 20 Mc. Read the generator dial at zero beat. If

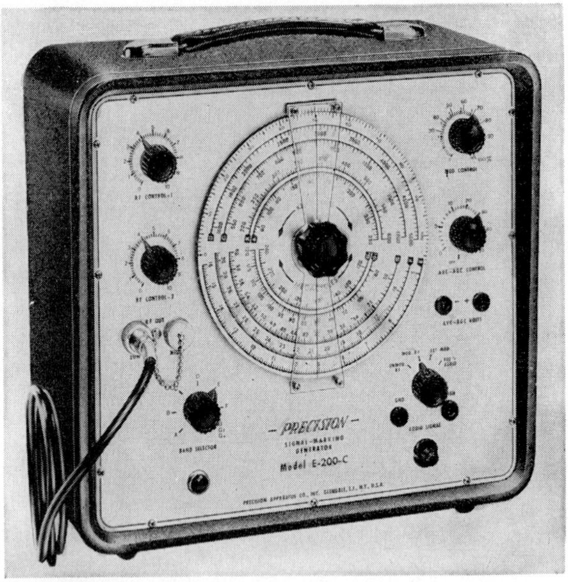

Fig. 12-19. Precision E-200-C signal generator.

the dial reads between 19.75 and 20.25 Mc, disregard the discrepancy. If the reading is off by a larger amount, the generator should be recalibrated by the manufacturer or a reliable laboratory which specializes in this work. Repeat the calibration at 35 Mc to check the accuracy of the generator at a frequency near 41.25 Mc.

Determining the Intermediate Frequency of the Receiver. If there is no information regarding the intermediate frequency of the receiver, hook it up as for alignment (see Fig. 12-23) and wobble the frequency control of the generator from 40 to 46 Mc. A receiver in the 20-Mc i-f band will not respond. Confirm this by repeating at 20 to 26 Mc.

Identifying and Locating Adjustment Screws. The signal circuit and chassis layout of a typical i-f amplifier are given in Fig. 12-20 as an aid in identifying the i-f adjustment screws. Note the last i-f transformer assembly. It has a removable cover for access to the crystal detector. In the schematic diagram, the arrow indicating the tuning slug for this transformer points downward. This is the convention employed to show that the adjustment screw is reached from the underside of the chassis.

Note also, that the i-f adjustment screw in the tuner is headless, in contrast to the r-f adjustments which are round-head screws. Mark the chassis near the traps and i-f coils with each adjustment frequency for easy identification in the alignment procedure. In the example given in Fig. 12-20, only the trap should be marked, since this is the only frequency given in the diagram.

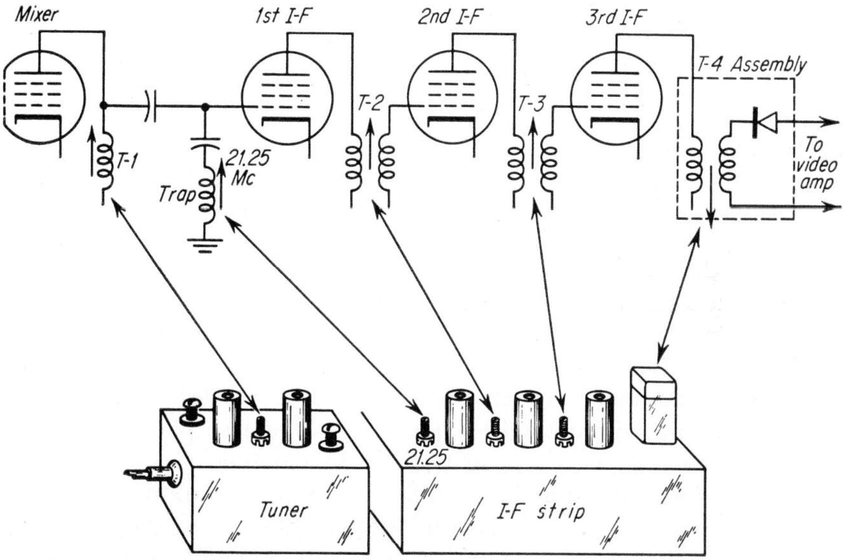

Fig. 12-20. Identifying i-f adjustment screws.

How to Connect an Output Indicator. The most reliable indicator for a peak or null adjustment is an a-c voltmeter connected as an output meter on the video output of the picture chain. Since the signal level is low, especially for the null adjustments of traps, the output meter is connected to the plate of the video output tube where the signal level is at its highest. The connections are shown in Fig. 12-21.

The 0.05-μf condenser in series with the voltmeter lead is for isolation purposes and to keep the direct voltage present at the plate of the video output tube from causing an erroneous reading on the output meter. The contrast control is left in the full on position throughout the alignment procedure so that the video amplifier is working at full gain.

The normal picture signal at the plate of the video amplifier has a peak-to-peak value of approximately 100 volts. This should not be exceeded during the alignment by the signal from the generator, but remember that it will register as about 35 volts on the output meter which reads rms values. For meter-protection purposes, keep the meter

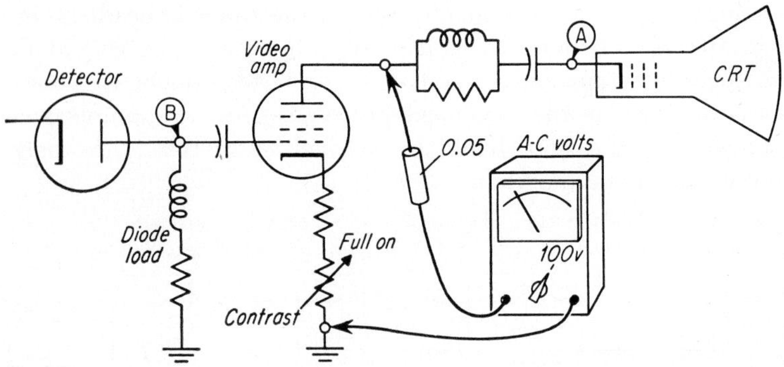

Fig. 12-21. How to connect an output meter for i-f alignment.

on a range which covers 100 volts, reducing to a lower range when needed.

How to Connect the Signal Generator. The signal injection point for aligning the i-f amplifier is the grid of the mixer tube. Since this point is usually inaccessible inside the tuner the connection is made as shown in Fig. 12-22. Locate the mixer tube, and refer to the schematic diagram for the pin number of the mixer grid (Fig. 12-22*a*). Wrap a small piece of hookup wire around the pin, using long-nose pliers to make a tight connection. Tape the wire securely to the tube as shown in the diagram. Reinsert the tube in its socket, and replace the tube shield, allowing the wire to come through the hole in the top, as shown in Fig. 12-22*b*. The hot lead of the signal generator is clipped to the insulation on the hookup wire. If greater signal input is desired, it may be clipped directly to the wire through a 200-$\mu\mu$f isolation condenser. The ground lead of the signal generator should be connected to the tuner chassis

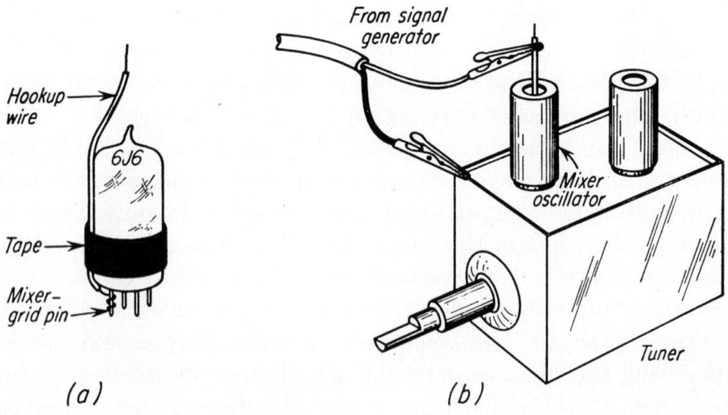

Fig. 12-22. How to connect the signal generator for alignment.

close to the mixer tube so as to keep the signal input path as short as possible.

Conditions for Alignment. The receiver should be insulated from the line. Refer to the schematic diagram, and check the type of low-voltage power supply. If it is of the transformer-tube rectifier type, it may be plugged directly to the line. In practically all other types, the chassis is connected to one side of the line. If this is the case, the receiver should be connected to the line through an isolation transformer.

To prevent interference from extraneous signals during the alignment, remove the antenna connection and short the antenna leads with a wire jumper at the connections to the tuner. Also disable the local oscillator in the tuner by allowing the channel-selector switch to rest at a between-channel point. If this cannot be done, set the channel selector to an unused high channel where the high frequency is not likely to interfere with the alignment.

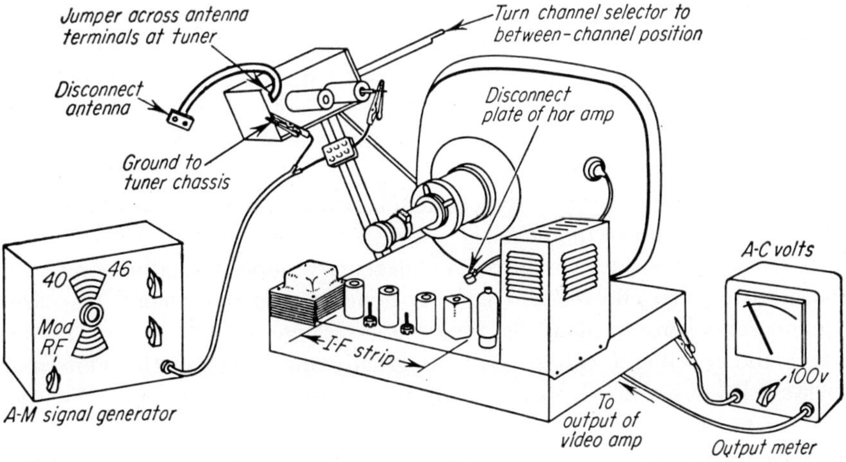

Fig. 12-23. Setup for i-f alignment.

If the receiver is equipped with agc or noise controls, these should be left in their normal reception positions. After the set has been realigned, these controls may require readjustment, because of altered gain characteristics of the receiver. Such readjustment is best performed in the consumer's home where the receiver is operating with its own antenna.

The complete alignment setup is shown in Fig. 12-23. The receiver is set on its side so that the connection points and adjustment screws are easily reached. Before starting the alignment, the receiver and signal generator are allowed a 15-min warm-up period. Check to see that the contrast control is turned fully on and that all tube shields are in place.

It is unnecessary to use a clamping battery for this alignment procedure. Traps are aligned first.

Adjustment of Wave Trap. If the receiver is equipped with a trap, it is adjusted at the frequency assigned to it in the manufacturer's service notes. Adjust the signal generator for a modulated signal at a moderate r-f output. Wobble the generator through the intermediate frequency of the receiver while watching the output meter reading, just to make sure that all connections are correct. It should rise and fall as the signal generator goes through the intermediate frequency.

Adjust the signal generator accurately to the trap frequency. If the output meter reading drops to zero, adjust the attenuator for higher output, and the range on the output meter to a lower scale. The meter response at the trap frequency should be low, and it will probably be necessary to utilize the full generator output by injecting the signal directly to the mixer grid wire rather than to its insulation in order to get a reading on the output meter.

When a readable indication is obtained on the output meter at the trap frequency, slowly adjust the tuning slug on the trap while watching the meter. Do not touch the frequency control on the signal generator. A point of minimum reading will be found. Turn the adjustment screw back and forth through this point to make sure that the reading increases on each side, and finally come to a stop at the lowest reading. This is called tuning for a null, and the receiver now has minimum response at the trap frequency.

If the receiver has two traps at different frequencies, either one may be adjusted first. Before shifting the generator to the second trap frequency, be sure to increase the voltage range of the output meter, since the receiver response will be greatly increased as the generator passes through the i-f band. Then repeat the procedure by adjusting the second trap for a null at its frequency.

How to Choose Trap Frequencies When Information Is Not Available. If the trap frequency is not marked on the diagram or mentioned in the service notes, it can be determined by assuming that its present setting is close to correct. First measure the height of the adjustment screw above the chassis as shown in Fig. 12-24. Make a note of the height so that you can return to the original setting. The three common trap frequencies for both i-f bands are also given in Fig. 12-24 for easy reference.

Connect the signal generator and output meter for alignment, and set the frequency control to the adjacent sound-trap frequency at 27.25 or 47.25 Mc as the case may be. Adjust the attenuator for a readable indication on the output meter, and rock the trap adjustment screw through a half turn on each side of the original setting while watching the result on the output meter. If the meter shows no change, the trap is ineffective at this frequency. Repeat the procedure at the associated sound and adjacent picture trap frequencies. The frequency at which the adjustment produces a sharp null, at a point close to the original setting, is the frequency to which the trap was adjusted originally.

If you have reason to believe the trap adjustment had been tampered with, the above procedure cannot be used since you cannot be sure of

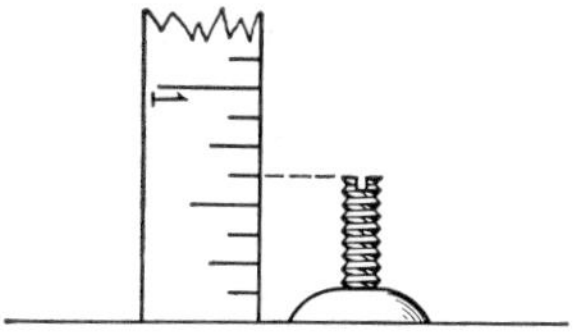

	20-Mc i-f	40-Mc i-f
Adjacent sound	27.25	47.25
Associated sound	21.25	41.25
Adjacent picture	19.75	39.75

Fig. 12-24. Common trap frequencies and how to mark an adjustment screw so that you can return to the original setting.

the original setting. The trap cannot be left at a random setting because this may be at a desired frequency in the i-f amplifier range. The random setting may therefore seriously distort the response curve and result in deterioration of the picture. You must choose a trap frequency, and adjust to it.

To choose a trap frequency, it would be wise to first find out the range of the trap. Set the generator to 27.25 or 47.25 Mc, and run the trap tuning slug all the way up and down while looking for the null on the output meter. It must produce a definite null, that is, the output reading must increase on each side of the null point. If the null is produced, the trap can reach the adjacent sound frequency. Repeat the fully up-and-down technique at the other two common trap frequencies to see if they can be reached also.

If the trap is effective at only one of the frequencies, there is no choice. Set the generator exactly at this frequency, and adjust for the null.

If two of the common frequencies can be reached (all three are not likely), the following considerations will guide your choice. If reception in your area is commonly marred by adjacent-channel interference, choose the adjacent trap frequency that can be reached, and adjust for the null. If there are no adjacent-channel problems in your area, choose the associated sound frequency, so that the receiver will have a wider pass band and greater freedom from intercarrier buzz.

If your problem is complicated by the presence of more than one trap and the absence of authentic information, it will yield to the same method. First determine the frequency range of each trap, and then choose advantageous trap frequencies. A final note to guide your choice is that it is unlikely that the same receiver will have two traps tuned to the same frequency, and that traps in the first i-f stage will be for adjacent-channel interference, whereas traps in a later stage are more likely to be meant for associated sound.

Peaking Frequencies. The manufacturer's information as to the peaking frequency for each coil is to be used if available and complete. If not, good results may be obtained by using the average frequencies listed in Fig. 12-25.

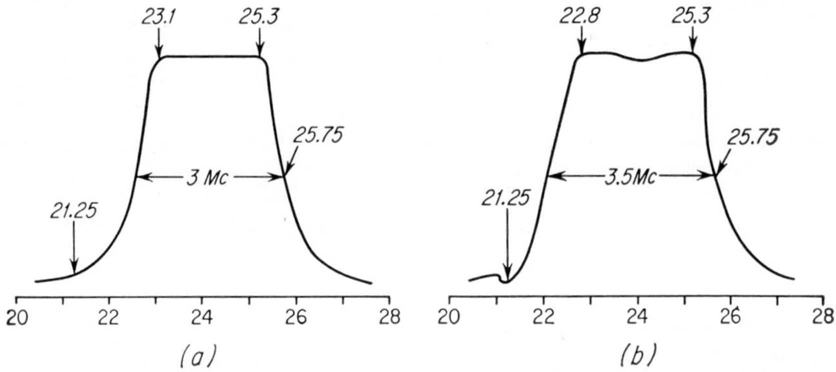

T-1, *T*-3	23.1	or 43.1
T-2, *T*-4	25.3	or 45.3

Trap	21.25	or 41.25
T-1, *T*-3	22.8	or 42.8
T-2, *T*-4	25.3	or 45.3

Fig. 12-25. Common peaking frequencies for aligning a three-stage i-f amplifier and the resulting response curves. (*a*) Without trap. (*b*) With trap for associated sound.

In all cases, the second and fourth i-f transformers are tuned to 25.3 or 45.3 Mc. This places the picture carrier (25.75 or 45.75) about halfway down on the response curve.

If the receiver has no trap at all, or a trap tuned to the interference from an adjacent channel, the first tuning circuit (mixer plate) and the third i-f transformer are peaked at 23.1 or 43.1 Mc. The approximate response which results from the use of these peaking frequencies is shown in Fig. 12-25*a*. Note that the curve is symmetrical and the bandwidth at half amplitude is approximately 3 Mc.

If the receiver uses a trap for associated sound at 21.25 or 41.25 Mc, the first tuning circuit (mixer plate) and the third i-f transformer are peaked at 22.8 or 42.8 Mc. The approximate over-all i-f response is as shown in Fig. 12-25*b*. The use of the trap makes the left slope sharper than the right and closer to the ideal i-f response curve. Note that the bandwidth at half amplitude is approximately 3.5 Mc.

Alignment of the I-F Transformers. The following procedure assumes the use of the average peaking frequencies given in Fig. 12-25. The connections of the signal generator and the output meter are the same as for the trap adjustment.

Set the signal generator to produce a modulated output at exactly 25.3 or 45.3 Mc, and adjust the attenuator for a reading of approximately 10 volts on the output meter. Adjust the last i-f transformer tuning slug for a peak reading. If the output reading goes over 15 volts during this adjustment, reduce the attenuator setting. As the alignment progresses, keep the output reading between 10 and 15 volts by reducing the input signal with the attenuator. If there is difficulty in reducing the output, change the signal-injection point to the insulation on the mixer-grid lead. Run back and forth through the peak position with the adjustment tool, finally coming to rest at the maximum reading. The adjustment must produce a true peak with the output voltage falling away on each side of the peak setting.

After the last i-f transformer has been adjusted, leave the generator at its 25.3- or 45.3-Mc setting, and proceed to adjust the second i-f transformer in the plate circuit of the first i-f tube. If the output meter should suddenly jump to a high reading while either the last or second i-f transformers are being adjusted, it is because the i-f amplifier has suddenly gone into oscillation. The unstable condition is caused by the other transformers' being close to the same frequency. Tighten the adjustment screw on either the third i-f transformer or mixer plate coil until the output meter drops to its original reading and then continue to peak the coil you were working on. Be careful not to touch a trap adjustment.

If any coil fails to come to a true peak, that coil is defective, or one of the condensers associated with it is open. This is more likely to be the agc bypass condenser (*C*-5 for the second i-f coil in the standard circuit of Fig. 12-10), since an open plate bypass condenser produces oscillatory effects which would have been found much earlier in the procedure. Solder in another agc bypass condenser at the coil lead, and try peaking again. If this is ineffective, repeat with the plate decoupling filter. If the coil cannot be adjusted to a peak after these changes, it must be replaced.

After the last and second i-f transformers have been peaked at 25.3 or 45.3 Mc, the signal generator is carefully set to the second peaking frequency. This is 23.1 or 43.1 Mc, unless the receiver has a trap tuned to the associated sound frequency, in which case it is set for 22.8 or 42.8 Mc. The third i-f transformer and the mixer-plate transformer are then peaked in the order mentioned. This completes the alignment of the i-f amplifier.

How to Check Frequency Response with a Signal Generator and Output Meter. The receiver, signal generator, and output meter are connected in the alignment setup of Fig. 12-23. The frequency control on the generator is wobbled through the intermediate frequency, and the attenuator is adjusted so that the greatest output voltage is within the scope of a low-voltage range of the output meter.

Provide yourself with some cross-section paper and tabulate a list of the intermediate frequencies in 0.5-Mc steps as shown in the first column of Fig. 12-26*a*. Then adjust the signal generator to the first figure, 21 or 41 Mc as the case may be. There may be a very small reading on the output meter. Record this as shown in the second column of Fig. 12-26*a*. Next adjust the frequency control on the generator 0.5 Mc higher, record the output meter reading, and continue through the list.

A graph of the figures is made by setting up the frequencies on a horizontal line and the output voltage on a vertical line as shown in Fig. 12-26*b*. Each point is marked on the graph by a dot or cross. Note that

f	Meter reading
21.0	0.2
21.5	0.2
22.0	0.5
22.5	1.5
23.0	4.5
23.5	5.0
24.0	4.5
24.5	4.5
25.0	5.0
25.5	3.0
26.0	0.5
26.5	0.2
27.0	0.2

(a)

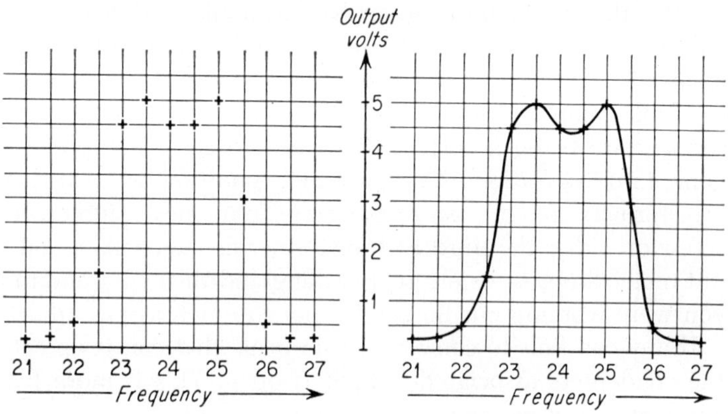

(b) (c)

Fig. 12-26. Steps in constructing a graph of the frequency response of an i-f amplifier.

the output voltage begins to climb at 22 Mc, and the cross for this frequency is one space up, denoting 0.5 volt of signal output. Continue placing the crosses, and then join them, using curved lines rather than straight, as shown in Fig. 12-26*c*. The result is the frequency-response curve of the amplifier.

The figures and curves of Fig. 12-26 are actual figures taken on a receiver which had just been aligned by the method given in the previous section. It misses perfection, but as we discuss these imperfections, we shall see that it is a very good response curve. Note the two peaks. This is usual when the amplifier is aligned at only two frequencies. The two peaks should be the same height, or nearly so, and the trough between should be a maximum of 30 per cent of the total amplitude. The actual trough is down 0.5 volt from a 5-volt peak. This is 10 per cent down. Moving to the half-amplitude point at 2.5 volts, the curve should be approximately 3 Mc wide. The curve crosses the 2.5-volt line at 22.7 and 25.6 Mc, making it 2.9 Mc wide. The picture carrier at 25.75 Mc should be at the half-amplitude point. This misses, since the curve shows about 2 volts for 25.75 Mc. This is permissible, since the function of the fine tuner is to move the picture carrier up and down the slope of the curve between 25 and 26 Mc until the point of best picture is found. At the same time, the fine tuner moves the sound carrier which is 4.5 Mc away from 20.5 to 21.5 Mc. Note that the curve is at an even low figure for these frequencies, so that sound amplitude is little affected as the point of best picture is found. The amplitude at the sound carrier should be 5 per cent of the total, which is 0.25 volt. The actual reading at 0.2 volt is close to correct.

How to Check the Frequency-response Curve with a Sweep Generator and Oscilloscope. If a sweep generator is available, the frequency-response curve can be seen directly on an oscilloscope. The setup is shown in Fig. 12-27. The r-f output of the sweep generator replaces the signal

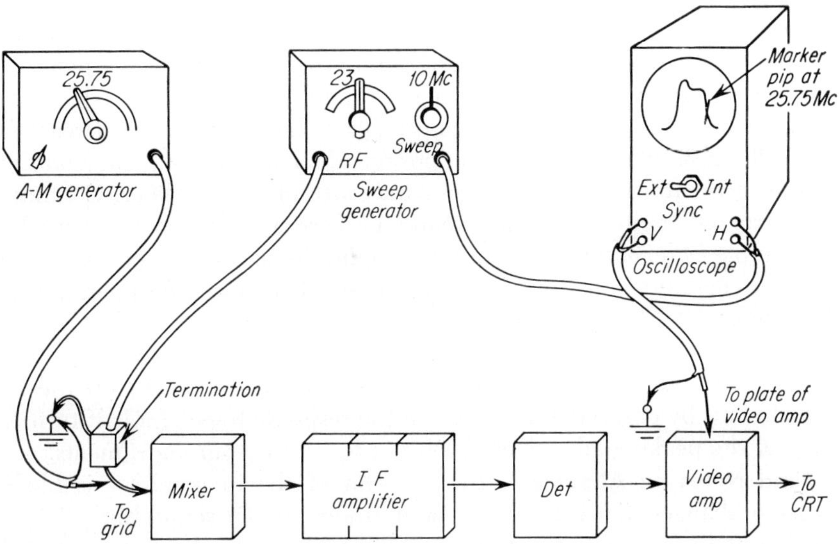

Fig. 12-27. Checking the frequency-response curve of an i-f amplifier with a sweep generator and oscilloscope.

generator output in the alignment setup, and the vertical input lead of the oscilloscope replaces the output meter. The synchronizing sweep output of the generator is connected to the horizontal input of the oscilloscope. The sync control switch on the oscilloscope is set for external horizontal sweep. Refer to the sweep generator instruction manual for connection details.

The center r-f frequency on the sweep generator is set at 23 or 43 Mc as the case may be, and the sweep range is adjusted for a 10- or 12-Mc sweep. The height of the trace on the oscilloscope can be controlled by the attenuator in the sweep generator and the vertical gain control in the oscilloscope. Use a low output from the sweep generator so as to prevent overloading the i-f amplifier, and compensate by increasing the vertical gain in the oscilloscope so that the trace is a proper height on the screen.

The trace shows the response curve of the i-f amplifier very much like the graph of Fig. 12-26*c*. However, the trace is unmarked as to frequency except that the center of the screen should correspond to the center frequency of the sweep generator. Your a-m generator may be used as a marker. Loosely couple it to the sweep generator by clipping its hot lead to the insulation on the hot lead of the sweep generator. The signal thus introduced causes an irregularity on the oscilloscope trace at a point corresponding to the frequency setting of the a-m generator. This irregularity is called a marker pip. Set the marker generator at the i-f picture carrier (45.75 or 25.75), and use just enough output to observe the pip. See if the pip is correctly located at the half-amplitude point of the trace. Then slowly bring the marker generator to lower frequencies while watching the progress of the pip across the trace. When it reaches the half-amplitude point on the opposite side, read the frequency on the marker generator to determine the bandwidth of the amplifier. Continue down to the frequency of the sound i-f carrier. The pip may not be discernible here because of the low amplitude at this frequency. Some sweep generators contain a built-in marker generator. The operation is the same.

If the trace shows an unsatisfactory frequency-response curve, the sweep generator and oscilloscope may be used for touching up the i-f adjustments by observing the effect on the response trace of a half turn of each of the peaking adjustments. Do not touch the trap adjustments.

Final Notes on Alignment. Realignment of the i-f amplifier is likely to have changed the ratio of the amplitudes of the sound and picture carriers with a resulting change in the 4.5-Mc sound beat. Low sound output or excessive intercarrier buzz may ensue. For this reason, it is best to realign the sound section after completing the alignment of the i-f amplifier. Alignment procedures for the sound section are given in Chap. 15.

When the receiver is tuned from channel to channel, it is desirable to have each station come in at the best picture position without having to readjust the fine tuner. To accomplish this, the final step in the alignment procedure is to touch up the oscillator adjustments in the tuner for each channel normally received in your area. Turn the selector to the highest channel which gives a station and set the fine tuner to its mid-range position, even though this is not at the best picture setting. Now rock the oscillator adjustment screw back and forth, finally leaving it at the position which gives best picture. Without touching the fine tuner, switch to the next lower station, repeat with its oscillator adjustment screw, and continue with all stations normally received. A more detailed description of this procedure is given in Chap. 14.

SUMMARY

Quick Check for the I-F Stages in the Picture Chain. Feed a strong modulated i-f signal to the input of the i-f section, and wobble the frequency control from 20 to 25 Mc. The appearance of modulation bars on the screen of the picture tube indicates that the i-f section is working.

Basic Circuit. The schematic diagram of a basic three-stage i-f section is given in Fig. 12-28. Normal voltage readings are included in the diagram.

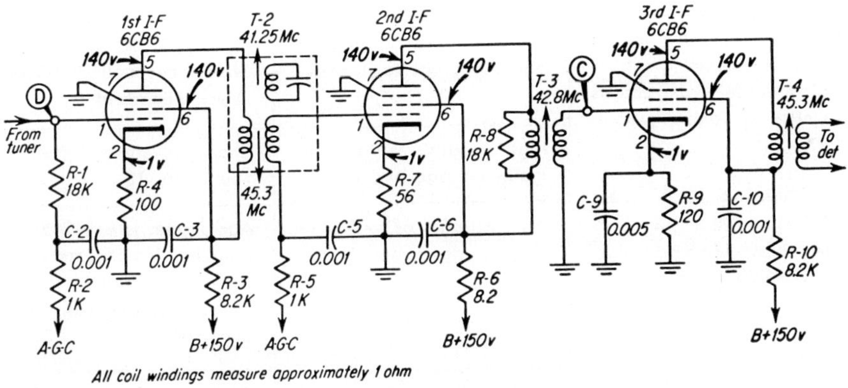

Fig. 12-28. Schematic diagram of a basic three-stage i-f amplifier.

Servicing Charts. Servicing procedures for loss of picture or weak picture without snow are given in outline form in separate charts. Another chart is given for the overload symptoms of unstable sync and negative picture. The symptoms of lack of definition, smearing, ringing, and trailing whites are caused by improper alignment when diagnosis places the fault in the i-f amplifier. A generalized realignment procedure is also given in chart form. Reference in the charts is to the basic i-f section of Fig. 12-28.

Servicing Procedure for Normal Raster—No Picture—No Snow

Procedure	*Result*	*Trouble and Subsequent Check*
1. Replace the first and second i-f tubes one at a time	Picture comes on and stays on	Original tube is defective
	None	Go to step 2
2. Tune to unused channel and check voltage on the agc bus	Positive voltage is found	Leakage from B plus to the agc line. Check for leakage in i-f coupling condensers, primary-to-secondary shorts in i-f transformers, or a shorted neutralizing condenser in the tuner
	High negative voltage is found	Oscillation in an i-f stage or improper operation of a keyed agc system. Go on to step 3
	Normal reading of 0 to —1 volt	Either i-f is inoperative, or agc is inoperative. Go on to step 4
3. Leave voltmeter on agc bus and connect a 0.001-μf condenser from chassis to each i-f control grid in turn	High negative voltage on agc bus drops to normal	I-f stage is oscillating. Check decoupling filters. Check alignment
	High negative voltage remains on agc bus	Check keyed agc system. See Chap. 13
4. Feed i-f test signal to the third i-f grid, and check resulting agc voltage	Agc voltage of —0.5 or more is developed	Agc system is working; i-f amplifier is not. Check first and second i-f stages by voltage and resistance analysis
	Agc voltage does not change	Agc circuit is not working, or there is a short on the agc bus. Go to step 5
5. Measure agc bus to ground with an ohmmeter	Normal resistance reading: Simple: 1 meg. Keyed: 50,000 to 150,000 ohms	Agc stage not working. See Chap. 13
	Low resistance reading	Check agc decoupling filter condensers for leakage.

Servicing Procedure for Weak Picture—No Snow. The complaint of weak picture or lack of contrast or definition without snow is diagnosed as a fault in the first or second i-f stages when an i-f test signal fed to the last i-f grid produces a normal response on the screen of the picture tube. Since there is no snow, the fault is not in the tuner.

1. Check i-f tubes by replacement.
2. Make a voltage check, looking for incorrect plate or screen voltage readings.
3. Check alignment. (An i-f adjustment that will not peak is an indication of an open coil or agc bypass condenser.)

Servicing Procedure for Overload Conditions of Unstable Sync or Negative Picture

An agc control, if present, is in the mid-range position.

Procedure	*Result*	*Trouble and Subsequent Check*
1. Check first and second i-f tubes by replacement	Normal picture	Original tube is defective
	None	Go to step 2
2. Tune to unused channel and check voltage on the agc bus	Positive voltage is found	Leakage from B plus to the agc line. Check for leakage in i-f coupling condensers, or primary-to-secondary short in i-f transformers
	Normal reading of 0 to —1 volt	Go to step 3
3. Measure agc bus to ground looking for abnormally low resistance reading	Normal resistance reading	Agc stage not developing sufficient control voltage. See Chap. 13
	Low resistance reading	Check agc decoupling filters.

Generalized Alignment Procedure for the I-F Section. Use isolation transformer for series filament receivers. Remove antenna connection, and short antenna terminals as close to the tuner as possible.

Set receiver controls as follows:

Channel selector—between channels, or to channel 12 or 13, whichever is not used in your area.

Contrast—full on.

Connect output meter from chassis to plate of video output tube. Use a 0.05-μf isolation condenser in the plate lead.

Connect wire to grid pin of the mixer tube, and clip hot lead of the signal generator to the insulation on this wire. Clip generator ground lead to tuner chassis.

Allow signal generator and receiver a 15-min warm-up period.

Adjust adjacent channel traps, if present, for minimum response at their assigned frequencies.

Alignment Procedure for Sets with an Associated Sound Trap

Signal Generator Frequency		*Adjust*	*Notes*
20-Mc I-F Band	*40-Mc I-F Band*		
1. 21.25	41.25	Trap for minimum	Clip generator lead to mixer grid wire for readable output indication
2. 25.3	45.3	*T*-4 and *T*-2 for maximum	Reduce attenuator for approximately 15-volt output. If attenuator is ineffective, receiver is in oscillation. Turn *T*-3 and *T*-1 slugs in until oscillation stops
3. 22.8	42.8	*T*-3 and *T*-1 for maximum	

Alignment Procedure for Sets without Sound Trap

Signal Generator Frequency		Adjust	Notes
20-Mc I-F Band	*40-Mc I-F Band*		
1. 25.3	45.3	*T*-4 and *T*-2 for maximum	Reduce attenuator for approximately 15-volt output. If receiver goes into oscillation, tighten *T*-3 and *T*-1
2. 23.1	43.1	*T*-3 and *T*-1 for maximum	

Final Steps. Realign sound section (see Chap. 15).

Set fine tuner at mid-range position, and adjust individual channel oscillator slugs in the tuner for best picture on all channels in your area (see Chap. 14).

QUESTIONS

1. What symptom is the result of an inoperative i-f stage in a television receiver?

2. Review the quick checks for the sections in the picture-signal chain:

a. Quick check for the video amplifier and picture tube.

b. Quick check for the detector stage.

c. Quick check for the i-f stages.

d. Which section of the receiver is at fault if positive results are obtained in checks *a*, *b*, and *c* above?

3. Describe the function of the i-f section. Include desirable bandpass characteristics of the i-f amplifier.

4. What is meant by a favorable ratio between the amplitudes of the video and sound carriers in the i-f amplifier?

5. Make a drawing of an idealized response curve for an i-f amplifier. Mark on the curve the positions of the video and sound carriers as well as the two adjacent trap frequencies.

6. Interference from adjacent channels is not a serious problem in large metropolitan areas. Why is this so?

7. Describe the system of stagger tuning to achieve wide-band characteristics in i-f amplifiers.

8. Describe a trap, and explain its function in an i-f amplifier. Give the three trap frequencies commonly employed.

9. What is meant by a bifilar winding in an i-f transformer? What is the purpose of this type of winding?

10. Draw a schematic diagram of impedance coupling between two i-f stages. Include a series trap in the grid circuit.

11. Draw a schematic diagram of transformer coupling between two i-f stages. Include an absorption trap.

12. What is the purpose of loading resistors when found across impedance or transformer windings in an i-f amplifier?

13. Describe the localizing checks which place the trouble in the agc system or the i-f amplifier for a no-picture symptom.

14. What is the probable cause of a high positive voltage on the agc bus?

15. What are the probable causes of a high negative voltage on the agc bus?

16. Give the important precautions to be observed when making replacements in the i-f section.

17. A smeared picture is caused by a distorted frequency response in the i-f or video amplifiers. Describe a check which places the trouble definitely in one of these sections.

18. If the information is not given in the manufacturer's service notes, how can you determine that an i-f amplifier has been designed to operate in the 20- or 40-Mc i-f band?

19. Describe the alignment procedure for the i-f amplifier using an a-m signal generator and an output meter.

20. If a test receiver is available, locate and identify the alignment adjustments, and go through the alignment procedure.

21. Provide yourself with a piece of cross-section paper, and construct a frequency-response graph of an i-f amplifier like the example shown in Fig. 12-26.

22. In the actual graph constructed in question 21, what are the allowable deviations from a perfect frequency-response curve?

23. If a sweep generator and oscilloscope are available, hook them up to the receiver and check the frequency-response curve of the i-f amplifier. Compare the trace with the graph obtained for question 21.

13

Servicing the Picture Chain: Further Notes on AGC Systems

Quick Check. Symptoms of overloading such as a too contrasty or negative picture, or loss of sync are usually laid to faulty agc action. It may be that the agc system itself is good but not operating properly because of trouble elsewhere in the set. For example, a leaky coupling condenser in an i-f amplifier stage will place a positive voltage on the agc bus, and thus cause the overload condition. Because of this, there is no quick check for the agc section. The serviceman is brought here by servicing procedures set down in previous chapters on the picture chain.

Troubles in the agc-controlled stages up to and including the agc feed line or bus have been dealt with in the preceding chapter. It now remains to locate and clear the troubles in the agc system itself.

Function of Automatic Gain Control. Broadcast radio receivers use an automatic volume control (avc) circuit to prevent blasting when tuning in a strong local station. The rectified signal at the second detector is used to establish a negative voltage which is filtered and fed back to the r-f and i-f tubes as a bias voltage. Strong stations produce a high bias voltage which decreases the gain of the controlled tubes, thereby decreasing volume. Weak stations produce a lower bias voltage. This permits more gain in the r-f and i-f tubes, thereby increasing volume. As a result, strong and weak stations come in at about the same volume level within the limits of the maximum sensitivity of the receiver.

In radio work this is called automatic volume control, since it brings in all stations at about the same volume. Actually, it is automatic gain control, since the constant volume level is attained by changing the gain characteristics of the controlled r-f and i-f stages. The similar system used in television receivers is more correctly called automatic gain control (agc).

A block diagram of a television agc system is shown in Fig. 13-1. It is called simple agc to distinguish it from other agc systems which are described later in this chapter. The rectified signal at the detector produces a negative voltage which is proportional to the strength of the received signal. This voltage is filtered and fed back to the r-f tube in the tuner and the first and second i-f tubes as a bias voltage. The gain of these tubes is controlled by the bias so that the signal level in the video amplifier is substantially the same for strong and weak stations.

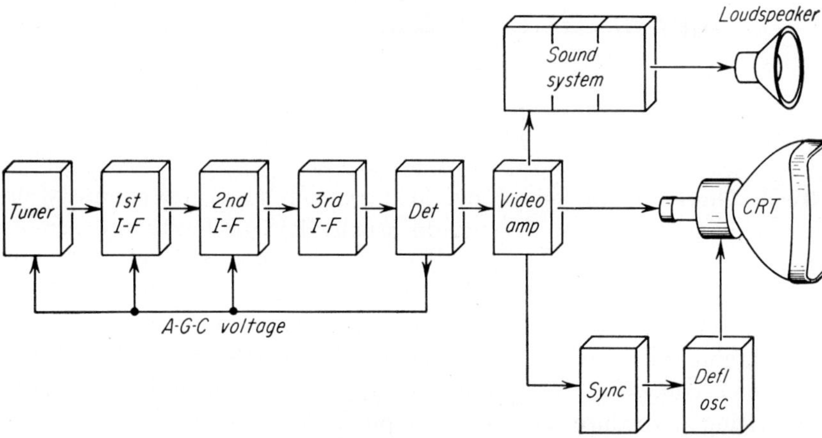

Fig. 13-1. Block diagram of a simple agc system.

In the television receiver, the constant signal level for strong and weak stations produces equal volume from the loudspeaker just as it does in the broadcast radio receivers. In addition, it produces the same degree of contrast in the picture, and a stable sync signal level to the sync section. This last item is of great importance, since correct operation of the sync section is dependent on correct signal level. The first sign of improper agc action is unstable syncing of the picture.

How the Control Voltage Is Developed in a Simple AGC System. The schematic diagram of a video-detector circuit is shown in Fig. 13-2. The input signal is obtained from the last i-f transformer *T*-4, and is an r-f voltage modulated by the sync pulses and video information. The waveform of the input signal is shown directly above transformer *T*-4 in the diagram. The diode detector is connected in such a way as to conduct on the negative phase of the input signal. Its output is therefore the negative half of the modulated r-f input signal. Next comes the i-f filter circuit consisting of *L*-4 and *C*-4. This removes the i-f pulsations, leaving the negative-going composite signal of sync pulses and video information across the diode load *L*-13, *R*-13. This is fed to the video amplifier, which in turn feeds the CRT and sync section. It is also fed to the agc filter, consisting of resistor R-12 and condenser C-12. Here condenser C-12 becomes charged at the peak negative voltage of the sync signal and has a large enough capacitance (0.25 μf) to maintain this potential until the arrival of the next sync pulse. The potential across condenser C-12 is therefore a negative d-c voltage which feeds the agc bus.

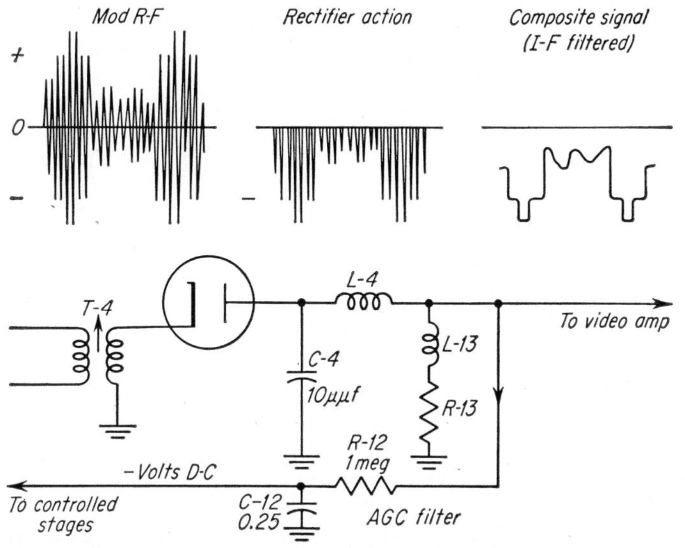

Fig. 13-2. Video-detector and filter circuits to develop the agc voltage

When the receiver is tuned to a weak station, it tends to produce a low-amplitude signal at the detector input, thereby creating a low negative voltage on the agc bus. The grid circuits of the controlled r-f and i-f tubes are returned to the agc bus as shown in Fig. 13-3. When these tubes are operated at a lowered bias from the agc bus, their gain is increased. This increases the signal level at the detector.

Similarly, when the receiver is tuned to a strong station, there is a high amplitude signal at the detector input. This time a high negative voltage is developed on the agc bus, which when fed back to the controlled tubes, decreases their gain. Thus the composite signal fed to the video amplifier remains substantially the same for strong and weak stations within the gain possibilities of the controlled r-f and i-f tubes.

In the diagram of Fig. 13-3, resistor R-5 and condenser C-5 provide a decoupling filter for the second i-f tube, and the resistor R-2, condenser C-2 combination serves a similar function for the first i-f tube. The r-f tube is on a separate chassis in the tuner, and so its decoupling filter is a two-section unit consisting of condensers C-11 and C-3 and resistors R-11 and R-3. Condenser C-11 is a component known as a feed-through capacitor. It consists of a metal cylinder attached to the tuner chassis. The feed wire runs through this cylinder and is insulated from it by a length of ceramic tubing. The wire and cylinder form the condenser plates and the tubing the insulator or dielectric. Feed-through capacitors are commonly used where it is desired to filter a lead as it enters a chassis or subchassis. They rarely cause any servicing difficulty.

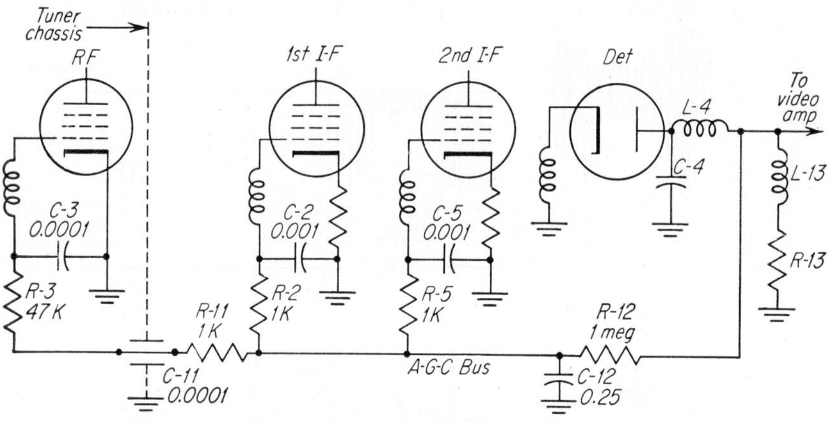

Fig. 13-3. Typical agc bus line in a simple agc system.

The diagram of Fig. 13-3 is typical for simple agc systems. In actual receivers there will be some differences in the values or arrangement of the decoupling filter. In some cases, the second i-f tube is not controlled. In almost all cases the last i-f tube is not controlled. Occasionally, a circuit will be found using two separate diodes or germanium crystals, one for the video detector, and one for the agc detector.

Troubles Caused by Faulty AGC Action. Assume a condition where the voltage developed on the agc line is just slightly less than it should be. The controlled tubes will operate at slightly increased gain. Weak stations in the area will come in as well as before, if not better, because of the increase in gain. On strong stations, however, the increased amplitude of the signal causes clipping of the amplitude peaks by limiting action in one of the tubes in the picture chain. The amplitude peaks are the sync signals, and when these are clipped off, the result is unstable sync. It is for this reason that the first trouble-shooting check for an unstable sync condition is to try operation on a weak station or to decouple the antenna. In either case, if sync stability is improved at the reduced signal level, the trouble is probably in the agc system.

Now consider the possibility of a larger loss of control voltage on the agc line. The increased signal that results will overload one or more tubes in the picture-signal chain. The symptom will vary depending on the degree of overload. One is called an overload or too contrasty picture where gray registers as black. A further advanced condition of overload is called negative picture, or reversed picture, where black registers as white and vice versa as in a photographic negative. The symptom may also be no picture. This happens when the high signal level at some tube in the picture chain develops a grid-leak bias high enough to block the tube completely, stopping the signal at that point.

Servicing the Simple AGC System. The voltage on the agc line is simply a supply voltage for the controlled r-f and i-f grid circuits just as the voltage on the B plus line is a supply voltage for their plate and screen circuits. Therefore, the i-f amplifier could not be serviced without taking into account the possible faults in the agc line. The procedure already set down in the previous chapter for servicing the i-f amplifier for no picture, negative picture, and unstable sync will have found most of the faults in a simple agc system. We shall quickly review these steps now to see how the agc system is affected, and go on to find any fault not previously covered. When servicing for no picture, remember that we do not know whether the break in the i-f chain is caused by a defective i-f component, resulting in no voltage on the agc bus, or whether no voltage on the agc bus has caused blocking in an otherwise normal picture chain.

A partial diagram of the picture chain which includes a simple agc system is given in Fig. 13-4. The first step in the procedure is to replace the i-f tubes. Obviously, a dead tube will stop the picture signal. There

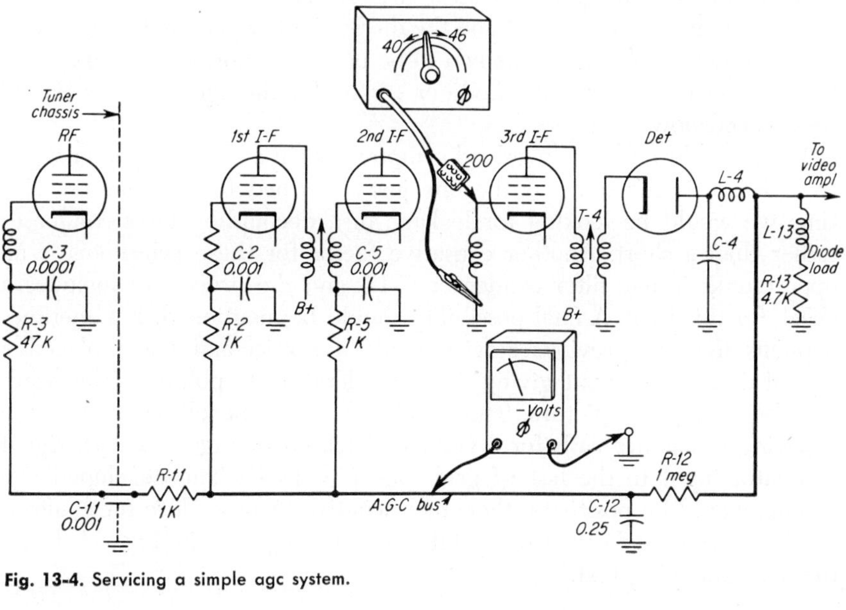

Fig. 13-4. Servicing a simple agc system.

is another tube trouble which might result in a stoppage. A tube with a small gas content or grid emission will cause a slight positive grid voltage. This will reduce the agc bias possibly to a point where the resulting overload will block a tube. Replacing the tubes eliminates these potential sources of trouble. The next step is to tune the receiver to an unused channel and make a voltage check on the agc bus. The normal voltage is zero, or a slightly negative reading caused by contact bias. Any positive voltage shows leakage into the agc bus through an i-f transformer or coupling condenser. A high negative reading is significant if the system is of the keyed agc type. This will be described later.

The next step is to feed a modulated i-f test signal to the last i-f grid, and measure the voltage produced on the agc bus. This step is illustrated in Fig. 13-4. Note that the last i-f stage is not controlled by the agc bus. It is therefore in working order regardless of the condition of the agc system.

If the generator produces an agc voltage of approximately −1 volt, we know that the agc system is working, and the fault is now definitely an inoperative first or second i-f stage. If the agc system fails to produce this voltage, it is either open or shorted.

The short is checked for by resistance reading from the bus to ground. The normal reading is 1 meg, through agc filter resistor *R*-12 and the diode load. A lower reading shows a short in one of the agc decoupling filter condensers: *C*-11, *C*-2, *C*-3, *C*-5, or *C*-12. These are checked.

If the resistance reading on the agc bus reads open rather than its normal value of 1 meg, filter resistor *R*-12 is open. The i-f filter consisting of coil *L*-4 and condenser *C*-4 and the diode rectifier need not be checked, since the detector circuit is known to be operating normally. However, if the receiver uses a separate diode or crystal for the agc line, this should be checked now.

The procedure for the overload symptoms such as loss of sync or negative picture is the same as the one just described for no picture. This time we would be looking for leakage in a decoupling filter condenser rather than a short. Another causative factor for these symptoms is an open circuit in agc filter condenser *C*-12, and the procedure includes a check for this item. A final possibility should be mentioned. It sometimes happens that, in previous service work, a serviceman has replaced a crystal detector without giving due consideration to polarity. The result may be a negative picture. This condition would be obvious when the servicing procedure calls for a voltage check on the agc bus with signal generator input to the last i-f grid (see Fig. 13-4). The developed agc voltage would be positive rather than negative. A procedure for confirming the correct polarity for crystal detectors is given in Fig. 11-6 and the accompanying text.

No new servicing techniques are required for the simple agc system. However, the keyed system is somewhat more complicated and will require additional checks. The keyed system will be described now.

KEYED AGC SYSTEMS

Simple agc systems have certain disadvantages. For example, a strong noise pulse would place a high bias on the agc bus, temporarily weakening the signal. The effect would remain for a perceptible length of time, because the noise pulse would charge the fairly large capacitance of the agc filter condenser *C*-12 and would take a long time to discharge through the high resistance of filter resistor *R*-12. This slow-acting agc system causes another type of picture defect especially prevalent in areas having considerable airplane traffic overhead. As an airplane is traveling near the line between transmitting and receiving antennas, it causes a series of multiple ghosts which periodically strengthen and weaken the signal arriving at the receiving antenna. The slow-acting agc system cannot change gain fast enough for these variations, and the picture is seen to strengthen and weaken while the airplane is moving in this vicinity. This symptom is known as airplane flutter.

The simple agc system could be made to act faster by reducing the capacitance of the agc filter condenser, but this would allow low-frequency video and sync currents on the agc line with regenerative or degenerative effects. A faster-acting agc supply must be obtained through a different method. This is the keyed agc system.

A block diagram of a keyed agc system is shown in Fig. 13-5. Here, the control voltage in the agc system is obtained from an extra stage

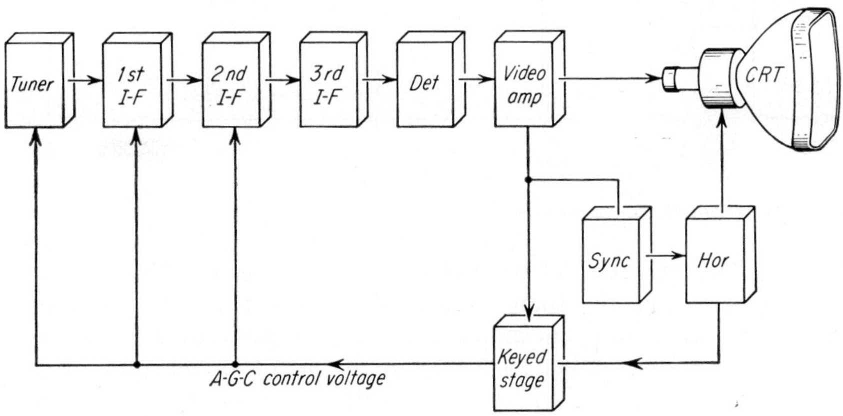

Fig. 13-5. Block diagram of a keyed agc system.

called the keyer or keyed stage, instead of from the detector tube, as in the simple agc system. In the keyed stage, the composite signal from the video amplifier and a pulse from the horizontal system combine to produce a control voltage which is fast-acting and so overcomes the disadvantages of the simple agc system.

How the Control Voltage Is Developed in the Keyed AGC Stage. Examine Fig. 13-6*a*, which shows the plate-cathode circuit of the keyed agc stage. Winding *W*-3 is a special agc winding on the horizontal output transformer which develops a potential of at least 300 volts peak

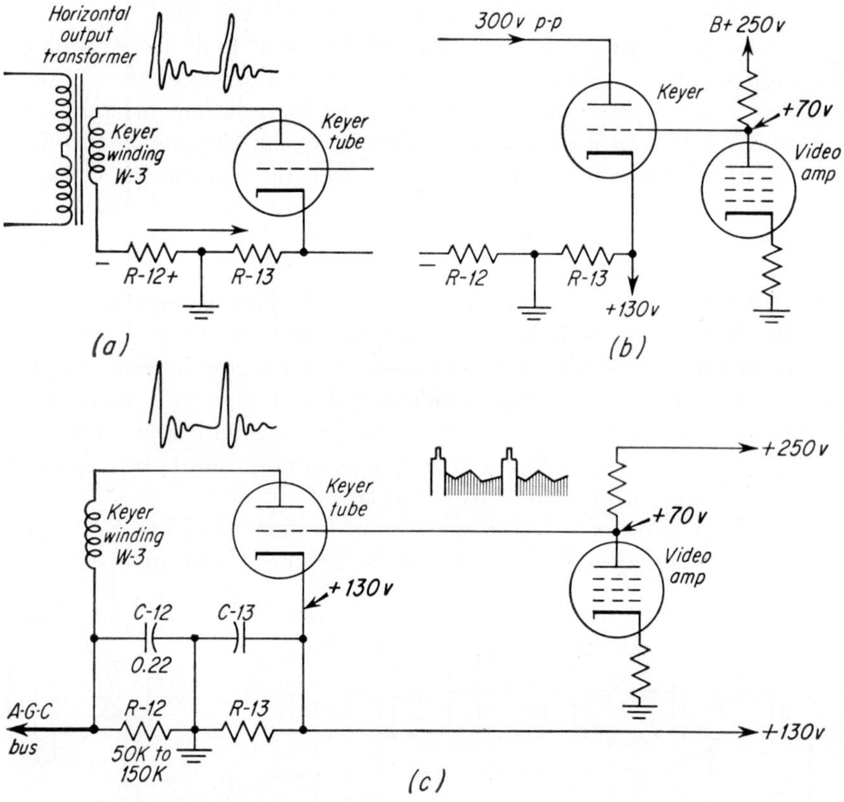

Fig. 13-6. How the keyed agc system works.

to peak. The waveform of this voltage is given over the keyer plate in the diagram. Note the sharp positive spikes which occur at the horizontal rate. As the diagram shows, this voltage is fed to the plate-cathode circuit of the keyer tube. Disregarding the action of the grid for the moment, the keyer tube will conduct on the positive spikes just like any diode rectifier. Resistor *R*-12 and *R*-13 in the cathode circuit acts as the diode load. The current through these resistors is a series of sharp d-c pulses at the horizontal rate. The junction of resistors *R*-12 and *R*-13 is grounded. Observe the direction of current flow through the resistors, and note that keyer conduction currents make the ungrounded end of resistor *R*-12 negative with respect to ground. This is the No. 1 key of the keyer circuit.

Next examine Fig. 13-6*b*. This shows the control-grid circuit of the keyer tube. It is directly coupled to the plate of the video-amplifier tube. In a no-signal condition this places the grid of the keyer tube at +70 volts, the normal voltage at the video-amplifier plate. Since the cathode is connected to the +130-volt bus in the low-voltage power supply, the control grid is −60 volts with respect to its cathode. At this bias voltage the tube is cut off.

When a positive-going composite signal is present at the plate of the video amplifier with an amplitude close to 60 volts, the bias is overcome, and the keyer tube conducts. This amplitude is reached only by the sync pulses which are the positive peaks of the composite signal. This is key No. 2 of the keyer tube.

Examine the full circuit in Fig. 13-6*c*. Note that the tube can conduct and produce a negative voltage across resistor *R*-12 only when both keys are closed, that is, when the plate is made positive by the sharp horizontal spike (key No. 1), and the grid is made more positive by the horizontal sync pulse (key No. 2). Since these pulses are simultaneous for a synced-in picture, a series of negative pulses is developed across resistor *R*-12 at the horizontal rate. The conduction current of the keyer tube is dependent on the supply voltages. The horizontal spike on the plate is about the same whether there is a station on the air or not. The grid voltage depends on the station. It will be at cutoff for no signal. Since strong stations tend to produce larger composite signals than weak, strong stations will make the grid more positive than weak stations. This gives more conduction for the strong station, resulting in higher negative pulses across resistor *R*-12. Resistor *R*-12 ranges from 50,000 to 150,000 ohms in various receivers, the value chosen so that the conduction currents will always develop a voltage suitable for its agc system. The voltage pulses across resistor *R*-12 are filtered by condenser *C*-12 and applied to the agc bus. Note that strong stations develop a high voltage to cut down the gain of the controlled tubes, and weaker stations produce a lower voltage to allow more gain.

Agc filter condenser *C*-12 is 0.22 μf, about the same size as is used in the simple agc system. However, in the simple circuit, the condenser is discharged by a 1-meg resistor, making the circuit slow-acting. The discharge circuit in the keyed system is resistor *R*-12, which discharges the condenser much faster because of its lower resistance (50,000 to

150,000 ohms). This gives the faster agc action to the keyed system, which overcomes the disadvantages of the simple system.

Condenser *C*-13 and resistor *R*-13 in the diagram of Fig. 13-6*c* may not be found in an actual receiver. This is because their function of completing the cathode circuit of the keyer tube is taken over by the low-voltage power-supply circuit.

Basic Keyed AGC Circuits. Two variations of a basic keyed agc system are shown in Fig. 13-7 for service analysis. In the simplified circuit of Fig. 13-6*c*, some components normally found in actual receivers are omitted because they are unnecessary for the explanation of how the control voltage is developed. These have been added to produce the basic diagram of Fig. 13-7*a*. They include resistor *R*-8 and condenser *C*-8, which together with condenser *C*-12 make up the agc filter. In addition,

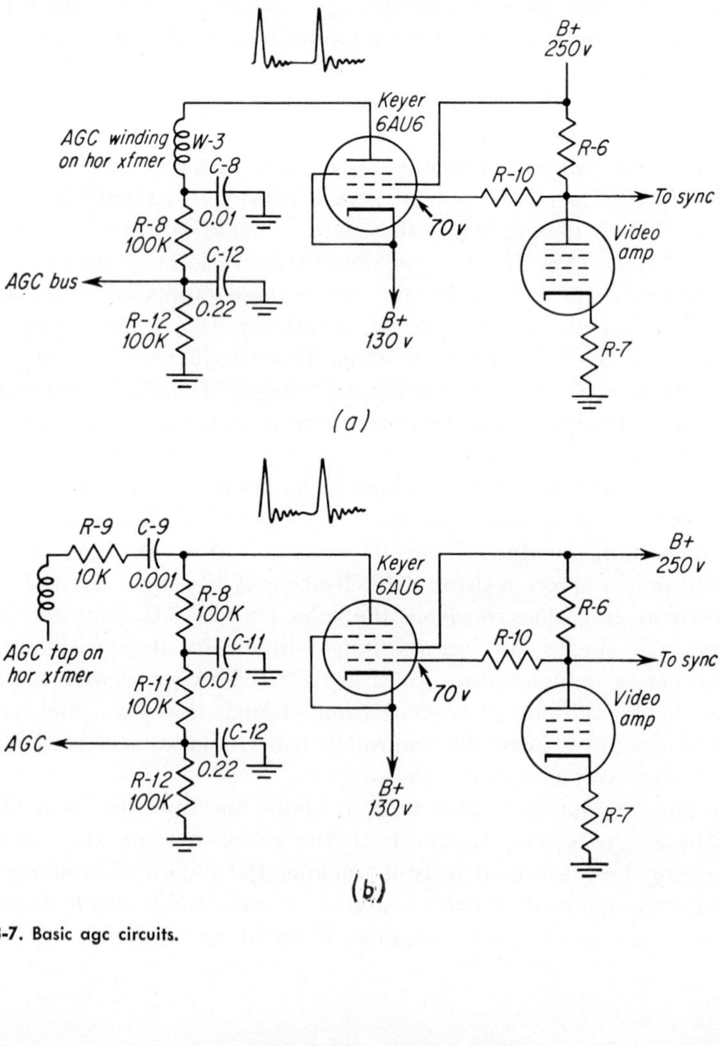

Fig. 13-7. Basic agc circuits.

actual receivers use one and sometimes two resistors in the direct-coupling feed line between the video-amplifier plate and the keyer grid. This is represented by resistor *R*-10 in the diagram. The same circuit is often used to feed the sync section, and resistor *R*-10 represents the de-coupling between the keyer and sync stages.

The circuit of Fig. 13-7*b* shows a different method of feeding the horizontal pulse to the plate of the keyer tube. The pulse is fed in through resistor *R*-9 and condenser *C*-9. This type of circuit is used where the agc pulse voltage is obtained from a tap on the horizontal output transformer. This makes a parallel feed to the keyer tube, rather than the series feed of Fig. 13-7*a*. In this case, the agc voltage is developed between the plate of the keyer tube and ground.

Note that the tube used in both circuits is a pentode rather than the triode of Fig. 13-6. This is usual in keyed agc circuits although triodes are sometimes found. The 6AU6 pentode and the pentode half of the 5BT8 are the most commonly used tubes in keyed agc systems. Normal voltages are indicated on the diagram. The figures given are typical, but there is a wide variance in this regard in different receivers, and it is best to use the manufacturer's service notes for more accurate data. In receivers using transformers for filament supply, the keyer tube is often connected to a separate ungrounded winding so as to reduce the cathode-heater potential and the possibility of cathode-heater breakdown.

Servicing Keyed AGC Systems. In the preceding chapter, you were told to investigate the agc system under two sets of conditions. These are:

1. There is a high negative voltage on the agc bus in a keyed agc system. Checks have proved that the voltage is not caused by oscillation in the i-f amplifier.
2. The agc bus has been found free of shorts and leakage to ground, and yet the full generator output, when connected to the last i-f grid, does not develop any agc voltage.

We shall take up the condition of a high negative voltage first.

Servicing for a High Negative Voltage on the AGC Bus in a Keyed System. Refer to Fig. 13-7 and note that, in either system, a high negative voltage developed across load resistor *R*-12 is most probably caused by too heavy a conduction current through the keyer tube. For the condition of no input signal to the receiver, the tube should be at cutoff. Signal current at the keyer grid causes plate current to flow. Because of the high resistance in the circuit, a very small plate-current flow is sufficient to develop the normal bias voltage on the agc bus. Obviously, the high agc voltage at no input signal to the system is caused by abnormally high plate current in the keyer tube. This in turn may be caused by a defective tube or by improper operating potentials.

First change the tube with one known to be good. If this does not restore normal operation, voltage analysis is the next step. Too much plate current in any tube is usually caused by too high a voltage on the plate or screen or by too low a bias on the control grid. Too high a plate voltage is not likely, since this is a pulse from the horizontal system, which is probably working normally because of the presence of the normal raster. The same applies to the screen voltage which is obtained from B plus bus or boost line, neither one of which is likely to be excessive. This leaves the control-grid bias as the likely offender.

The manufacturer's service notes may give the normal voltage reading at the keyer control grid, but do not depend on this reading, since there will be large variations due to signal conditions at the time of the test. The basic diagram of Fig. 13-7 gives a normal reading of 130 volts for the keyer cathode, and 70 volts for the keyer grid. This is for a no signal input condition, making the control grid 60 volts more negative than the cathode, and the tube is cut off. Then a 60-volt peak-to-peak composite video signal will drive the keyer tube to conduction on the positive sync peaks. If the cathode is below 130 volts or the grid is above 70 volts, the cutoff balance is upset, and the tube may be conducting with or without the input signal.

Measure the cathode voltage in the defective receiver, and compare it with the voltage given in the manufacturer's service notes. If the reading is incorrect, trace back along the cathode line to the source. If the cathode reading is normal, the trouble must be an abnormally high voltage on the control grid of the keyer tube.

Because of the direct coupling between the keyer grid and the video-amplifier plate, there will also be an abnormally high voltage on the video-amplifier plate, and this is what is causing the trouble. Abnormally high plate voltage on the video-amplifier tube may be caused by weakness of the tube or incorrect resistance values in its cathode or plate circuits. The most common offender is the tube; so try a new video-amplifier tube. If this does not restore the proper voltage balance, check the cathode and plate resistors of the video amplifier. In the basic diagram of Fig. 13-7, these are resistors *R*-6 and *R*-7. Check the equivalent resistors in the set you are working on.

Servicing a Keyed AGC System That Does Not Develop a Control Voltage. Let us review the checks described in previous chapters which bring us to this point. The i-f–video–keyer chain is shown in Fig. 13-8. The symptom is normal raster, no picture, or one of the overload symptoms such as negative picture.

The signal generator is set for maximum output and is connected to the last i-f grid. The frequency control is wobbled through the intermediate frequency of the receiver, and the production of bars on the face of the picture tube proves the set is operative to this point. When the generator is connected to the tuner output, the nonappearance of bars places the trouble in the i-f–agc system. Then the receiver is tuned to an unused channel, and the voltage on the agc bus is measured. When this reads a normal 0 to −1 volt for no signal input, the generator is

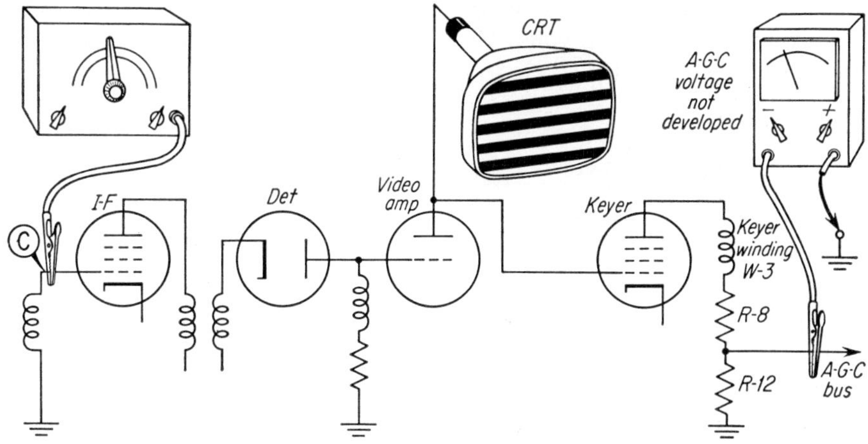

Fig. 13-8. Test conditions which point to improper keyer action.

reconnected to the last i-f grid. Since the last i-f grid is not controlled from the agc bus, the full output of the generator should develop a voltage on the agc bus. When the voltmeter remains at its 0 to −1 volt reading, either the keyer stage is not producing the voltage or the voltage is being shorted out at the agc bus. The last check is to measure the resistance on the agc bus to ground. When this shows the normal 50,000 to 150,000 ohms it indicates that there are no shorts or leakage and that the trouble is in the keyer stage.

The trouble can be a bad keyer tube, improper operating potentials, or the failure of the keying pulses to arrive at either the plate or grid of the keyer tube. These possibilities are checked in accordance with the following procedure.

Step 1. Change the keyer tube. The receiver is kept in the setup of Fig. 13-8, with bars on the picture tube, and the voltmeter on the agc bus. If the new tube causes the agc voltage to increase, the original tube is bad. Check operation to make sure the set is now operating normally. If the new tube causes no change, go on to step 2.

Step 2. Check the keyer plate for the appearance of the horizontal pulse (key No. 1). The pulse at the keyer plate has a peak-to-peak value of 300 volts or more, and the waveshape shown in Fig. 13-9. The waveshape and normal peak-to-peak voltage are often given in the service notes for the receiver.

To calibrate the oscilloscope for a 300-volt peak-to-peak reading, set its horizontal sweep frequency to 30 cps, reduce its vertical gain to zero, and connect its vertical leads to an a-c outlet. Then adjust the vertical gain control to bring the 2-cycle trace within the horizontal tape markers on the face of the oscilloscope screen. The peak-to-peak value of the a-c line is 2.8×117, or 328 volts. Mark this setting for future reference on the vertical gain control.

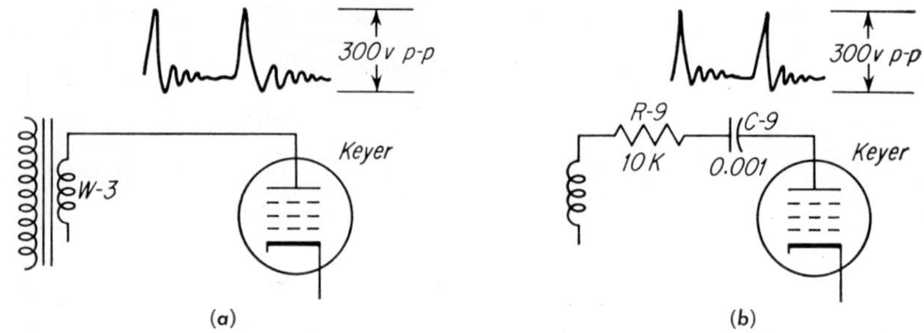

Fig. 13-9. Waveform at the plate of the keyer tube. (*a*) Keyer winding on flyback transformer. (*b*) Keyer tap on flyback transformer.

To check the waveform at keyer plate, set the oscilloscope to the markers used for observing the horizontal trace, and connect the vertical probe to the keyer plate through a 0.05 condenser. The waveform can now be observed and its amplitude judged by the height of the trace between the markers on the screen. The waveform is the same for both keyer-plate circuits of Fig. 13-9. If the trace is missing or abnormal, check back along the keyer-plate wiring to the flyback transformer.

In the separate-winding type of Fig. 13-9*a*, the trouble is probably an open keyer winding. If the break in the winding cannot be found and repaired, the flyback transformer must be replaced with an exact duplicate. The keyer winding is often part of the width coil. In this case, only the pulse-width coil need be replaced. Replacement units of this type are available. In either case, when installing the replacement, it is necessary to check the polarity of the keyer winding. This is done with an oscilloscope. The vertical lead of the scope is touched to one lead of the winding with the set in operation. If the trace is positive-going, this is the correct polarity for connection to the plate of the keyer tube.

In the keyer circuit of Fig. 13-9*b*, the trouble is probably an open in resistor *R*-9 or condenser *C*-9. If condenser *C*-9 is at fault, make sure the replacement condenser has a rating of at least 1,600 volts. These condensers are subjected to sharp pulse voltages and break down easily.

If the normal waveform is observed, go on to step 3.

Step 3. Check the keyer grid for the appearance of the signal.

Under normal operating conditions, the signal at the keyer grid is the composite video signal. When the symptom is no picture or reversed picture, the composite signal at the keyer grid and consequently the agc voltage cannot be correct. Stages in the picture chain before the last i-f stage are controlled by the agc voltage and therefore cannot be adequately checked. From the last i-f stage on, the chain is uncontrolled and can be checked regardless of the condition of the agc line. This is the basis of these checks in the picture chain illustrated in Fig. 13-8. With signal generator input to the last i-f grid, the normal signal at the video plate is the amplified 400-cycle modulation note of the generator. This is present, as we know by the appearance of the modulation bars across the face of the picture tube. It now remains to be seen if the modulation signal is reaching the keyer grid. This check is illustrated in Fig. 13-10.

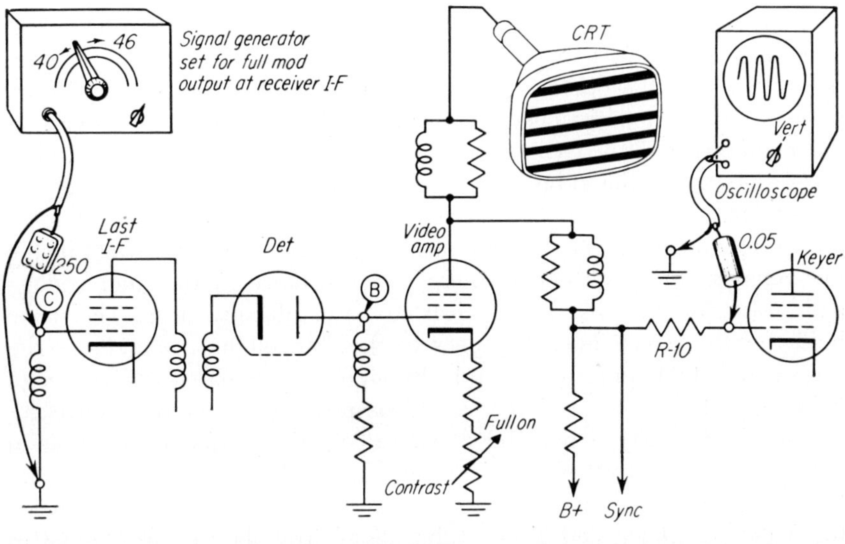

Fig. 13-10. Normal waveform at the keyer grid under test conditions of a modulated r-f signal fed to the last i-f grid.

With the oscilloscope set at the vertical rate, the 400-cycle modulation signal appears as a series of 10 or more sine waves. The vertical probe of the oscilloscope connected to the video-amplifier plate will show this 400-cycle trace. Now shift the vertical probe to the keyer grid. The nonappearance of the trace shows that the keyer tube is not receiving its normal unlocking signal (key No. 2). Trace back along the keyer grid line to the video plate, looking for an open resistor. The diagram of Fig. 13-10 shows a single resistor *R*-10, but the actual receiver circuit may contain more components. This line also connects to the sync section, and there may be a decoupling network.

If the sine wave appears at the keyer grid with approximately the same amplitude as at the video plate it may be assumed that the coupling line is good, and a normal video pulse, if present, can arrive at the keyer grid. Go on to step 4.

It should be mentioned at this point that, although the sine-wave trace does not have the same waveform or amplitude as the normal composite video signal which reaches the keyer grid, practical experience with many receivers shows that the 400-cycle note provides sufficient grid excitation to unlock the keyer tube and develop 0.5- to 2.0-volt agc bias.

Step 4. Check operating potentials on the keyer tube. Refer back to Fig. 13-8. We now have the condition of no voltage on the agc bus, a known good tube in the keyer socket, a normal horizontal pulse on the keyer plate, and a normal signal capable of arriving at the keyer grid. The only possibility of stopping operation of the keyer tube is incorrect potentials on the cathode, grid, or screen. Since the normal operating potential places the operation of the tube near cutoff, these voltages are critical and want to be checked and held close to tolerances. Refer to the manufacturer's service notes, and if normal voltages are not indicated directly at the tube prongs, check back along the cathode and screen lines to their supply source, where the voltage reading is more apt to be indicated. The cathode usually goes to B plus 130 volts in the low-voltage supply system. If there is a resistance in the line, voltage and resistance analysis should clear up any difficulty here. The same holds for the screen line, if it goes to the 250-volt bus. Sometimes the screen goes to B plus boost, and although the normal raster seems to indicate that the boost voltage should be normal, it should be checked, because boost voltage may be down sufficiently to stop conduction in the keyer tube, and yet provide sufficient high voltage to light the picture tube. Finally, as mentioned before, because of direct coupling between the video amplifier and keyer tube slight irregularities in the video stage can affect the control-grid voltage on the keyer tube. Remember that the normal grid voltage for the keyer tube should be about 60 volts below its cathode voltage for conditions of no input signal.

Delayed AGC. The servicing procedure just described is formulated for a basic or abbreviated keyed agc system. Actual circuits used in receivers usually include some refinements like delay voltages, clamps, and manual controls in the agc system. These variations do not interfere with the basic servicing procedure, which may be applied to all of them. The difference, however, may lead to some confusion in tracing circuits, and so some typical agc systems will be examined to see how the servicing procedure will be applied.

The keyed agc stage of the Magnavox 106C chassis is given in Fig. 13-11, as a typical example of a keyed agc system that incorporates a delay voltage. This is used to delay agc action until the signal reaches a certain predetermined strength. In the ordinary agc system, even a weak

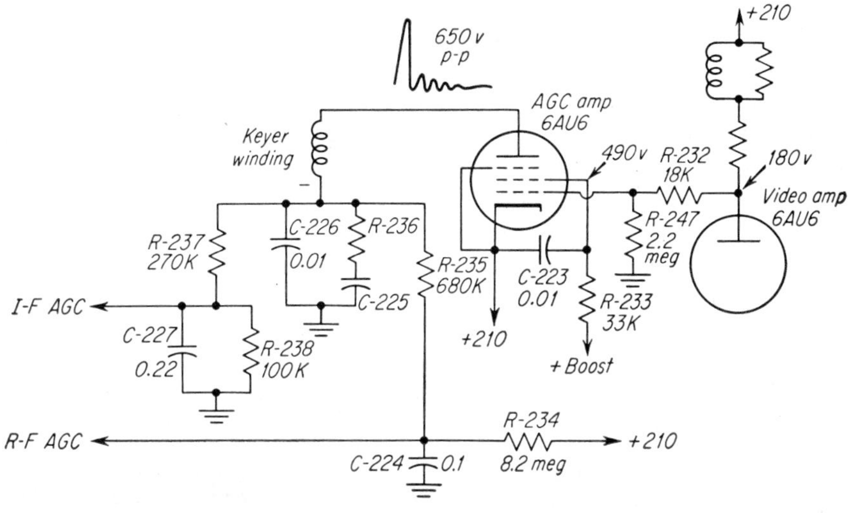

Fig. 13-11. Keyed agc circuit in Magnavox 106C.

signal will produce some agc voltage. This tends to reduce the gain of the controlled stages, and a weak station needs all available gain. The matter is of importance in fringe areas, and so in some receivers, the control voltage is not applied until the signal reaches a certain level. This is accomplished by applying a small positive voltage to the agc line through a high resistance. The developed agc voltage cannot therefore be applied until it overrides this small positive voltage. The small positive voltage is called a delay voltage, and it is usually applied only to the r-f tubes in the tuner.

In the circuit of Fig. 13-11, the agc voltage starts at the negative end of the keyer coil. Note the combined filter and agc load consisting of resistors *R*-237 and *R*-238 and condensers *C*-226 and *C*-227. The agc bus for the i-f tubes is taken from the junction of these resistors. The negative end of the keyer coil also feeds agc voltage to the r-f tubes through resistor *R*-235. However, the same point is fed a positive voltage through resistor *R*-234. The circuit constants are worked out so that the r-f bias will be approximately zero for weak stations, thereby allowing maximum gain in the r-f section. On stronger stations, the developed agc voltage overrides the positive potential through resistor *R*-234, and the r-f tubes are biased accordingly. Note further the high pulse voltage fed to the plate of the keyer tube, which is called AGC AMP. in the diagram. This is due to the design of the keyer winding and is the reason for the higher voltages on the cathode and screen. Also note resistor *R*-247 added to the grid circuit of the keyer tube. This will drop the grid voltage slightly.

All the basic tests will operate satisfactorily on this circuit with due reference to the different voltages used. The delay voltage causes no interference with the checks beyond adding some extra components to check.

AGC Clamp and Automatic Gain Control. In the delay system just described, a small positive voltage is fed to the r-f section of the agc bus. When the receiver is tuned to an unused channel there is no developed agc voltage to overcome the positive bias, and the r-f tube may draw excessive plate current. Some manufacturers include a circuit known as a "clamp" to prevent this.

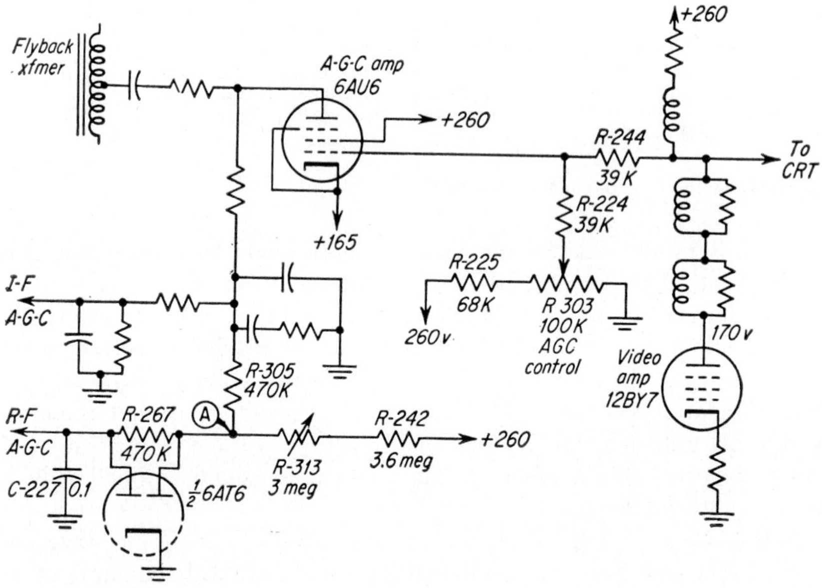

Fig. 13-12. Keyed agc system in the Dumont RA-301 chassis.

Such a circuit is shown in Fig. 13-12 which gives the keyed agc system used in the Dumont RA-301 chassis. In this circuit the plate of the agc tube is fed from a tap on the flyback transformer so that the control voltage is developed between plate and ground. The agc feed line for the i-f stages is conventional. The control voltage is fed to the r-f line through resistor *R*-305. Feeding the same point, labeled *A* in the diagram, is a positive voltage from B plus through resistors *R*-313 and *R*-242. A diode is connected to point *A*. In cases of no input signal, point *A* becomes positive because of the voltage from the B plus source. However, the diode conducts on a positive potential, shorting out this voltage. The delay voltage operates in the usual way, but diode conduction will keep (clamp) the r-f bus at zero potential. The diode used is the diode section of the first audio tube. In this circuit, resistor *R*-313 is made variable so as to give a control of the delay voltage. Since the r-f stages operate at full gain until the signal is strong enough to override the delay voltage, the variable delay voltage permits a means of attaining more gain from the receiver if needed. The control is of particular importance in fringe-area work. It is adjusted for maximum signal strength on weak stations to just before the point where the reception is marred by snow.

This delay voltage control is not often found in receivers. A control which is found more often regulates the over-all agc action.

A control of the amount of signal fed to the keyer grid or the cathode or grid potentials of the keyer tube determines the operating point and amount of agc voltage developed for a given level of signal. In the circuit of Fig. 13-12, the manual control in the agc system combines two of these variables in one unit. Potentiometer 303 is so connected in the grid circuit of the keyer tube as to control simultaneously the amplitude of the video signal and the potential on the control grid.

The manual control in the agc system is usually set as follows: Set the control to a mid-range position and tune the receiver for a good picture on the strongest station in your area. The manual control in the agc system is then varied through its range. Toward one end, it allows heavy conduction in the keyer tube. This develops a high agc voltage and reduces gain in the controlled stages and the picture begins to lose contrast. Rotating the control toward the other end, the bias on the keyer tube is increased, its conduction decreased, and the agc voltage reduced. This allows more gain, and the contrast comes back. Continued rotation in this direction allows still more gain, until the system overloads. The first sign of overloading is loss of sync due to clipping by limiting action in one of the tubes in the picture chain. The control is turned up until the picture begins to slip and then is backed off until it once more becomes stable. Since this is done on the strongest station in the area, it allows maximum gain without overloading the receiver. The adjustment, of course, is best made in the consumer's home.

A keyed agc system of this type is serviced by the same procedure already set forth. An inoperative clamping diode may cause one complication. There will be no direct symptom, since the circuit is normally negative on a station and the diode is not conducting. Off-station operation allows a positive bias on the r-f tubes with consequent overheating. Frequent breakdown of r-f tubes in sets of this type calls for a check of the clamping diode. This is done by checking grid voltage on the r-f tube. If it is positive on an unused channel, the clamping diode is likely to be inoperative.

Keyed-pulsed AGC Systems. The circuits just discussed are sometimes called keyed, sometimes gated, and sometimes keyed-gated agc systems.

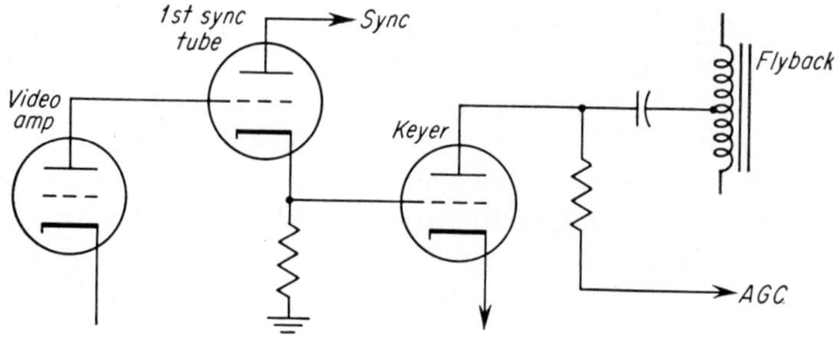

Fig. 13-13. Skeletonized diagram of a keyed-pulsed agc system.

There is another variation of the keyed agc system used mainly by RCA in some of their receivers, and called keyed-pulsed agc. A skeletonized diagram of the keyed-pulse agc system is given in Fig. 13-13, so that its operation may be understood without the complications caused by the presence of all the components. A complete diagram taken from an RCA KCS72 chassis is given in Fig. 13-14.

Referring to Fig. 13-13, the composite signal from the video amplifier is fed to the grid of the first sync tube, the plate of which feeds the rest of the sync section. The cathode of this same tube feeds the control grid of the keyer tube which then develops the agc voltage in the usual way. The first sync tube, therefore, acts as a cathode-follower type of amplifier between the video amplifier and the keyer tube.

From the serviceman's point of view, we now have another stage in the agc chain which may fail to deliver the proper signal to the keyer grid, and so it must be checked. The servicing procedure previously presented must be altered slightly to include the extra stage.

Since the circuit is quite common in RCA receivers and is also used in their color receivers, we shall quickly review the servicing procedure, making the extra provisions for the added stage. Refer to Fig. 13-14 for voltage readings and components which must be checked.

The agc system is proved inoperative when a modulated r-f signal at the last i-f grid produces bars on the picture tube but fails to develop agc voltage. Resistance from the i-f–agc bus to ground measures its normal value of 39,000 ohms (*R*-142).

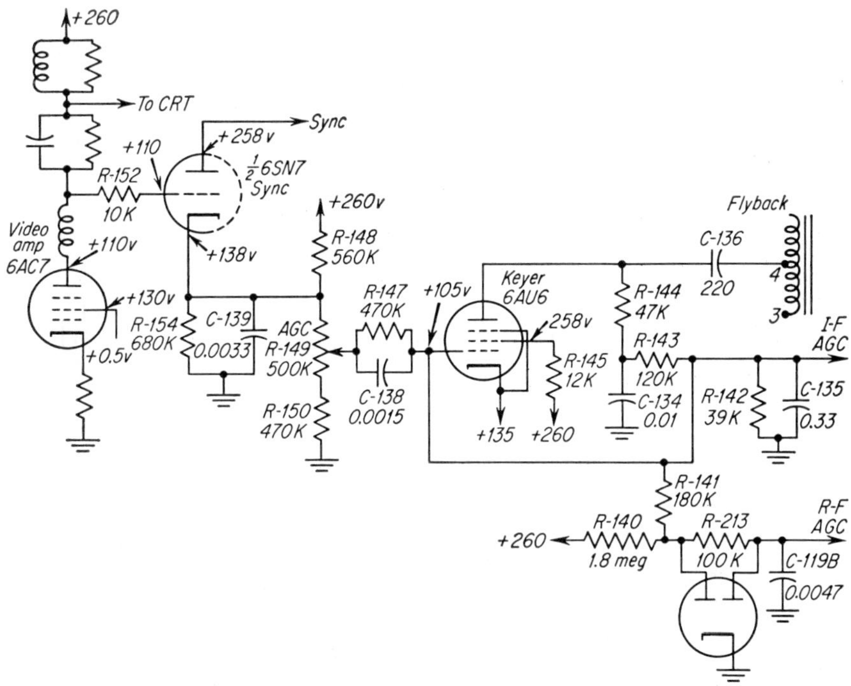

Fig. 13-14. Keyed agc system in RCA KCS72.

First change the keyer tube, and also the first sync tube. If this does not restore operation, check the keyer plate for the appearance of the horizontal pulse. Since the diagram does not give the peak-to-peak voltage which is normal for this point, look for a trace of at least 300 volts. If the pulse is not present, condenser *C*-136 is probably open.

If the normal pulse is present, check the grid of the first sync tube for the 400-cycle sine-wave trace of the modulation note of the signal generator. If this does not appear, resistor *R*-152 is open. If it is present, the trouble must be in the components and feed voltages to the sync and keyer tubes. Note that all pertinent voltages and resistance data are given on the diagram to aid in voltage and resistance analysis.

SUMMARY

Symptoms of overloading such as negative picture or loss of sync are usually laid to faulty agc action. While this is probably true, the trouble is not necessarily in the agc system itself but may be elsewhere in the set. Because of this, there is no quick check for the agc section. The serviceman is brought here after checks in previous chapters tell him that the agc section specifically is at fault.

These previous checks and the subsequent servicing procedure will have cleared up all troubles in a simple agc system, and some of the troubles in a keyed system. The procedure outlined in this summary is for the keyer stage itself.

Standard Circuit. The schematic diagram of a basic keyed agc system is given in Fig. 13-15 as the standard circuit. Normal voltages are indicated on the diagram.

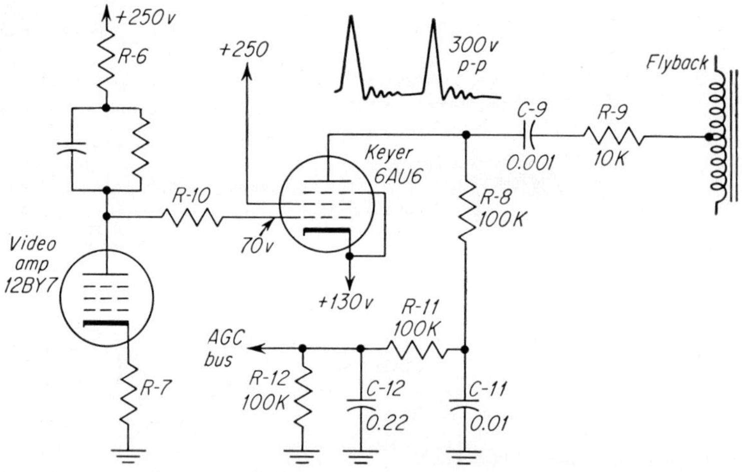

Fig. 13-15. Schematic diagram of a basic keyed agc system.

Servicing Charts. In the preceding chapter, when checking the i-f–agc system, the keyer stage was to be checked under two conditions:

1. There is a high negative voltage on the agc bus.
2. A test signal fed to the last i-f grid does not develop an agc voltage on the bus.

A servicing chart is given for each of these conditions. Reference in the charts is to Fig. 13-15.

Servicing Chart for a High Negative Voltage on the AGC Bus

Procedure	*Result*	*Trouble and Subsequent Check*
1. Replace keyer tube	Voltage on agc bus returns to normal	Original tube is defective
	Voltage on agc bus remains high	Go on to step 2

Servicing Chart for a High Negative Voltage on the AGC Bus (Continued)

Procedure	*Result*	*Trouble and Subsequent Check*
2. Make a voltage check of the keyer tube	Cathode voltage is low	Check cathode back to its source in the voltage-distribution system of the low-voltage power supply
	Grid voltage is high	Check for weak video-amplifier tube or incorrect value of cathode or plate resistors in the video-amplifier circuit

Service Chart for a Keyer Stage That Does Not Develop AGC Voltage

Throughout these checks, a signal generator is feeding a modulated r-f signal to the last i-f grid at the correct intermediate frequency to produce bars on the picture tube. A meter on the agc bus registers zero or slight negative. Previous checks have shown there is no short or leakage on the agc bus.

Procedure	*Result*	*Trouble and Subsequent Check*
1. Replace keyer tube	Agc voltage increases to —1 or —2 volts	Original tube is bad
	No change in agc voltage	Go on to step 2
2. Make oscilloscope check at plate of keyer tube	Horizontal pulse is missing or abnormal	Open keyer winding Open condenser C-9 Open resistor *R*-9
	Normal trace	Go on to step 3
3. Make oscilloscope check at grid of keyer tube (vertical rate)	Modulation trace (400 cps) is missing	Open resistor *R*-10
	Modulation trace is present	Go on to step 4
4. Make voltage check of keyer tube	Screen voltage is low	Check screen circuit back to its source in low-voltage power supply or boost circuit

QUESTIONS

1. Describe the function of automatic gain control (either simple or keyed) in a television receiver.

2. A radio receiver without automatic volume control would blast when tuning from a weak to a strong station. Early television receivers were not equipped with automatic gain control. Describe what might happen if such a receiver, adjusted for a good picture on a weak channel, were suddenly switched to a stronger channel.

3. A radio receiver without automatic volume control would lose volume when tuning from a strong to a weak station. Similarly, a television receiver without automatic gain control would lose volume and contrast when changing from a strong to a weak station. What other condition may manifest itself? Explain.

4. Starting with a symptom of raster normal, no picture, no snow, list the procedural checks in the picture chain which places the trouble in the agc system.

5. What are some of the disadvantages of the simple agc system that are corrected in the keyed agc system?

6. Describe the signals present at the plate of the keyer tube and the grid of the keyer tube. Explain why both must be present simultaneously to develop the agc voltage.

7. A high negative voltage measured on a keyed agc bus with the receiver on an unused channel can be caused by either of two conditions. Name them, and explain how you would determine which of the two conditions prevails.

8. A weak video-amplifier tube causes a symptom of insufficient contrast in a receiver with a simple agc system. The same tube placed in a receiver with a keyed agc system results in no picture. Explain.

9. Outline a procedure for checking a keyer stage which is not developing an agc voltage.

10. A test signal fed to the last i-f grid produces bars on the picture tube. Feeding the same test signal to the first i-f grid produces nothing, placing the trouble in the i-f–agc system. Either an i-f stage is inoperative, or the agc system is inoperative, causing blocking in the picture-signal chain. Explain how you can determine which system is at fault.

11. Describe a delayed agc system.

12. What is meant by clamp action on the agc line? Explain its operation.

13. Describe the function of a manual control in the agc system.

14. Why is the manual control in the agc system left in the mid-range position when working in the i-f–agc section or sync section of the receiver?

15. Describe a procedure for setting up a manual control in the agc system. Why should this adjustment be performed in the customer's home?

14

Servicing the Picture Chain: Tuners

The tuner in a television receiver is sometimes called the "front end." It requires no quick check for operation, since it is the last link in the picture-signal chain. If all previous quick checks showed normal operation, the tuner section is at fault. Of course, we must not overlook the possibility that the entire receiver is good and the trouble is in the antenna, which is the true beginning of the signal. However, this will not occur on the service bench, where the antenna condition is known. In field servicing, where this is a factor to be considered, checking the antenna installation is a routine part of the home-service call.

As mentioned in a previous chapter, the appearance of snow on the picture tube is a short cut eliminating the need for intermediate quick checks in the picture chain and making it possible to go directly to the tuner for the trouble. It will be remembered that snow is the result of noise pulses, and if the i-f amplifier, detector, and video sections can carry through these noise pulses, they are probably in good working order. The tuner is simply putting out more noise than picture signal.

The tuner in a television receiver is a complete subassembly chassis like one of those illustrated in Fig. 14-1, containing the r-f, mixer, and oscillator stages. It is usually mounted in a cutout section of the main chassis. Sometimes the mounting is outboard for cabinet arrangement, since the tuner controls must be accessible for front-panel operation of the receiver. In either case, the unsoldering of five or six connections and the removal of a few screws allows the entire tuner chassis to be removed for servicing or replacement.

Tuner repairs are complicated and time-consuming. When the trouble is obvious or easily found and repaired, like a defective tube, a misaligned oscillator slug, or dirty contacts, the repair is usually made in the shop. When the trouble is not immediately obvious, general practice is to invest in a complete replacement tuner rather than an unknown amount of bench time.

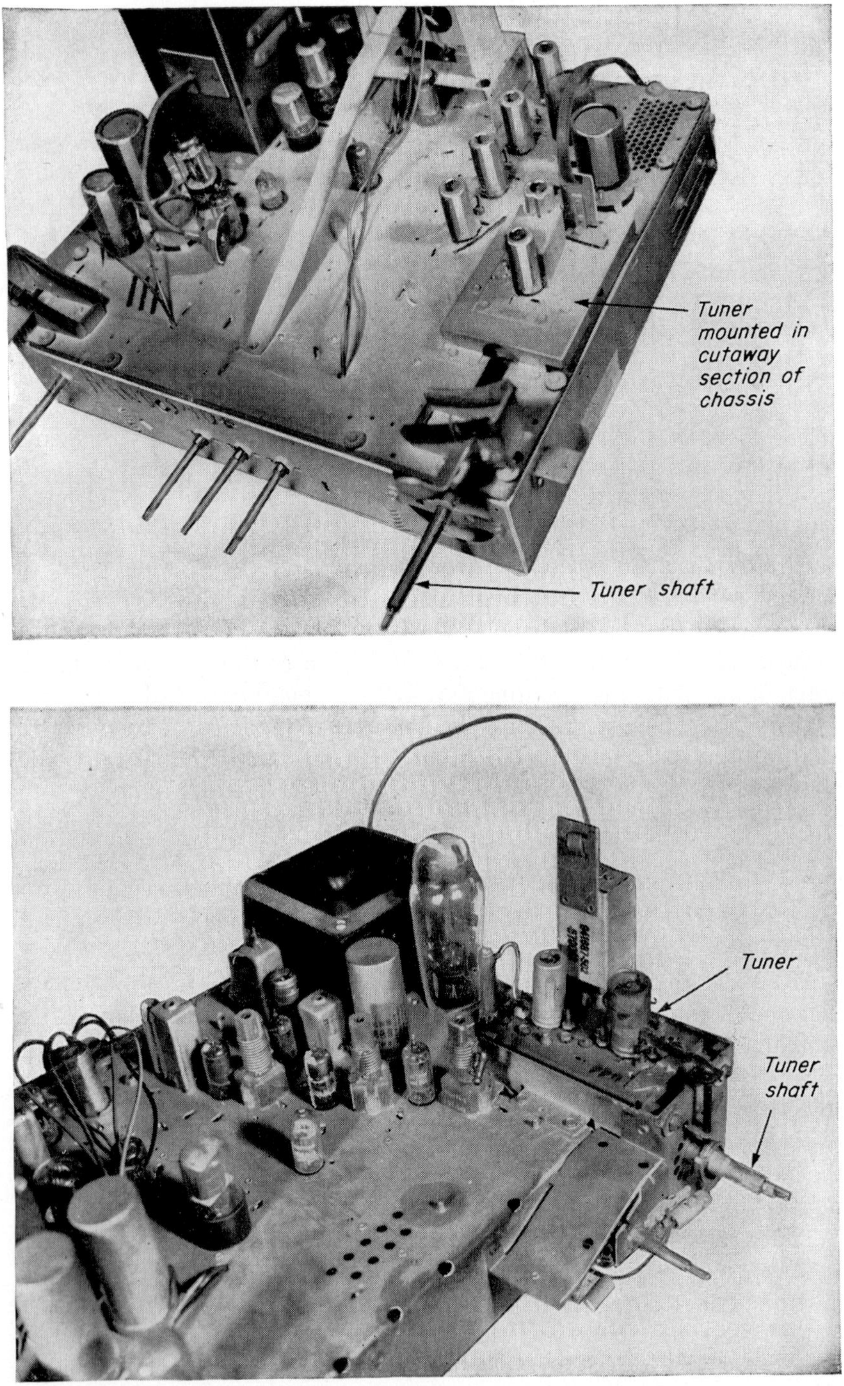

Fig. 14-1. Typical tuners.

The servicing information given in this chapter will describe servicing techniques as far as they are likely to be carried out in any shop. However, the beginning serviceman must be warned not to exceed his mechanical skill when disassembling a tuner to get at the defective component, because it is often necessary to return the old tuner intact in order to obtain a replacement.

Function of the Tuner. A diagram showing the functional position of the tuner in relation to the rest of the receiver is shown in Fig. 14-2.

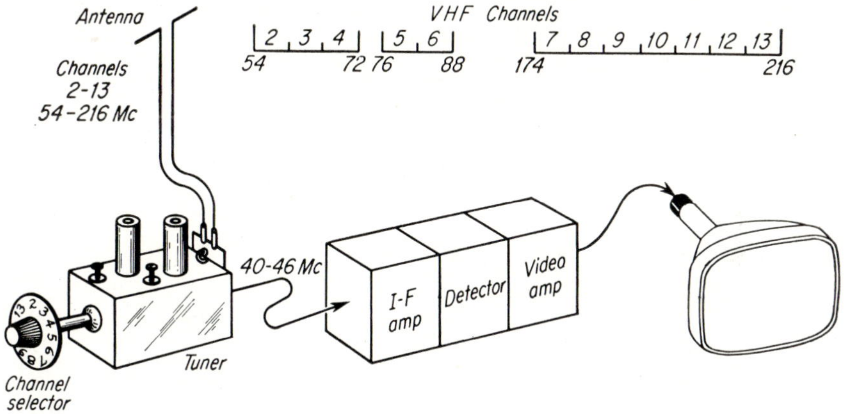

Fig. 14-2. Vhf television channels and functional location of the tuner in the picture chain.

Since it is between the antenna and i-f amplifier, its job is to select a station and change the frequency of the signal to 40 to 46 Mc or 20 to 26 Mc for those receivers which operate with an i-f amplifier in the 20-Mc band.

In the vhf band, television stations operate on assigned frequencies on Channels 2 to 13. This represents a frequency spread from 54 Mc, the low end of Channel 2, to 216 Mc, the high end of Channel 13. As shown in the upper part of Fig. 14-2, there are two breaks in this frequency spread. These are assigned to other services. Note that each channel occupies 6 Mc.

The antenna picks up energy from all stations broadcasting in its area. One function of the tuner is to select only one channel and amplify the picture and sound signals that are present on that channel. The other function is connected with superheterodyne action. The frequency of the received signal must be changed to the intermediate frequency of the receiver.

In most receivers, tuning is accomplished by turning a channel-selector switch to any one of 12 preset positions. A fine tuner is added to compensate for slight inaccuracies in the presetting.

Stages in the Tuner. A block-diagram representation of the stages in the tuner is given in Fig. 14-3. Signals picked up by the antenna are fed to an r-f stage. The diagram shows one r-f stage for simplicity, although two stages are often found. Here amplification and tuning take place at signal frequency, but the tuning circuits must have wide bandpass characteristics to pass the full 6 Mc of a television channel.

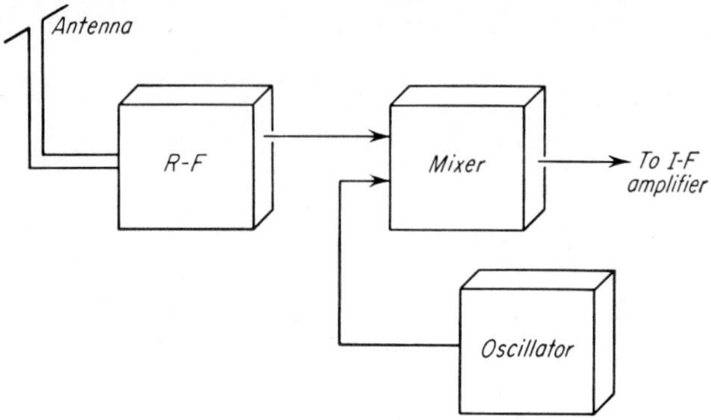

Channel No.	Frequency limits	Oscillator frequency	Heterodyne beats	
			Picture i-f signal	Sound i-f signal
2	54–60	101	45.75	41.25

Fig. 14-3. Stages in the tuner.

The r-f amplifier feeds the signal to the mixer stage of a superheterodyne frequency changer or converter. An unmodulated r-f voltage from a local oscillator operating at a frequency 40 Mc higher than the desired station is also fed into the mixer stage and produces heterodyne beats in the 40-Mc i-f band. If the receiver uses an i-f amplifier with a 20-Mc i-f band, the local oscillator is adjusted to a frequency 20 Mc higher than the desired station.

The important frequencies to be considered for Channel 2 are included in the diagram of Fig. 14-3. Channel 2 extends from 54 to 60 Mc, and the bandpass circuits in the r-f stages must pass at least this spread when the receiver is tuned to this channel. The oscillator is tuned to 101 Mc. This beats with the picture carrier at 55.25 Mc to produce a picture i-f signal at 45.75 Mc since

$$101 - 55.25 = 45.75$$

At the same time, the oscillator signal beats with the sound carrier at 59.75 Mc to produce the sound i-f signal at 41.25 Mc since

$$101 - 59.75 = 41.25$$

When the receiver is tuned to Channel 3, the r-f stages are switched to cover 60 to 66 Mc, and the oscillator is simultaneously switched to 107 Mc. The picture carrier at 61.25 Mc and the sound carrier at 65.75 Mc beat with the oscillator at 107 Mc to produce the same i-f signals as before since

$$107 - 61.25 = 45.75$$

and

$$107 - 65.75 = 41.25$$

The tuner chassis generally employs only two tubes, even though there are three or four stages. A single r-f stage uses an r-f pentode like the 6CB6 or the 6BC5, or their 600-ma equivalents the 3CB6 or 3BC5. Tuners with two r-f stages use twin triodes like the 6BZ7, 6BQ7, or 4BQ7. The oscillator and mixer stages are also contained in one envelope, twin triodes like the 6J6 and 5J6 and triode-pentodes like the 6X8, 5U8, and 5AT8 being the types most commonly found. As a general rule, the mixer-oscillator tube is mounted toward the front of the tuner chassis and the r-f tube toward the rear.

The R-F Amplifier. The r-f amplifier most commonly found in the front end of television receivers is a two-stage unit called a "cascode" amplifier. A basic schematic diagram of this amplifier is given in Fig. 14-4.

Coils *L*-3 and *L*-4 make up the antenna coil assembly, feeding signal energy from the antenna to the first r-f grid circuit. The coils are wound to cover the frequency range of one channel, and the arrowheads at the coil contacts indicate the switching arrangement whereby the coil assembly is changed for another channel. Resistor *R*-4 is across coil *L*-4 on all channels, its purpose being to broaden the response. Coils *L*-3 and *L*-4 are wound with very close coupling for the same purpose. The grid circuit is returned to the agc bus for control.

The tube used in the first r-f amplifier stage is one triode section of a 6BQ7, 6BZ7, or 4BQ7 twin triode. Feedback through the interelectrode capacitance of the tube may cause oscillation and a high noise level for the circuit. These effects are overcome by using a neutrodyne circuit where condenser *C*-4 returns some of the energy in the plate circuit back to the grid. Condenser *C*-4*A* is a screw-adjusted variable trimmer condenser. It is connected so as to bypass some of the feedback energy to ground, thereby providing a control of the neutralization.

The plate-supply circuits of the two r-f stages in the cascode amplifier are connected in series. Trace the B plus 250-volt line through decoupling filter *R*-16 and plate coil *L*-5 to the plate of the second r-f tube. Consider the plate-cathode circuit of this tube as a conductor, and the series circuit becomes evident by the connection from the cathode of the second r-f tube to the plate of the first. Since the tubes are in series, they divide the available plate-supply voltage between them. Note the reading of

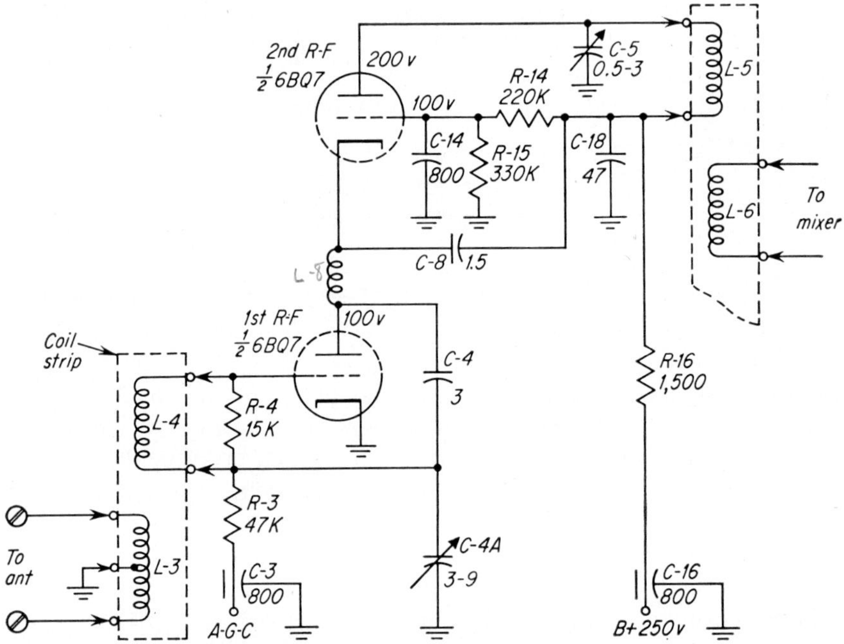

Fig. 14-4. Schematic diagram of a cascode r-f amplifier.

200 volts at the second r-f plate, and 100 volts at the second r-f cathode and first r-f plate. For proper operation of the second r-f tube, its control-grid voltage must be close to its cathode at +100 volts. This is accomplished by the voltage divider consisting of resistors *R*-14 and *R*-15.

The circuit used for the second r-f stage in the cascode tuner is known as a grounded grid amplifier. The control grid is at ground potential for r-f signals because of the capacitance of condenser *C*-14 which is connected between grid and ground. At the same time the grid is at +100 volts d-c because of the voltage divider. A signal fed to the cathode causes the cathode voltage to vary above and below its normal 100-volt point, thereby varying the bias and consequently the plate current of the second i-f tube. The shielding effect of the r-f grounded grid between the cathode input and plate output of the tube avoids the need for neutralization. The two r-f stages are directly coupled through coil *L*-8.

The signal in the output circuit of the second r-f tube is coupled to the mixer grid through coils *L*-5 and *L*-6. Again, the arrowheads indicate the channel-selector switch which throws in a different set of coils for another channel. Condenser *C*-5 is another screw-adjusted variable trimmer condenser which tunes the r-f plate in the bandpass tuning circuit.

The Pentode R-F Amplifier. Another r-f amplifier often found makes use of an r-f pentode like the 6CB6 or 6BC5. It is a single-stage amplifier, a basic circuit of which is given in Fig. 14-5. The antenna coil feeds signal energy to the grid circuit of the r-f pentode. Although the coil-switching arrangement of Fig. 14-4 could be used, this circuit shows a different method of tuning in the different channels. Coil *L*-4 which tunes the grid circuit is a tapped inductance, with a tap brought out for each channel. The tap switch S-1 is ganged with similar tap switches for the mixer and

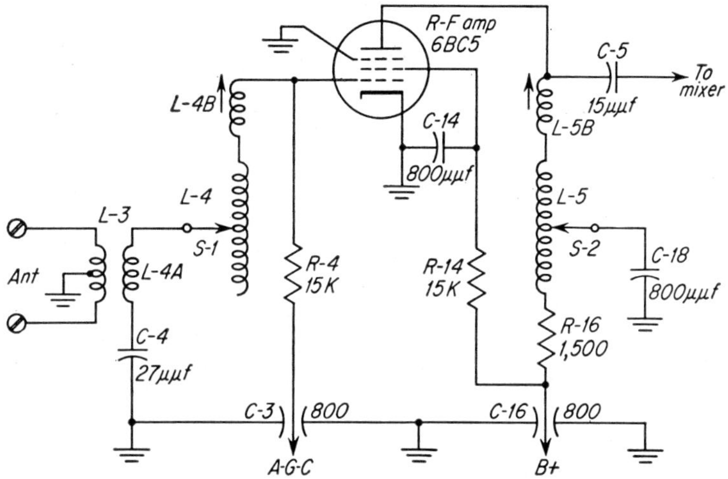

Fig. 14-5. The pentode r-f amplifier.

oscillator tuning circuits to become the channel selector for the receiver. Note that coil *L*-5 which tunes the r-f plate circuit also tunes the mixer grid circuit. In all other respects the circuit is a standard r-f amplifier.

Antenna-matching Devices. Antenna winding *L*-3 of the antenna coils in Figs. 14-4 and 14-5 is center-tapped so that a balanced 300-ohm line feeds the picked-up signal energy to the first r-f grid. Some receivers include an additional antenna-matching transformer for more efficient coupling of the antenna energy to the first r-f grid. The matching transformer, or "balun," as it is sometimes called, is enclosed in a separate shielded container mounted outboard from the tuner chassis. The schematic diagram and mounting arrangement of a typical balun are shown in Fig. 14-6.

The matching transformer is a dual unit consisting of two closely coupled sets of coils, *T*-1 and *T*-2, one in each leg of the input line. The windings are connected in series with the centers grounded for balancing purposes. The output of the two secondaries is connected across grid coil *L*-4. The arrowheads on this coil indicate that different grid coils can be switched in for the different channels. Coil *L*-4 could also be the tapped inductance coil *L*-4 of Fig. 14-5.

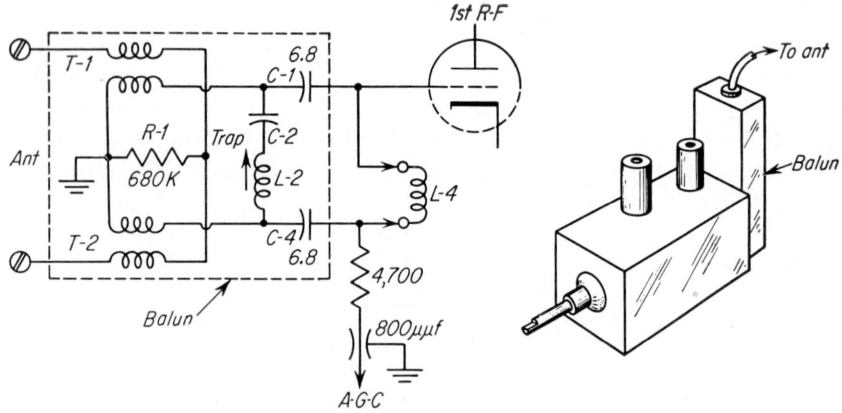

Fig. 14-6. Antenna-matching unit, or balun.

The balun often includes one or more wave traps. This is represented by coil *L*-2 and condenser *C*-2. The trap may be set to remove interference at the intermediate frequency of the receiver, or for some station in the f-m band (88 to 108 Mc) which may cause image-frequency interference. Traps for f-m stations are usually found on receivers with an intermediate frequency in the 20-Mc i-f band.

The Frequency-changer Section of the Tuner. The schematic diagram of the mixer and oscillator stages as used in a typical tuner is given in Fig. 14-7. Coils *L*-5, *L*-6, and *L*-7 are, respectively, the r-f plate, mixer grid, and oscillator coils wound to cover the frequency range of one channel. All three coils are in inductive relationship on one coil form so that mixer coil *L*-6 is fed signal energy from r-f plate coil *L*-5 and local oscillator energy from oscillator coil *L*-7. Condenser *C*-6 is an adjustable condenser which tunes coil *L*-6. Acting in conjunction with condenser *C*-5 which tunes the primary coil, we have a tuned-plate–tuned-grid bandpass circuit for the needed broad-band response. In some receivers, a loading resistor of about 6,800 ohms is included across coil *L*-6 to broaden the response further. Condenser *C*-12 and resistors *R*-12 and *R*-12*A* provide grid-leak bias for the mixer tube. The grid load is tapped as shown in the diagram to provide a test point for the mixer grid. This test point is called the "looker point" and is included on most tuners as a convenient connection for test equipment. Resistor *R*-12*A* is included in the circuit so that the looker test point is provided with an isolation resistor.

The oscillator circuit is a Colpitts, where the junction of condensers *C*-7 and *C*-7*A* across oscillator coil *L*-7 provides the feedback tap. This type of oscillator is very stable at the high frequencies used in the tuner. Condenser *C*-11 and resistor *R*-11 form the oscillator grid leak and condenser.

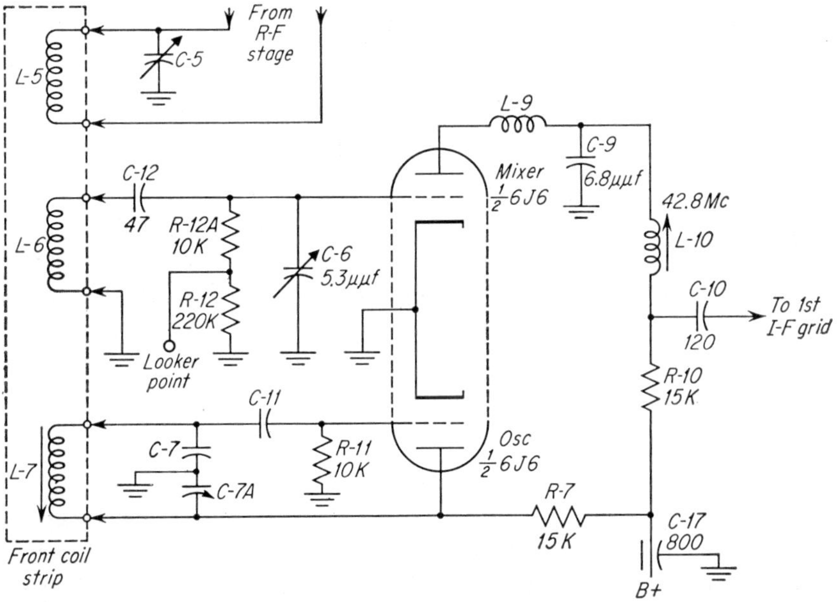

Fig. 14-7. Schematic diagram of the mixer and oscillator stages in a tuner.

The oscillator frequency adjustment is very critical in the operation of any superheterodyne receiver. There is usually a tuning slug for each channel coil as indicated by the arrow on coil *L*-7 in the coil assembly. These are set for optimum performance when the receiver is installed. In addition, condenser *C*-7*A* is made variable and controlled from the front panel as the fine-tuning adjustment. Since condenser *C*-7*A* is on the tube side of the channel-selector switch, it is operative on all channels.

The signal in the plate circuit of the mixer tube contains components which include the original signals, the oscillator signals, additive beats, and difference beats. The signals to be fed to the i-f amplifier are the difference beats, and since these are at the lowest frequency, the circuit consisting of coil *L*-9 and condenser *C*-9 in the mixer plate acts to filter out the signals at the higher frequencies. Coil *L*-10 is made to resonate at the intermediate frequency and is the first tuned circuit of the i-f amplifier. The adjustment of this coil is described in the i-f alignment procedure of Chap. 12.

The circuit of Fig. 14-7 shows a triode for the oscillator and a triode for the mixer—both tubes being the two halves of the twin triode 6J6. As mentioned previously, some tuners use a triode-pentode like the 6X8, where the triode section is used as the oscillator and the pentode as the mixer. The circuit and its operation would be very similar to the twin-triode circuit just described with the addition of a dropping resistor and bypass condenser for the screen grid of the pentode-mixer. This variation is shown in Fig. 14-10*d*.

SERVICING PROCEDURE FOR NO PICTURE

When a symptom of no picture is traced to the tuner, the first step is to change the two tubes one at a time. If this does not help, the next step is to isolate the trouble to one of the stages in the tuner. This is done with a signal check.

Signal Checks for the Mixer, Oscillator, and R-F Stages in the Tuner. A review of the signal checks in the picture chain is illustrated in Fig. 14-8. The symptom is normal raster—no picture. An audio signal applied to the diode load at test point *B* produces bars on the picture tube, showing that the video amplifier and picture tube are working. The test signal is changed to modulated r-f and fed to the last i-f grid at test point *C*. The generator frequency is wobbled back and forth around the intermediate frequency of the receiver, and the appearance of bars shows that the last i-f tube and detector stage are working. The modulated r-f signal is next fed to the input of the i-f amplifier at test point *D*. The appearance of the modulation bars over a 3- or 4-Mc swing of the generator shows normal operation of the i-f amplifier. This places the trouble in the tuner.

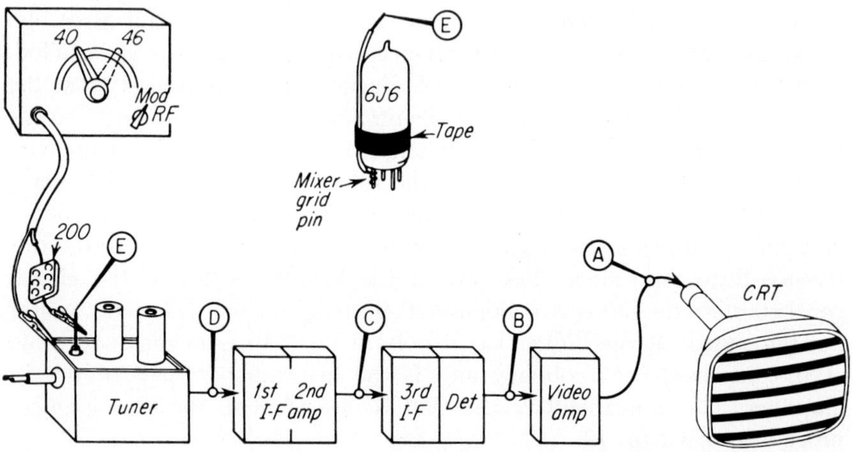

Fig. 14-8. Signal checks in the picture chain.

The same signal-check setup of Fig. 14-8 may be used to localize the trouble further to the mixer, oscillator, or r-f stages in the tuner. Without changing any settings, feed the test signal to the looker point (test point *E*) in the tuner. As previously described, the looker point is a connection to the mixer grid which is provided in most tuners. If the receiver is not so equipped, refer to its schematic diagram for the mixer-grid pin number. Wrap a piece of hookup wire around this pin, and secure it to the tube with a piece of tape as shown in the upper part of Fig. 14-8. Use this wire as test point *E*.

Turn the station selector to Channel 2 or 3, whichever is not used in your area, and check to see that the controls have not been disturbed. The contrast control should be in the full on position, and brightness should be adjusted for a normal raster. With the generator connected to the mixer grid at test point *E*, rotate the frequency control from 40 to 46 Mc (20 to 26 Mc if the receiver uses the 20-Mc i-f band).

If modulation bars do not appear on the picture tube, the mixer stage is not working. If the bars appear and remain steady through most of the i-f band, the mixer stage is working as an i-f amplifier and will presumably work as a mixer.

Continue to rotate the frequency control on the signal generator toward 60 Mc. As the generator dial leaves 46 Mc (26 Mc for receivers with a 20-Mc i-f amplifier) the bars disappear. With the receiver tuned to Channel 2, and a working oscillator, the bars should appear again at approximately 54 Mc. If the receiver is tuned to Channel 3, the bars should reappear at approximately 60 Mc. This check localizes the trouble to the oscillator or the r-f amplifier. If the bars do not reappear, the oscillator is not working. If the bars do appear at proper frequency, the r-f amplifier is not working.

Making Voltage Checks in the Tuner. The signal check just given places the trouble in the mixer, oscillator, or r-f stages. The next step in trouble shooting the tuner is to make voltage checks. However, unlike the other stages in the receiver, it is best to make these checks from the top of the chassis. Not only is it difficult to get at the underside of the tuner chassis, but it is unwise to move components about to wedge in a test prod, because even a slight misplacement may affect operation in these high-frequency circuits.

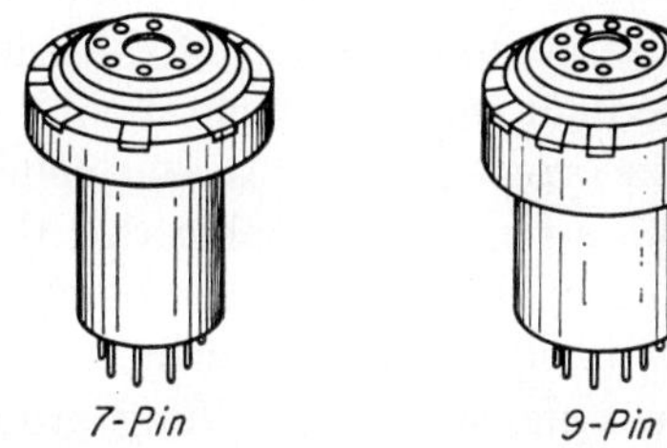

Fig. 14-9. Tube-adapter sockets.

Tube-adapter plugs are available for making voltage checks. Plugs for seven- and nine-pin tubes are illustrated in Fig. 14-9. The tube is placed in the adapter, and the adapter is plugged into the socket. The adapter is provided with numbered tabs which are connected to the tube elements. Voltage measurements are then made at the tabs.

Refer to the schematic diagram of the receiver for normal voltage readings. A set of typical readings for the most common tuner tubes is given in Fig. 14-10 for voltage analysis. Do not take a missing oscillator grid voltage as being too significant, since the use of the adapter plug may have stopped the oscillator. The voltage check is primarily to help in the quick location of defective components in the plate and screen circuits. When a plate voltage is found to be missing, before tearing the tuner apart, check to see that this voltage is coming in to the tuner at the connection strip. It may be missing because of an open decoupling filter resistor in the feed line.

The circuit of Fig. 14-10*a* shows a 6BZ7 twin triode in a cascode r-f amplifier. No voltage at any pin is most likely to be caused by an open plate resistor *R*-16. Incorrect voltage at the No. 7 grid pin calls for a check of resistors *R*-14 and *R*-15 and condenser *C*-14. A positive voltage at the No. 2 grid pin may be caused by a defective tube or a leakage in neutralizing condenser *C*-4. Since the tube has already been changed, condenser *C*-4 is more likely to be at fault.

In the twin-triode mixer-oscillator circuit of Fig. 14-10*b*, incorrect voltages at either plate pin indicate trouble in resistors *R*-7 or *R*-10 or mixer plate coil *L*-10. As mentioned previously, the voltages at the grid pins may be incorrect because the adapter plug stopped operation. Incidentally, lack of proper bias may cause a considerable change in the plate reading. The check is primarily for a missing plate voltage and an open resistor or coil.

The circuit of Fig. 14-10*c* shows an r-f amplifier using a pentode tube. This is a fairly familiar circuit and requires no special information.

The pentode section of the triode pentode of Fig. 14-10*d* is similarly a familiar circuit which should give no trouble in voltage analysis of an incorrect plate or screen voltage. When checking the triode section, you must again remember that the oscillator may be inoperative, causing incorrect plate and grid voltages. A missing plate voltage, however, is most likely to be caused by an open resistor *R*-10.

In some tuners, shielding around the socket bases may prevent the use of the adapter plugs. In this case, plate and screen voltage readings may be taken by removing the tube and taking them directly at the socket pins. When doing this, remember that, in working on the top of the socket, pin numbers are counted counterclockwise. Furthermore, be careful to merely touch the test prod to the socket contact, since any pressure may spread the springs so that they will no longer contact the tube pins. Another point to remember is that a voltage reading will be observed only at the second r-f plate pin in a cascode tuner. This is normal, because the series plate circuit is broken by the removal of the tube.

The purpose of making the voltage checks is to locate the defective

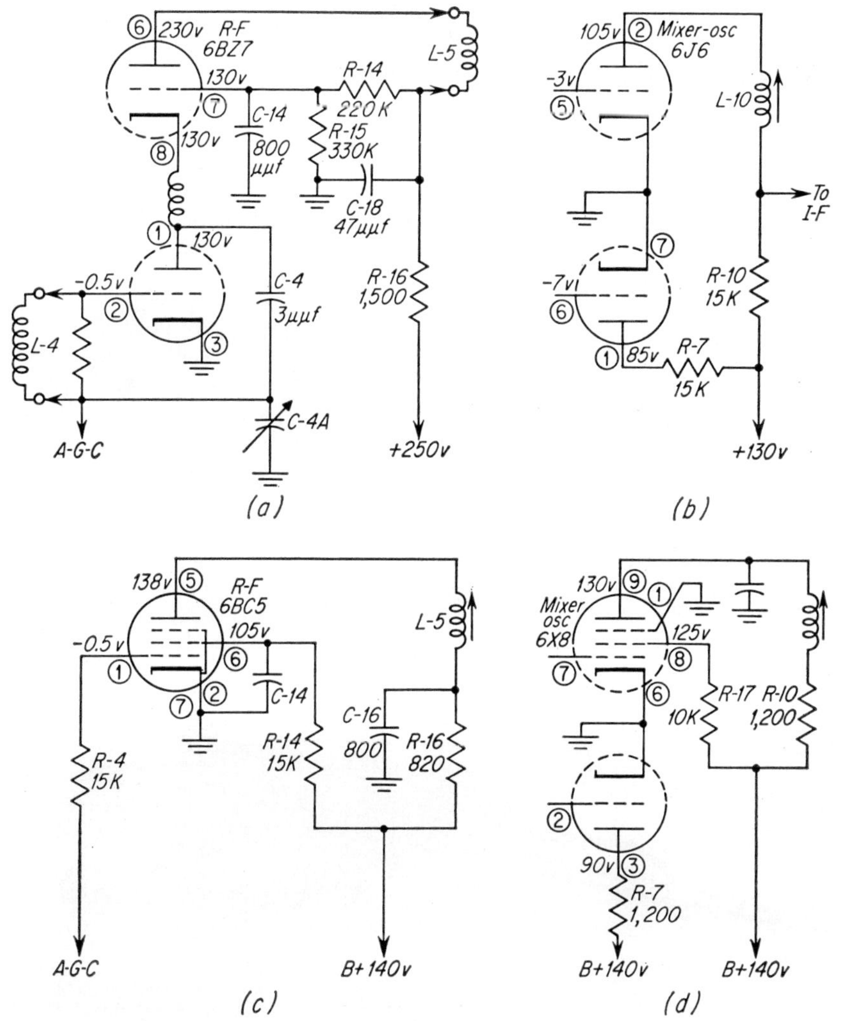

Fig. 14-10. Voltage analysis of tuners.

component, if possible, before opening the tuner. If this is accomplished, the defective part can be replaced with a minimum of disturbance to other components.

Removing the Outer Shielding from the Tuner. In order to gain access to the inside of the tuner subchassis, the cover which acts as the outer shielding must be removed. Examine the tuner to see how the shielding is fastened. It may be held by self-threading screws, or it may fit into notches. If it is a notch fit, the cover is pried loose with a screwdriver.

After completing repairs in the tuner subchassis, remember to replace the cover, since oscillation may result from the lack of shielding. Figure 14-11 shows two typical tuners with covers removed for servicing.

Drum and Wafer Tuners. Tuners fall into two general classifications depending on the arrangements for switching channels. They are called drum and wafer types, both of which are illustrated in Fig. 14-11.

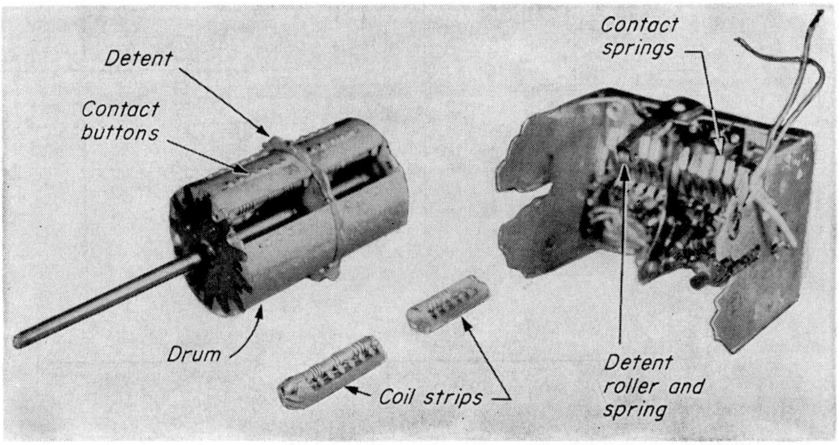

(*a*)

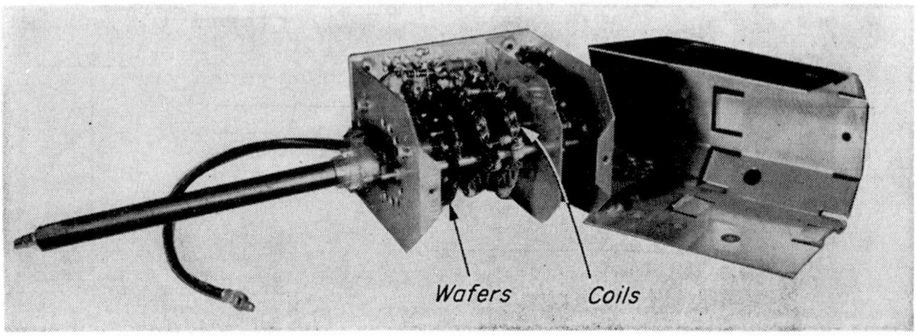

(*b*)

Fig. 14-11. Types of tuners. (*a*) Drum. (*b*) Wafer.

The drum, or turret, tuner of Fig. 14-11*a* has a set of coils for each of the 12 channels mounted on a fiber strip with the coil-contact buttons on the outside of the strip. The coil strips are clipped into the drum assembly as shown in the photograph. A set of stationary switch springs makes contact to the buttons on one strip. As the drum is rotated, the set of coils for the next channel wipes into the contact springs. A roller spring against the detent plate causes each channel to snap into its correct contact position.

The wafer tuner of Fig. 14-11*b* is built around a multisection switch. The sections consist of fiber wafers, each one containing a 12-point switch. The coils are mounted directly on the wafers in a series or incremental arrangement so that the switch points act as taps. Rotating arms connected to the tuner circuits take off the proper tap or short out successive sections of the main coils for each channel. The switch has a ball-type detent arrangement so that the switch snaps into place for each channel.

In both tuners, the fine tuning is controlled by a sleeve around the drum shaft or switch shaft. Rotating the sleeve changes the capacitance of a small variable condenser in the oscillator-tuning circuit.

When the outer shield can is removed from the tuner, the type of tuner is immediately evident, and defects like a broken switch shaft or detent spring may be seen. In the case of the wafer type of switch, further disassembly is not practicable, and the underside of the tuner chassis is serviced with the switch in place. In the drum type, the entire drum assembly may be removed for further inspection and servicing.

Dropping a Drum Assembly. The following procedure is used in dropping the drum or turret assembly for cleaning, or to gain access to components on the underside of the chassis in drum-type tuners. Refer to Fig. 14-12 for the appearance and location of components.

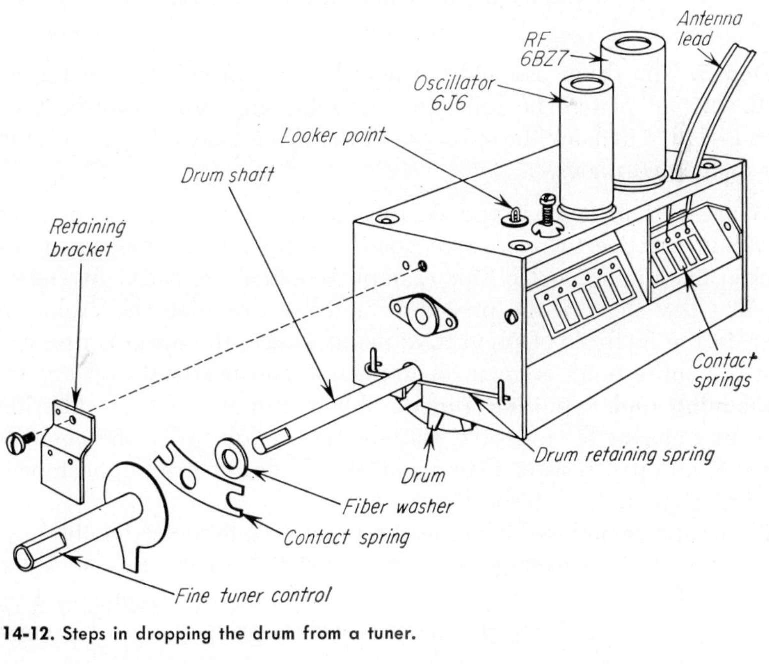

Fig. 14-12. Steps in dropping the drum from a tuner.

Step 1. This step is needed in receivers where the tuner is mounted toward the center or rear of the chassis, and there will be long shafts to the front panel. In these receivers, the shaft is supported by a collar on the front apron of the chassis. This collar should be removed first. The tuner illustrated in Fig. 14-12 is of the short-shaft type, and so the collar is not shown.

Step 2. Remove the screw holding the retaining bracket, and then the bracket.

Step 3. Slide off the fine tuner, contact spring, and fiber washer.

When reassembling, make sure that the parts are replaced in the proper sequence, with the fiber washer going back first.

Note the button behind the retaining bracket. This button and the bracket form the two plates of a condenser. The cam-shaped fiber disk at the back of the fine-tuner control is the dielectric for this condenser. Turning the control changes the ratio of fiber to air in the space between the plates because of the cam shape. This varies the capacitance of the condenser. Since the condenser is in the oscillator-grid circuit, the fine tuner controls the oscillator frequency. Its size permits a change of 1 to 3 Mc.

Step 4. Lift the drum-retaining spring out of its tab in the front plate of the tuner chassis, and remove it. There is another similar spring on the rear plate to hold the back end of the drum shaft. Lift and remove this spring also.

Step 5. The drum assembly is now held in place only by the roller in the detent plate. The roller and its retaining spring may be seen in Fig. 14-11*a*. Lift back the spring, and the drum assembly can be slipped out of the end plates.

When a drum tuner is opened for almost any reason, it is a standard servicing practice to remove the entire drum for cleaning purposes. A broken fine-tuning control, for example, could be replaced by following the first few steps of the procedure and not dropping the drum. However, in the interests of preventive maintenance, the drum is dropped so that the entire tuner is open for inspection and overhaul.

Cleaning and Repairing Tuners. When brought to the underside of the tuner chassis by a missing plate or screen voltage, as measured from the top, an open resistor may reveal itself by a charred appearance. So the first step is visual inspection.

If this shows nothing, begin at the tuner voltage point for the missing reading, and check voltage along the feed line to the tube-socket connection. If voltage is present all along the line, the trouble is lack of contact between the tube pin and the socket spring. The socket contacts can be tightened or resprung by inserting a sharply pointed tool like a pick between the socket hole and contact, and prying, as shown in Fig. 14-13. This spring-tightening process can then be performed on all socket contacts in the tuner.

If the receiver uses a drum tuner, the r-f plate circuit will be broken by the removal of the drum. If you are checking such a tuner for loss of voltage at the r-f plate, and the voltage is found to be present from the B plus feed point up to the drum contact, it is not likely that all 12 plate coils have been burned open. The more likely trouble is that the plate or B plus stationary contact spring has become bent or dirt-encrusted, so that the contact is lost. This can be determined by visual inspection. All contact springs should be at the same level, and bright. Of course, contacts other than those carrying plate voltage may likewise lose contact and stop operation, and these will be found by the same visual inspection. It is therefore a standard procedure when servicing drum tuners to look for dirt, corrosion, and depressed springs.

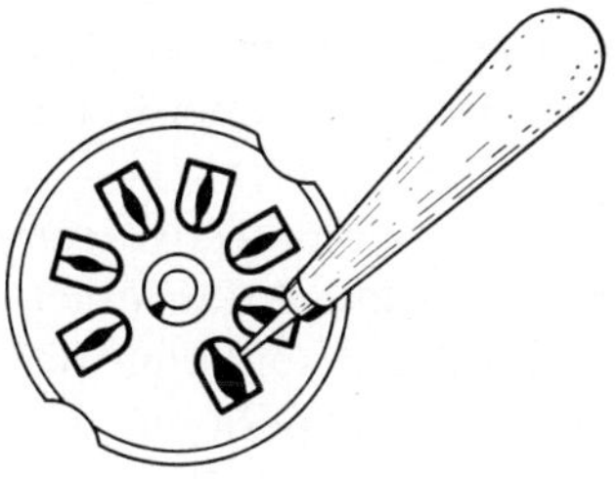

Fig. 14-13. Tightening socket contacts.

Dirt or corrosion is removed by rubbing with a mild abrasive. The eraser on the end of a pencil is ideal for this purpose. The springs are then lubricated with a touch of petroleum jelly. A slight film of lubrication is all that is required. Excessive amounts will absorb dirt, which may be conductive and therefore cause point-to-point leakage. Do not use Lubriplate as the lubricant because it is conductive. Any springs that have been depressed are lifted to the same height as the others. Then press in gently on each to make sure spring tension is present. If one of the contact springs is broken, a replacement contact spring is obtainable. The contact buttons on the coil strips in the drum are also cleaned with an eraser and lubricated with a thin film of petroleum jelly.

Wafer-type switches are also subject to poor contact. However, the contacts here are not too accessible for a cleaning job. The best procedure for these is to use a commercially prepared cleaner like Quietrole or Walsco Contactene. These are equipped with a dropper-type stopper which will inject a drop of the cleaning fluid to the wiping switch arm on each wafer deck. The switch is then rapidly rotated through all channels so that the cleaning solution is worked through all contacts. If the inspection shows a broken wafer, replacement wafers are not usually obtainable, and the entire tuner is replaced.

Visual inspection on either type of tuner will also reveal mechanical defects like a broken switch shaft, a missing detent ball, or a broken detent spring. Detent springs complete with roller are available for drum tuners and are easily replaced. Detent shafts for many wafer-type tuners are also available. A procedure for replacing the shaft is given later in this chapter.

When neither the voltage checks nor the visual inspection reveals the fault in the tuner, we finally resort to resistance checking of all components. Since the signal check has indicated the faulty stage, we are down to a very few components to be checked. Care must be exercised in this procedure. If a component must be moved slightly to insert a test prod, remember to return it to its original position after the check has been made.

When the defective component has been found, it should be replaced with as close to an exact duplicate as is obtainable. Note its physical location and soldering points, and be sure to place the new component in the same position, with the same lead length, and soldered to the same connective points.

Before reassembling the tuner, it is a standard servicing practice to clean and respring selector-switch contacts by the procedure just described as a preventive-maintenance measure, even though the trouble was found elsewhere.

SERVICING THE TUNER FOR OTHER FAULTS

The procedure just described is for a completely inoperative tuner. Troubles in the tuner can be responsible for other faults. These are weak picture on all channels and weak or no picture or sound on some channels, although others give normal reception.

There are other tuner faults of a type where the trouble is immediately obvious when making the initial check of the receiver's operation. In this type of trouble, when tuning from station to station, it is observed that pressure on the station-selector knob causes the picture to come in; or the picture may become weak and snowy with or without this pressure; or there may be flashes caused by noise pulses which come on or disappear as the pressure on the selector knob is varied. All these troubles point to faulty contacts in the selector switch. The symptoms depend upon which contact is making and breaking, and the degree of the break. The contact-cleaning and respringing procedure described in the previous section usually clears up faults of this nature.

Servicing Procedure for Weak Signals on All Channels. Weak picture coupled with snow is a direct indication of trouble in the tuner. Weak picture without snow is caused by trouble in the picture chain after the tuner and has been dealt with in previous chapters.

The weak picture may be caused by a defective tube, and since this is easiest, the first step is to change the tuner tubes one at a time.

If new tubes do not clear up the trouble, a check of the antenna is the next step in a field-servicing procedure. Antenna checks are described in Chap. 16. On the bench, the antenna is known to be good; so the next step is to localize the trouble to the r-f stages or the mixer-oscillator section. This is done by connecting one lead of the bench antenna to the looker point. If the mixer-oscillator stages are working, there will be some reception. It will probably be below normal, because the r-f amplifier is not contributing its gain and tuning qualities to the final picture. Try this check first on a number of good receivers. Connect your bench antenna to the looker point, and check operation until you are familiar with the results to be expected. Then you have a basis for determining if any receiver is operating normally when the antenna is connected to its looker point.

If it turns out that the r-f stage is at fault, and the receiver is equipped with a balun, or antenna-matching unit, we should next determine whether the trouble is in the balun or the r-f stages. This is best done by connecting the antenna to the grid of the first r-f tube. Refer to the schematic diagram of the receiver, and determine the pin number of the first r-f grid. Remove the tube, wrap a piece of wire around this grid pin, and replace the tube in its socket, bringing the wire up. Connect one of the antenna leads to this wire through a 25-$\mu\mu$f condenser. If a normal picture without snow comes on, the trouble is in the balun. If the picture remains weak, the r-f stage is at fault.

If the balun is at fault, open it to see if the defective condition may be repaired. If this cannot be done, a new balun must be obtained and installed. Use an exact replacement.

If the trouble is within the tuner, in the mixer, oscillator, or r-f stages, the servicing procedure is the same as for no picture. A voltage check from the top of the chassis may help to narrow down the search for the defective component. Next visual inspection below the chassis may reveal a charred resistor or a defective switch contact. Finally, resistance checking will locate the defective component.

Servicing for Poor Reception on Some Channels. A condition of normal reception on some channels and poor reception, that is, weak picture or sound on one or more channels, may be due to several factors. If abnormally poor results are obtained on stations known to be weak in the area, the weakness may be anywhere in the picture chain, since agc action will build up the stronger station signals so that they seem at normal strength. If the weak picture is accompanied by an abnormal amount of snow, the trouble is probably in the r-f stage of the tuner. But in the absence of snow, all stages in the picture chain must be subjected to a routine check for weak picture.

If the symptom is a weak picture or no picture on a strong station channel, whereas weaker stations are received with normal results, tune the receiver to this strong channel and adjust the fine tuner. If overload symptoms of loss of sync and negative picture appear, the trouble is in the i-f–agc system, and the receiver is serviced as described in Chap. 12. If the fine tuner cannot quite bring in good picture and sound, the trouble is likely to be a misadjusted oscillator-coil slug. Procedures for aligning oscillator-tuning slugs for the different types of tuners is given in the next section. If the fine tuner seems to have no effect on this channel, although it is operative on others, either the oscillator coil is mistuned so far that the fine tuner is ineffective or there is an open coil on this channel. To determine which condition exists, go through the oscillator-coil adjustment. If this adjustment is also ineffective, the tuner must be checked for an open coil or switch contact on this channel. This is done by visual inspection and an ohmmeter check.

We have been speaking of poor reception on a station normally weak or normally strong because, with these, we may encounter factors outside the tuner, like a weak stage elsewhere in the picture chain or improper operation of the agc circuit. If poor reception is experienced with an average station, while stronger and weaker stations give normal reception, the trouble is sure to be in the tuner, either a mistuned oscillator slug or an open coil.

If the trouble is an open coil in a drum tuner, a repair may be possible. First rotate the drum to a position where the coil is accessible. The coils are marked with the channel number and a letter. Refer to Fig. 14-14 for the location of the coil-retaining spring. Pull this back, and remove the coil. Examine the coil connections, and if one is poorly

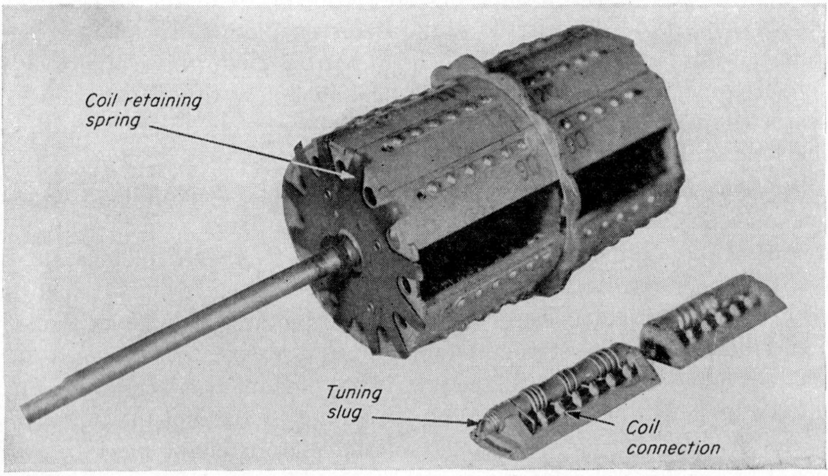

Fig. 14-14. Coil-retaining spring on a drum tuner.

soldered, resolder it. If the coil is beyond repair, a new one is obtainable. Make sure the replacement has the same number and letter, since coils differ, depending on the circuit and the intermediate frequency of the receiver. If you have any discarded tuners, check the coil number and letters. A usable coil may be salvaged in this way. Finally, it may be possible to use the coils from adjacent channels, if there is no reception on them in your area. Remove both adjacent coils, and compare them with the broken coil. Replace it with the one which is closest in size of wire, number, and spacing of turns. Then readjust the oscillator slug for best reception on this channel.

Adjusting Oscillator Slugs in Drum Tuners. Since the drum tuner uses individual coils for each channel, the frequency of the oscillator may be varied for any one channel without affecting the settings for any other. As a result, the channels may be adjusted in any order, and unused channels need not be adjusted at all.

Connect the receiver to an antenna, and turn the channel-selector switch to a channel needing adjustment. In the absence of the marked knob, in order to determine which coil is making contact, rotate the drum until the flat on the inner tuning shaft is level and points down. This is the position shown in Fig. 14-15. Note that in this position the coils for Channels 3 and 4 are at the bottom of the drum, the coils for Channel 7 are making contact, and the adjustment screw available through the hole in the front tuner plate is also for Channel 7. Thereafter one notch clockwise brings Channel 8 in for adjustment, etc. Remember that Channel 2 follows Channel 13.

Set the fine tuner to the center of its rotation range even though this detunes the picture. The mid-range position is shown in Fig. 14-15. Note that the flat side of the fine tuning shaft points upward. Keep the fine tuner in this position throughout the adjustment.

An enlarged view of an oscillator coil is also shown in Fig. 14-15. A slotted brass screw acts as the tuning slug. The inside of the coil form is not threaded. A piece of spring wire fitted into the slot in the coil form rests on the screw and acts as the thread. Turning the screw against this spring forces the slug farther into the form, thereby tuning the coil. If the screw is turned in too far, the last thread will go beyond the retaining spring, and then the screw will slide around inside the coil form. This is prevented by using a special alignment tool. The General Cement No. 9087 alignment screwdriver has a shoulder about $\frac{1}{4}$ in. back from the tip which allows full control and prevents the screw from being driven beyond the retaining spring. If the screw has been driven in too far, it can be retrieved with a Walsco No. 2565 tool which has been specially made for the purpose. If one of these tools is not available, the coil is removed from the drum, and the screw drops out when the coil form is tapped against the bench while holding back the screw-retaining spring.

Turn the channel selector to any channel normally received in your area. See that the fine tuner is in the center of its range, regardless of what this does to the picture, and adjust the oscillator slug for best picture. The point of best picture should coincide with good sound

Fig. 14-15. Making oscillator adjustments in a drum tuner.

and minimum intercarrier buzz. If there is a slight intercarrier buzz at the best picture setting, a compromise setting is permissible. If the buzz is too pronounced, service the receiver for excessive intercarrier buzz as described in Chap. 15. Switch to another channel, do not touch the fine tuner, and repeat the oscillator-coil adjustment. Continue on all channels normally received in your area.

Adjusting Oscillator Slugs in Wafer Tuners. Wafer tuners may be classified into several types as regards oscillator slug adjustments. Three common wafer tuners are illustrated in Fig. 14-16. In the type illustrated in Fig. 14-16*a*, the oscillator inductance is provided with screw adjustments on all channels. In the type illustrated in Fig. 14-16*b*, every other inductance is provided with a screw adjustment, making seven adjustment screws in all. The tuner illustrated in Fig. 14-16*c* has only two adjustments, one for the high-frequency channels and the other for the low-frequency channels. In all cases, the screws may be reached from the front of the tuner chassis and are available for adjustment regardless of which channel is being tuned in.

We must remember that, in wafer tuners, the coils are arranged in series. On Channel 13, only the smallest section of the coil is operating. As each lower channel is switched in, another section of coil is added.

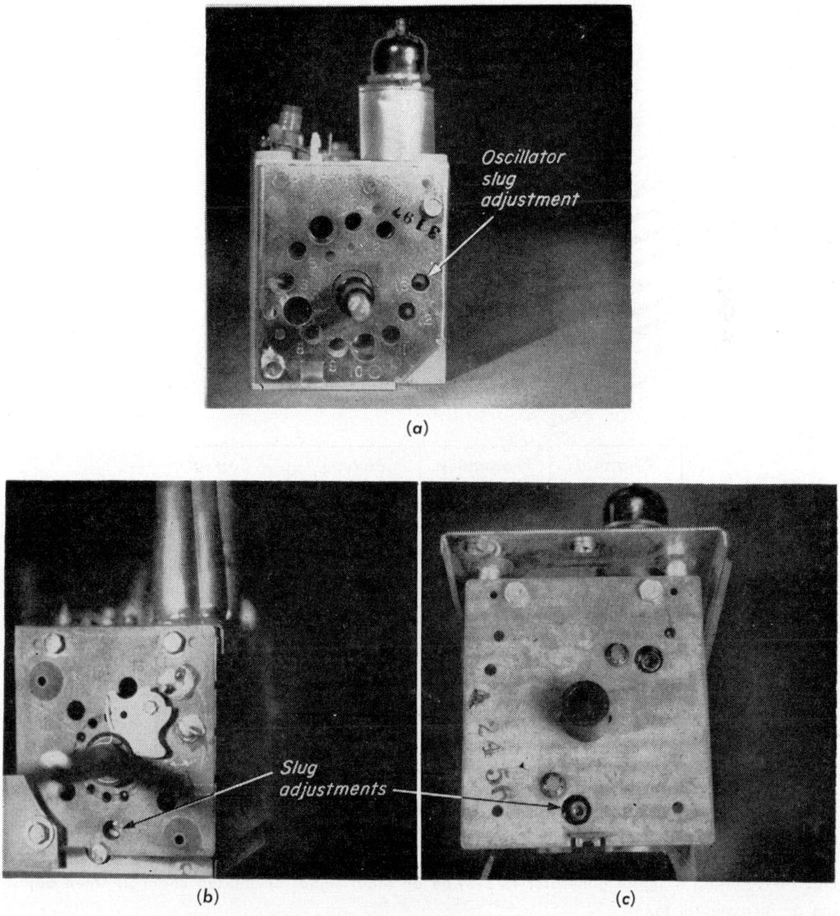

Fig. 14-16. Oscillator-slug adjustments on wafer tuner.

It follows therefore that the adjustment of Channel 13 will have an effect on all following channels, and alignment must start at the highest channel. Each adjustment thereafter must be accurately made before starting on the next lower channel.

Since no one area will have stations on all 12 channels, the alignment must be carried on with the aid of a signal generator. Where a station signal is available, it is to be preferred, the signal generator being used to adjust those channels which are unused in your area.

The alignment setup is shown in Fig. 14-17. The receiver is connected to an antenna, and the modulated r-f output of the signal generator is loosely coupled in by clipping the hot lead to the insulation on the 300-ohm line. The following procedure is for a tuner that provides tuning adjustments for all 12 channels.

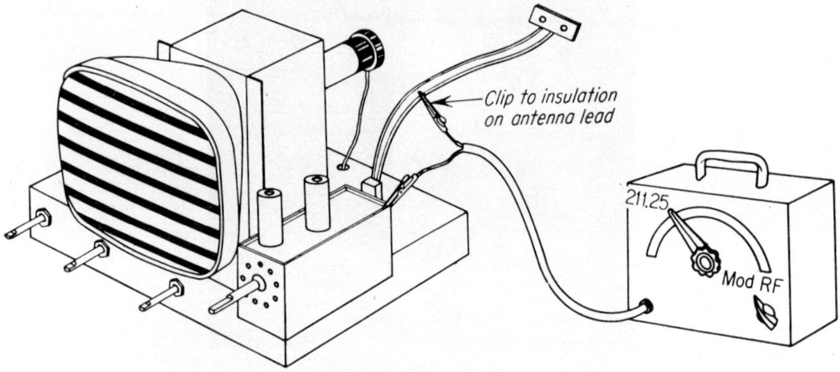

Channel No.	Frequency Range Mc	Picture Carrier, Mc	Sound Carrier, Mc
2	54–60	55.25	59.75
3	60–66	61.25	65.75
4	66–72	67.25	71.75
5	76–82	77.25	81.75
6	82–88	83.25	87.75
7	174–180	175.25	179.75
8	180–186	181.25	185.75
9	186–192	187.25	191.75
10	192–198	193.25	197.75
11	198–204	199.25	203.75
12	204–210	205.25	209.75
13	210–216	211.25	215.75

Fig. 14-17. Using a signal generator to provide an input signal on unused channels when adjusting an incremental tuner.

Turn the channel selector to Channel 13, and set the fine-tuning adjustment to the mid-point of its rotation. Leave the fine tuner at this position throughout the alignment. If you have a station transmitting on Channel 13 in your area, detune the generator away from 210 to 216 Mc, and adjust the Channel 13 screw for best picture. In making this adjustment, first drive the screw until it is fully in. Do not use too much pressure since this will crack the wafer behind it. Then loosen for two or three turns, leaving the adjustment at the best picture setting. Do not loosen beyond four or five turns, since the screw will fall out and become lost. Always follow this technique in making all adjustments so as to avoid breaking the wafer or losing the adjustment screw. If there is no station on Channel 13 in your area, set the frequency control of the signal generator to 211.25 Mc, the picture carrier for Channel 13, and adjust for black bars on the picture tube. Repeat this procedure for Channel 12, and so on down to Channel 2, always adjusting for best pictures where there is a station, or black bars where there is none. A list of frequencies of the picture carrier for the various channels is given in Fig. 14-17 for reference.

When the tuner provides an adjustment screw for every other channel, like the one shown in Fig. 14-16*b*, use the same setup of Fig. 14-17, tune to Channel 13, and adjust this slug either on station or generator signal as before. Then, if there is a station on Channel 12 or 11 in your area, tune to that station, and adjust the 11 to 12 screw for best picture. If you have neither station, tune to the lower-frequency channel, and feed in a modulated r-f signal at 199.25 Mc, the picture carrier for Channel 11. Adjust the 11 to 12 screw for bars. Continue through the channels in the same way, adjusting for best picture where you have a station or, where there is none, tuning the lower-frequency channel for best bars on the generator signal.

Where the tuner has only two adjustments like the one shown in Fig. 14-16*c*, set the station selector to the channel of the poorest station in your area from Channels 7 through 13. Adjust the Channel 13 screw for best picture on this station. Check the operation of any other stations on Channel 7 through 13, and touch up the adjustment, if needed, finally leaving the screw at the adjustment which gives best over-all operation from Channel 7 to 13. Then set the channel selector to the weakest station from Channel 2 to 6, and adjust the low-frequency screw for best picture on this station. Check the operation of the receiver on any other stations you may have on these channels, touching up the adjustment, if needed. On these tuners, the oscillator-slug adjustment is a compromise, but the fine tuner should give optimum performance on all channels.

The adjustment screws on all the tuners illustrated in Fig. 14-16 are accessible regardless of which channel is being tuned in. It is therefore possible to touch up the adjustment screw of a higher channel if this is desired, to improve reception or the range of the fine tuner on a lower one. In some tuners this cannot be done, since a rotating plate blocks off all adjustment screws except for the channel being tuned in. This creates no problem if adjustments are provided for all 12 channels. Each adjustment is made in order from Channel 13 down, using station signal or generator signal. When there is one adjustment for two channels, however, it may not be possible to adjust on station signal.

A tuner of this type is illustrated in Fig. 14-18. Adjustment screws are provided for Channels 2, 4, 6, 8, 10, 12, and 13, the others being dependent upon correct adjustment of the existing screws. Channel 3, for example, has no adjustment of its own and is brought to correct frequency by the previous adjustment of Channel 4. If, in your area, you have a station on Channel 3, and none on 4, the adjustment cannot be

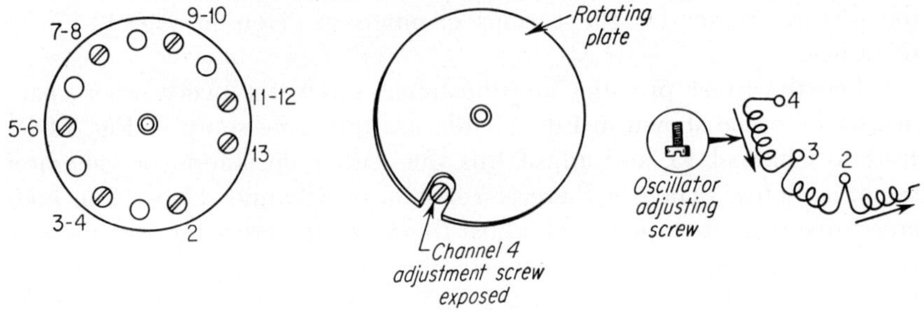

Fig. 14-18. Adjustment screws covered by rotating plate.

touched up while on Channel 3, because it is made inaccessible by the cover plate. The notch in the cover plate gives access to the empty spot on Channel 3.

When adjusting a tuner of this type, start on Channel 13, adjusting for best picture or bars, depending on whether you have a station. Continue by adjusting the 11 to 12 screw on Channel 12. Now assume you have a station on Channel 11 but none on 12. The 11 to 12 screw must be adjusted on Channel 12 feeding in 205.25 Mc from the signal generator. Then turn to Channel 11, and try the operation of the fine-tuning control. If the point of best picture is near the center of rotation, the adjustment is correct. If it is too far to one side, return to Channel 12 and tighten the screw a quarter turn. Then check operation on Channel 11. If the best picture now comes in near the center of rotation the adjustment is correct, but if it is even farther to one side, the quarter turn was in the wrong direction. Go back to Channel 12, loosen a quarter turn to bring back the original setting, and then loosen another quarter turn for the new correction setting. Check operation again on Channel 11, to make sure that best picture comes in near the rotation center of the fine tuner.

Continue by adjusting the 9 to 10 screw on Channel 10, the 7 to 8 screw on Channel 8, etc. Always adjust for the point of best picture on a station when available, using the generator when it is not. If you should again encounter the condition of a station on an uncontrolled position, use the quarter-turn technique described above for Channel 11.

Replacing Detents and Shafts. A broken tuner shaft is so common a fault that replacement detents and shafts are obtainable for most tuners. This is true even for the drum tuner where the shaft is metal and not easily broken.

In the case of the drum tuner the replacement unit consists of the shaft and the drum and detent plates. The old drum is dropped in accordance with the instructions given earlier in this chapter, and the contact springs are cleaned and adjusted. The coils are removed from the old drum, cleaned, and inserted in the new one. When changing the coils, set the two drums alongside each other, with the flat sides of both shafts pointing downward. Remove one coil, and insert it in its correspondingly correct position in the new drum, using the flat of the shaft as a guide. Continue with the rest of the coils until all are transferred. If the coils are transferred incorrectly in relation to the flat of the shaft, the channel-marking system on the knob indicator will be incorrect. The correct relationship is that the coils for Channels 3 and 4 are below the flat side of the shaft, as shown in Fig. 14-15.

The shafts in wafer tuners are usually made of fiber which is comparatively fragile and so require replacement quite often. Replacement detents and shafts are available for the more popular tuners. If a replacement is available for your tuner, the following pointers should be kept in mind when making the change:

The shaft is usually broken somewhere in the center, and subsequent turning of the channel selector in the hope of getting a station has misaligned the switch rotors, since some have turned with the shaft and some have not. Therefore, after removing the outer cover of the tuner, turn the selector shaft and note which switch sections are turning with it. Set

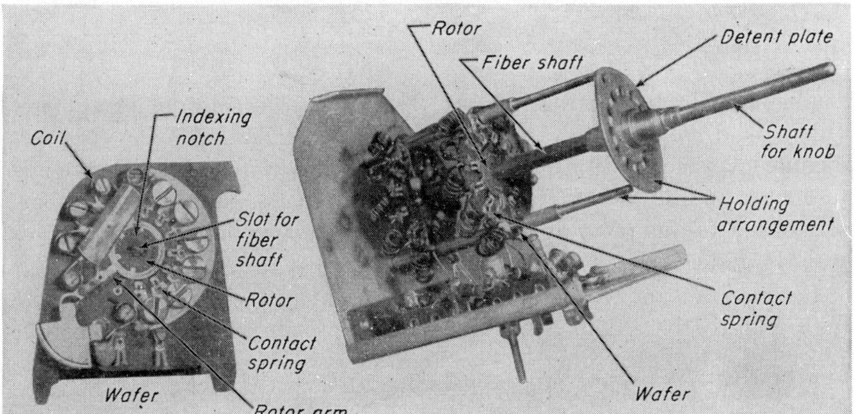

Fig. 14-19. Replacing a fiber shaft in a wafer tuner.

the shaft to an easily remembered position, with the flat section facing upward, let us say. Examine the holding arrangement for the shaft, and remove it. It usually pulls straight forward after removing two or three holding screws. Examine the wafers of the switch sections that moved with the shaft, looking for the position of the indexing notch labeled in Fig. 14-19. The remaining switch sections have to be turned so that the notch is in the same position. If the broken end of the shaft extends through the rear of the tuner, it may be possible to rotate the remaining switch sections to the same position as the front wafers by turning it. The broken end is removed by gently pulling it out. Check to see that all switch sections have the indexing notch in the same position. Any out of place are corrected by hand, being careful to see that the rotor arm positions itself correctly between the switch-contact springs. Note that the replacement shaft can be installed in either of two positions. Determine the correct position by rotating the flat side of the shaft upward, and insert the shaft carefully through the wafers. Finally, the shaft is locked into position with its holding screws.

There is the possibility that the shaft is broken close to the detent plate and none of the switch sections turn with the knob. In this case, remove the old shaft, position the wafers so that the identifying notches are aligned, and insert the new detent and shaft in either of its two possible positions. Check the receiver for normal operation, and complete any other service work which may be needed. Before reinstalling the receiver in its cabinet, operate the receiver on a known station, and check the channel number which will be indicated on the indicator knob. If it is incorrect by 180°, note the location of the shaft flat for the knob. Carefully remove the detent and shaft, rotate it so that the knob flat is in the opposite location, and reinsert the shaft into the wafers.

UHF TUNERS

Channels 14 to 83 from 470 to 890 Mc in the ultrahigh-frequency (uhf) band are allocated to television broadcasting, and many receivers include provision to pick up stations on these frequencies. Channels 2 to 13 are in the very-high-frequency (vhf) band, and the tuners described up to this point are of the vhf type. Each channel in both bands is 6 Mc wide.

Receivers meant to operate on both the uhf and vhf bands are essentially vhf receivers which include an extra tuner for operation on the uhf band. The extra tuner is a separate chassis similar to the vhf tuner. It is usually attached to the vhf tuner and operated from the one-dial assembly like the unit shown in Fig. 14-20. If the uhf tuner chassis is omitted, the remainder is a standard vhf receiver.

Fig. 14-20. A typical uhf-vhf tuner.

Types of UHF Tuners. Uhf tuners may be classified into two types. Both are illustrated in Fig. 14-21.

The two types are the same in that they contain mixer and oscillator stages in a frequency-changer circuit. They operate from a separate uhf antenna. Note the antenna symbol in the diagram. It represents the triangular dipole often used for uhf reception and commonly called a bow tie. There is no r-f amplifier preceding the mixer stage, because amplification at these frequencies is impractical. The tuning device covers the entire range from 470 to 890 Mc in one sweep of a continuous single control.

The two uhf tuner types differ in the frequency of the output signal and the provision in the vhf receiver for accepting the uhf station signal. In the system shown in Fig. 14-21*a*, the uhf frequency converter changes the received station signal to 82 to 86 Mc. This is the frequency range of vhf Channel 6, and the output is simply fed to the antenna terminals of the unchanged vhf receiver. The station selector is placed on Channel 6, and uhf stations are tuned in. If there is a local station on Channel 6, Channel 5 is used instead and the output frequency is changed to 76 to 82 Mc. In order not to transpose the video and sound carriers, the frequency of the oscillator circuit is kept below that of the received signal. This system works as a double-conversion type of superheterodyne receiver.

The uhf system illustrated in Fig. 14-21*b* is a single-conversion type of superheterodyne receiver. Here the oscillator frequency is higher than the received signal and produces in its output a video carrier at 45.75 Mc and a sound carrier at 41.25 Mc—the standard intermediate frequencies

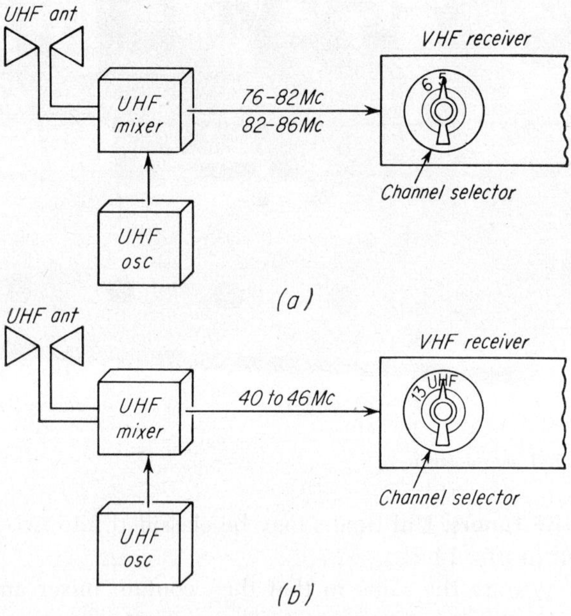

Fig. 14-21. Types of uhf tuners.

of the television receiver. An extra position on the channel-selector switch is marked UHF. This switches 40-Mc coils into the r-f amplifier circuits. These now operate as a two-stage preamplifier for the i-f section, making five i-f stages in all. At the same time, the vhf oscillator is switched off.

Tuning Circuits at Ultrahigh Frequencies. The coils and condensers we are accustomed to in tuning circuits cannot reach ultrahigh frequencies. Shorted parallel conductors show resonant characteristics at a quarter wavelength. At ultrahigh frequencies, where wavelength is low, a parallel line-tuning circuit is possible in a comparatively short length and is the tuning circuit commonly employed in uhf tuners.

The tuning circuit is made variable by a sliding shorting bar, as shown in Fig. 14-22. The lines may be straight as shown in Fig. 14-22*a* or can be formed into an arc as shown in Fig. 14-22*b* for ease in ganging. Most tuners use three such tuning circuits: one as a preselector, one in the mixer grid, and one in the oscillator stage. Some provision must be made so that all three circuits are operating at their correct frequencies

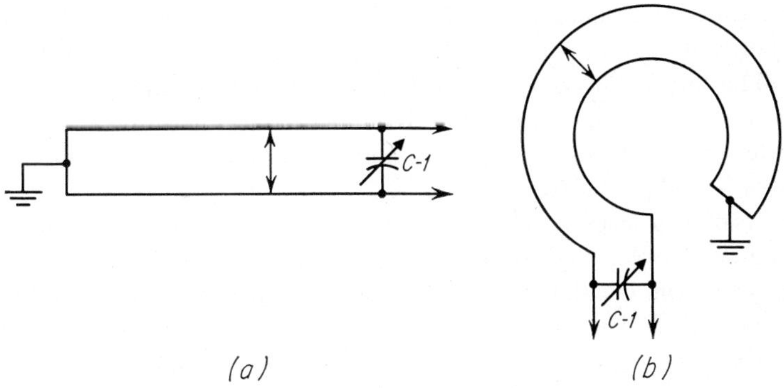

Fig. 14-22. Quarter-wave tuning circuit at ultrahigh frequencies.

when ganged. This is accomplished by the small variable condensers *C*-1 in Fig. 14-22.

Circuit of the UHF Tuner. In a dual-conversion superheterodyne system, the uhf tuner produces an i-f signal in the Channel 5 band; in a single-conversion superheterodyne system it produces an i-f signal in the 40-Mc band. In either system the uhf tuner itself has the same basic circuit: a preselector, mixer, and oscillator.

The circuit of the Magnavox 700459 tuner is given in Fig. 14-23 as a typical example of a uhf tuner. The uhf station selector is the ganged shorting bar on the three tuned lines *L*-601A, *L*-601*B*, and *L*-601*C*. Line

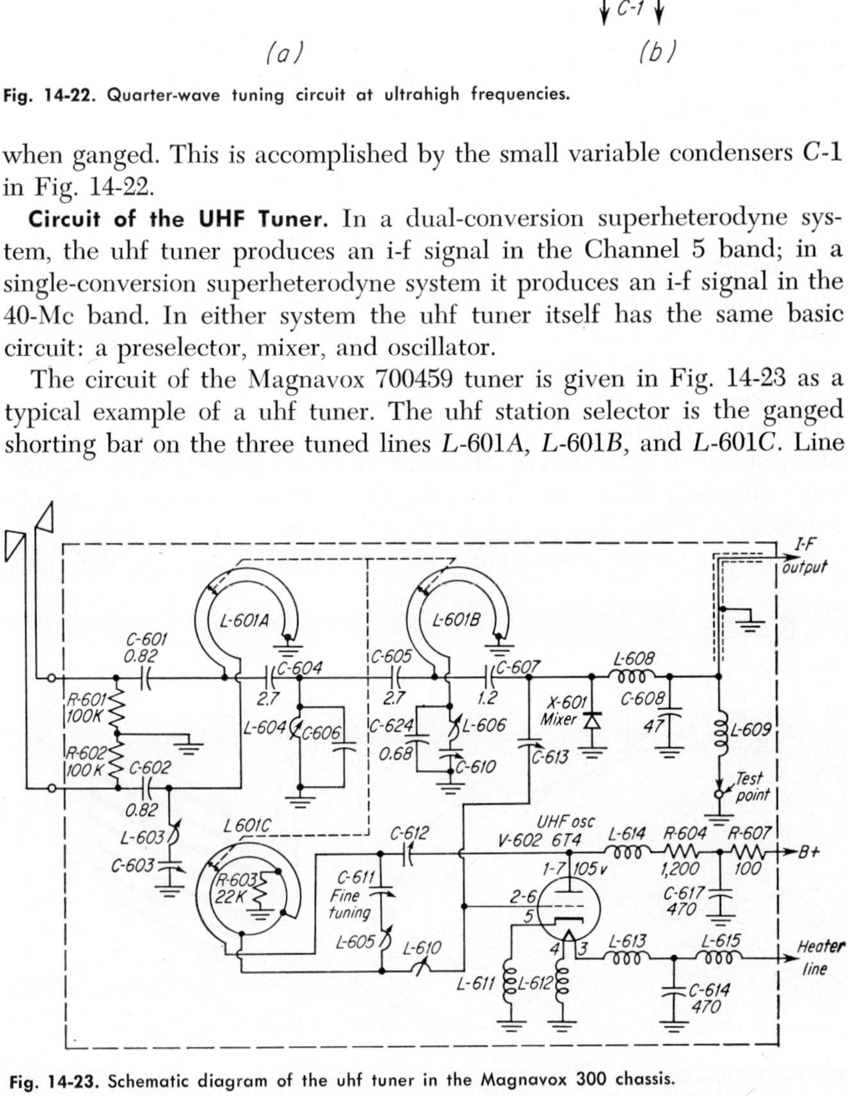

Fig. 14-23. Schematic diagram of the uhf tuner in the Magnavox 300 chassis.

L-601*A* tunes the antenna circuit, and condensers *C*-604 and *C*-605 feed this energy to line *L*-601*B* which acts as the tuning circuit for the mixer stage. The latter does not use a tube, crystal X601 acting as the first detector, or mixer for the superheterodyne receiver. Tuned line *L*-601*B* is the equivalent of the mixer-grid circuit. Some uhf tuners use a triode for mixing.

The oscillator uses a 6T4 triode tube in a modified Hartley circuit. Tuned line *L*-601*C* determines the frequency of oscillation. Condenser *C*-611 across the oscillator-tuning circuit is a front-panel control and acts as the fine tuner for the uhf receiver. Condenser *C*-606 and coil *L*-604 across the preselector and condensers *C*-610 and *C*-624 and coil *L*-606 in the mixer tuning circuit act as tracking and bandpass adjustments. Note the r-f filter network in the heater circuit to prevent radiation of the oscillator current through the common heater line.

Oscillator energy is fed to the mixer circuit through condenser *C*-613, where it beats with the received station signal to produce a 40-Mc i-f signal in the mixer output circuit. This is fed to the vhf tuner which has a thirteenth position labeled UHF on the channel selector. In this position the vhf antenna is shorted, the vhf oscillator is disabled, and the r-f tubes in the vhf tuner are operated through 40-Mc coils to act as the first stages of the i-f amplifier.

Servicing UHF Tuners. When a receiver gives normal reception on vhf stations but poor results on the local uhf channels, the trouble is obviously in the uhf system. However, it is not necessarily in the uhf tuner chassis itself. It may be in the vhf switching system or the uhf antenna.

First refer to the service notes to determine whether the uhf tuner

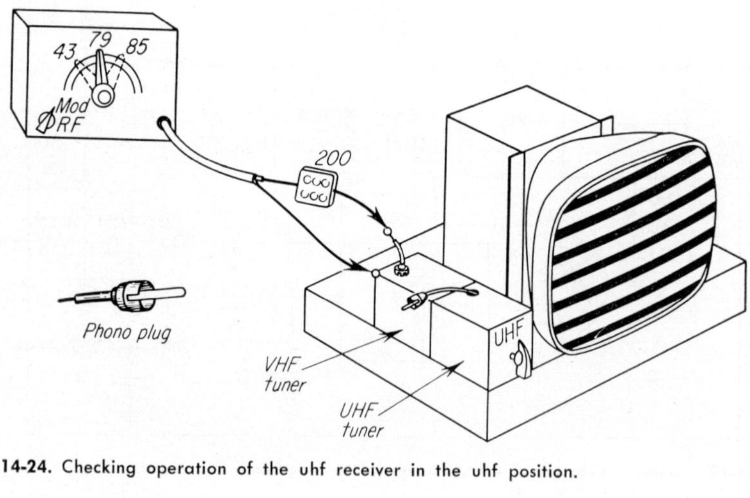

Fig. 14-24. Checking operation of the uhf receiver in the uhf position.

converts to Channel 5, or the 40-Mc i-f band. Then switch to the uhf position, and disconnect the uhf tuner output lead from the vhf input terminal. The connection is usually in the form of a phonograph plug. Make up a similar plug with a short insulated lead, like the one shown in Fig. 14-24, and connect the output of the signal generator to this lead. Feed in a modulated signal at 79 or 43 Mc, depending on which system is in use, and rock the generator frequency control while looking for bars on the picture tube. Couple in a test signal at 85 Mc if the tuner converts to Channel 6. The appearance of the modulation bars shows that the vhf tuner is functioning normally and places the trouble in the uhf tuner or antenna. Conversely, the nonappearance of the bars places the trouble in the vhf tuner. Since local vhf stations give normal reception, the trouble is limited to the coils, their alignment at 43, 79, or 85 Mc as the case may be, or the switch contacts in the uhf position. These are then serviced as described earlier in this chapter.

If the above check places the fault in the uhf tuner or antenna, the antenna possibility should be investigated before dismantling the uhf tuner. Reception at ultra high frequencies is not so reliable as vhf broadcasts, and it would be wise to establish that a signal is being delivered by the antenna. This, of course, is a field-servicing problem. It is best solved by checking the antenna with a known good uhf receiver. A good choice for the latter is a self-powered uhf converter like the Mallory TV-101 which delivers a strong output signal in the frequency range of vhf Channels 5 and 6. The converter is connected to the uhf antenna, and its output is connected to the vhf antenna terminals of the receiver being tested, or any receiver. The receiver is tuned to Channel 5 or 6, and the converter to the channel number of the local uhf station. The setup is shown in Fig. 14-25. Normal reception under these conditions

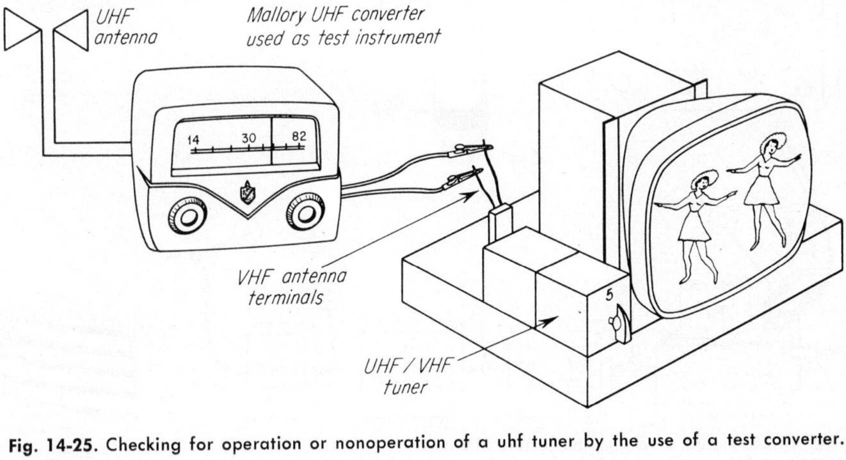

Fig. 14-25. Checking for operation or nonoperation of a uhf tuner by the use of a test converter.

proves that the uhf antenna is picking up signals, and the trouble is therefore in the uhf tuner of the receiver.

The following procedure is designed to clear up troubles in the uhf tuner.

Step 1. Change uhf oscillator tube.

Since tubes used in uhf oscillators are more critical than those used in vhf work, several should be tried. The tube should be seated firmly in its socket and the shield replaced. Check operation on the highest uhf station received in your area. If a tube which gives good over-all results cannot be found, realignment according to manufacturer's specifications will be necessary.

Step 2. Replace the mixer tube or crystal.

Crystals as well as tubes vary widely in performance. Try operation with several crystals. Select the one which gives the best results. Avoid using excessive heat on those crystals which have pigtail leads. Grasp the crystal lead with long-nose pliers as a heat shunt when soldering. Use extreme care so as not to disturb the wiring. A suitable replacement for uhf crystal mixers is 1N82A. When replacing, observe polarity.

Step 3. Make voltage and resistance checks to locate any off-value components.

SUMMARY

Quick Check for the Tuner. Snow on the screen is an indication of trouble in the tuner. In the absence of snow, normal responses at all previous test points in the picture chain place the trouble in the tuner. The check points are illustrated in Fig. 14-26.

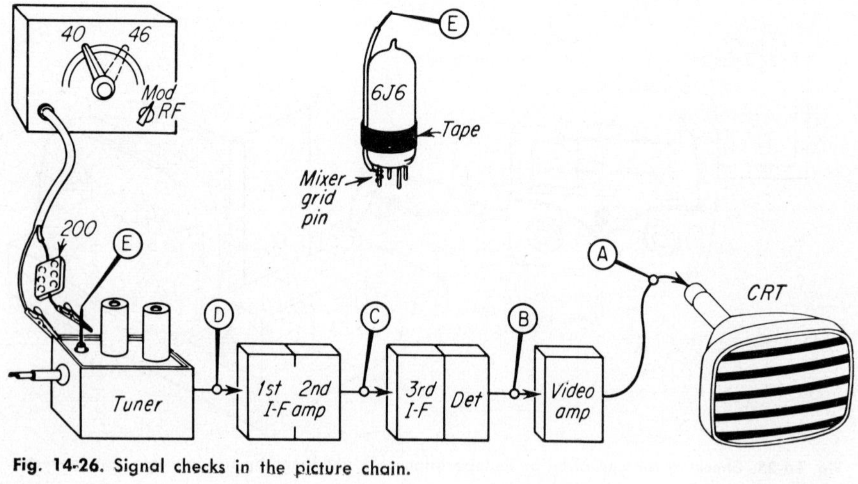

Fig. 14-26. Signal checks in the picture chain.

Signal Checks for the Mixer, Oscillator, and R-F Stages in the Tuner. Set the receiver on Channel 2, and disconnect the antenna. Connect the modulated output of an a-m signal generator to the grid of the mixer tube, or the looker point, and rotate its frequency control through the intermediate frequency of the receiver. The appearance of modulation bars on the screen of the picture tube indicates that the mixer tube is working. Advance the frequency of the signal generator toward 60 Mc. As the frequency goes above the i-f range of the receiver, the modulation bars on the screen of the picture tube die out. The reappearance of the bars at 54 to 60 Mc, the frequency range of Channel 2, proves that the oscillator is working.

If both the mixer and oscillator stages are working, the trouble is in the r-f stage or stages.

QUESTIONS

1. Describe the procedural checks by which the symptom of normal raster —no picture is isolated to the tuner.

2. Explain how the appearance of snow is an indication of tuner trouble.

3. Name the three stages in the tuner, and give the function of each.

4. Where is the fine-tuner control located in the tuner circuit, and what is its function?

5. Assume a receiver with an i-f amplifier in the 40-Mc band. Give the band range of the tuning circuits in the r-f, mixer, and oscillator stages when tuning in Channel 2.

6. Draw a diagram of a two-stage cascode r-f amplifier.

7. Describe a balun and give its function.

8. Describe the purpose of the wave traps which are often found in the antenna circuit of tuners.

9. Draw the diagram of the mixer-oscillator stage in a tuner. Use a twin-triode tube, and label the fine tuner.

10. What is the frequency range of the tuned coil in the plate circuit of the mixer tube?

11. Describe the signal checks which isolate trouble to the mixer, oscillator, or r-f stages in a tuner.

12. What is a looker point? Draw a partial diagram of a mixer stage, and label the looker point.

13. In the diagram of Fig. 14-10*a*, list the possible causes of a voltage reading of zero at the No. 1 pin of the 6BZ7 tube.

14. Describe the main differences between the drum and wafer types of tuners.

15. What is the purpose of a detent in a switch?

16. Discuss the cleaning of contacts in drum and wafer switches.

17. Explain how you would determine that tuner-oscillator slugs need readjustment.

18. Describe the oscillator-slug adjustment procedure for a drum tuner.

19. Describe how to adjust the two oscillator slugs provided for the tuner of Fig. 14-15*c*.

20. What precautions should be observed when adjusting oscillator slugs?

21. Describe the two types of uhf tuners.

15

Servicing the Sound Section

A complaint of "normal picture–no sound" is the indication of trouble in the sound section of a television receiver. Sound and picture signals are common to the tuner, i-f, and video sections. Since a normal picture is present, these sections may be assumed to be working properly, thereby limiting the trouble for "normal picture–no sound" to the sound section.

The above must be true for all channels normally received. If picture and sound come in on even one channel, the sound section is working. It may be weak, but it is likely that absence of sound on most stations is due to other factors. This will be taken up in the localizing procedures given later in this chapter.

The Sound-signal Path of a Television Receiver. Practically all television receivers use the intercarrier type of sound system. A block diagram of the picture- and sound-signal paths in such a receiver is shown in Fig. 15-1.

A television transmitter sends out two signals, one for picture and one for sound. Both carriers are within the 6 Mc allotted to a station

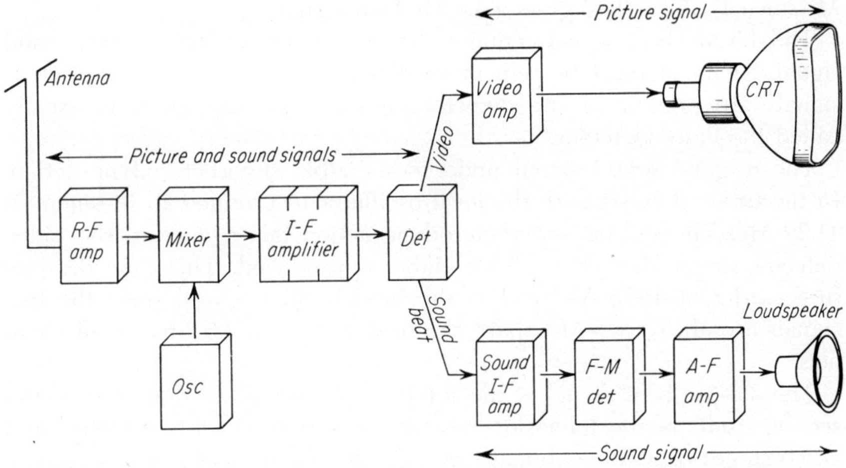

Fig. 15-1. The picture-sound signal path of a television receiver of the intercarrier type.

channel. In each channel, the FCC regulation places the picture carrier 1.25 Mc above the lower frequency limit and the sound carrier 4.5 Mc higher, or 0.25 Mc below the upper frequency limit. Thus, in Channel 2, which is 54 to 60 Mc, the picture carrier is at 55.25 Mc and the sound carrier is at 59.75 Mc. The picture carrier is amplitude-modulated; the sound system uses frequency modulation.

These are the two signals which are picked up by the receiving antenna and selected by the tuner when it is adjusted to receive Channel 2. At the same time, the channel selector switches the local oscillator in the receiver to 101 Mc. The r-f section is broad enough to pass both the picture and sound carrier. By superheterodyne action the picture carrier beats with the oscillator signal to produce a video i-f signal at 45.75 Mc, and the sound carrier beats with the same oscillator signal to produce a sound i-f signal at 41.25 Mc since:

$$101 - 55.25 = 45.75 \quad \text{(i-f for picture)}$$
$$101 - 59.75 = 41.25 \quad \text{(i-f for sound)}$$

The two i-f signals are then fed to the common i-f amplifier which usually contains three stages. These are broadly tuned so as to pass both i-f signals. The presence of both signals in the i-f amplifier must be kept in mind, even though it is often called the video i-f amplifier.

The amplified i-f signals are then fed to a detector. This is a simple diode or crystal detector working in a conventional a-m detector circuit. It demodulates the a-m picture i-f signal, extracting the video information and sending it through the video amplifier to the picture tube. At the same time, the detector will act as a superheterodyne detector for the beat between the picture and sound i-f signals. This beat will be at 4.5 Mc since 45.75 − 41.25 gives a 4.5-Mc beat signal.

The 4.5-Mc beat signal contains the modulation of the original sound signal. Again, it must be remembered that both the sound and picture signals are present in the detector circuit, even though it is usually called the video detector.

The original sound signal undergoes double superheterodyne action. In the tuner, it beats with the local oscillator to produce an i-f signal at 41.25 Mc. The second superheterodyne action takes place in the video detector stage where the 4.5-Mc signal is produced. This time the picture carrier at 45.75 Mc acts as the local oscillator, and since the two signals are always 4.5 Mc apart, the beat will be at 4.5 Mc for all channels.

The 4.5-Mc beat signal is the input to the sound section. The sound section itself, in the television receiver, is simply the i-f, detector, and audio stages of a conventional f-m receiver which has been designed to operate at an intermediate frequency of 4.5 Mc. There may be one or two sound i-f stages. If there are two, the second stage is usually a limiter; if only one, the following f-m detector is of the limiting type, like the ratio detector. Behind the detector are one or two a-f stages to drive the loudspeaker.

Receivers with Sound Take-off Behind the Video Amplifier. The input to the sound section is called the sound take-off point. In the block diagram of Fig. 15-1, the sound take-off point is behind the video detector. In some receivers, the sound signal is taken off behind the video amplifier, so as to get the gain of one more stage before the signal is fed to the sound section. A receiver of this type is shown in block-diagram form in Fig. 15-2.

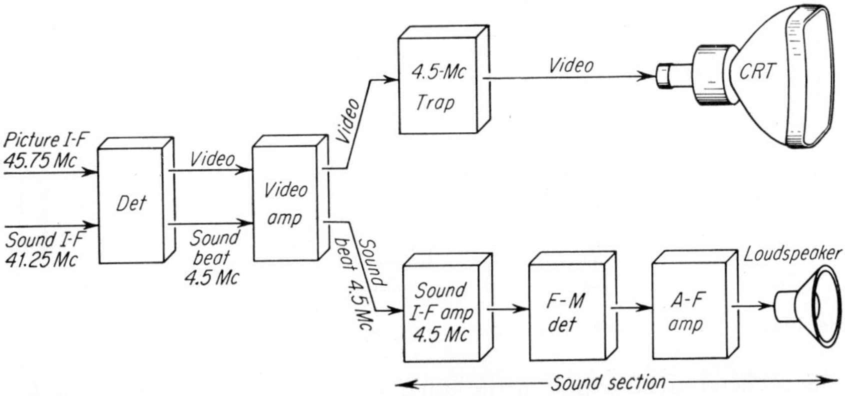

Fig. 15-2. Intercarrier receiver where the sound is taken off behind the video amplifier.

The second detector demodulates the a-m picture signal and acts as the double converter for the sound signal as was just described. The detector output, therefore, contains the video signal and a 4.5-Mc beat signal, both of which are fed to the video amplifier. This is an untuned stage, which must give good amplification over a wide frequency range to give good picture detail. Video amplifiers are designed to give reasonably flat amplification out to 5 Mc or higher. It will therefore pass and amplify the 4.5-Mc sound signal. This is then fed to the sound section, with the added gain of the video-amplifier tube.

If the 4.5-Mc sound beat is fed to the picture tube along with the video signal, it will give the picture a herringbone pattern or pebbly appearance. The picture-tube feed line therefore contains a 4.5-Mc trap, as shown in Fig. 15-2. This removes the sound signal from the picture-tube circuit. The 4.5-Mc trap is not limited to receivers where the sound take-off point is behind the video-amplifier stage. The sound signal is likely to be present in the video chain of any type of receiver, and some provision is made for its removal behind the sound take-off point in practically all receivers.

If the herringbone pattern is noted, it is likely to be caused by a misadjusted trap allowing the sound beat to get to the picture tube. The indication for correct adjustment is a minimum of the herringbone pattern in the picture.

Split-sound Systems. In some few receivers, the sound take-off point is before the video detector. This type is called a split-sound system, as distinguished from the more commonly used intercarrier system. A block diagram of a split-sound receiver is shown in Fig. 15-3.

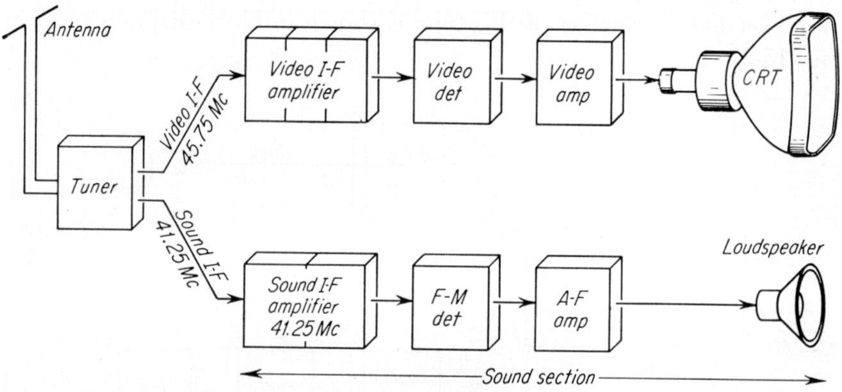

Fig. 15-3. Picture- and sound-signal chains in a split-sound system.

Superheterodyne action in the tuner produces two i-f signals, one for sound and one for picture. The sound signal picked up in the antenna beats with the local oscillator to produce a 41.25-Mc sound i-f signal (21.25 Mc in older receivers). The picture carrier beats with the same oscillator to produce a 45.75-Mc picture i-f signal (25.75 Mc in older receivers). The two i-f signals are channeled to separate i-f systems, one for sound and one for picture.

The sound signal at the output of the tuner is of low amplitude and is followed by as many as three i-f stages in some of the early split-sound receivers. Other receivers amplify both the sound and picture signals in a common i-f stage or stages before the signals are split to the two separate channels. These receivers have fewer stages. As long as the sound take-off point is before the video detector, it is regarded as a split-sound system. When the sound signal is taken off behind the video detector, we have an intercarrier system.

In the split-sound system, even though it has been taken off in a previous stage, some of the sound signal may be present at the video detector. Dual conversion takes place, and a 4.5-Mc beat signal appears in the video-amplifier stages. This will cause the herringbone pattern of the sound beat signal to appear on the picture tube. As a result, split-sound systems often include a 4.5-Mc trap in the video amplifier to reduce sound interference.

Types of Sound Section. Although the sound-signal chain in a television receiver begins at the antenna, the sound take-off point is usually considered as the beginning of the sound section. As was mentioned previously, the sound section is an f-m receiver including a 4.5-Mc i-f amplifier, a limiter, an f-m detector, and an audio amplifier. In the few cases of a split-sound system, the i-f amplifier is tuned to 21.25 or 41.25 Mc rather than the 4.5 Mc of the intercarrier system.

Sound sections in television receivers may be classified into three general types. One uses a limiter-discriminator circuit, another a ratio detector, and the third a 6BN6 or 3BN6 gated-beam detector. All three types are illustrated in block-diagram form in Fig. 15-4.

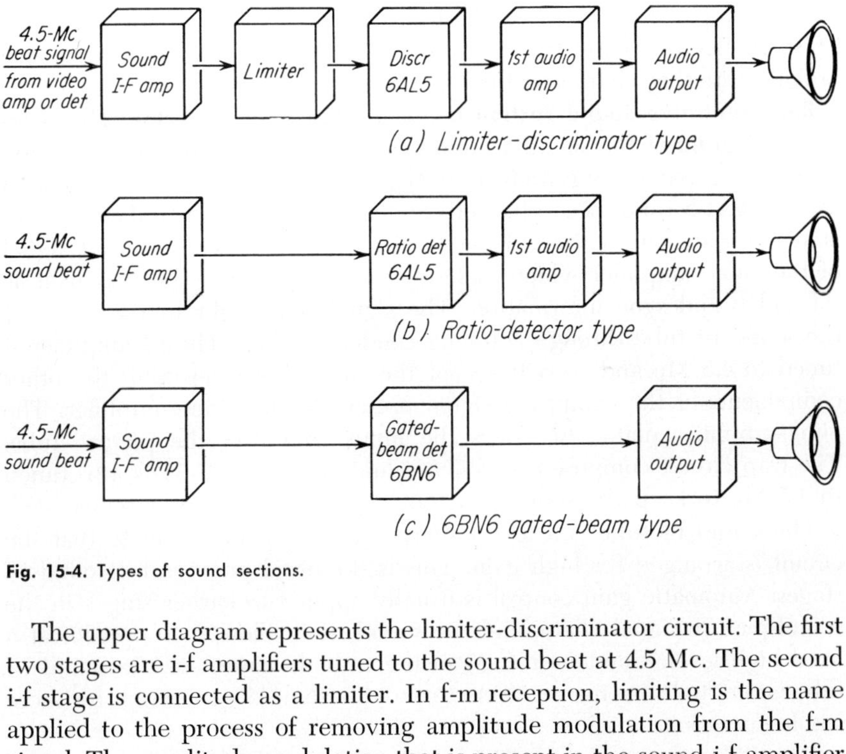

Fig. 15-4. Types of sound sections.

The upper diagram represents the limiter-discriminator circuit. The first two stages are i-f amplifiers tuned to the sound beat at 4.5 Mc. The second i-f stage is connected as a limiter. In f-m reception, limiting is the name applied to the process of removing amplitude modulation from the f-m signal. The amplitude modulation that is present in the sound i-f amplifier is primarily noise pulses and the sync peaks of the video signal. These may cause noise and buzz. A limiter stage tends to keep its output signal at a predetermined level, thereby removing a-m peaks which are greater than this level or limit. The limiter stage is followed by a discriminator-type f-m detector and two stages of audio amplification.

The center diagram of Fig. 15-4 represents the ratio-detector type of sound system. A ratio-detector circuit smoothes out amplitude variations and so is self-limiting. This removes the need for the limiter stage, which has been omitted from the diagram. However, the over-all gain of the system is lowered because of the absence of the limiter stage, and some ratio-detector circuits incorporate a second sound i-f stage to make up for this lack.

Aside from the limiter stage and some circuit differences, the discriminator and ratio-detector systems are very similar and may be serviced by the same procedures. Even in the differing detector circuits, both systems use a dual diode. This is usually a 6AL5 type. Sometimes two germanium crystals are used.

The lower block diagram of Fig. 15-4 shows a common arrangement where a 6BN6 or 3BN6 gated-beam detector tube is used. This is a special f-m detector which has a very high efficiency. There is usually only one i-f stage before it and only one a-f stage behind it. The circuit also provides limiting characteristics.

Representative Sound Section Using a 6AL5 Ratio Detector. The schematic diagram of the sound section in the Pacific Mercury 201 television chassis is given in Fig. 15-5 as a typical example of a sound section using a 6AL5 tube in a ratio-detector circuit. The sound take-off point is at the plate pin of the video-amplifier tube. This is the output signal of the video amplifier which includes the 4.5-Mc sound beat as well as the video and sync information. The signal is brought to the input of the sound i-f tube through coupling condenser *C*-23. The i-f amplifier is tuned to 4.5 Mc and so will accept the sound beat, rejecting the other components of the composite signal present in the video amplifier. The video-amplifier plate also feeds the input circuit of the picture tube. The trap circuit composed of coil *L*-4 and condenser *C*-22 is also tuned to 4.5 Mc to keep the sound beat out of the picture-tube circuits.

The sound i-f tube is a 6AU6 sharp-cutoff r-f pentode. Note that the circuit is arranged for high gain. This is the usual practice for sound i-f stages. Automatic gain control is usually applied to earlier stages in the sound-picture signal chain, and the sound i-f is allowed to operate at a fixed high level. Condenser *C*-25 in the cathode circuit of the sound i-f tube is omitted in some receivers to provide some degeneration for more stable operation.

Discriminator transformer *T*-5 couples the sound i-f stage to the 6AL5 tube which is hooked up as a ratio detector. Condenser *C*-28 is connected across the output of the two diodes and will be charged to a voltage dependent on the signal level. Its capacity is large enough to absorb changes in signal level caused by amplitude modulation, thereby providing the limiting action of this circuit. It is called a stabilizing condenser. The

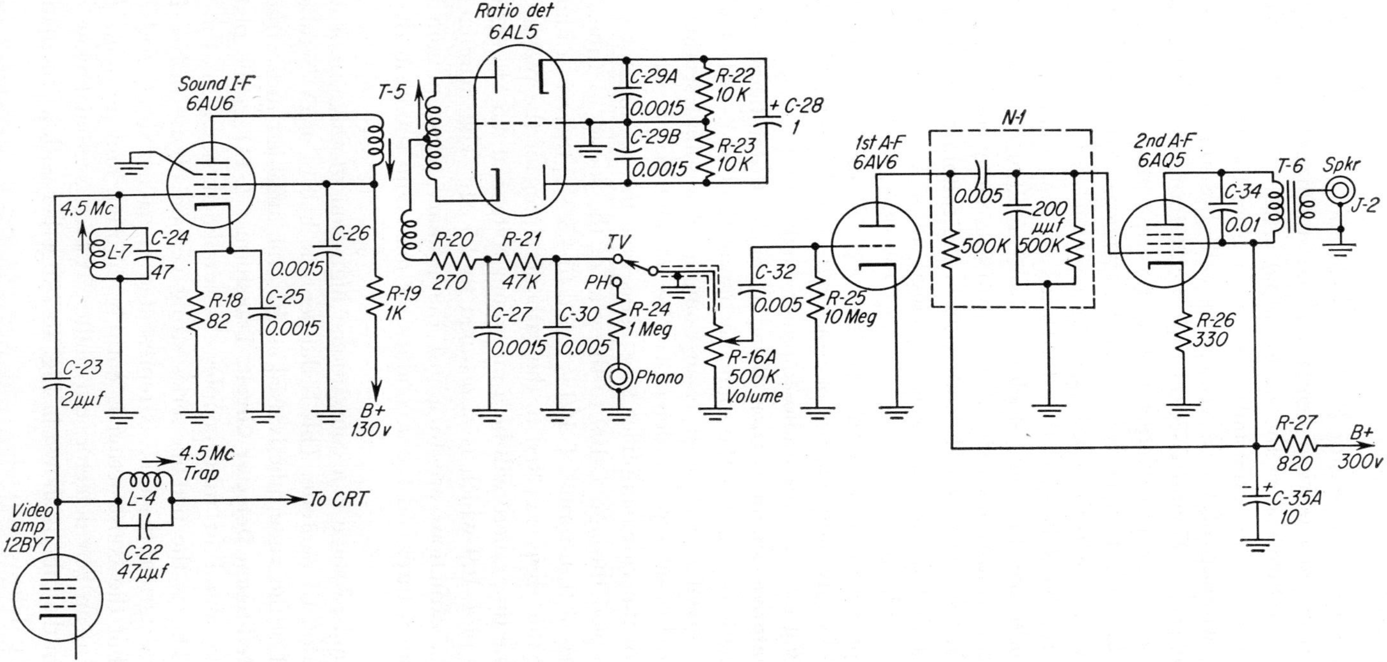

Fig. 15-5. Sound section in the Pacific Mercury 201 chassis.

network consisting of resistor *R*-22 and *R*-23 and condensers *C*-29*A* and *C*-29*B* provide an electrical center for the stabilizing condenser. This center is the return end of the audio output of the discriminator diodes. The high end of the audio output is taken from the center tap of the secondary winding of the discriminator coil. Resistors *R*-20 and *R*-21 and condensers *C*-27 and *C*-30 in this circuit form an i-f filter for the audio output of the ratio detector. Note how the phonograph input and discriminator output come together at the TV-PHONO switch so that either signal may be fed to the hot end of the volume control. The layout of television receivers is such that the leads to the volume control and phono switch may be long. Shielded wiring is often used in these circuits to prevent unwanted signal pickup in these long leads.

Note that the resistors and condensers in the coupling circuit between the two audio-amplifier tubes is a printed-couplate unit. In other respects, the audio amplifier is a standard two-stage unit. Also note the use of a speaker outlet jack *J*-2, and that one side of the speaker line is grounded. Finally, note the use of resistor *R*-27 and condenser *C*-35*A* as a decoupling filter for the audio-amplifier stages.

Representative Sound Section Using a 6BN6 Gated-beam Detector. The schematic diagram of the sound section of the Crosley 441 television receiver is given in Fig. 15-6 as an example of a sound section which uses a gated-beam type of detector. The sound take-off point is again the plate of the video-amplifier tube. This time the sound beat is transformer-coupled through *L*-109 to the grid of the sound i-f tube. When the primary of transformer *L*-109 is tuned to 4.5 Mc, there is a minimum current at this frequency fed to the cathode circuit of the picture tube. At the same time, there will be a maximum voltage at 4.5 Mc across the secondary of *L*-109 which is also tuned to 4.5 Mc. The manufacturer's alignment instructions are to adjust the primary for minimum sound beat in the picture, and the secondary for maximum sound in the loudspeaker.

The sound i-f stage is a conventional high-gain r-f stage built around a sharp-cutoff r-f pentode. This is the pentode half of a 5AV8 tube. The triode half of the same tube is used as the sync amplifier in this receiver.

The Gated-beam Detector Circuit. Transformer *L*-111 in the plate circuit of the sound i-f tube couples the amplified signal to the 3BN6 detector. This is the 600-ma series string version of the 6BN6. In the schematic diagram, the tube is represented as having three grids like a pentode, but the construction is entirely different and accounts for its operation as an f-m detector and limiter. A cross-sectional view of the elements in the tube is shown in Fig. 15-7 together with a schematic diagram of the detector circuit.

The first grid at pin No. 2 is called the limiter grid. The f-m i-f signal

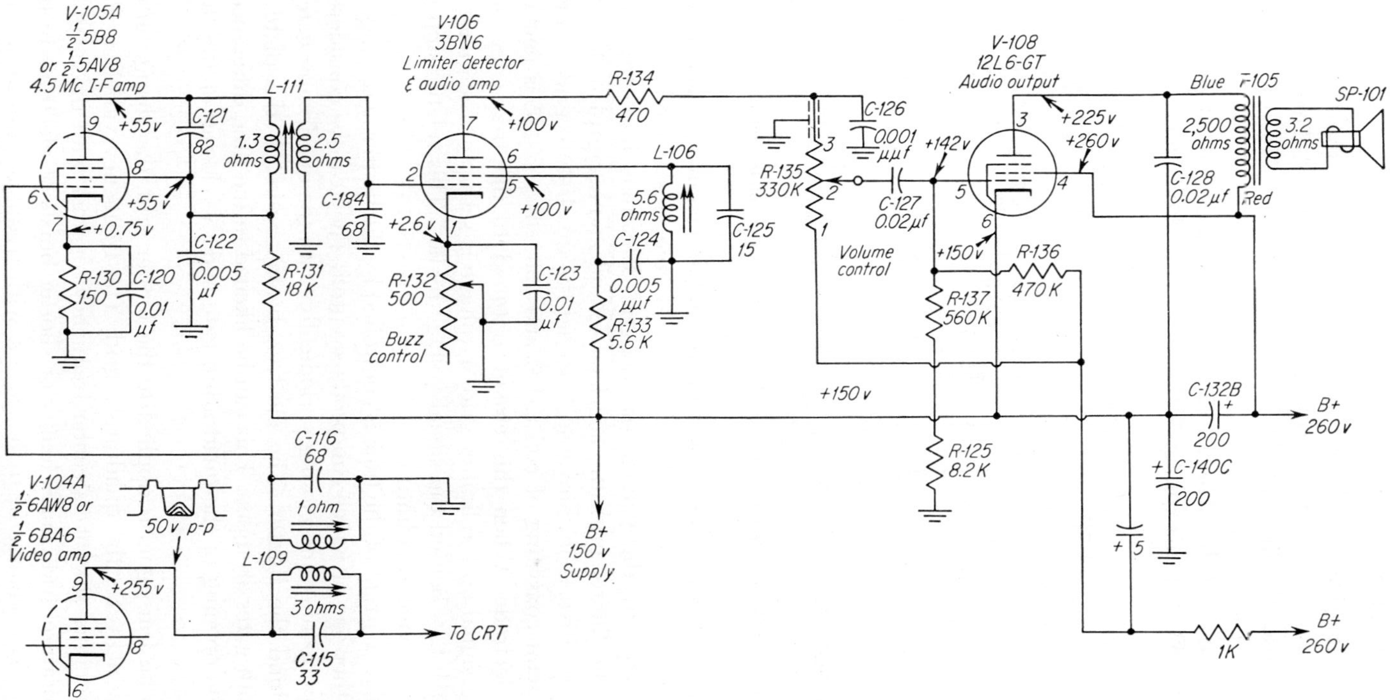

Fig. 15-6. Sound section of Crosley chassis 441.

is fed to this grid from the secondary winding of i-f transformer *L*-111. The connection at pin No. 5 is called the accelerator. This is an anode which encloses the limiter grid and together with the focusing electrode around the cathode forms a sort of electron gun similar to the one in a cathode-ray tube. The accelerator is connected to B plus through resistor *R*-133, and this voltage is the main control of the cathode current.

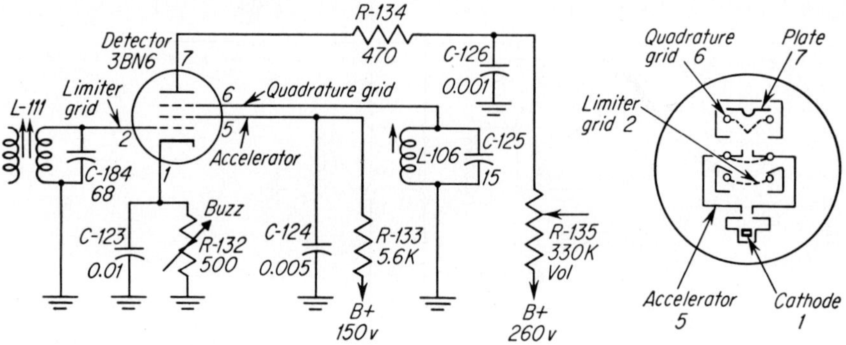

Fig. 15-7. The gated-beam detector circuit and a cross-section view of the tube.

The gunlike construction of the cathode and accelerator projects a steady stream of electrons upon the limiter grid, and this cathode current is little affected by the voltage on the grid. Electrons from the gun go through the limiter grid, the accelerator, and another grid to the plate, which is positive. The third grid is called the quadrature grid. A high-Q tuned circuit consisting of coil *L*-106 and condenser *C*-125 is connected to this electrode. When this circuit is tuned to the frequency of the incoming signal on the limiter grid, a voltage is set up at the quadrature grid which lags behind the signal voltage by 90°. This relationship gives the grid its name, quadrature.

The plate current of the tube is controlled by the limiter and quadrature grids, either one of which can produce cutoff. They can be considered as two gates opening and closing periodically, with the quadrature gate lagging behind the limiter. The electron beam can reach the plate only when both gates are open. This can be likened to the time difference between the opening of the quadrature gate and the closing of the limiter gate.

When an f-m signal is applied to the limiter grid, the change in electron flow activates the quadrature grid. The frequency variation of the signal on either side of its center frequency together with the lagging of the quadrature grid causes longer or shorter bursts of electrons to reach the plate. This change in plate current with frequency deviation of the applied signal is f-m detection. Resistor *R*-124 and condenser *C*-126 act as the i-f filter, applying the audio component across potentiometer *R*-135, which is the manual volume control for the receiver.

If the bias on the limiter grid is adjusted so that the operating point is between plate-current cutoff and plate-current saturation, a small positive voltage on the grid achieves saturation, and a small negative voltage produces cutoff. Larger excursions of signal voltage such as those caused by a-m components can have no greater effect on the plate current, and thus limiting is accomplished. The correct operating point for the limiter grid is critical and is controlled by variable resistor *R*-132 in the cathode circuit of the tube. Adjustment of this control sets up the limiting points for the circuit. It is called the noise rejection, or buzz control.

Audio-output Stage. The audio output of the gated-beam detector tube is large enough to drive the power audio tube without an intervening voltage-amplifier stage. The audio-output tube is a 12L6-GT which is the 600-ma controlled heater equivalent of the 25L6-GT beam power tube. The Crosley receiver makes use of a shelf-type voltage-distribution system in the power supply, where the last audio tube is used as a voltage-dropping resistor to provide plate supply for other tubes in the receiver.

The circuit of the last audio stage is shown in Fig. 15-8. First examine Fig. 15-8*a*. The B power supply produces 260 volts which is applied to

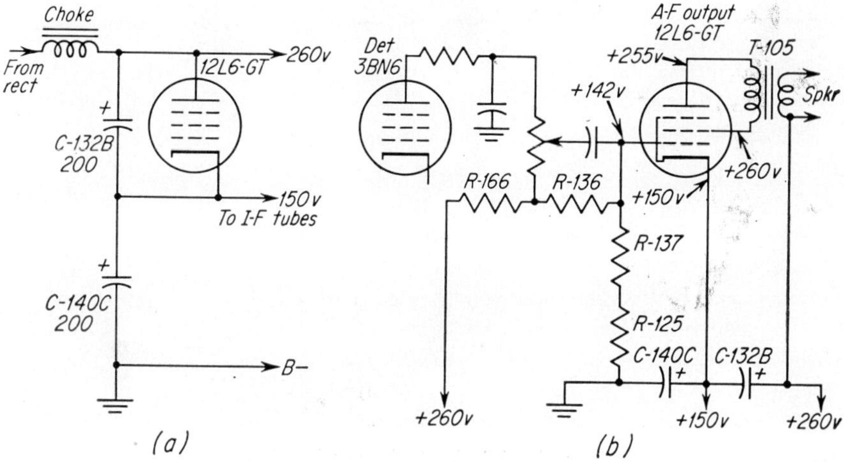

Fig. 15-8. The shelf-type circuit of the audio-output stage in the Crosley 441 receiver.

the plate of the 12L6-GT tube. The plate-cathode circuit of the tube acts as a voltage-dropping resistor, and the cathode is therefore the source of the B-plus 150-volt line for the rest of the receiver. Condensers *C*-132*B* and *C*-140*C* filter the two halves of the voltage-distribution system.

With the cathode of the 12L6-GT at +150 volts, and the plate at +260 volts, the actual B voltage applied to the tube is the difference, or 110 volts. The grid circuit is shown in Fig. 15-8*b*. The grid is returned to ground through resistors *R*-137 and *R*-125. This would place a bias of 150 volts on the grid of the tube, where it would be cut off. To avoid this condition, resistor *R*-136 connects the grid to the high B plus point. The grid is thus placed at +142 volts, which makes it 8 volts more negative than the cathode at +150, and the tube is therefore operating at its normal bias of 8 volts.

Shelf-type power supplies may be used in receivers using ratio detectors. Similarly, 6BN6 detectors are often used with conventional audio amplifiers and power supplies.

How to Identify the Sound Section. Although set layouts vary widely, they usually follow a general pattern. The pattern which helps to identify the sound section in any receiver is illustrated in Fig. 15-9. The i-f

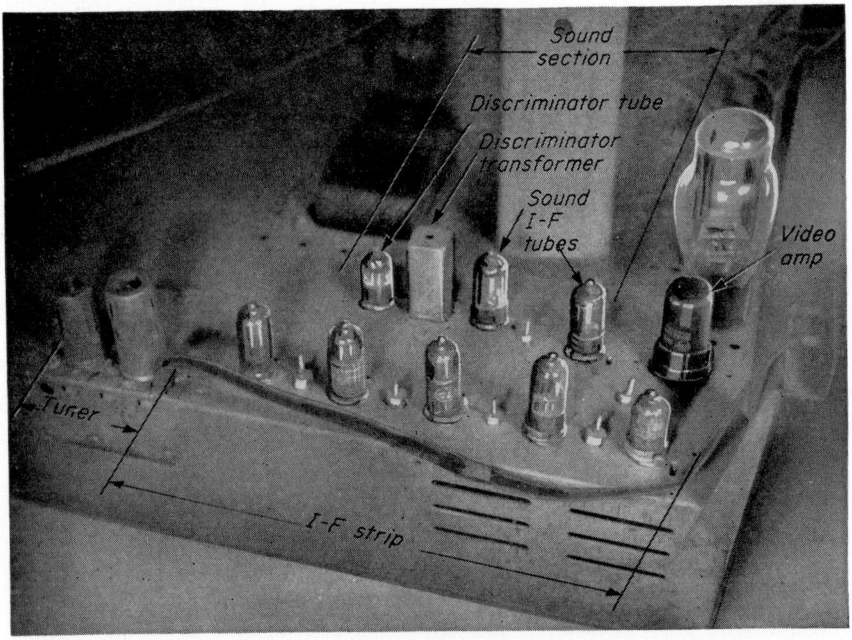

Fig. 15-9. Identification pointers that help locate the sound section.

amplifier usually leads out from the tuner in a straight line. The i-f adjustments and tubes in a row make the i-f strip easily recognized. The first stage will be closest to the tuner, and the last stage farthest. The video-amplifier stage is close to the last i-f stage, and nearby will be the adjustment screws and tubes of the sound i-f amplifier. This will be followed by the discriminator transformer. Behind the discriminator transformer comes the 6AL5 ratio detector, easily recognized by its small size. The set may have two 6AL5 tubes, because this tube is often used as a phase detector in horizontal deflection systems. It may also be that the detector tube may not be found at all, since some receivers use crystal diodes, usually mounted in the discriminator can.

Now look for the volume control which is usually combined with the on-off switch. The volume control may be at some distance from the detector tube, in which case there will be a shielded cable between the two. The audio tubes can be traced back from the loudspeaker, jack, and output transformer.

If the set uses a 6BN6 gated-beam discriminator, the layout and identification pointers are exactly the same except for the tube itself. The 6AL5 is easily recognized because of its small size; the 6BN6 is a little taller than the 6AU6 sound i-f tube.

Servicing Procedure in the Sound Section: A-F Stages. When a complaint of normal picture and no sound throws suspicion on the sound section, the first step in a servicing procedure for this complaint is to localize the trouble to the audio stage or stages, or the sound i-f and detector stages.

The volume control is the connecting link between these two subdivisions of the sound section. The output of the detector is impressed across the volume control and becomes the input to the audio stages. If the audio stages and loudspeaker are in working order, any audio signal fed to the volume control will be heard in the loudspeaker. This is the basis for checking the audio stages.

The test setup is illustrated in Fig. 15-10. The audio output of a standard signal generator is used as the test signal. The top diagram is a pictorial representation of the test signal being fed to the volume control. The next diagram shows a conventional two-stage amplifier behind a ratio detector, and the bottom one is the one-stage amplifier usually found behind a gated-beam detector. The test point is the arm of the volume control which is labeled *A* in all three diagrams. The volume control is an easily recognized component, since it has the off-on switch. Point *A* is the center of the three volume-control terminals. The volume control is turned to the full on position when this check is made.

As shown in the diagram, the hot generator lead is connected through a 0.05-μf condenser for isolating purposes. The audio-frequency output of most generators is near 400 cps. When the above connections are made, if the audio amplifier is working, a 400-cycle note will be heard in the loudspeaker. In receivers where there are two a-f stages behind the volume control, a much louder test note will be heard than in single-stage

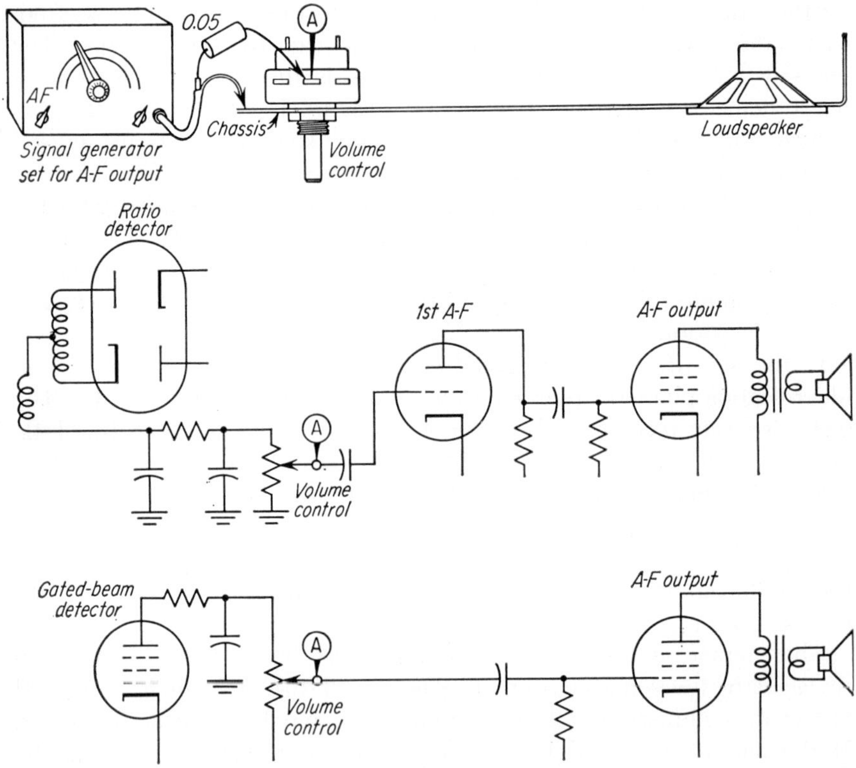

Fig. 15-10. Audio signal check to locate trouble in the sound section.

amplifiers. If the note is heard, the audio amplifier and loudspeaker are working, and the trouble is in the sound i-f and detector stages.

If a signal generator with a 400-cycle output is not at hand, there are other ways of obtaining an audio test signal. The tip of a soldering iron or gun or even a moist finger when applied to the arm of the volume control will produce a low-pitched growl from the loudspeaker to show that the audio amplifier is working.

Servicing the Audio-amplifier Stages. A diagram of a basic two-stage amplifier commonly used in television receivers is given in Fig. 15-11. The arm of the volume control is labeled point *A* as in the previous check. An audio test signal at point *A* places trouble in the audio stages when no response is heard from the loudspeaker.

The next step is to apply the test signal to test point *C*, the grid of the power audio tube. If the test note is heard in the loudspeaker, the last audio stage and speaker are working, and the trouble is between points *A* and *C* in the receiver. If the test note is not heard, the trouble is either in the last audio stage or in the loudspeaker. The latter is easily checked with an ohmmeter on the $R \times 1$ range. Placing the ohmmeter test prods across the voice coil will produce a distinct click in the loudspeaker if the voice coil is good. The transformer secondary does not have to be disconnected for this check.

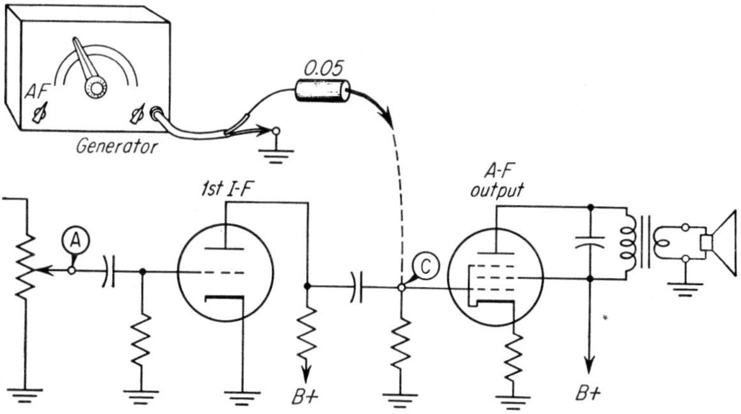

Fig. 15-11. Isolation check to locate faulty stage in the audio amplifier.

The same three check points: the arm of the volume control, the grid of the power audio tube, and the voice coil would be used in receivers having only one audio stage. However, in these sets the only component between points *A* and *C* is the coupling condenser.

These checks limit the fault closely to a very small area. Thereafter, voltmeter, ohmmeter, and substitution checks point out the defective component.

It is recommended that you run through these signal checks on several working receivers. In this way, you will become familiar with the amount of response that may normally be expected from any test point with your test equipment. This will be a considerable help when you are working on a set for a complaint of weak sound. When making the trial runs, operate the set without an antenna, and on an off-station channel so that extraneous sounds will not interfere with the test.

Servicing Procedure in the Sound Section: Sound I-F and F-M Detector Stages. The block diagram of the picture-sound signal chain has been repeated in Fig. 15-12 as an aid in localizing the fault to a particular subsection. The complaint is normal picture—no sound. The normal picture indicates that the picture chain from the antenna to the CRT is working, and the lack of sound places the trouble in the sound section. In the previous check, an audio test signal fed to the volume control (test point *A* in the diagram) produced a note in the loudspeaker, proving that the a-f amplifier and speaker are working. The trouble now should be in the sound i-f or f-m detector stages. This may be confirmed as follows:

Check to see that the receiver is on an unused channel, and the contrast and volume controls are in the full on position. Feed a modulated r-f signal at 4.5 Mc to the output of the video detector. This is test point *B*, at the diode load. The check is illustrated in Fig. 15-12. Note that it

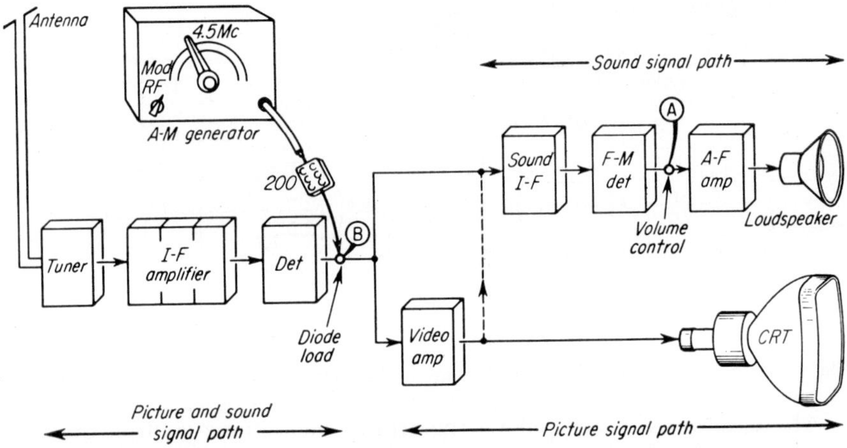

Fig. 15-12. Placing the cause of no sound in the sound i-f and f-m detector stages.

feeds a test signal to the sound section regardless of whether the sound take-off point is before or after the video amplier. We know the video amplifier is operative because of the normal picture. The frequency control on the signal generator is wobbled back and forth between 4 and 5 Mc.

Since we are checking for a suspected break in the sound-signal chain, there should be no response from the loudspeaker. If a response comes through at an off-frequency setting of the generator, realignment of the sound section is indicated. But if a signal comes through at 4.5 Mc, the sound section is definitely working. Now the absence of sound must be due to the inability of the receiver to produce the 4.5-Mc intercarrier beat in the video-detector stage. Since we have a normal picture, this can only be due to suppression of the sound carrier or the sound i-f signal in the tuning circuits in the common tuner or i-f stages before the sound take-off point. These are now checked in accordance with the manufacturer's alignment instructions, or the generalized alignment procedure given in Chap. 12.

Servicing the Sound I-F and Detector Stages. The check just given places the trouble in the sound i-f and detector stages, and now the procedure is continued to localize the fault to just one stage.

The diagrams of Fig. 15-13 illustrate the localizing check. An audio test signal at point *A* in either of the two diagrams produces a normal response in the speaker, proving that the audio section is operative. Next, a 4.5-Mc test signal at point *B*, the diode-load test point gives no response, indicating trouble in the sound i-f and sound detector stages.

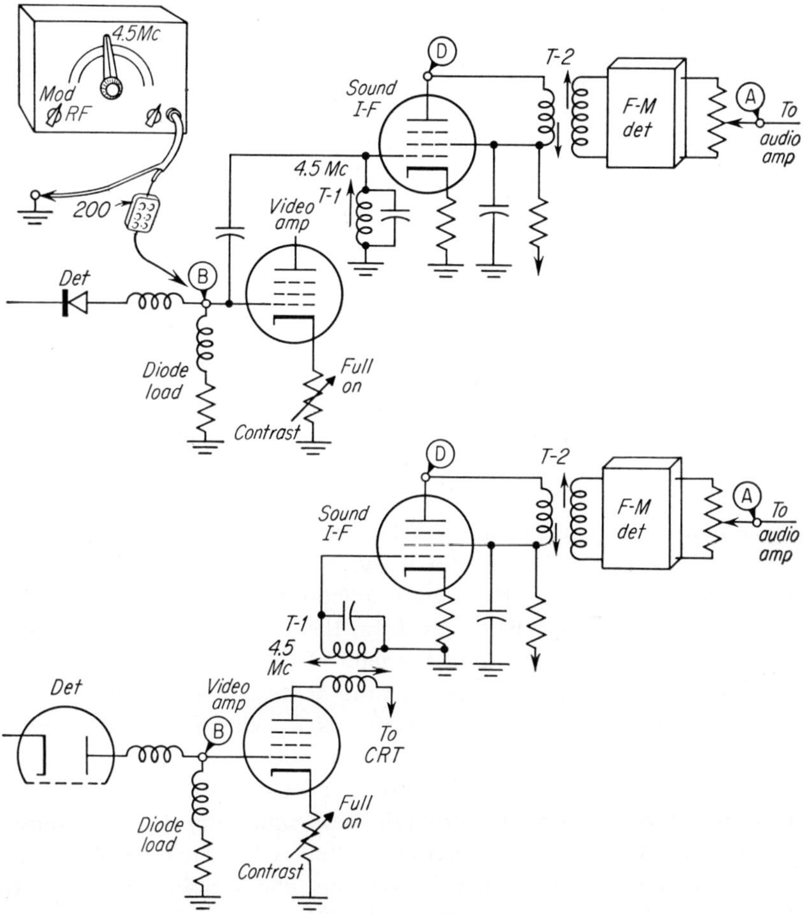

Fig. 15-13. Isolation checks to locate faults in the sound i-f and f-m detector stages.

Two diagrams are given to illustrate different designs commonly found in receivers. The upper diagram shows the sound take-off point at the output of the video detector; the lower gives the take-off point at the output of the video amplifier. Also one shows a crystal as the video detector, and the other uses a diode. A particular receiver may use any combination of these variations. However, in all cases, point *B* may be used to feed a 4.5-Mc signal to the sound section. As mentioned before, no response places the trouble in the sound i-f or f-m detector stages.

The servicing procedure continues as follows: The a-m generator is adjusted to produce a strong modulated r-f signal at 4.5 Mc and is fed to the plate pin of the sound i-f tube. The receiver is kept on an unused channel, and the volume control is turned to the full on position. The frequency control on the generator is then wobbled back and forth between about 4 and 5 Mc while listening for the test note in the speaker. If the f-m detector stage is working, the signal-generator note will be heard in the loudspeaker as the frequency control passes 4.5 Mc.

Now reduce the attenuator on the signal generator so that the note in the speaker is just about audible, and wobble again. You will now be able to observe the exact frequency at which the note is heard. A gated-beam discriminator should produce peak sound as the generator passes 4.5 Mc, and the ratio detector should produce a null at 4.5 Mc, with loud points on each side of the null. If the generator note is heard at an off-frequency setting, realignment is indicated. If the note is not heard at all, the indication is an inoperative detector stage which is then thoroughly tested for the fault. A ratio detector is checked by tube substitution and resistance checks. Since the gated-beam discriminator is supplied with operating potentials, voltage checking is helpful when working on this detector.

If the normal response is heard, the detector stage is working, and the test clip from the signal generator is moved back to the sound i-f grid. The frequency control on the signal generator is again wobbled while listening for the characteristic note from the loudspeaker. The response should now be louder because of the gain of the tube. If the note is not heard, the trouble is in the sound i-f tube or its associated circuits.

If the note is heard, the sound section is working from the sound i-f grid through to the loudspeaker. Now the trouble must be in the coupling devices between the sound take-off point and the sound i-f grid.

You are again urged to run through these signal checks on a number of receivers that are operating normally. You will then be familiar with the amount of test signal needed from any check point in any of the various types of sound sections encountered.

Servicing Split-sound Systems. An occasional split-sound system may be encountered in service work. It will be remembered that these are the receivers where the sound take-off point is before the video detector and where the sound section contains a separate i-f channel tuned to 41.25 or 21.25 Mc.

The serviceman will recognize that he is working on this type of receiver by observing the separate i-f channel on the chassis, or by the schematic diagram, or by the operation of the fine tuner on the receiver.

In an intercarrier receiver, the fine tuner is adjusted for best picture and is relatively ineffective on the sound signal. In the split-sound system, the fine tuner is relatively ineffective on the picture and tunes the sound sharply. This effect is so marked that, in the early days of television, when practically all receivers had split-sound systems, the fine tuner was named the sound-sensitivity control.

The servicing procedure for the split-sound system is exactly the same as the one just described for the intercarrier system except that the test frequency in the sound i-f stages is 41.25 or 21.25 Mc instead of 4.5 Mc.

Sound Section in a Color-Television Receiver. Color-television receivers employ a sound section that is very similar to the sound section in black-and-white receivers. Some color receivers are somewhat different in that a separate detector is used to produce the 4.5-Mc sound beat. The sound section of the CBS 205 color-television chassis is given in Fig. 15-14 as an example of this type of circuit.

Some of the energy in the plate circuit of the fifth i-f amplifier is fed to a separate crystal detector through a 1.5-$\mu\mu$f coupling condenser, as shown in the diagram. Note that the 1N60 crystal detector is located within the *TB*-3 sound-transformer assembly. Here the sound and picture i-f signals beat to produce a 4.5-Mc signal which becomes the input to the sound i-f amplifier. In all other respects the sound section is conventional.

Service work in this sound section is the same as in any other, except that a 4.5-Mc test signal could not be fed into the sound section from the video amplifier. Furthermore, the detector producing the 4.5-Mc sound i-f signal has to be checked. In the usual receiver, where one detector is used for production of the sound beat as well as demodulation of the video signal, it is not necessary to check the detector when servicing the sound section. This is because the normal picture shows that the detector is working. The sound-section test procedure is therefore adapted to this variation in the following way.

For a complaint of normal picture—no sound or weak sound, feed an audio test signal to the arm of the volume control. If there is no response or a weak response, check the audio-amplifier stages and loudspeaker. If there is a normal response, these stages are working, and an a-m modulated r-f test signal at 4.5 Mc is fed into the plate of the last sound i-f amplifier. The frequency of the signal generator is wobbled from 4 to 5 Mc, and if the normal response is heard at 4.5 Mc, the ratio detector is working. The 4.5-Mc test signal is then fed grid by grid to the two sound i-f stages to show whether these stages are working.

Finally, the 4.5-Mc test signal is fed to the grid of the fifth i-f tube and is wobbled from 4 to 5 Mc. If the normal response is heard as the frequency control on the generator passes 4.5 Mc, the entire sound section

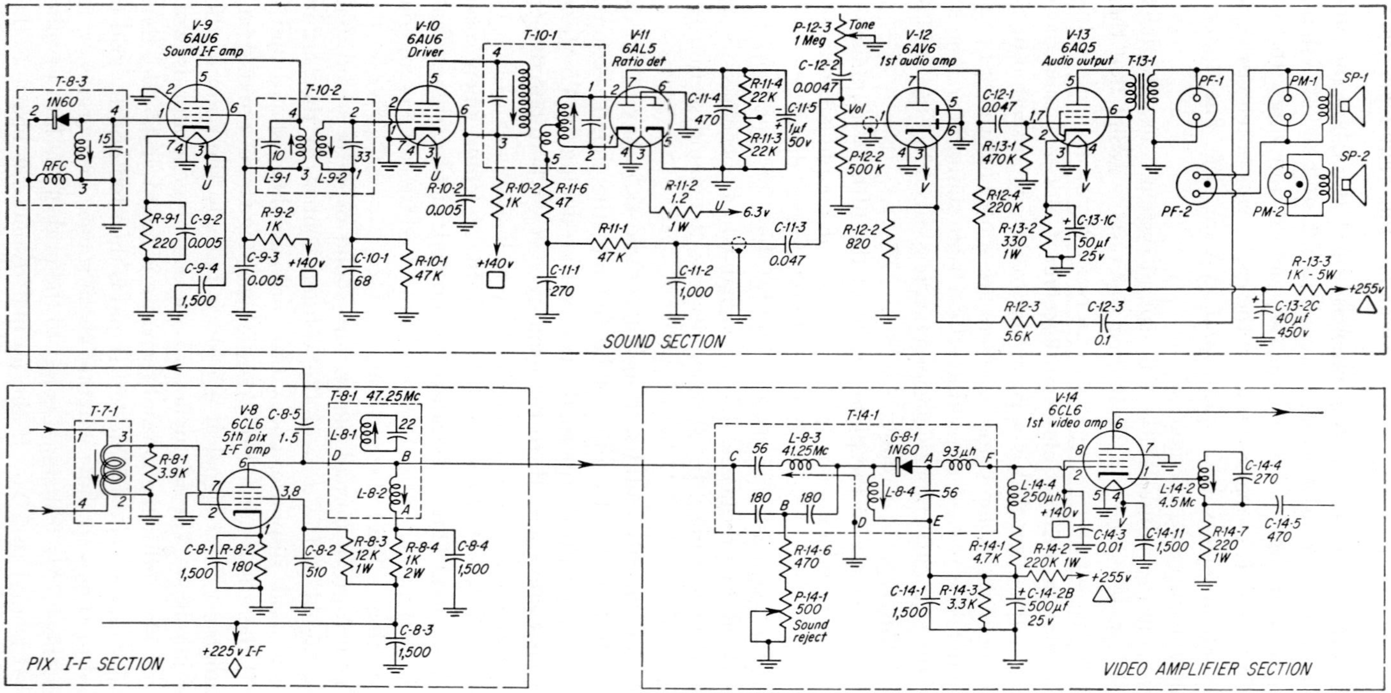

Fig. 15-14. Sound section in the CBS 205 color-television receiver.

is working, and the sound i-f signal is lost in the tuner or video i-f stages. If the signal is not heard, the defect is an open in coupling condenser *C*8-5, or crystal assembly *TB*-3. The condenser and crystal are checked by substitution, and the associated components with an ohmmeter.

ALIGNMENT OF THE SOUND SECTION

The servicing procedure indicates a need for realignment when the test signal produces a response at an off-frequency setting. Alignment is also necessary when a component in one of the tuned circuits proves defective and requires replacement. The replacement part needs to be tuned, and the best procedure is realignment of the entire section. When servicing for intercarrier buzz, in a set using a gated-beam detector, the setting of the buzz control wants to be checked, and this is also best done by a complete realignment of the sound section.

When performing the realignment, it is, of course, always preferable to follow the alignment instructions given in the manufacturer's service notes. In cases where these are not available, generalized procedures are given in the next sections. Two procedures have been prepared, one for use with ratio-detector receivers, and one for sets using the 6BN6 or 3BN6 gated-beam detector.

On-station alignment methods are given in the service notes offered by most manufacturers and are used in this procedure. For those receivers where the complaint is no sound due to off-frequency operation of the tuned circuits in the sound section, the procedure is first to tune roughly to 4.5 Mc by the aid of the signal generator. The station signal will then come through strongly enough to start the accurate sound alignment.

Location of Aligning Screws in a Ratio-detector Sound Section. In any alignment procedure, a first requisite is to locate and identify the various adjustment screws. A basic diagram of a ratio-detector type of sound section, and the layout of a representative sound strip are given in Fig. 15-15. The 6AL5 tube is easily recognized because of its small size. Discriminator transformer *T*-2 is usually larger than the other i-f cans, if present. The adjustment screw labeled *A*-1 is as shown, or accessible through the hole in the top of an input-transformer assembly can.

When aligning the ratio detector, it is important to distinguish between the adjustments for the primary and secondary of the discriminator transformer. The diagram of Fig. 15-15 places *A*-3, the secondary adjustment in the top of discriminator can *T*-2, and A-2, which tunes the primary winding, in the bottom. This slug is adjusted from the underside of the chassis. Check the service data carefully to determine the location of these adjustments.

If the service data do not include location diagrams, another convention

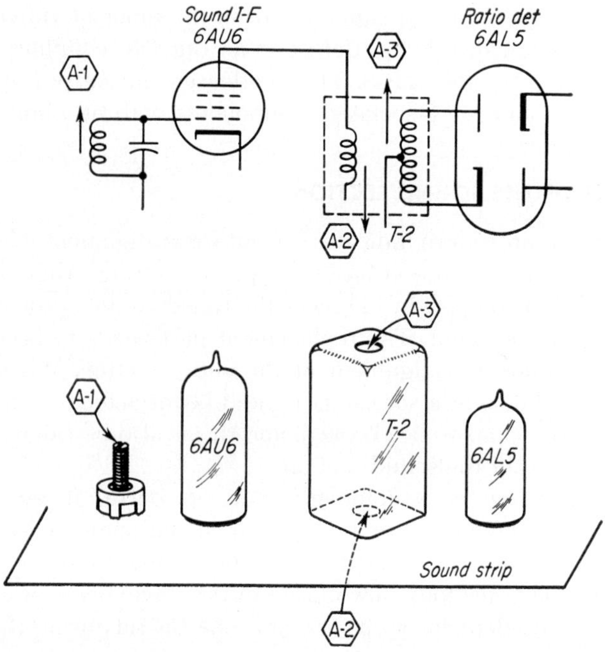

Fig. 15-15. Location of aligning adjustments in a sound section of the ratio-detector type.

may be an aid in identifying the adjusters. In the schematic diagram, note that the arrow which represents the primary tuning slug A-2 points down, and the arrow for secondary tuning slug A-3 points up. This convention is used in many schematic diagrams. Top arrowheads indicate top adjustments, and bottom arrowheads indicate bottom adjustments. If there is no other way of determining the primary and secondary adjustments of the discriminator transformer, it will be necessary to remove the shield can and check the positions of the windings. The primary coil has two leads, and the secondary three.

In some discriminator-transformer assemblies, both adjustments are made from the top of the chassis. A coil of this type is shown in Fig. 15-16. The primary and secondary coils are wound on an internally threaded fiber form. The tuning slugs are threaded on the outside. The adjusting slots are hexagonal in shape and extend through the cores. The aligning tool is also shown in Fig. 15-16. This has a hexagonal-shaped head mounted on a thin nylon shank. When adjusting the lower core, the aligning tool is slipped through the upper core slug to engage the lower. Then turning the tool causes the lower core slug to go up or down in the threaded fiber form. The position of the upper core slug is unaffected, because the thin nylon shank will not engage its hexagonal

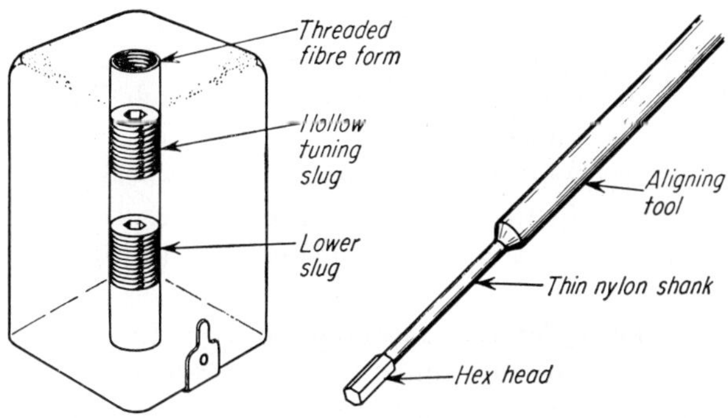

Fig. 15-16. Two-coil assembly where both slugs are adjustable from the top.

hole. To adjust the upper core, the tool is withdrawn so that its head engages the upper hole.

I-F Peaking Adjustments in a Ratio-detector Sound Section. The test signal for alignment of the sound section is obtained from a regular television program. Use the strongest channel in your area, and tune the receiver for the best picture. Use a vacuum-tube voltmeter as the output indicator. Set it up on its lowest range as a d-c voltmeter to measure negative potential. The common terminal is connected to chassis, and the d-c probe is connected to the negative end of stabilizer condenser *C*-5, point *E* in Fig. 15-17.

Adjust *A*-2, the primary slug of the transformer, for maximum indication of the output meter. Then adjust *A*-1, the tuning slug in the grid of the sound i-f tube, for peak indication. If the sound take-off is by means of a transformer with a tuned primary and secondary, the primary is adjusted now for maximum sound. This adjustment may be retouched

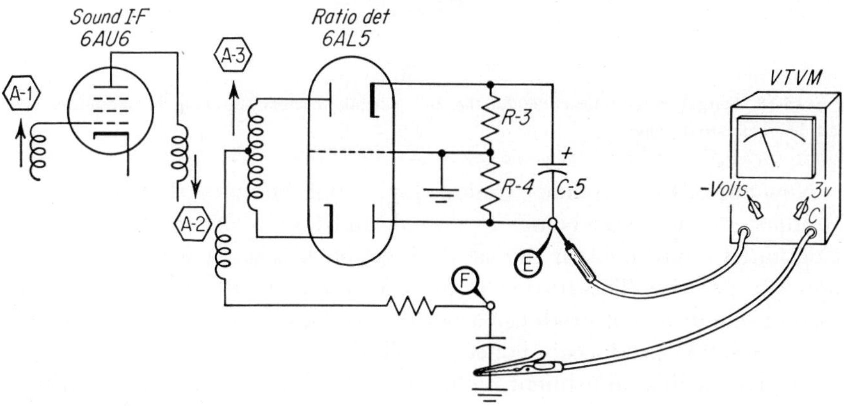

Fig. 15-17. Output meter connections for peaking adjustments in the sound i-f stage.

later, if needed, for minimum sound in the picture. In some receivers, the tuned primary of the sound take-off transformer also acts as the sound trap for the picture. Recheck the adjustment of each slug until a maximum indication is obtained.

If the receiver has two sound i-f stages, there will be one or two more peaking adjustments in the i-f interstage coupling. All peaking adjustments are made at the same time, followed by the discriminator null adjustment.

Adjusting the Ratio-detector Secondary Circuit. In the ratio detector, the secondary coil is properly tuned when the voltages in the two diode circuits are balanced. This condition may be measured by zero voltage between point *F*, the audio feed point, and the mid-point of the diode-load resistors *R*-3 and *R*-4 which is at ground potential (see Fig. 15-17). Some ratio-detector circuits use a single load resistor like *R*-3 in Fig. 15-18. In these sets, to get the balance reference point, the diode load is bridged with a pair of resistors of about 100,000 ohms each, and the junction of these is used. The connections are shown by the dotted lines in Fig. 15-18. The two resistors are removed after the alignment is completed.

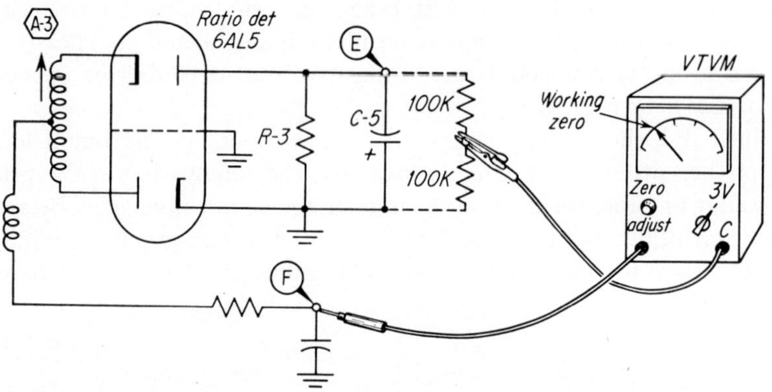

Fig. 15-18. Output meter connections for the null indications when adjusting the secondary of a discriminator transformer.

Now adjust the secondary tuning slug. It will be noted that, as the slug is adjusted, the meter comes to a peak and then drops sharply to zero. Continued adjustment in the same direction brings on a peak with the opposite polarity. The correct position is a zero reading, where a slight turn in one direction produces a positive voltage and a slight turn in the other direction produces a negative voltage.

In making this adjustment, it is desirable to have a voltmeter with a zero in the center of the scale. If this scale is not available on your meter, set the zero adjust control to a convenient marker on any scale, with the meter disconnected. Then this no-voltage marker becomes the working zero for the alignment. The meter pointer in Fig. 15-18 is adjusted for a convenient working zero. Tuning slug *A*-3 is adjusted so that the pointer is at the working zero, and a slight turn in either direction causes the pointer to swing either positive or negative.

If it is found that tuning slug *A*-3 was far off from its correct setting, repeak *A*-1 and *A*-2, the tuning adjustments in the grid and plate of the sound i-f tube.

Alignment of a Sound Section Having a Gated-beam Discriminator. First locate and identify each adjustment in the sound section. If a location diagram is not available, find the 3BN6 or 6BN6 tube. Then trace leads so as to identify the adjustments.

Refer to Fig. 15-19, which shows a basic diagram and sample layout for a gated-beam sound section. The No. 6 pin of the 6BN6 connects to the quadrature coil which establishes the position of this coil adjustment. The cathode at the No. 1 pin leads to the a-m rejection or buzz control *R*-4. This control is sometimes mounted on the rear apron of the chassis,

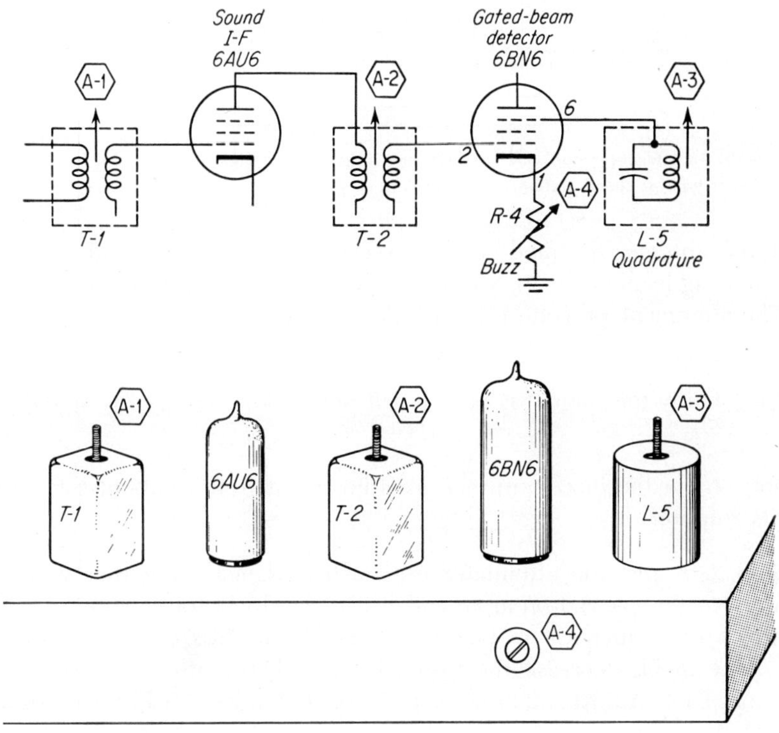

Fig. 15-19. Identifying and locating alignment adjustments in a gated-beam detector system.

as shown in Fig. 15-19. The No. 2 pin is connected to the interstage transformer between the sound i-f and detector tubes. The sound take-off coil *A*-1 will be somewhere between the video-amplifier and sound i-f tubes.

A transmitted television signal is used for alignment. No output meter or indicating devices are needed. However, correct alignment of the input coils *T*-1 and *T*-2 calls for an input signal that is below the limiting level of the gated-beam detector tube. Various methods may be used to reduce the input signal to this level. One is to disconnect the antenna leads and to loosely couple the 300-ohm lead-in wire to the antenna leads in the receiver. A better method is to use an adjustable attenuator pad like the one illustrated in Fig. 15-20. This is inserted between the antenna

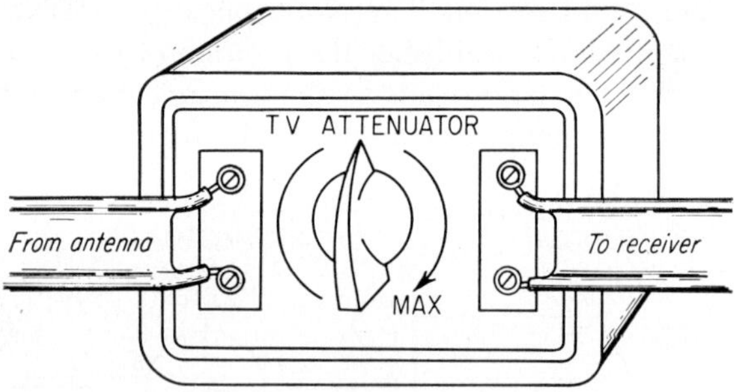

Fig. 15-20. An attenuator pad which reduces the signal fed to a receiver is useful in aligning a gated-beam type of sound section.

and the antenna terminals on the receiver. When the signal is below the limiting level of the 6BN6 tube, a hiss is heard in the sound output.

The alignment procedure is as follows:

Step 1. Tune in a station, and adjust for best picture. A test pattern with a steady tone signal is best, but if none is on the air, any signal will do.

Step 2. Set the buzz control *R*-4 (adjustment *A*-4) to about the center of its rotation.

Step 3. Adjust the attenuator until a hiss is heard together with the sound. Next adjust *A*-1 on input coil *T*-1 in Fig. 15-19 for maximum sound. Disregard any increase in buzz level, but if the hiss disappears during the adjustment, decrease the signal level with the attenuator pad until it comes back. Adjust *A*-2 on interstage coil *T*-2 in Fig. 15-19 for maximum sound. Again disregard increase in buzz, and reduce the signal if the hiss disappears. Then repeat both adjustments for maximum sound output.

Step 4. Readjust the attenuator so that the signal level is increased and the hiss disappears. The 6BN6 is now limiting, which is the correct condition for adjusting the quadrature coil and buzz control.

Step 5. Adjust *A*-3 on the quadrature coil *L*-5 in Fig. 15-19 for maximum sound output. Next adjust the buzz control *A*-4 until the intercarrier buzz disappears. Then touch up both the quadrature coil and buzz adjustments.

Proper alignment of the sound section can be checked by turning the buzz control first to the right and then to the left of the final adjustment. The buzz level should come up with each change. Also, if the sound level goes down in one direction, it should go down in the other. If this does not happen, repeat the alignment.

Adjustment of the 4.5-Mc Sound Trap. As was previously mentioned, the video amplifier includes a sound trap to remove the effect of the 4.5-Mc sound signal on the picture. If this trap is misadjusted, the 4.5-Mc beat interference will appear in the picture as very fine lines close together in a vertical or diagonal pattern. The pattern of the fine lines will change with the sound modulation, forming a changing herringbone pattern.

The diagrams in Fig. 15-21 show two common sound-trap systems. In Fig. 15-21*a*, transformer *T*-1 provides the take-off for the sound section.

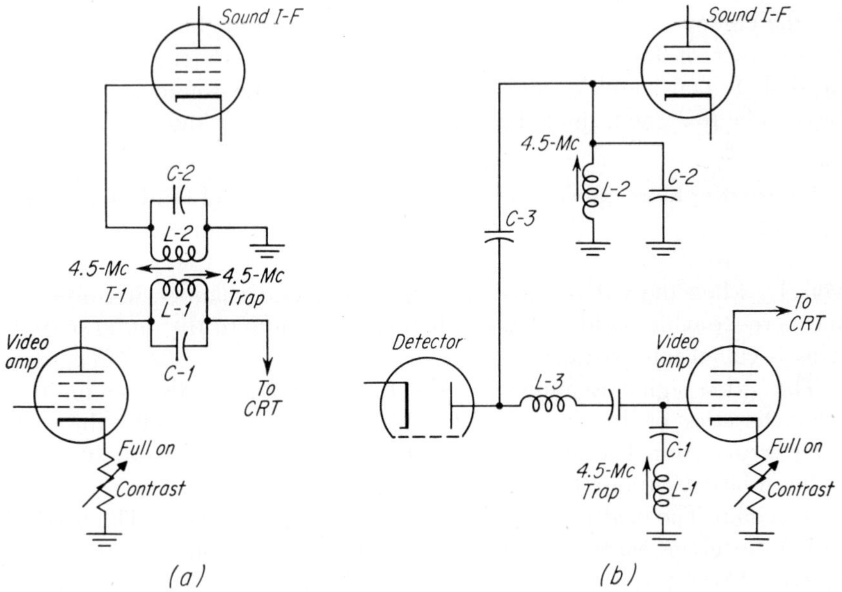

Fig. 15-21. Sound take-off systems showing 4.5-Mc sound trap. In both diagrams, the sound trap is composed of coil *L*-1 and condenser C-1.

When the primary winding *L*-1 is accurately tuned to 4.5 Mc, it will provide maximum signal voltage to the sound section. At the same time, 4.5-Mc current through the plate circuit is at a minimum, thereby removing the sound beat from the picture-tube circuit. Coil *L*-1 is adjusted for maximum sound during the sound-alignment procedure. It is therefore at correct adjustment, or very close to it for sound elimination from the picture.

In the circuit of Fig. 15-21*b*, the 4.5-Mc sound beat is fed through coupling condenser *C*-2 to the sound section. It is also fed through coil *L*-3 along with the video signal to the video-amplifier tube and so to the picture tube. Now coil *L*-1 and condenser *C*-1 in the grid of the video-amplifier tube are tuned to 4.5 Mc to act as a sound trap. The adjustment of coil *L*-1 will not affect the sound output since it is separated from the sound section by the impedance of coil *L*-3.

Regardless of which of the circuits is being used, the 4.5-Mc sound trap is adjusted in the same way. The procedure is as follows:

Step 1. After the sound alignment, check for normal picture and sound on all channels usually received in your area, while looking for the herringbone pattern which is characteristic of sound interference. If this is not evident, the trap adjustment need not be disturbed.

Step 2. If the pattern is observed, adjust the fine tuner on the receiver for maximum beat interference in the picture, and then adjust the trap for the elimination of the beat pattern. In the case of the sound take-off transformer of Fig. 15-21*a*, no more than a slight touchup should be needed. If considerable adjustment is required, the sound section will have to be realigned, since the adjustments of coils *L*-1 and *L*-2 are interdependent.

Intercarrier Buzz. In the intercarrier system, the sound- and picture-signal paths are closely allied. We have just seen that sound signals in the picture circuit cause a herringbone interference on the picture. Conversely, when the video signal gets into the sound channel, it causes an interference which adds a low-pitched staccato note to the audio output. This is called intercarrier buzz.

The video signal is amplitude-modulated, and normally the limiting characteristics of the ratio or gated-beam detectors handle this a-m interference. But if a large amount of the video signal is brought in, or if the limiting circuits become ineffective, the video signal gets into the audio chain. The peaks of the video signal are the sync pulses. The vertical pulses, occurring 60 times a second, have the greatest effect on the audio system. These produce the low-pitched staccato note.

Hum or Buzz? The vertical sync pulse is at 60 cps, and the hum frequency from the power line is also at 60 cps, or 120 cps in the case of a full-wave rectifier. It is easy to confuse these sounds, and a serviceman can waste considerable time trouble shooting for buzz when the trouble is really hum.

Hum and buzz can be readily recognized, but word pictures are inadequate for the purpose. Take a receiver which is operating normally, and tune it to an unused channel. Then take a clip lead containing a 0.05-μf isolating condenser, and clip one end to the arm of the volume control. Now touch the prod end to a hot 6-volt filament lead. The resulting sound is 60-cycle hum. Touch the prod to the filament pin of the rectifier tube. The resulting sound is 120-cycle hum. Note that these two sounds are very much alike, and it is quite difficult to distinguish between 60- and 120-cycle hums by ear, since they are an octave apart. Finally, touch the prod to the plate pin of the video-amplifier tube, where the peak signal is the sync pulse. The resulting sound is buzz and is quite different from the hum sounds and therefore easily recognized.

By way of confirmation, intercarrier buzz can occur only when on a station channel and will disappear when the set is switched to an unused channel. Hum is generally present on or off station and is usually accompanied by hum bars in the picture.

Servicing Procedure for Intercarrier Buzz. In addition to hum, there are other sounds which may be mistaken for intercarrier buzz. To see if this is the case, turn the channel selector to an unused channel. If the disturbance is still present, it is not intercarrier buzz, since this requires the presence of a station.

Now turn down the volume control. If the noise continues, it is probably of a mechanical nature. Check the power transformer and vertical output transformer for loose laminations. These have been known to create a noise very similar to intercarrier buzz. If the laminations are loose, you will be able to feel the vibration by touching the transformers. Usually tightening the core-holding bolts will take care of this trouble.

If the noise is not mechanical, turn up the volume control, and try varying the vertical-hold control. If this changes the pitch, there is coupling between the vertical and sound sections. The vertical deflection section operates at 60 cps, and produces the same staccato note, which is similar to the intercarrier buzz. The coupling is probably caused by the field around the vertical yoke leads which are operating at a high pulse voltage. Try rerouting these leads. Another possibility which may occur in some sets is the failure of a decoupling condenser in the B plus lead to either the vertical or audio sections.

A noise similar to intercarrier buzz may be caused by coupling between the high-voltage lead and the sound section. You will be keyed to this trouble by varying the brightness control. This control varies the current drawn by the picture tube and so affects the current in the high-voltage lead and therefore the field around it. If the brightness control varies the noise, try rerouting the high-voltage lead. Check also the aquadag coating and the spring contact on the coating. If these are at fault, there is likely to be a concurrent complaint of vertical jitters.

Sounds similar to buzz are treated in this section because they are often mistaken for buzz. Assume the buzz sound appears only on station and is eliminated on an unused channel. This is a bona fide case of intercarrier buzz.

First a check should be made of the limiting devices in the sound section of the receiver. If the set uses a gated-beam discriminator, check the setting of the buzz control in the cathode of the 6BN6 tube. Incidentally, if a receiver is equipped with some kind of noise control, do not adjust this control for intercarrier buzz. Noise controls are meant to stabilize the operation of the sync circuits and have no effect on the sound section.

If the receiver uses a ratio detector, check the stabilizer condenser across the diode loads (*C*-5 in Fig. 15-18) which controls the limiting in this type of circuit. This is done by connecting a test 4-μf condenser across the original condenser with due regard for polarity, and observing results on the buzz. Experienced servicemen suspect this condenser after making a fast swing of the fine-tuning control. In order to gain this same experience, open the stabilizing condenser in a normally operating receiver, operate the fine-tuning control, and observe the sharp tuning effect on the buzz.

Another factor which may affect buzz in a ratio-detector type of receiver is the setting of the tuning slug on the secondary winding of the discriminator transformer. This is correctly set during the sound alignment by a zero reading as indicated by an output meter. It may be possible to obtain better buzz elimination by a touchup of this adjustment. First tune the set for normal picture and sound. Then readjust the secondary slug of the discriminator transformer for minimum buzz. A very slight rotation in either direction should be all that is required. The correct adjustment point is located between the two maximum buzz peaks that will be noticed when tuning the slug back and forth about a quarter to a half turn. Another note on buzz elimination in the ratio detector is to be sure that the shield covering the 6AL5 tube is in place. If the set uses a limiter stage, this should be checked for proper operating voltages.

In sets where the sound take-off is in the output circuit of the video amplifier, the contrast control will have a considerable effect on intercarrier buzz. In strong-signal areas, if the contrast control is advanced too far, there will be a large increase in buzz. When servicing a set for intercarrier buzz under these conditions, first set the contrast control for a normal picture. If this reduces the buzz to a negligible amount, the customer probably prefers a very contrasty picture and should be instructed in the proper use of the contrast control.

Intercarrier buzz may also be caused by too strong a signal in the video chain. To check for this, decouple the antenna, or use an attenuator pad like the one shown in Fig. 15-20, to cut down the signal strength to the receiver. If this clears up the buzz while still providing good picture and sound, the video interference is greater than can be handled by the normally operating limiting circuits in the sound section. First check the operation of the manual control in the agc system as described in Chap. 8. If this is normal, check the tubes, operating voltages, and alignment of the video i-f stages as described in Chap. 12.

Excessive intercarrier buzz is a considerable problem in television reception. Very often, later runs of the same set incorporate design changes which are intended to cut down the buzz. It would be a good idea, therefore, when working on a set for this complaint, to check manufacturers' service notes on later runs to see if some provision has been made for excessive buzz.

The setting of the fine-tuning control has some effect on buzz, and the customer should be instructed to adjust it for best picture with minimum buzz. Normally, the position for best picture is not very sharp, and the fine tuner allows a considerable change in sound level without too much effect on the picture. In the video detector, the changing sound level changes the amount of video a-m components in the 4.5-Mc beat signal, and therefore the intercarrier buzz.

On some channels the oscillator setting is such that the range of the fine tuner does not change frequency sufficiently for good buzz elimination. A final step in servicing for buzz is to set the fine tuner to the center of its rotation, and tune the set to each channel normally received. If the buzz appears on any station, adjust the oscillator slug in the tuner for best picture with minimum buzz.

Service Notes on the Sound Section in a Receiver with a Stacked Power Supply. The sound section in a receiver with a shelf or stacked power supply is subject to the same faults and serviced in the same way as any other, except for troubles in the last audio stage. Here, the voltage-distribution system of the stacked power supply must be taken into account.

A basic voltage-distribution diagram of the last audio stage in a stacked-power-supply receiver is given in Fig. 15-22. Normal voltage readings are given at the pin terminals of the tube. Note that the tube is acting as the voltage-dropping resistor for the B plus 150-volt line.

In most cases, trouble in this stage would result in a no-sound no-picture condition. For example, if the primary of the output transformer were to open, there would be no sound. At the same time, the B plus 150-volt line would go dead because of the open to the 250-volt line, and there would be no picture. Servicing for no sound, no picture, normal raster calls for an investigation of the common sound-picture r-f and i-f stages. The lack of voltage on the video i-f tubes would lead to the open voltage-dropping resistor, in this case the second a-f stage.

However, there are some defective conditions which would result in the usual sound-section symptom: picture with no sound or poor sound. One of these is a short in filter condenser *C*-1 or plate bypass condenser

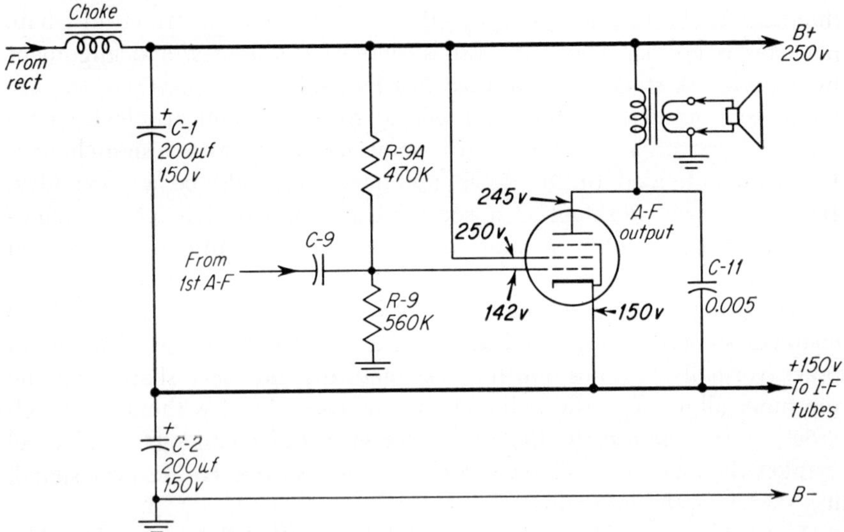

Fig. 15-22. Voltage analysis of the last audio stage in a stacked- or shelf-type power supply.

C-11. If either of these condensers were to become shorted, B voltage would be removed from the last audio stage, resulting in no sound. Voltage on the 150-volt line would go up, producing an overload symptom. When servicing the i-f–agc system for this complaint, excessive plate voltage on the i-f tubes would be noted. This leads to a check of the voltage-distribution system, and the discovery of the shelf-type power supply. Voltage and ohmmeter checks in the audio output stage then find the shorted condenser.

Another possible complaint is poor tone quality. Again, a careful voltage check of the final audio stage must be made. Note the normal readings of +142 volts at the grid and +150 volts at the cathode. Here, it would be best to measure the bias voltage directly from cathode to grid. The normal reading should be −8 volts at the grid. Any variance in the bias is likely to cause the poor tone. If the bias voltage is incorrect, check coupling condenser *C*-9 for leakage. If the condenser is good, grid resistors *R*-9 and *R*-9A are checked for correct value. These resistors form a voltage divider which establishes the correct potential at the grid, and small differences may be enough to spoil the tone. As a general rule, gold-band resistors with a 5 per cent tolerance rating are used in this circuit and should be used for replacement purposes.

Common Troubles and Replacement Notes in the Sound Section. The most common troubles in the sound section are defective tubes, loudspeakers, output transformers, volume controls, and the usual resistors and condensers. The sound i-f transformers rarely cause any trouble other than misalignment. If one should become defective, a satisfactory replacement can usually be chosen from catalogues, where they are listed by operating frequency and use. Discriminator transformers are different for the ratio-detector and limiter-discriminator types. Be sure to use the correct replacement if this component should become defective. Replacements for output transformers are likewise a simple problem. Here it is safe to standardize and stock a transformer designed to couple a beam output tube to a 3-ohm voice coil for all receivers except those using push-pull output tubes. A spare push-pull output transformer will complete your replacement stock. As for volume controls, the usual complaint is noisy operation. It is generally considered good servicing practice to replace the control, since a cleaning may be only a temporary repair.

Service Notes on Loudspeakers. Most receivers use a p-m type of loudspeaker. In these, the voice coil somtimes goes off center, and sometimes it opens. An off-center voice coil causes raspy sounds and poor tone. An open voice coil results in no sound. Check for an off-center voice coil by operating the set and applying a little finger pressure at various spots on the paper cone near the voice coil. Pressure at the right spot will recenter the coil, clearing up the trouble. Most speakers have no provision for recentering; so the entire unit must be replaced. To check for an open voice coil, open the lead to the output transformer, and check the voice coil on the $R \times 1$ range of your ohmmeter. A good voice coil will read about 3 ohms, and a click will be heard as the ohmmeter leads are disconnected. If the coil is open the entire speaker should be replaced. When replacing a p-m loudspeaker, any 3.2-ohm unit with the same outside diameter as the original will work satisfactorily.

Some receivers use electrodynamic loudspeakers where the field coil is used as a filter choke in the low-voltage power supply. These also develop off-center and open voice coils, but this time the replacement loudspeaker should be an exact replacement, since the filter-choke circuit will not permit a wide tolerance.

If an exact or close replacement for the electrodynamic loudspeaker cannot be obtained, the following expedient will effect a repair: Assume the field coil is good. Then choose a p-m speaker of a size comparable

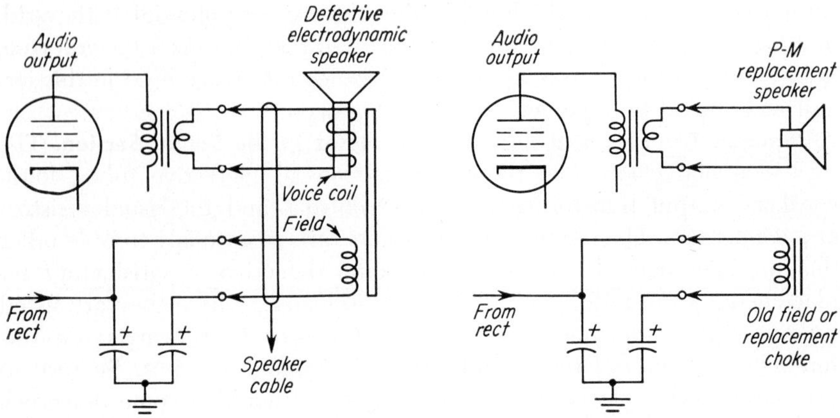

Fig. 15-23. Replacing an electrodynamic loudspeaker with a p-m type.

with the original, and use this for the replacement, wiring in its voice coil and disconnecting the old. The old speaker is then mounted somewhere in the cabinet with its field wires untouched, and the voice coil is disconnected. The replacement scheme is shown in Fig. 15-23. It may be possible to remove the frame holding the voice coil and cone, to conserve space.

If the field coil is open, or if it is impractical to mount two speakers in the cabinet, the field coil must also be replaced. A 1-henry 300-ma replacement filter choke will generally serve for this purpose. It may be mounted on the chassis or in the cabinet, using the original speaker leads. When mounting the old field or choke in the cabinet, it should be mounted toward the rear, where its magnetic field will have the least effect on other sensitive circuits in the receiver.

Replacement Note on Audio Plate-bypass Condenser. A fault very often encountered is a shorted condenser in the plate circuit of the sound output tube. This is condenser *C*-11 in Fig. 15-24. One lead of the condenser is always connected to the plate; the other sometimes goes to B plus as shown in the diagram and sometimes is connected to chassis or the cathode of the audio output tube. In all cases it is subject to peak audio signal voltages and often breaks down as a result. When replacing

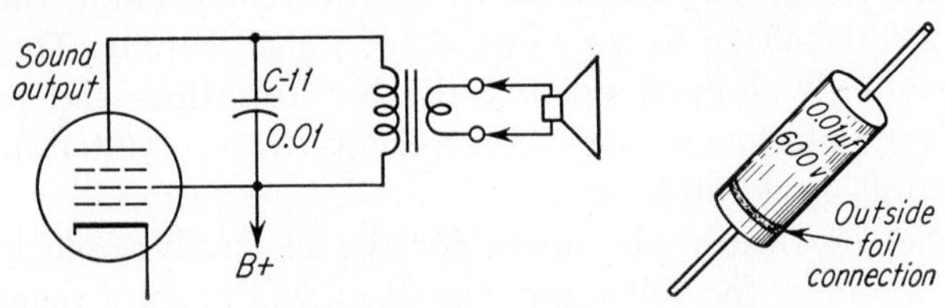

Fig. 15-24. Audio plate-bypass capacitor and its position in the circuit.

the condenser, use the same capacitance rating as the original, but use a voltage rating of 600 or 800 volts, going higher than the original rating which has already proved itself inadequate.

SUMMARY

When there is a normal picture, and the sound is weak or dead, there is trouble in the sound section.

Typical Sound Sections. A basic diagram for a sound section with a ratio detector is given in Fig. 15-25*a*; a sound section using a gated-beam detector is shown in Fig. 15-25*b*. These are the two circuits which are most commonly used in television receivers.

Servicing Procedure for a Complaint of Normal Picture, with No Sound or Weak Sound.

Step 1. Feed an a-f test signal to the arm of the volume control (test point *A* in Fig. 15-25). If there is no response, check the a-f stage or stages for the fault. If there is a normal response or weak response, go on to step 2.

Step 2. Feed a strong r-f amplitude-modulated signal to the plate pin of the last sound i-f tube (test point *D* in Fig. 15-25). Wobble the frequency control between 4 and 5 Mc. If there is no response, check the detector stage. If there is a response, go on to the next step.

Step 3. Feed the 4.5-Mc a-m test signal to the grid of the first sound i-f tube. If there is no response, check the sound i-f stage or stages. If there is a normal response, go on to step 4.

Step 4. Feed the 4.5 Mc a-m test signal to the output of the video detector (test point *B*, the diode load). See Fig. 15-13. If there is no response, check the coupling between the sound take-off point and the first sound i-f grid. If there is a normal response, check the alignment of the tuner and video i-f stages.

Alignment Procedure: Ratio Detector

1. Tune the receiver for best picture on the strongest channel in your area.
2. Connect a vacuum-tube voltmeter to chassis and the negative end of stabilizer condenser *C*-5 (test point *E* in Fig. 15-25*a*).
3. Beginning with the primary slug of the discriminator transformer, adjust all tuning slugs back to the sound take-off point for maximum indication on the vacuum-tube voltmeter. Repeat the adjustments.
4. Connect the probe of the vacuum-tube voltmeter to the discrimi-

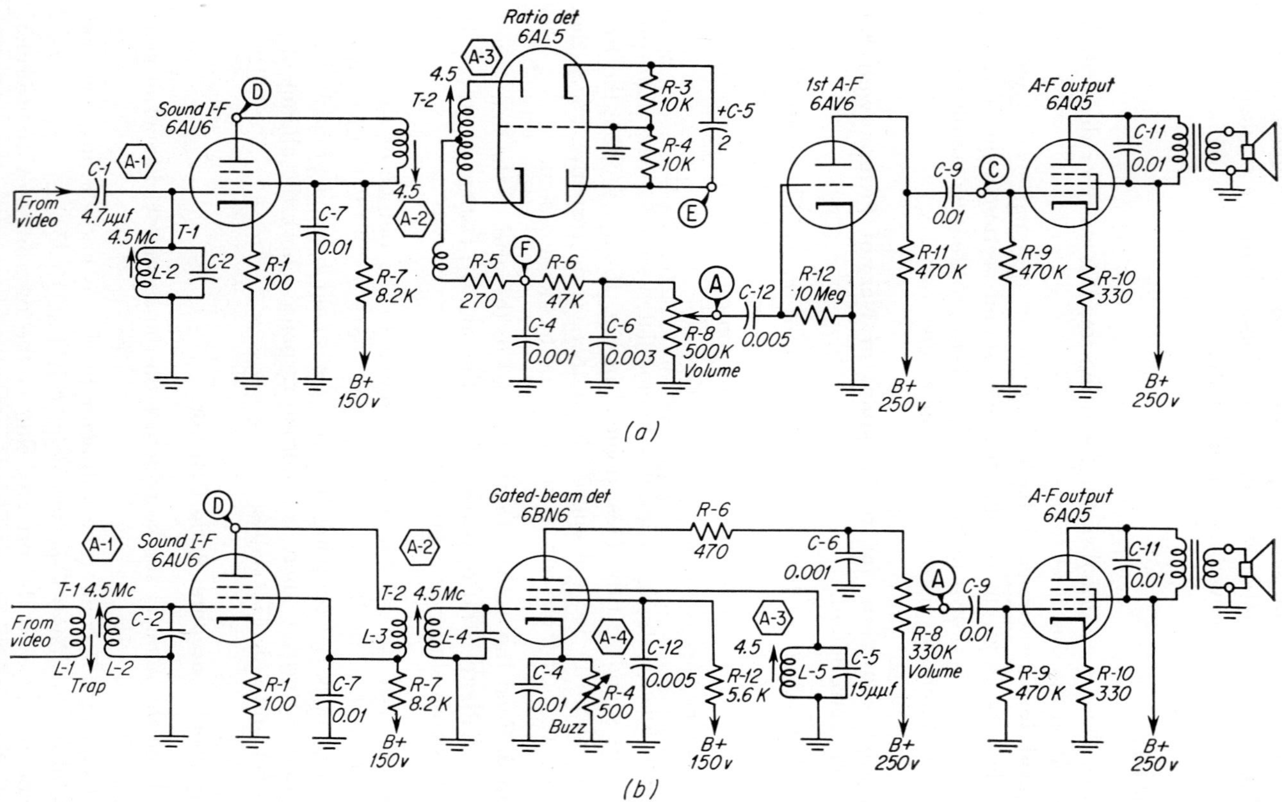

Fig. 15-25. Basic diagrams for sound sections. (*a*) Ratio-detector type. (*b*) Gated-beam type.

nator sound output (test point *F* in Fig. 15-25*a*). If the diode loads are not center-grounded as shown in Fig. 15-25*a*, provide a meter-return point, as shown in Fig. 15-18. Use a zero center scale on the meter.

5. Adjust the secondary tuning slug of the discriminator transformer for a null. The correct zero point is located between a positive and a negative maximum. Disconnect the output meter, and retouch this adjustment, if needed, for minimum buzz.

Alignment Procedure: Gated-beam Detector

1. Tune the receiver for best picture on a strong channel. Connect a variable attenuator pad in the antenna lead.
2. Set the buzz control to the center of its rotation.
3. Adjust the attenuator for a hiss in the sound.
4. Beginning with the last sound i-f transformer (*T*-2 in Fig. 15-25*b*), adjust all tuning slugs back to the sound take-off point for maximum sound. If the hiss disappears during the alignment, reduce the attenuator setting. Repeat the adjustments.
5. Reset the attenuator for strong sound with no hiss.
6. Adjust the quadrature coil (*L*-5 in Fig. 15-25*b*) for maximum sound. Then adjust the buzz control for minimum buzz. Repeat both adjustments.

Sound-trap Adjustment. If the herringbone pattern of the 4.5-Mc sound beat is observed on any channel, adjust the fine-tuning control to exaggerate the beat interference. Then adjust the 4.5-Mc sound trap for minimum beat pattern in the picture.

Servicing Chart for Intercarrier Buzz

Procedure	*Result*	*Trouble and Subsequent Check*
1. Check manufacturer's service notes for run changes		Incorporate changes designed to reduce intercarrier buzz
2. Tune set to unused channel	Buzz continues	Check: (*a*) laminations of power transformer and vertical output transformer, (*b*) lead dress of yoke and high-voltage wiring, (*c*) grounding of aquadag coating
	Buzz stops	Go on to step 3
3. Tune to station where buzz is present and decouple the antenna	Buzz stops	Check agc adjustment Check i-f alignment (Chap. 12)
	Buzz continues	For 6BN6 detector, check sound alignment especially of buzz control (*R*-4 in Fig. 15-25*b*) For ratio detector, (*a*) check stabilizing condenser C-5 in Fig. 15-25*a*, (*b*) check limiter stage in sound section, if used, (*c*) shield 6AL5 tube, (*d*) touch up alignment of secondary winding of discriminator transformer For both types, check setting of oscillator-tuning slug in tuner

QUESTIONS

1. What is the symptom which leads to investigation of the sound section?

2. Draw a block diagram of the sound-signal path in a television receiver of the intercarrier type beginning with the antenna and ending at the loudspeaker. Label the sound take-off point.

3. The sound-signal path in a television receiver begins at the antenna. Yet the sound section is considered as beginning at the sound take-off point. Explain why this is so.

4. Describe the double superheterodyne action by which the original sound signal picked up by the antenna is changed to a 4.5-Mc beat at the video detector in an intercarrier system.

5. In some receivers, the sound take-off point is in the output of the video detector; in others, it is in the output of the video amplifier. Explain how the video amplifier passes the 4.5-Mc sound signal. What is the advantage of this arrangement?

6. What is the purpose of the sound trap in the picture-signal path? At what frequency does it operate?

7. Use block diagrams to explain the difference between intercarrier and split-sound systems.

8. Explain why split-sound receivers often include a sound trap in the video-amplifier chain. At what frequency does it operate?

9. Draw a block diagram of the necessary stages in the sound section of an intercarrier system. Begin at the sound take-off point. Name the stages.

10. Draw a schematic diagram of a basic ratio-detector circuit.

11. Draw a schematic diagram of a basic gated-beam detector circuit.

12. Define limiting as it applies to an f-m sound system.

13. What factors provide limiting in a ratio-detector circuit? in a gated-beam detector circuit?

14. In the intercarrier system, what would be the result of a breakdown of the limiting function?

15. Describe a servicing procedure which localizes trouble in the sound section to a particular stage.

16. An audio test signal is applied to the arm of the volume control. The response is generally louder in a set with a ratio detector than in a set with a gated-beam detector. Why is this the case?

17. A receiver comes in for repair with the symptom "normal picture–no sound." A test signal applied to the arm of the volume control gives a normal response. In which stages is the trouble likely to be? How would you proceed to localize the trouble further?

18. A receiver comes in for repair with the symptom "normal picture–no sound." An audio test signal at the volume control gives a normal response. An

r-f test signal applied to the sound i-f plate gives a response when the generator is wobbled to 5.1 Mc. What is at fault?

19. A receiver comes in for repair with the symptom "normal picture—no sound." An audio test signal at the volume control gives a normal response. An r-f test signal applied to the sound i-f plate gives a normal response. The same thing happens at the sound i-f grid and at the video-detector output. What is likely to be at fault?

20. Describe the alignment procedure for a ratio-detector type of sound section.

21. Describe the alignment procedure for a gated-beam type of sound section.

22. If a television receiver is available, identify all tubes and alignment controls in the sound section. Go through the sound-section alignment for practice.

23. Describe how you would distinguish between the primary and secondary tuning slugs of the discriminator transformer in a ratio-detector type of receiver.

24. Describe in detail the output meter connections for the null adjustment of the secondary tuning slug in a ratio detector.

25. Describe the adjustment of the 4.5-Mc sound trap.

26. How could you distinguish between hum and intercarrier buzz?

27. If a noise similar to intercarrier buzz is present on an off-station channel, what are some of the likely causes? How do you check for each?

28. What checks should be made in a ratio detector for buzz elimination? in a gated-beam detector?

29. How do you check for buzz caused by too strong a signal?

30. In a receiver with a shelf or stacked power supply, what will be the result on reception as the last audio tube becomes weak and loses emission? Describe the servicing procedure which will reveal the weak tube.

16

Antenna Installation and Service Notes

The antenna is the first link in the receiver signal chain. It picks up signal energy from the transmitting station and feeds it to the receiver. The television antenna is generally some form of the basic dipole shown in Fig. 16-1*a*. This type is known for its directional characteristics and the fact that it has good pickup qualities over the frequency range.

In areas close to transmitting stations, the signal level is high, and an indoor antenna like the rabbit-ear type illustrated in Fig. 16-1*b* may

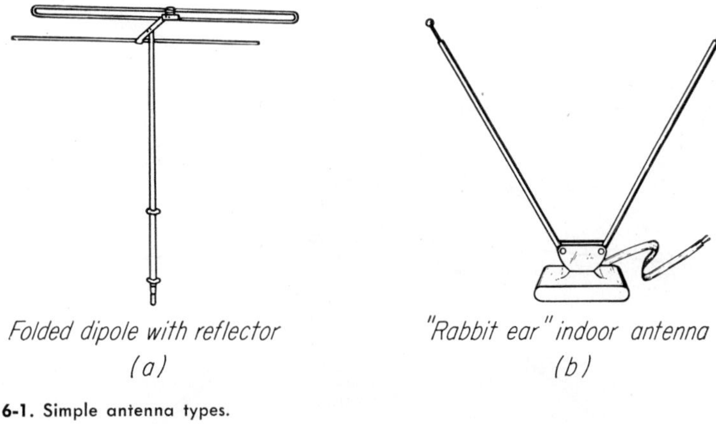

Fig. 16-1. Simple antenna types.

be adequate to provide good reception of sound and picture. In some receivers, an antenna is built into the cabinet, and these also may be successfully used in strong-signal areas.

As the distance from the transmitting station increases, signal strength decreases, and more efficient antennas are required. As a general rule, increased height from ground level increases signal strength, and the antenna is installed on the roof. Even in the strong-signal area very close to the transmitter, shielding effects caused by the use of metal in the building or r-f interference caused by proximity of the antenna to the house wiring may make the built-in antenna inadequate and require the installation of a roof antenna.

At even greater distances from the transmitting station, the signal level becomes very low and reception very poor. This is called the fringe area. In order to get favorable reception conditions, masts and towers are used for added height, together with complex arrays designed to give more signal voltage. Even with the best antenna equipment, fringe-area reception is often only passable. This chapter will describe antenna-installation problems and techniques relative to obtaining good reception in both strong- and weak-signal areas.

Antenna Selection. There is a very large variety of antennas available, each one of which serves a slightly different purpose, so that an inexperienced serviceman may be easily confused as to which type to choose for a particular installation. The following discussion is intended to help clear up some of this difficulty.

In a strong-signal area close to transmitting stations, almost any television antenna will prove adequate. As a general rule, strong-signal areas are built-up city residential sections, with a dense population, and serv-

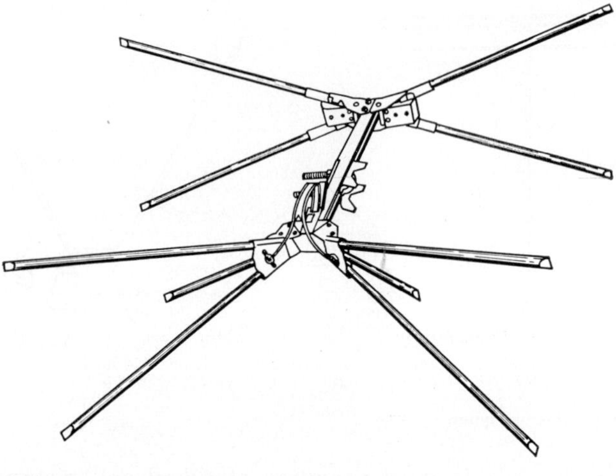

Fig. 16-2. Conical antenna used in strong-signal areas.

iced by several broadcast stations. The tendency is for the transmitting stations to be very close to each other in the center of such an area so that at each receiver installation the stations are in one general direction. When a built-in or indoor antenna does not give satisfactory reception, a roof antenna is installed. The antenna is sometimes installed in the attic of a home, but for this classification, the attic antenna is considered as a roof antenna. There will be more information on this subject later.

Probably the most popular antenna in strong-signal areas is the conical antenna illustrated in Fig. 16-2. This is generally a good choice because the broad-band characteristics of this type produce good over-all results on all channels.

In fringe areas, where the signal level is low, the efficiency of the antenna is of the utmost importance. However, the problem is an individual one for each area, and about the only general rule that can be applied is to take advantage of the experience gained in previous installations. The local parts distributor will know and recommend the type of antenna which gives good results. He will also have some ideas about the mast height needed. This information is easily confirmed by making a survey of existing installations. Two commonly used fringe-area antennas are illustrated in Fig. 16-3.

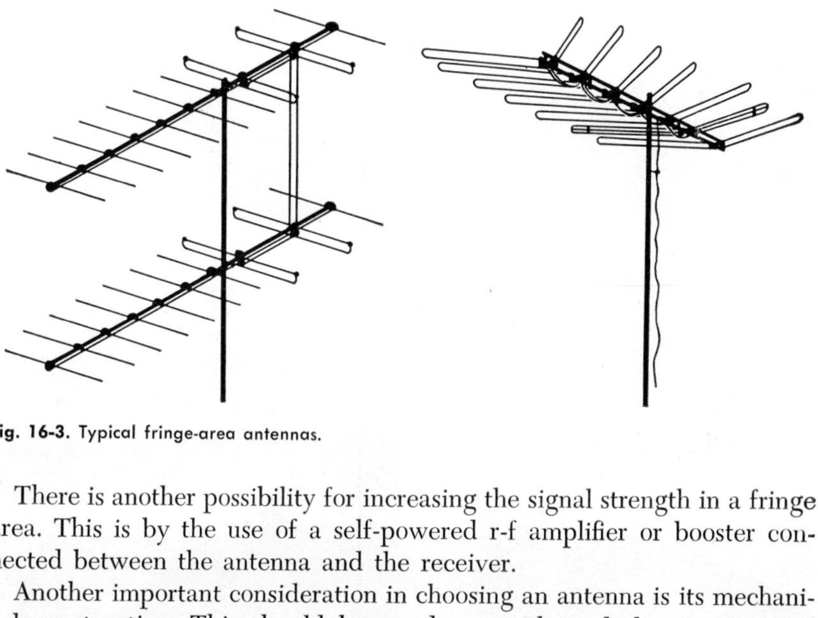

Fig. 16-3. Typical fringe-area antennas.

There is another possibility for increasing the signal strength in a fringe area. This is by the use of a self-powered r-f amplifier or booster connected between the antenna and the receiver.

Another important consideration in choosing an antenna is its mechanical construction. This should be sturdy to withstand the pressures of high winds and ice loading. In addition, the materials used should be able to withstand the corrosive action of weathering.

Planning the Antenna Installation. General reception conditions in the area determine the type of antenna to be used and the height needed. There are other factors to be considered in making the installation. Some of these are the type of mounting, the location of the receiver in the home, and how to run the transmission line from the antenna to the receiver.

The type of mounting used depends on the type of house. Some common installation systems for outdoor antennas are shown in Fig. 16-4. In the case of an apartment house, the antenna mast is mounted on brackets which are fastened to the inside of the parapet which usually surrounds the roof. This is called a parapet or wall mount and is illustrated in Fig. 16-4*a*. In a private house with a hip roof, the mast is usually fastened to the chimney, as shown in Fig. 16-4*b*.

Fig. 16-4. Types of antenna mounts. (*a*) Parapet. (*b*) Chimney. (*c*) Gable. (*d*) Side wall.

The chimney cannot always be used for this purpose. Sometimes it is in the center of the house and does not extend high enough over the roof to space the two mounting clamps sufficiently for a stable installation. Or it may be made of fieldstone to which it is very difficult to attach a clamp. In these cases, the antenna is mounted as shown in Fig. 16-4*c* or *d*. The eave clamps of Fig. 16-4*c* can be used on a gable roof, and the side-wall clamps of Fig. 16-4*d* are fitted with extensions so that the mast clears the overhang of the roof. This type can be mounted on any upright wall.

There may not be too much choice as to the location of the mast on a parapet roof. The apartment-house parapet is likely to be crowded with other antennas, leaving you as little room on the roof as you found for parking your car in front of the house. When making this type of installation, place a newspaper on the outside sill of the window which is to take the transmission line, and close the window on it so that it will not blow away. This will serve to identify the window from the roof. Then, if possible, the mast is erected directly in line over the window, so as to avoid horizontal runs with the transmission line. In a private-house installation, try to keep the mast toward the rear of the house. This has a dual advantage. One is that it is less likely to spoil the appearance of the house, and the other is that the house will shield the antenna from r-f interference emanating from automobile traffic in the street. If a line-of-sight view of the transmitting tower is obtainable at any portion of the roof, that is the preferred location. In all cases, permission to erect the antenna must be obtained, and local ordinances pertaining to antenna installations should be consulted.

Inside the house, the receiver should be placed where the consumer wants it. The installation man should limit himself to a tactful suggestion about changing the location away from the glare of a bright window, or to shorten inside runs of line cord or transmission lines. As for the latter, the preferred method is to bring the line into a basement window and then up through the floor behind the receiver.

The antenna installations illustrated in Fig. 16-4 show self-supporting masts attached at the bottom. This type is sufficiently strong for a mast length of 10 to 12 ft at most. In fringe areas, where greater height is desirable, the masts are extended and supported by guy wires. A two-section mast supported by a set of guy wires can increase the antenna height to 20 ft above the roof. In extreme fringe areas, even greater heights are used, but this calls for towers and special rigging. Step-by-step instructions are included with the purchase of a tower or high mast. These should be followed carefully.

The Extension Ladder. An important piece of equipment in antenna-installation work is a ladder. The type preferred by most men is a 36-ft extension ladder made of magnesium alloy and equipped with safety feet. These handle easily because of the light weight, will reach to the peak of most private homes, and in the collapsed state, will not extend too far beyond the extremes of most automobiles. It is usually carried on the roof of the delivery car or station wagon by means of a car-top carrier.

The conductivity of the metal ladder introduces some hazards. When placing the ladder against a house, be especially careful to avoid any overhead wires. Contact to even low-tension wires may be very dangerous. It would also be wise to avoid the use of any electric equipment when working on the ladder, because of short and shock hazards.

Other specialized materials and tools are needed in installation work. These will be introduced when the installation procedures are described.

Installing Wall Brackets. The different types of antenna mounts illustrated in Fig. 16-4 show a common characteristic in that the installations are based on securely fastening two mast brackets one above the other in concrete, brick, or wood. The antenna is then assembled and fastened to the mast; the mast is raised and then made fast to the brackets. A typical wall bracket and the necessary tools and hardware are shown in Fig. 16-5.

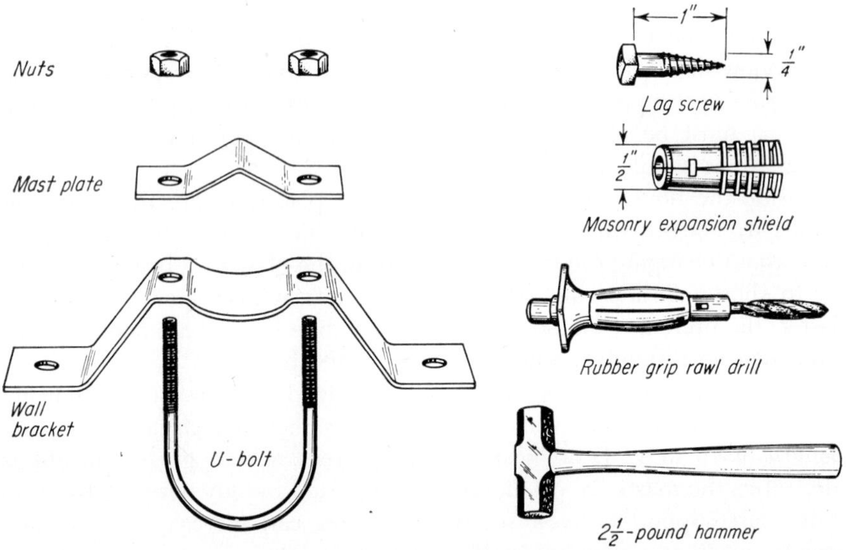

Fig. 16-5. Antenna-mast wall bracket with hardware and tools needed for installation.

The brackets are made of heavy steel, treated to withstand weathering, and come in various sizes. The size refers to the distance the bracket extends from the wall. A common size is 4 in. The U bolt and top plate are used to fasten the mast. The bracket is held to the wall with ¼-in. lag screws. When the brackets are mounted on wood, as in the case of the gable mount of Fig. 16-4*c*, the lag screws go directly into the wood after drilling a pilot hole. On a brick chimney or concrete wall, the lag bolts are screwed into masonry expansion shields. The shields for $\frac{1}{4}$-in. lag bolts requires a $\frac{1}{2}$-in. hole 1 in. deep in the masonry.

In drilling these holes, some mechanics prefer to use a heavy-duty slow-speed electric drill fitted with a $\frac{1}{2}$-in. carbide masonry bit. This tool has its advantages where there is sure footing but is to be avoided when you are working on a ladder. The trailing power cord is awkward and dangerous, and pressure on the drill tends to push the ladder away from the wall. For ladder work, a safer method is to use the rubber-grip rawl drill and short-handled $2\frac{1}{2}$-lb hammer shown in Fig. 16-5. The rawl drill handle can be fitted with cutting drills of different sizes. The $\frac{1}{2}$-in. size will require the heavy $2\frac{1}{2}$-lb hammer. A lighter one will bounce back off the drill without making very much impression on the masonry.

Having decided on the approximate location of the mast brackets, hold the top bracket in position, and mark off one of the mounting holes with a pencil. When mounting on brick, try to place the bracket so that the holes come in the approximate centers of two adjoining bricks as shown in Fig. 16-6*a*.

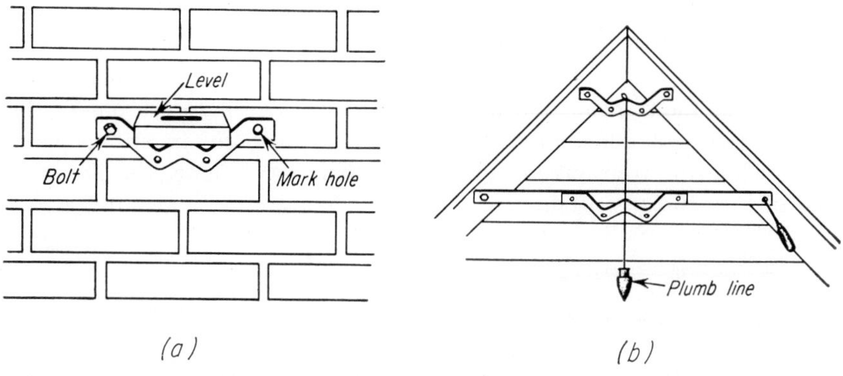

Fig. 16-6. How to get the brackets level and in line for a perpendicular mast mounting.

When pounding the hole in brick or concrete, the rawl drill is twisted slightly between hammer blows to prevent wedging and to facilitate removal of chips. Insert the shield in the hole, and if it protrudes, pound the hole a little deeper so that the shield is flush with the surface. Thread the lag bolt through the bracket, and screw it down, but not fully tight, so that the bracket is free to swing on the one mounting bolt. Swing the bracket so that it is perfectly level, using a small carpenter's level for the purpose, and mark off the second hole. Then swing the bracket out of the way, cut the second hole, insert the shield, and now drive both lag bolts fully tight so that the upper bracket is securely fastened to the wall. A ratchet type socket wrench makes short work of tightening the lag bolts.

The installation of an eave mount on the peak of a gable roof is illustrated in Fig. 16-6*b*. Note that the upper bracket is similar to the upper bracket of the wall mount of Fig. 16-6*a*. It is leveled and mounted by the same procedure, except that this time only a pilot hole is needed for the lag bolt.

In both mounts, the lower bracket should be placed approximately 18 in. below the upper for a sturdy mast support. On a brick wall, seven courses down will be about correct spacing and again will provide adjacent bricks for the two mounting holes. In eave mounts, the bottom brackets are furnished in different lengths so as to accommodate different roof pitches and yet maintain adequate spacing between the brackets.

The lower bracket should be exactly in line with the upper so as to keep the mast perpendicular. To achieve this end, drop a plumb bob from the center of the mounted upper bracket as shown in Fig. 16-6*b*. Center the lower bracket on the plumb line, and holding it level, mark off one mounting hole. Drill and fasten this end of the bracket. Check leveling and centering before marking the second hole. Then drill and fasten the last bolt.

Transmission Lines. The wire connecting the antenna to the receiver is known as the transmission line. Common practice is to use one of the various forms of 300-ohm twin lead for the purpose. In strong-signal areas, when the transmission line is less than 100 ft long, the flat twin lead of Fig. 16-7*a* is used.

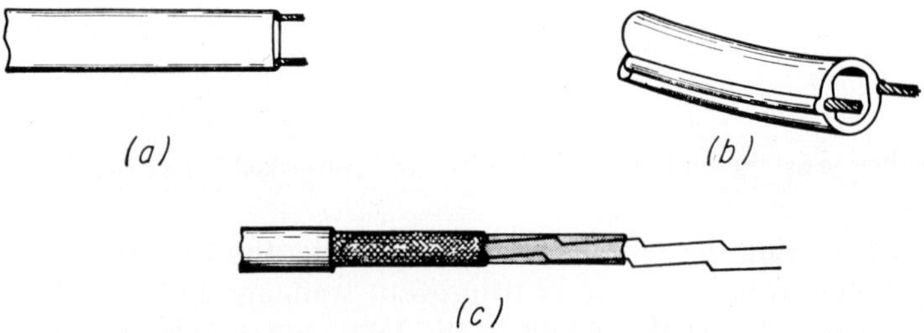

Fig. 16-7. Transmission lines. (*a*) Flat twin lead. (*b*) Tubular twin lead. (*c*) Shielded twin lead.

The tubular lead of Fig. 16-7*b* has heavier wire and low-loss characteristics. This makes it more suitable for use on long transmission lines or in fringe areas where losses should be avoided. Another advantage of the tubular twin lead is its resistance to weathering effects. In the moist salt-air atmosphere present in a seacoast area, flat twin lead develops high losses in a comparatively short time, whereas tubular twin lead retains its good qualities. In a problem area of this type, if the tubular twin lead is unobtainable, the ordinary flat twin lead can be treated with a moisture-repellent seal like Walsco silicone compound.

Some installations are troubled by manmade static interference. This is usually a strong r-f field around the electric wiring in the house. The antenna on the roof is usually in the clear as far as this interference is concerned, but the transmission line picks some up on the way to the receiver. This is prevented by the use of the shielded 300-ohm twin lead shown in Fig. 16-7*c*. This type of transmission line has heavy losses, and its use is generally avoided in fringe-area installation unless compensated for by a booster.

Splicing 300-ohm Twin Lead. Splices in transmission lines are generally avoided. However, it is sometimes necessary to make a splice, and a careful job should be done. A procedure for making a satisfactory splice that will not impair strength or electrical characteristics is illustrated in Fig. 16-8.

The two ends are stripped, leaving staggered lengths, as shown in

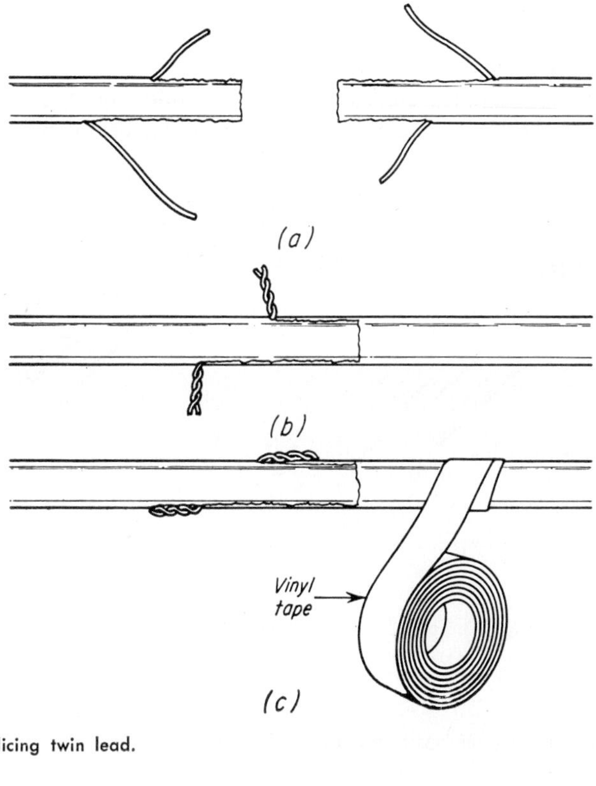

Fig. 16-8. Splicing twin lead.

Fig. 16-8*a*. The center insulation pieces are allowed to overlap as shown in Fig. 16-8*b*, and the splices are made. Splices made outdoors are not easily soldered. Clinching the joints with heavy pliers is considered satisfactory for a permanent connection. The ends are then tucked in and the joint covered with vinyl plastic electrical tape as shown in Fig. 16-8*c*. Do not use friction tape since it is a poor insulator at high frequencies. Tape No. 33 made by the Minnesota Mining & Mfg. is satisfactory. Splices in tubular and shielded twin lead are best made by means of special connectors.

Connecting the Transmission Line to the Antenna. After the brackets are mounted, the next step in an installation is to assemble the antenna and fasten it to the mast. Most antennas are of the quick-rig type with assembly instructions on the package. The mast fastening is generally a U bolt. The assembly is usually simple enough to require no special instructions, and we can go on to connect the transmission line.

As a first step in preparing the transmission line, study the installation to see if there will be any points like the edge of a parapet where the line is likely to touch and be frayed by normal swinging. A 12-in. piece of loom or Walsco twin-lead Flexitube for each contact point of this type is slipped over the end of the line to be used as a protection against fraying. Later, as the line is fastened into place, the loom is slid into position and secured to the line with vinyl tape.

The antenna is usually provided with a connection block with binding screws for the twin lead. It is preferable to use crimp-type lugs under

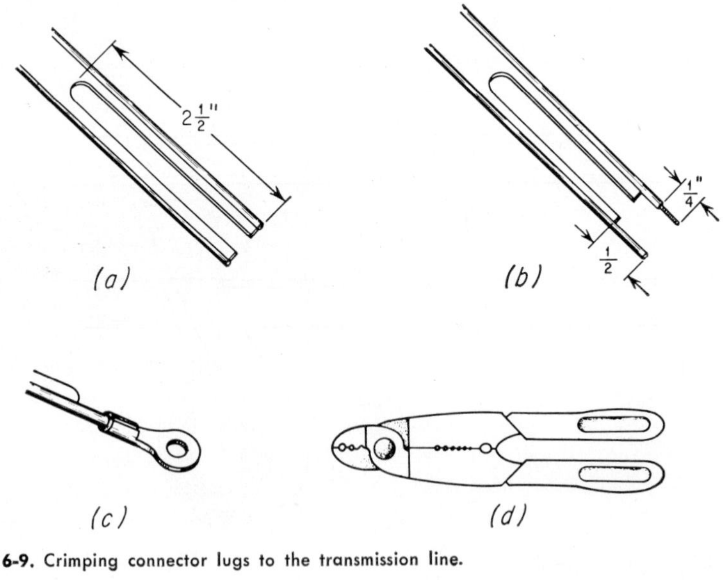

Fig. 16-9. Crimping connector lugs to the transmission line.

these screws for secure connections. Refer to Fig. 16-9 for the method of attaching the lugs to the transmission line. The center insulation of the line is cut out for a distance of about 2½ in., as shown in Fig. 16-9*a*. The General Cement No. 8386 tool is specially designed to make this cut. The rest of the center insulation is then cut away for about ½ in. from the end, and the final ¼ in. is bared as shown in Fig. 16-9*b*. The wire end is placed into the barrel of the lug as shown in Fig. 16-9*c*. Some of the insulation goes into the barrel as well as the stripped end. A crimping tool like the one shown in Fig. 16-9*d* secures the connection. Be sure that the crimping tooth is opposite the seam, in the barrel of the lug. Another lug is crimped to the second wire end, and the two lugs are securely fastened under the binding heads of the antenna connection block. Spray the connections well with Krylon to prevent corrosion.

Erecting the Mast. The transmission line is secured close to the top of the mast to prevent strain and breakage near the connection block.

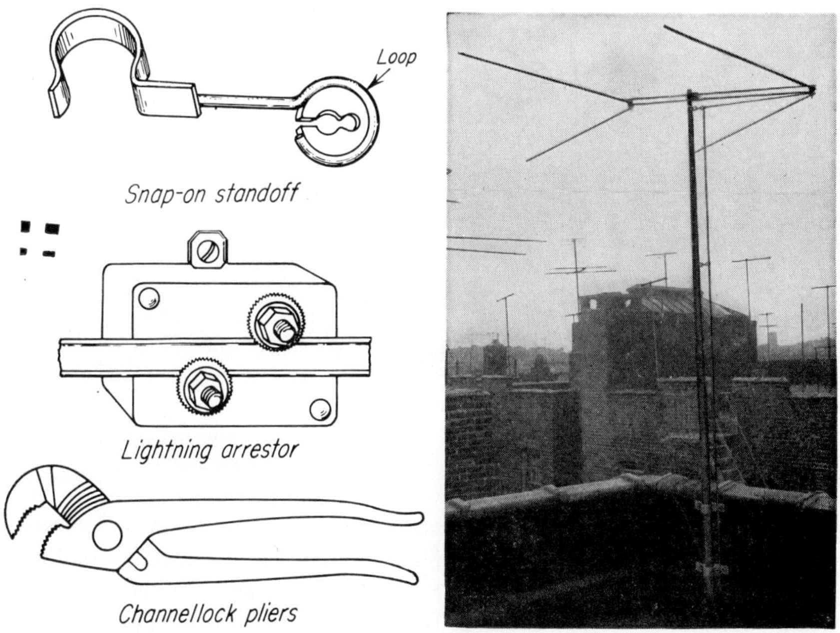

Fig. 16-10. Standoff insulators attached to the mast start the transmission line on its way.

A snap-on standoff like the one illustrated in Fig. 16-10 is used for this purpose. The clamp is attached to the mast right below the antenna, and the line is slipped into the slot in the insert. The insert is then turned so that the slot is away from the opening in the loop; the wire is pulled taut and fastened by squeezing the wire loop together. Channel-lock pliers, also illustrated in Fig. 16-10, are a good tool for this job.

A lightning arrester is clamped to the mast at about its center. The transmission line is run through the groove in the top of the arrester and the holddown nuts are tightened so that the teeth of the serrated washers below them bite into the line, thereby making contact. The ground connection on the arrester is run to an outside ground. This may be a vent pipe on the roof, or a grounding rod driven into moist earth. The ground wire is usually bare aluminum. It grounds the mast at the same time.

The mast is now erected into position against the wall brackets and is fastened securely. Generally, a second snap-on standoff is needed near the bottom of the mast to start the transmission wire down the side of the house. A typical mounting showing the arrangement of clamps is shown in Fig. 16-10.

Running the Transmission Line from the Antenna to the Receiver. Refer to Fig. 16-11 for illustrations of the transmission-line accessories men-

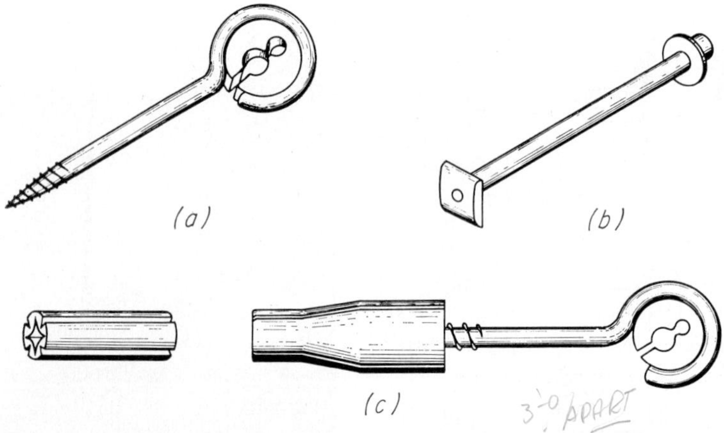

Fig. 16-11. Transmission-line accessories. (*a*) Screw-eye standoff. (*b*) Watertight wall bushing. (*c*) Screw anchor.

tioned in this section. The transmission line is held to the wall by means of screw-type standoff insulators. There should be at least two of these, one at the roof and one at the point of entry. If possible, others should be installed at points spaced about 10 ft apart on vertical runs. The insulators are screwed directly into wood when it is available.

On a brick or masonry wall, the standoff insulators are fastened with plastic screw anchors. A hole is cut into the masonry with a No. 14 rawl drill, the screw anchor is inserted, and the standoff is screwed in. In order to drive the standoff, the polyethylene insert is removed by compressing the slot and a heavy screwdriver is inserted in the wire loop and is used as a T handle. The insulation is then reinserted, the line slipped into the slot, the insert turned, and the wire loop squeezed tight.

The transmission line is usually brought into the house through a window near the receiver. The 300-ohm lead is given a protective coating of vinyl tape and stapled to the window sill after first cutting away a ½-in. section of the metal weather stripping so that the wire can lie flat. The window is then closed on the line. A better installation is to drill through the trim below the window sill and use a polystyrene feed-through bushing. These are designed to make a weathertight seal.

Windows of the casement type do not lend themselves to these methods. In a private home, it is best to bring the line into the basement through a hopper window, and then up through the floor to the receiver. Where this is not practical, one method is to remove a small pane of glass from a fixed section of the casement window, and replace it with a lucite plate which has been drilled to accommodate a short weathertight feed-through bushing. Another method is to use a capacitance-type feed like the Walsco type 1555 Window-Thru. This consists of two sets of metal disks which adhere to both sides of a fixed glass pane. Connection posts for the transmission line are provided. The disks form the plates of two condensers with the glass as the dielectric. There is therefore a condenser in each side of the line, giving a capacitance feed without cutting through the glass.

Inside the house, the transmission line is run to the receiver by stapling the wire to wood trim. A stapling gun is handy for this purpose. Be sure the staple is driven parallel to the wires in the twin lead and is centered between them. Carelessness in this regard may short the twin lead. At the receiver, crimp on a pair of connectors for convenient connection to the antenna posts. Leave enough slack in the lead so that the receiver may be moved from the wall for cleaning or servicing. Do not leave too much slack, since this may be coiled, and the inductance thus introduced may have a detrimental effect on reception.

Installing a 20-ft Mast. A 20-ft mast properly mounted on the roof of a house which is 25 ft or so above ground level gives the antenna sufficient height for reception in most fringe areas. An installation of this type does not require special equipment and is within the scope of the average television technician.

A diagram showing a typical installation on a gable roof is given in Fig. 16-12. The diagram also shows close-up views of some of the hardware used. A mast of this height is self-supporting, but it must be guyed to keep it from swaying or bending in the wind. A 20-ft mast consists of two 10-ft sections. Depending on the manufacturer, the sections are telescoped one inside the other or are two separate pieces of 1¼-in. tubing with fitted or crimped ends. The mast is attached to the eave or

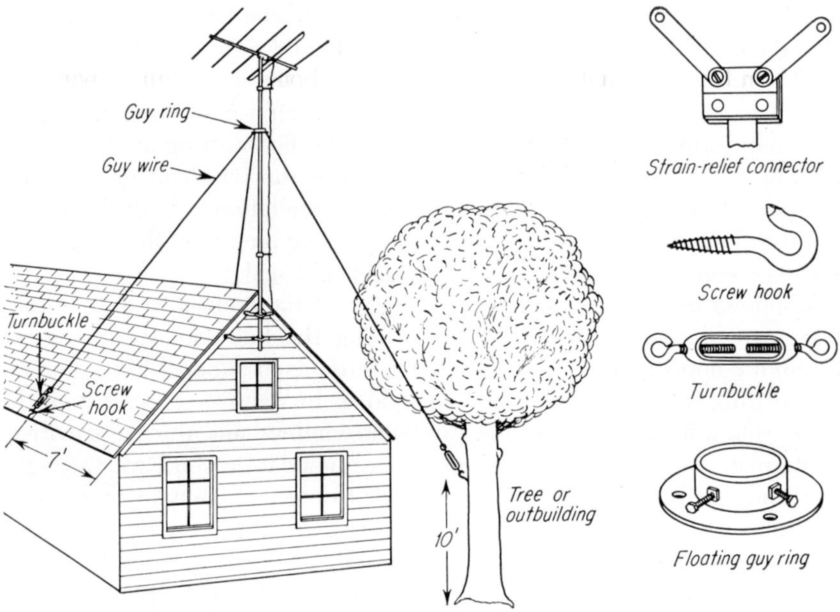

Fig. 16-12. Installation details of a 20-ft mast supported by guy wires.

chimney, using the same brackets described before for a 10-ft mast. The guying is done with stranded galvanized wire. The usual size is four strands of No. 20, which comes in coils in lengths of 25 or 50 ft.

The guy wires in a three-wire system as shown should be 120° apart and dropped at a 30° angle to the mast. These conditions are roughly met in a house of average proportions by fastening two guy wires at the roof eaves about 7 ft back from the gable, as shown in the diagram. The third wire is sent out on a line with the peak and is fastened to an outbuilding or tree at least 10 ft above ground so as to be out of the way. If there is no such structure available, the third guy wire is extended out by means of an outrigger strut attached to one of the mast brackets like the setup shown in Fig. 16-13.

The wall brackets and screw hooks are mounted first. The wall brackets are mounted in the usual way, except that the lower bracket should be a little more than 18 in. below the upper to take care of the added strain. Extra care should be taken to level and plumb the brackets, since irregularities will be exaggerated by the greater height. The screw hooks should be driven in so as to be approximately in line with the direction of pull.

The antenna is assembled and attached to the mast on the ground. If the mast is telescopic, the two sections cannot be separated. The assembly is made with the upper section extended about halfway. If the mast is of the fitted-end type, the two halves are separated and the assembly is made with the upper section alone. The guy ring is slipped over the top of the upper section first. The guy ring should be of the "floating" type, so that the mast may be rotated inside the floating washer. Fringe-area conditions require antenna orientation to take full advantage of the directional characteristics of the high-efficiency antennas.

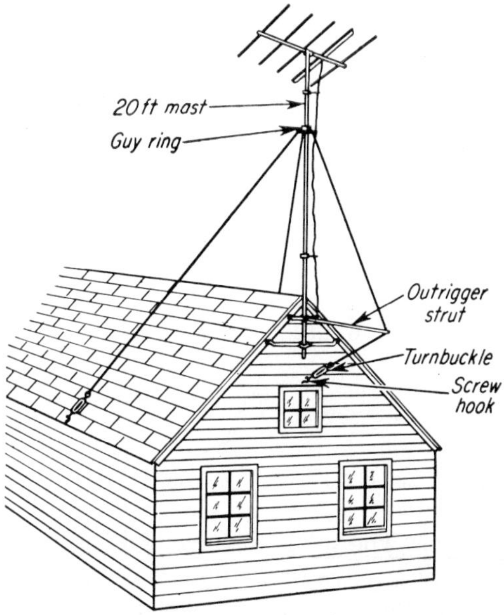

Fig. 16-13. Use of an outrigger strut to support the third guy wire.

The antenna is assembled and attached to the top of the mast, the transmission line is connected, and the top clip-on standoff is installed. Considering that this antenna does not come down easily for servicing or inspection, it would be wise to remove the weak link which exists at the antenna connection. The transmission line often breaks near the point of attachment to the antenna because of no strain relief. An illustration of a connector which provides strain relief is shown in Fig. 16-12. A device of this type should be used.

Another device to facilitate future servicing is to connect a noninductive resistor of about 100,000 ohms across the antenna connection if the antenna is not of the folded-dipole type. Continuity of the entire antenna system is easily ascertained with an ohmmeter at the receiver terminals when the antenna is folded, thereby establishing that the transmission line is good. This cannot be done on any of the open-circuit antennas. The 100,000-ohm resistor at the antenna connection provides the same easy check for the transmission line.

After the transmission line is clamped below the antenna, the guy ring is secured to the mast just below the standoff. Determine the approximate best direction for the antenna, and bring the transmission line to a temporary clamp below the guy ring so that the mast may be rotated back and forth through this approximate setting without having the transmission line cross a guy wire.

Three rolls of guy wire of proper length are chosen, and one end of each is fastened to one of the holes in the floating washer of the guy ring. When fastening a guy wire, the end is looped through the hole twice as shown in Fig. 16-14, brought back upon itself, and wrapped

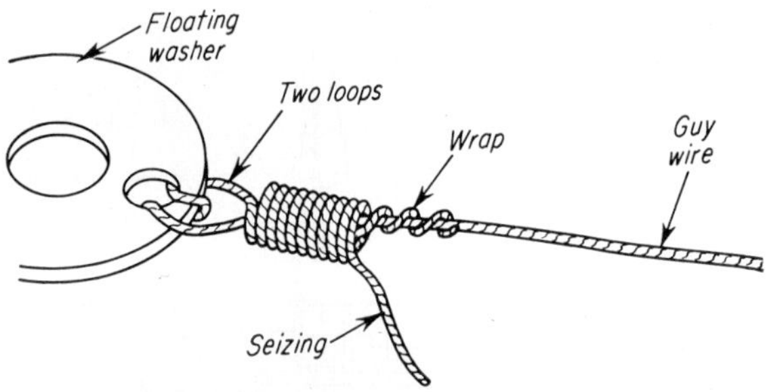

Fig. 16-14. Fastening the guy wire to the floating guy ring.

tightly for several turns. The splice thus created is seized by a tight wrap with an excess piece of guy wire. The wire is comparatively soft and can be tightly wound with the aid of heavy pliers.

It is dangerous to work while standing on the peak of the roof. In the following procedure, the work can be done from a kneeling or sitting position on the peak. The technique must be varied for the fitted-end and telescopic masts.

In the fitted-end type, the bottom section alone is first fastened to the wall brackets below its normal position, so that the top is about level with the peak of the roof. The assembled top section is hoisted or carried up to the peak, turned to the predetermined best direction, and inserted into the bottom mast section. If a lock is provided for the two mast sections, it is tightened. Locking devices are not always used with fitted end sections, especially if tapered, since the weight of the antenna and the downward pull of the guy wires will ensure a tight fit between the sections. If the dangling guy wires and transmission line have become twisted together, they are straightened and sent in their proper directions. The transmission line is clamped to the mast with a standoff near the joint of the mast sections. Check again to see that the mast can be rotated around the approximately correct position without interference between the transmission line and guy wires. The mast plates on the wall brackets are loosened just enough to permit motion of the mast, and it is pushed up to its correct position. The antenna is now at full height, and the mast plates are tightened again.

When working with a telescopic mast, the antenna and guy ring have been attached to the inner section as described previously. The sections are closed until stopped by the fastened guy ring. The entire assembly is carried or hoisted to the roof, and the outer section is fastened to the wall brackets so that the guy ring is about level with the peak. The top telescoping section is then extended to full height rotated to approximate best direction, and locked. Locking pins are always provided for telescopic masts. The rest of the procedure is the same. The wires are untangled and sent off in their correct direction, and the transmission line is clamped near the lock, after making sure that it will not cross a guy wire when the final orientation is made for best direction. Finally, the bottom section is pushed up to full height and is tightened in position.

To attach the guy wires to their fastening points, choose any one of the three points, loosen the turnbuckle all the way, and slip one eye over the open screw hook. The guy wire is threaded through the other eye and pulled while watching the mast. When the mast begins to bend toward the fastening point, loosen up so that it straightens and temporarily hold this position by making a couple of hitches with the loose end of the guy wire. Repeat the procedure at the other two fastening points.

Before making permanent ties at the turnbuckles, check to see that the mast is perpendicular and that there is about the same amount of sag in each of the three guy wires. Loosen or tighten the temporary hitches, if necessary, recheck, and then make permanent splices using the same wrap previously described for the guy ring. Finally, take up the sag by tightening the turnbuckles, using a screwdriver as a lever in the center section. Do not use too much pressure. The guy wires should be just tight enough to hold the mast against swaying. Excess pressure puts a strain on the system without improving the stability. Check that the mast remains perpendicular throughout the tightening process.

Installing Attic or Window Antennas. In an area of good reception, when the antenna built into the cabinet of the receiver is not quite adequate, the obvious solution is the installation of a good outdoor antenna on the roof. However, this is not always feasible. The owner of a private house may object on the grounds that the outside antenna will spoil appearances, and an apartment house may have restrictions against antennas.

When the private house has a little-used attic, the installation in this space of a standard antenna on a short mast is generally very effective. An antenna inside the attic is not very much lower in height than if it were mounted outside and is still one or two floors higher than the inside antenna at the set.

The antenna can be hung from a single wall bracket mounted high on a roof rafter, as shown in Fig. 16-15*a*, or it may be fastened to a floor joist by using a floor flange as shown in Fig. 16-15*b*. The transmission line may

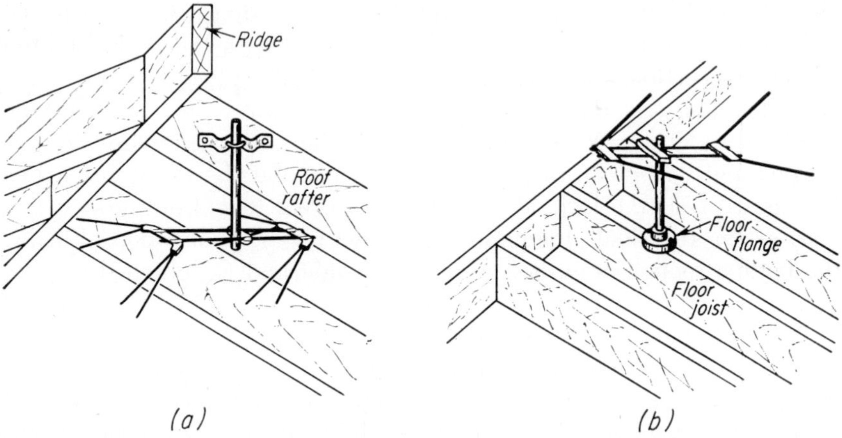

Fig. 16-15. Methods of mounting an antenna in an attic.

be fed through an attic ventilator or window to the outside of the house, and carried to the receiver in the usual way.

The apartment-house dweller who is denied the use of the roof can usually improve reception over his indoor antenna by mounting an antenna outside a window with the setup illustrated in Fig. 16-16. The window is not necessarily in the same room as the receiver, since greatly improved reception may be obtained by using a window in another room which faces toward the transmitter. The window antenna consists of a jack-type mount which wedges across the sides of the window frame. The jack supports a short mast and the antenna. Small squares of rubber cut from an old inner tube and inserted under the legs prevent slipping while positioning and tightening the jack. Orienting a window antenna for best reception is likely to be critical.

Orienting the Antenna. In strong-signal areas, the antenna is pointed in the general direction of the transmitting stations, the positioning being generally determined by observing the direction of other antennas already installed in the vicinity. The receiver is tried out, and unless there is a serious condition of multiple images or ghosts, there is little to be gained by rotating the antenna for better reception.

In fringe areas, antennas with highly directional characteristics are often used in order to obtain maximum signal strength. This type may also be used to overcome a serious ghost condition. In either case, in order to take full advantage of the directional characteristics of the antenna, it must be oriented, that is, rotated while observing results on the receiver.

Essentially, orientation of an antenna is a two-man job. One man tunes the receiver and watches results, while the other rotates the antenna. The two men maintain communication by means of two sets of sound-

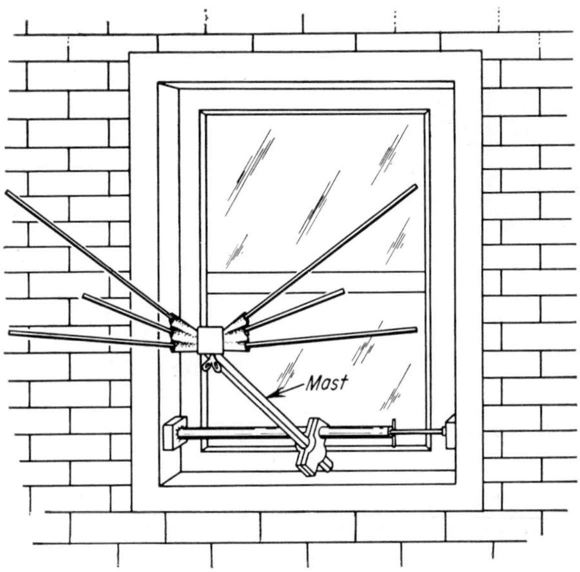

Fig. 16-16. A window antenna.

powered phones interconnected by a separate length of transmission-line or extension-cord cable. When the man at the receiver finds the position of best all-round reception, he informs the man at the antenna, who then ties down the mast-holding bolts to retain this position.

In some areas, where reception is possible from stations in two different directions, it is not easy to achieve a satisfactory compromise and obtain good reception from both directions. One method is to install two separate antennas and transmission lines, with a switch at the antenna terminals of the receiver. One antenna is oriented for best results from one direction, and then the receiver is switched to the other antenna, which is rotated and fixed at the position of best results.

Another method is to make use of an antenna rotator, like the one illustrated in Fig. 16-17. An electric motor is mounted on the wall brackets and rotates the antenna mast as directed by a control switch and indicator which are installed at the receiver. The installation man controls the antenna direction from the receiver, noting the positions of best recep-

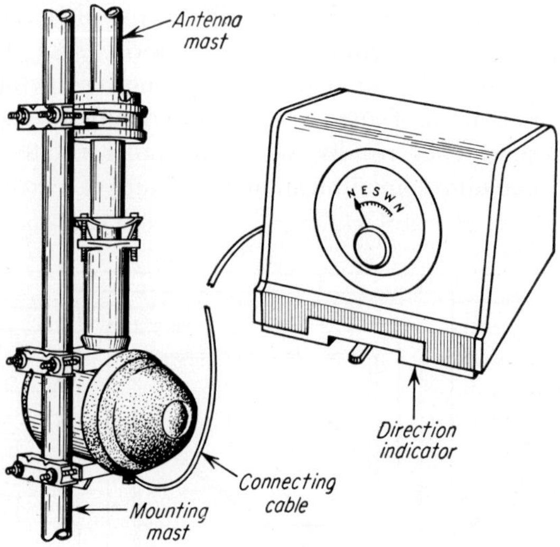

Fig. 16-17. An antenna rotator.

tion for each station, and instructs the consumer in the proper use of the device.

Improving Reception on a Given Channel. When trying out a receiver on installation, it sometimes happens that normal results are obtained on all channels but one, and this one is known to give generally good reception in the area. Normal reception from other stations indicates that the set is good. Poor reception on one channel may be caused by a mistuned oscillator slug or a mismatch between the antenna and receiver. Readjust the oscillator-tuning slug. This procedure is given in Chap. 14. If this

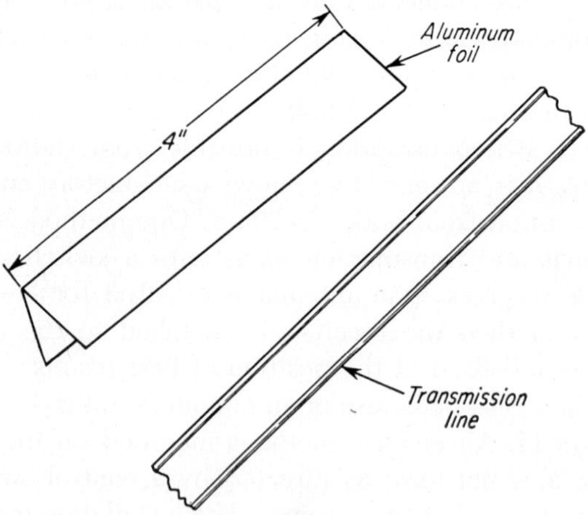

Fig. 16-18. Tuning the transmission line.

does not improve the condition, try tuning the transmission line to overcome the mismatch.

This is done as follows: A piece of heavy aluminum foil about 2 by 4 in. is folded lengthwise as shown in Fig. 16-18 and is slipped over the transmission line about 2 ft back from the antenna terminals of the receiver. The receiver is tuned to the station giving poor results, and the piece of foil is slid back and forth along the transmission line. If improvement is noted, check reception on the other channels to make sure that reception has not been impaired, and tape the foil securely in the position of best results.

Cross Modulation (Windshield-wiper Effect). A single tuned circuit is comparatively broad. In receiver circuits, tuning becomes efficient after the received signals have passed through the successive tuned circuits of several stages. It is therefore to be expected that signals from more than one station may be present in the early stages.

Assume we have a receiver tuned to one station and that, in addition to the desired station, signals from a station on another channel are also present in the first r-f stage. If the signal level is high, the first r-f tube may be overloaded and act as a detector, demodulating some of the signals which are present. Now, although the output signal is primarily at the frequency of the desired station, some of the modulation from the second station will have been added to the desired station signal. This interference cannot be eliminated by subsequent tuning circuits because it is now part of the modulation of the desired signal. Such a condition is called cross modulation.

In general television reception, cross modulation is not too much of a problem because signal strength is generally low, and the agc circuits tend to reduce strong signals below the overload point of the tubes. However, in very strong signal areas close to transmitting stations, the received signal may be so strong that the agc system is inadequate, and cross modulation occurs.

In radio reception, the effect of cross modulation is that music from the interfering source is added to the program being received as a sort of soft background music, most noticeable in pauses and often drowned out by the sounds of the desired program. Similarly in television reception, most of the interfering signal is drowned out by the desired signal. The most noticeable effect is the vertical dark bar of the horizontal blanking of the interfering signal shifting from side to side across the screen like a background effect. The deflection system is locked to the horizontal frequency of the desired station, and the interfering bar appears and moves as the horizontal frequency of the interfering signal shifts from the locked frequency. The shifting bar gives somewhat the impression of a windshield wiper, which accounts for the name often applied to this condition—windshield-wiper effect. The symptom is illustrated in Fig. 16-19*a*.

The windshield-wiper effect will be noted when trying out the receiver after the installation of the antenna. If the receiver has a manual adjustment for the agc system, this is tried. If there is no control, or if its adjustment does not remove the interference, the input signal strength must be reduced. This is best done with an antenna attenuator like the one illustrated in Fig. 16-19*b*.

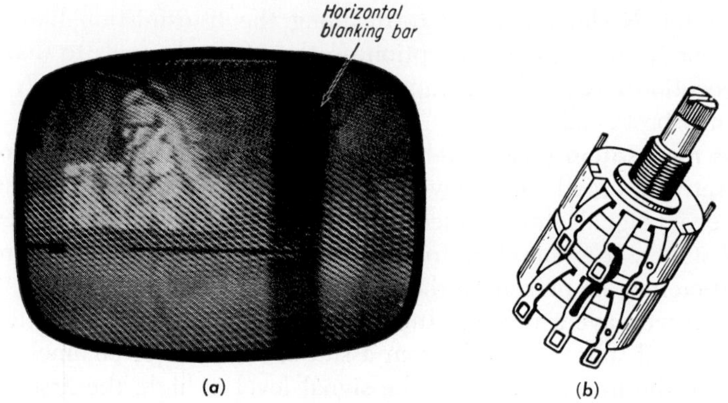

Fig. 16-19. (*a*) Windshield-wiper effect and (*b*) an antenna attenuator designed to correct it.

Interference. Cross modulation is one of many forms of interference which mar television reception. Others are interference from stations on adjacent channels, and strong signals from nearby f-m and amateur stations. Interference from other stations produces a beat signal in the video amplifier which appears on the screen as a mesh of slanting lines like the illustration of Fig. 16-20*a*.

Remedial measures for interference in any one installation vary, depending on the source. One expedient, however, may be tried in all cases where signal strength is comparatively high. If the desired signal is stronger than the interfering signal, and also stronger than needed, an antenna padder similar to the one used for cross modulation will cut down both the desired signal and the interference. In this way, the interference may be reduced to a point where it is not objectionable before the picture becomes too weak for good reception. This method is successful often enough to be worth trying.

If attenuation of the input signal does not help, the interfering signal must be trapped out. If the frequency of the interfering signal is known, a proper wave trap is chosen, connected in the antenna line at the receiver, and tuned until the interfering lines are at a minimum, or disappear. The trap illustrated in Fig. 16-20*b* comes in the following frequency ranges:

13.5 to 31.6 Mc for amateur or commercial short-wave transmission.
30 to 60 Mc for amateur or diathermy interference.
90 to 112 Mc for f-m transmission.
140 to 170 Mc for the commercial transport band.

(a)

(b)

Fig. 16-20. (*a*) Effect of r-f interference and (*b*) a wave trap designed to reduce it.

If the frequency of the interfering signal is unknown, it is necessary to get a set of traps and connect them in one at a time, tuning each one through its range while watching the interference pattern for improvement. The traps are provided with a ground lead, which should be used. A convenient ground is the wall plate of an electric outlet.

Before connecting any traps, refer to the service notes of the receiver. Some manufacturers include one or two wave traps for the reduction of interference in the antenna circuit. These are usually designed to cover the i-f band of the receiver and the f-m band. If these traps are included, proper adjustment may clear up the reception.

Troubles Common to the Antenna. The interference problems described in the previous sections are antenna troubles, but they are primarily installation problems in that the symptoms are evident immediately on making the installation, or soon after. We shall now consider troubles that develop after the antenna has been in operation for some time.

The antenna is subject to two common troubles. One is that the antenna comes down as the result of a severe windstorm or the rusting through of mounting components. The other is that the transmission line opens. In either case, the input signal to the receiver is reduced, resulting in weak picture, or no picture, coupled with snow.

When called in for such a complaint, the alert serviceman looks to see that the antenna is up before even entering the building. But let us assume that the call is to an apartment house, where the maze of antennas rules out this quick check. After verifying the complaint of weak picture, or no picture, coupled with snow, the first step will probably be to change the tubes in the tuner. When this does not make any improvement, the trouble is limited to the antenna or a circuit fault in the tuner. The latter is usually a pull job to the shop; so the antenna installation is checked next, and even in the apartment house, a dismasted antenna is now easily found.

Even though the antenna is still standing, the installation may still be at fault because of the possibility of an open transmission line. This should be checked before the set is taken to the shop. Since a folded dipole is a one-turn loop at the end of the transmission line, this type of antenna system is easily checked for continuity with an ohmmeter at the receiver terminals. The transmission line in an open type of antenna is not so easily checked. It is for this reason that the suggestion was made earlier in the chapter to connect a 100,000-ohm resistor across an open-line antenna. If such a resistor is in place, an ohmmeter check can still be made for continuity of the line. However, if there is no resistor, the antenna system may be checked by other means.

In a strong-signal area, the serviceman should carry with him a portable indoor antenna, like the rabbit-ear type illustrated in Fig. 16-1*b*. This is connected to the receiver instead of the outdoor antenna, and if the signal is strengthened and the snow cleared up, the outdoor antenna is at fault. In a fringe area, this method may not work satisfactorily because the indoor antenna may not deliver enough signal to the receiver to make a reliable check. Instead, it is recommended that the serviceman carry in his car an extra working chassis or table-model receiver. Traded-in receivers have little resale value, and one of these, kept in repair and brought along for just this purpose, is a good investment. The old receiver is tried on the antenna, and if it works, the antenna is good. It might be a good idea to leave the old trade-in at the customer's house, so that he has reception until his own set is repaired. This makes for good consumer relations.

If the transmission line proves to be open, look for the break at one of the wear points by flexing the lead. The break will be obvious on inspection. Start at the antenna posts on the receiver, and check back along the line to the window. Too much moving of the receiver for cleaning purposes may have broken the line in this area. Check at the window, where the closing of the window may have caused the break. Check at any bends in the transmission line between the window and the roof. If the line goes over a parapet wall, this again is a likely spot. Finally, it may be necessary to pull down the mast to check for breaks at the connection point to the antenna. Repair the break, or replace the transmission line, whichever is most practical.

Visual Aids

The motion pictures and filmstrips in the following list can be used to supplement the material in this book. Lengths of motion pictures (MP) are given in minutes and of filmstrips (FS) in frames. The motion pictures are 16 mm and have sound, and the filmstrips are 35 mm and silent; both are black-and-white.

Addresses of producers or distributors are given at the end of the list. In many cases, the films can be borrowed or rented from state or local film libraries.

The Cathode-ray Tube: How It Works (MP, UWF, 15 min). Demonstrates the construction and function of various parts of the cathode-ray tube and explains electrostatic and electromagnetic deflection.

How Color Television Works (FS, McGraw, 45 fr). Explains the basic principles of television tubes and shows how television colors are formed. Compares the three-electron-gun tube and the single-electron-gun Chromaton tube. Describes color television cameras and transmission.

The How of Television (MP, UWF, 10 min). Explains the elementary principles of television and shows the work of the television cameraman and other technicians and the operation of the television camera, amplifying and receiving systems, and antenna.

Magic in the Air (MP, GM, 10 min.). Goes behind the scenes at the television studio to show the construction of the transmitter.

Radio, Television, and Electrical Appliance Repairmen (FS, SVE, 47 fr). Points out the various kinds of jobs available, the qualifications required, and the advantages and disadvantages of each occupation.

Story of Television (MP, DuMont, 14 min). Discusses research which led to the development of both black-and-white and color television, traces the history of television broadcasting, and shows the operation of television cameras, transmitters, and receivers.

Television: How It Works (MP, Coronet, 11 min). Shows the fundamentals of television broadcasting and reception, including operation of the cathode-ray tube, image-orthicon tube, and sync generator.

This Amazing World of Television-Electronics (MP, De Forest, 22 min). Shows uses of electronics in television manufacturing and broadcasting and in military service and opportunities for trained men in the field.

Magic in Television (MP, De Forest, 10 min). Traces the development and shows the operation of the cathode-ray tube.

Vital Link (MP, Assn, 10 min). Explains the function of a television antenna and how it determines the quality of the picture received.

Antenna Is the Pay-off (MP, Channel-Master, 20 min). Explains various aspects of antenna performance. Primarily intended for those who sell and install television antennas.

How to Use the Signal Generator (FS, McGraw, 38 fr). Explains the operation of the controls as used in a basic generator and identifies similar controls on various commercial instruments.

Procedure of TV Alignment (MP, Libra, 18 min). Demonstrates the alignment of television sets, utilizing various pieces of standard testing equipment as recommended by the manufacturer.

Television (McGraw). A set of six films and six correlated filmstrips which demonstrate the basic principles and servicing of television receivers. Titles and running times are:

Television System (14 min)
Antenna Installation (12 min)
Television Receivers (7 min)
Localizing Troubles (11 min)
Deflection Circuits (11 min)
Practical TV Alignment (16 min)

Typical Television Interference Patterns (MP, UWF, 11 min). Shows the various types of television interference patterns, what causes them, and how they can be corrected.

Film Sources

Assn—Association Films, Inc., 347 Madison Ave., New York 17

Channel Master—Channel Master Corp., Naponoch Rd., Ellenville, N.Y.

Coronet—Coronet Films, Coronet Bldg., Chicago 1

De Forest—De Forest's Training, Inc., Film Service Dept., 2533 N. Ashland Ave., Chicago 4

DuMont—Allen B. DuMont Laboratories, Inc., 750 Bloomfield Ave., Clifton, N.J.

GM—General Motors Corp., Department of Public Relations, Film Section, 3044 W. Grand Blvd., Detroit 2

Libra—Libra Film Distributors and Producers, 6525 Sunset Blvd., Hollywood 28, Calif.

McGraw—McGraw-Hill Book Company, Inc., Text-Film Dept., 330 W. 42 St., New York 36

SVE—Society for Visual Education, Inc., 1345 W. Diversey Pkwy., Chicago 14

UWF—United World Films, Inc., 1445 Park Ave., New York 29

Index